INNOVATIVE APPROACHES IN DIAGNOSIS AND MANAGEMENT OF CROP DISEASES

Volume 2
Field and Horticultural Crops

Innovative Approaches in Diagnosis and Management of Crop Diseases
Volume 1: The Mollicutes / hardback ISBN: 978-1-77463-024-2

Innovative Approaches in Diagnosis and Management of Crop Diseases
Volume 2: Field and Horticultural Crops / hardback ISBN: 978-1-77463-025-9

Innovative Approaches in Diagnosis and Management of Crop Diseases
Volume 3: Nanomolecules and Biocontrol Agents / hardback ISBN: 978-1-77463-026-6

**Innovative Approaches in Diagnosis and Management of Crop Diseases,
3-volume set**
hardback ISBN: 978-1-77463-027-3

INNOVATIVE APPROACHES IN DIAGNOSIS AND MANAGEMENT OF CROP DISEASES

Volume 2
Field and Horticultural Crops

Edited by
R. K. Singh, PhD
Gopala, PhD

APPLE ACADEMIC PRESS

First edition published 2022

Apple Academic Press Inc.
1265 Goldenrod Circle, NE,
Palm Bay, FL 32905 USA

4164 Lakeshore Road, Burlington,
ON, L7L 1A4 Canada

CRC Press
6000 Broken Sound Parkway NW,
Suite 300, Boca Raton, FL 33487-2742 USA

2 Park Square, Milton Park,
Abingdon, Oxon, OX14 4RN UK

© 2022 Apple Academic Press, Inc.

Apple Academic Press exclusively co-publishes with CRC Press, an imprint of Taylor & Francis Group, LLC

Library and Archives Canada Cataloguing in Publication

Title: Innovative approaches in diagnosis and management of crop diseases / edited by Rakesh Kumar Singh, PhD, Gopala, PhD.

Names: Singh, Rakesh Kumar (Plant pathologist), editor. | Gopala (Plant pathologist), editor.

Description: First edition. | Includes bibliographical references and index. | Contents: Volume 2. Field and horticultural crops.

Identifiers: Canadiana (print) 20210171766 | Canadiana (ebook) 20210172266 | ISBN 9781774630259 (v. 2 ; hardcover) |
    ISBN 9781774639535 (v. 2 ; softcover) | ISBN 9781774630273 (set) | ISBN 9781003187837 (v. 2 ; ebook)

Subjects: LCSH: Phytopathogenic microorganisms—Control. | LCSH: Plant diseases. | LCSH: Mycoplasmatales.

Classification: LCC SB731 .I56 2021 | DDC 632/.3—dc23

Library of Congress Cataloging-in-Publication Data

Names: Singh, Rakesh Kumar (Plant pathologist), editor. | Gopala (Plant pathologist), editor.

Title: Innovative approaches in diagnosis and management of crop diseases. Volume 2, Field and horticultural crops / edited by Rakesh Kumar Singh, Gopala.

Description: First edition. | Palm Bay, FL, USA : Apple Academic Press, [2021] | Includes bibliographical references and index. | Contents: Paradigm Shift in Detection of Rice Diseases and Their Management / Manas Kumar Bag, Prahlad Masurkar, Anuprita Ray, and Rakesh Kumar Singh -- Diagnosis and Management of Fungal Diseases of Rice Prevalent in Telangana State, India / S. Kiran, M. Surekha, and S. M. Reddy -- New Insights into the Identification and Management of Wheat Diseases / T. L. Prakasha, A. N. Mishra, J. B. Singh, S. V. Sai Prasad, and Suresh Chand -- Detection and Management Approaches of Bakanae (Foot Rot) Disease in Rice / Sachin Kumar Jain, Kamal Khilari, and Mukesh Dongre -- Biotic and Abiotic Stresses in Cotton Crop in Punjab, India / Rupesh Kumar Arora and Paramjit Singh -- Modern Approaches for Management of Sesame Diseases / N. Ransingh, B. Khamari, and N. K. Adhikary -- Approaches for Diagnosis and Management of Banded Blight in Small Millets / A. K. Jain, S. K. Tripathi, and R. P. Joshi -- Maize Diseases and Their Sustainable Management in India: Current Status and Future Perspectives / M. K. Khokhar, K. S. Hooda, P. N. Meena, R. Gogoi, S. S. Sharma, Rekha Balodi and M. S. Gurjar -- Recent Advances in Detection, Diagnosis, and Management of Finger Millet Diseases / Pardeep Kumar, Shrvan Kumar, Jiwan Paudel, and D. P. Singh -- Recent Approaches for Diagnosis and Management of Economically Important Diseases of Field Pea (Pisum sativum L.) in India / Sonika Pandey, R. K. Mishra, Monika Mishra, A. K. Parihar, and G. P. Dixit -- Recent Advances in Bio-Intensive Management (BM) of Major Diseases of Pigeon Pea in India / Monika Mishra, R. K. Mishra, Sonika Pandey, U. S. Rathore, Rajesh K. Pandey, and Manjul Pandey -- Innovative Approaches in Diagnosis and Management of Diseases in Ginger (Zingiber officinale Roscoe) and Turmeric (Curcuma longa L.) / Ajit Kumar Singh, Devendra Kumar Choudhary, Shrikant Sawargaonkar, and Rakesh Kumar Singh -- Diagnosis and Diversity Analysis of Alternaria brassicae and A. brassicicola in Vegetable Cruciferous Crops / Pratibha Sharma, Shaily Javeria, Swati Deep, Manika Sharma, and Raja Manokaran -- Recent Advances in Diseases Management of Aonla (Emblica officinalis) / R. K. Prajapati, C. S. Pandey, P. K. Gupta, V. K. Singh, and S. R. Singh -- Exploitation of Biofumigation and Biocontrol Agents for the Management of Soil-Borne Diseases / G. Bindumadhavi and R. Gopi. | Summary: "This book is the second of the 3-volume Innovative Approaches in Diagnosis and Management of Crop Diseases, which provides an abundance of new research and information on major diseases of various crops along with new techniques and technology for the detection of plant pathogens along with appropriate management strategies. Divided into three volumes and with chapters written by renowned and expert scientists working in different areas of plant pathology, the volumes cover important diseases of crops, incited by bacteria, fungi, viruses, viroids, phytoplasma, and nematodes. It addresses these disease challenges to commercial field and horticultural crops and their management. Innovative Approaches in Diagnosis and Management of Crop Diseases, Volume 2 focuses on recent advances in diagnosis, detection, and management of diseases of specific crops, such as cotton, sesame, rice, wheat, millet, maize, field pea and pigeonpea, ginger and turmeric, guava, aonla, and vegetable cruciferous crops. Key features: Presents diverse research of leading plant pathologists on detection, diagnosis, and management of crop diseases Shares innovative and emerging techniques for diagnosis and management of major plant diseases Covers a vast array of important crops and their diseases"-- Provided by publisher.

Identifiers: LCCN 2021016910 (print) | LCCN 2021016911 (ebook) | ISBN 9781774630259 (v. 2 ; hardback) | ISBN 9781774639535 (v. 2 ; paperback) | ISBN 9781003187837 (v. 2 ; ebook)

Subjects: LCSH: Phytopathogenic microorganisms--Control. | Plant diseases.

Classification: LCC SB731 .I56 2021 (print) | LCC SB731 (ebook) | DDC 632/.3--dc23

LC record available at https://lccn.loc.gov/2021016910
LC ebook record available at https://lccn.loc.gov/2021016911

ISBN: 978-1-77463-025-9 (hbk)
ISBN: 978-1-77463-953-5 (pbk)
ISBN: 978-1-00318-783-7 (ebk)

About the Editors

R. K. Singh, PhD

Head, Plant Pathology, Rajmata Vijayaraje Scindia Krishi Vishwa Vidyalaya, College of Agriculture, Indore, M.P., India

R. K. Singh, PhD, is Head, Plant Pathology at Rajmata Vijayaraje Scindia Krishi Vishwa Vidyalaya (RVSKVV), College of Agriculture, Indore, M.P., India. He is esteemed member of national and international research societies and a fellow of the Indian Society of Pulses Research and Development. He has been honored with four awards from national societies. Dr. Singh has guided 25 MSc (Ag) and 3 PhD students on modern integrated areas in biology and has published over 65 research papers along with book chapters, instructional manuals, 1 text book and popular articles. He has experience in teaching, research, and extension work in agriculture in various capacities for over 20 years. He was PI of a project funded by the Japan International Cooperation Agency (JICA), ICAR, and evaluated several newly evolved agrochemicals. He is actively working in detection of plant pathogens and identification of sources of genetic resistance in chickpea and mungbean. Dr. Singh has set up a molecular plant pathology laboratory, wilt sick plot for chickpea and mushroom cultivation and value addition laboratory at college.

Gopala, PhD

Rajmata Vijayaraje Scindia Krishi Vishwa Vidyalaya (RVSKVV), College of Agriculture, Indore, M.P., India

Gopala, PhD, is affiliated with Rajmata Vijayaraje Scindia Krishi Vishwa Vidyalaya (RVSKVV), College of Agriculture, Indore, M.P., India. He earned his MSc and PhD degrees in Plant Pathology from IARI, New Delhi, India. He has cleared his exams such as ICAR PGS-JRF and had a IARI merit scholarship for his PhD. He also cleared ICAR NET in 2014. He has received three awards from national societies. During his MSc work, he has developed a new screening technique and rating scale for stalk rot of maize caused by *Macrophomina phaseolina*. During his PhD work, he has reported new phytoplasma diseases in *Cucurbita pepo*, bougainvillea, dianthus, petunia, and ornamental kale along with its associated vectors. He has submitted more than 50 sequences to the National Center for Biotechnology Information and received accession numbers. Dr. Gopala has published more than 15 research papers in journals of national and international repute and 1 text book.

Contents

Contributors

N. K. Adhikary
Junior Pathologist, ICAR-AICRP on Sesame and Niger, Institute of Agriculture Science, University of Calcutta, West Bengal, India

Rupesh Kumar Arora
Punjab Agricultural University (PAU), Regional Research Station, Bathinda – 151001, Punjab, India, Mobile: 9646687131, E-mail: rkarora@pau.edu

Manas Kumar Bag
Crop Protection Division, ICAR-National Rice Research Institute, Cuttack, Odisha – 753006, India, E-mail: manas.bag@gmail.com

Rekha Balodi
ICAR-National Research Center for Integrated Pest Management, Pusa Campus, New Delhi – 110012, India

G. Bindumadhavi
Regional Agricultural Research Station, Lam, Acharya N.G. Ranga Agricultural University, Guntur, Andhra Pradesh, India, E-mail: bindugopireddy@gmail.com

Suresh Chand
School of Life Sciences, Devi Ahilya University, Indore – 452 001, Madhya Pradesh, India

Devendra Kumar Choudhary
College of Agriculture and Research Station, IGKV, Raigarh, Chhattisgarh, India

Swati Deep
Technology and Innovation Center, International Panacea Limited, Gurugram, Haryana – 122003, India

G. P. Dixit
ICAR-Indian Institute of Pulses Research, Kanpur – 208024, Uttar Pradesh, India

Mukesh Dongre
Department of Plant Pathology, RVSKVV, College of Agriculture, Indore, Madhya Pradesh, India

R. Gogoi
Indian Agriculture Research Institute, New Delhi – 110012, India

R. Gopi
ICAR-Sugarcane Breeding Institute Research Center, Kannur – 670002, Kerala, India

P. K. Gupta
Scientist (Plant Protection and Technical Officer), JNKVV, Directorate of Extension Service, Jabalpur, Madhya Pradesh, India

M. S. Gurjar
Indian Agriculture Research Institute, New Delhi – 110012, India

K. S. Hooda
ICAR-Indian Institute of Maize Research, PAU Campus, Ludhiana – 141 004, Punjab, India

A. K. Jain
Department of Plant Pathology, JNKVV, College of Agriculture, Rewa – 486001, Madhya Pradesh,
India, E-mail: akjagcrewa@gmail.com

Sachin Kumar Jain
Department of Plant Pathology, Amar Singh College, Lakhaoti, Bulandshahr, Uttar Pradesh, India,
E-mail: sachinjain1115@gmail.com

Shaily Javeria
Division of Seed Science and Technology, ICAR-Indian Agricultural Research Institute,
New Delhi – 110012, India

R. P. Joshi
Department of Plant Breeding, JNKVV, College of Agriculture, Rewa – 486001, Madhya Pradesh, India

B. Khamari
Assistant Professor, Department of Plant Pathology, IAS, SOADU, Bhubaneswar, Odisha, India

Kamal Khilari
Department of Plant Pathology, S.V.P. University of Agriculture and Technology, Modipuram,
Meerut, Uttar Pradesh, India

M. K. Khokhar
ICAR-National Research Center for Integrated Pest Management, Pusa Campus, New Delhi – 110012,
India, E-mail: khokharmk3@gmail.com

S. Kiran
Department of Botany, Satavahana University, Karimnagar, Telangana, India

Pardeep Kumar
KVK, Sohna (ANDUA&T, Ayodhya), Siddharthnagar, Uttar Pradesh, India,
E-mail: drpardeepviro@gmail.com

Shrvan Kumar
Rajiv Gandhi South Campus, BHU, Barkachha, Mirzapur, Uttar Pradesh – 231001, India

Raja Manokaran
Department of Plant Pathology, SKN Agriculture University, Jobner, Jaipur – 303329, Rajasthan, India

Prahlad Masurkar
Department of Mycology and Plant Pathology, Institute of Agricultural Sciences,
Banaras Hindu University, Varanasi, Uttar Pradesh – 221005, India

P. N. Meena
ICAR-National Research Center for Integrated Pest Management, Pusa Campus,
New Delhi – 110012, India

A. N. Mishra
ICAR-Indian Agricultural Research Institute, Regional Station, Indore – 452 001,
Madhya Pradesh, India

Monika Mishra
ICAR-Indian Institute of Pulses Research, Kanpur – 208024, Uttar Pradesh, India

R. K. Mishra
ICAR-Indian Institute of Pulses Research, Kanpur – 208024, Uttar Pradesh, India,
E-mail: rajpathologist@yahoo.com

C. S. Pandey
Assistant Professor (Horticulture), JNKVV, College of Agriculture, Jabalpur, Madhya Pradesh, India

Manjul Pandey
KVK, Banda University of Agriculture and Technology, Banda, Uttar Pradesh, India

Rajesh K. Pandey
Bundelkhand University, Jhansi, Uttar Pradesh, India

Sonika Pandey
ICAR-Indian Institute of Pulses Research, Kanpur – 208024, Uttar Pradesh, India

A. K. Parihar
ICAR-Indian Institute of Pulses Research, Kanpur – 208024, Uttar Pradesh, India

Jiwan Paudel
Rajiv Gandhi South Campus, BHU, Barkachha, Mirzapur, Uttar Pradesh – 231001, India

R. K. Prajapati
Scientist (Plant Protection), JNKVV, Krishi Vigyan Kendra, Tikamgarh, Madhya Pradesh, India,
E-mail: rkiipr@yahoo.com

T. L. Prakasha
ICAR-Indian Agricultural Research Institute, Regional Station, Indore – 452 001,
Madhya Pradesh, India, E-mail: prakash7385@gmail.com

S. V. Sai Prasad
ICAR-Indian Agricultural Research Institute, Regional Station, Indore – 452 001,
Madhya Pradesh, India

N. Ransingh
Associate Professor, College of Agriculture, OUAT, Bhawanipatna, Odisha, India,
E-mail: nirakar.ranasingh@gmail.com

U. S. Rathore
ICAR-Indian Institute of Pulses Research, Kanpur – 208024, Uttar Pradesh, India

Anuprita Ray
Crop Protection Division, ICAR-National Rice Research Institute, Cuttack, Odisha – 753006, India

S. M. Reddy
Department of Botany, Kakatiya University, Warangal, Telangana, India

Shrikant Sawargaonkar
College of Agriculture and Research Station, IGKV, Raigarh, Chhattisgarh, India

Manika Sharma
North American College of Pharmaceutical Technology, Mississauga, Ontario, Canada

Pratibha Sharma
ICAR Emeritus Scientist, Department of Plant Pathology, SKN Agriculture University, Jobner,
Jaipur – 303329, Rajasthan, India, E-mail: psharma032003@yahoo.co.in

S. S. Sharma
Maharana Pratap University of Agriculture and Technology, Udaipur – 313001, Rajasthan, India

Ajit Kumar Singh
College of Agriculture and Research Station, IGKV, Raigarh, Chhattisgarh, India,
E-mail: ajitspices8@gmail.com

D. P. Singh
KVK, Maharajganj (ANDUA&T, Ayodhya), Uttar Pradesh, India

J. B. Singh
ICAR-Indian Agricultural Research Institute, Regional Station, Indore – 452 001,
Madhya Pradesh, India

Paramjit Singh
Punjab Agricultural University (PAU), Regional Research Station, Bathinda – 151001, Punjab, India

Rakesh Kumar Singh
Department of Mycology and Plant Pathology, Institute of Agricultural Sciences,
Banaras Hindu University, Varanasi, Uttar Pradesh – 221005, India; RVSKVV, College of Agriculture,
Indore, Madhya Pradesh, India

S. R. Singh
Scientist (Plant Protection), CSUA&T, Krishi Vigyan Kendra, Mahamayanagar, Uttar Pradesh, India

V. K. Singh
Associate Professor (Horticulture), JNKVV, College of Agriculture, Tikamgarh, Madhya Pradesh, India

M. Surekha
Department of Botany, Kakatiya University, Warangal, Telangana, India

S. K. Tripathi
Department of Plant Pathology, JNKVV, College of Agriculture, Rewa – 486001,
Madhya Pradesh, India

Abbreviations

2,4-D	2,4-dichlorophenoxyacetic acid
AFLP	amplified fragment length polymorphism
APR	adult plant resistance
ARDRA	amplified 16S ribosomal DNA restriction analysis
AUDPC	area under disease progress curve
B.s.	*Bacillus subtilis*
B/CYDVs	barley and cereal yellow dwarf virus
BCSM	*Brassica carinata* seed meal
BGL	β-glucosidase
BIP	backward inner primer
BLP	backward loop primer
BM	bio-intensive management
BYDV	barley yellow dwarf virus
CBH	cellobiohydrolases
CDA	Czapek's dox agar
CLCuD	cotton leaf curl virus disease
CLCuMuB	cotton leaf curl Multan beta-satellite
CLFD	chromatographic lateral flow device
CLS	Cercospora leaf spot
CMA	corn meal agar
CNN	convolution of neural network
Co-PCR	co-operational PCR
CPG-TTC	casamino acid-peptone-glucose-triphenyl tetrazolium chloride
CWDEs	cell wall degrading enzymes
DGGE	denaturing gradient gel electrophoresis
DIBA	dot immunobinding assay
DMI	demethylation inhibitor
DSBs	double-strand brakes
dsRNAs	double-stranded RNAs
EDR1	enhanced disease resistance 1
EG	endoglucanase
ELISA	enzyme-linked immunosorbent assay
EPPO	European and Mediterranean Plant Protection Organization

EPS	extracellular polysaccharide
ETS	transcribed external spacer
FCM	flow cytometry
FI	fluorescence imaging
FIP	forwarding inner primer
FISH	fluorescence in situ hybridization
FL	fluorescence
FLP	forward loop primer
FRET	fluorescent resonance energy transfer
FYM	farmyard manure
GLS	glucosinolates
HIGS	host-induced gene silencing
hpRNA	hairpin RNA
HR	homologous recombination
HRCA	hyper-branched RCA
HT	hyperspectral technology
ICT	information and communication technology
IDM	integrated disease management
IF	immunofluorescence
IGS	intergenic spacer
ISSR	inter simple sequence repeat
ITCs	isothiocyanates
ITS	internal transcribed spacer
LAMP	loop-mediated isothermal amplification
LBPH	local binary patterns histograms
LISS	linear imaging self-scanner
LVQ	learning vector quantization
LW	leaf wetness
MAbs	monoclonal antibodies
MAS	marker-assisted selection
MCRI	modified chlorophyll absorption ratio index
MGA	malachite green agar
MLB	maydis leaf blight
MLO	mildew resistance locus
MLP	multilayer perception
MR	moderately resistant
MYMV	mung bean yellow mosaic virus
NASBA	nucleic acid sequence-based amplification
NDVI	normalized-differenced-vegetation index

NHEJ	non-homologous end-joining
NIR	near-infrared
NMR	nuclear magnetic resonance
NN	neural networks
NPCI	normalized pigments chlorophyll ratio index
NRPS	non-ribosomal peptide synthase
NSKE	neem seed kernel extract
OM	otitis media
OSAVI	optimized soil-adjusted vegetation index
P.f.	*Pseudomonas fluorescens*
PAM	protospacer adjacent motif
PCR	polymerase chain reaction
PCR-RFLP	polymerase chain reaction-restriction fragment length polymorphism
PDA	potato dextrose agar
PDI	percent disease index
Pfl	*Pseudomonas fluorescent*
PG	polygalacturonase
PLPs	padlock probes
PMTs	photomultiplier tubes
PNN	probabilistic neural networks
PSB	phytophthora stem blight
QoI	quinine oxidation inhibitor
qPCR	quantitative PCR
QTL	quantitative trait loci
R/FR	red/FarRed
RAPD	random amplified polymorphic DNA
RBF	radial based functions
RCA	rolling circle amplification
RCR	rolling circle replication
rDNA	ribosomal DNA
RGAs	resistance gene analogs
RISA	radio-immunosorbent assay
RISC	RNA inducing silencing complex
RLH	relative lesion height
RNAi	RNA interference
rRNA	ribosomal RNA
RT	reverse transcriptase
RT-PCR	reverse transcription-polymerase chain reaction

SA   salicylic acid
SBCMV   soil-borne cereal mosaic virus
SBWMV   soil-borne wheat mosaic virus
SCAR   sequence-characterized amplified region
SDA   step-wise discriminate analysis
SDW   sterile distilled water
SF   sodium fluoride
SIBA   seed immunoblot binding assay
siRNA   small interfering RNA
SLP   single layer perception
SMD   sterility mosaic disease
SMSA   selective medium South Africa
SNP   single nucleotide polymorphism
SOP   self-organizing map
SPR   surface plasmon resonance
SSC   scattered laser light
SSEM   serologically specific electron microscopy
SSR   simple sequence repeats
ssRNA   single-stranded RNA
SSU   small subunit
SVT   small vein thickening
SWMV   soil-borne wheat mosaic virus
T.h.   *Trichoderma harzianum*
T.v.   *Trichoderma viride*
TBIA   tissue blot immunoassay
TCARI   transformed chlorophyll absorption ratio index
TLB   *turcicum* leaf blight
UAV   unmanned aerial vehicle
VAM   vesicular arbuscular mycorrhizae
VC   vermicompost
WSMV   wheat streak mosaic virus
WSSMV   wheat spindle streak mosaic virus
WYMV   wheat yellow mosaic virus
YMD   yellow vein mosaic disease

Foreword

Over the years, the science of plant diseases has grown by leaps and bounds, and the contributions made by scientists all over the world have immensely enhanced our understanding of the subject and helped humanity in facing the challenges posed by plant pathogens to various crops. But further, we need to bring in the advanced technologies such as host-induced gene silencing against many pathogens through RNAi, gene silencing for virus disease management, the science of omics in pathogenicity, and several such other molecular approaches for our enhanced understanding of the subject. Pathogenomics may show the way for the management of many plant diseases. The exploitation of safe secondary metabolites from beneficial microbes and integration of the same in disease management is another area to explore. With climate change a reality now, as well as the host-plant agrochemical resistance, food safety issues are more concerning than ever before.

On one side, we are proud of the advances made in molecular taxonomy, state-of-the art diagnostic techniques, identification of newer fungicide molecules, nanotechnology, understanding the mechanism of host-plant resistance, specifically effector genes, whole-genome sequencing of important pathogens, deciphering the hitherto unknown functions of many genes in plant pathogens, developing new formulations of biocontrol agents including plant growth promoters, etc. On the other side, we have the challenges to be addressed *viz.*, invasion and emergence of new diseases, biosecurity, inappropriate use of fungicides or antibiotics and development of fungicides resistance, lack of fool-proof management practices for virus and phytoplasma diseases, especially the viruses transmitted by sucking pests cum vectors.

Further, the principles of quarantine need to be practically implemented in letter and spirit. In spite of our enhanced understanding of the subject, some of the emerging plant diseases. I am aware that the editors Dr. R. K. Singh and Dr. Gopala are editing a book titled, *Innovative Approaches in Diagnosis and Management of Crop Diseases*, published by Apple Academic

Press, Inc. I congratulate both editors and hope that the book would provide knowledge on plant diseases for the benefit of faculty, students, farmers, and society at large.

—**Prof. S. K. Rao**
Vice-Chancellor
Rajmata Vijayraje Scindia Krishi Vishwa Vidyalaya
Raja Pancham Singh Marg, Gwalior (M.P.)–474002, India

Preface 1

There are many kinds of plant pathogens ranging from ultramicroscopic entities to well-defined multicellular organisms, with wide variations in pathogenic potential. They are known to be the principle causes of destructive diseases of many economically important crops cultivated all over the world. Many of them are widely distributed and survive in varied habitats. It is well recognized that development of effective crop disease management depends on the rapid detection and precise identification of the pathogen(s) causing the disease in question. In this context, knowledge of the different methods available for the detection and identification of pathogens is a basic requirement for the successful management of disease(s) affecting the various crops in many location. This volume provides this vital information on currently applied methods of detection. It is our hope that this book will serve the science fraternity as well, and equally hope that the book will stimulate young scholars to work on the paradigm shift in the detection of rice, maize, small millets, wheat, sesame, field pea, pigeon pea, ginger, guava, and vegetable diseases and its management.

— **R. K. Singh, PhD**
Gopala, PhD

Preface 2

Diagnosis and management of biotic stresses play a pivotal role in the foundation of agriculture production. The changing scenario of agriculture from localized technology to modern innovative scenarios has focused our attention on ways to manage plant diseases, reduce crop losses, avoid wide fluctuations in production, and sustain higher levels of productivity.

During the last few decades, the science of crop production has developed rapidly, resulting in many diverse techniques evolved to manage a variety of menace incited by pathogenic microbes and nematodes. Currently, the prospects of chemical control do not appear good because of accidental and incidental trauma encountered in the field of chemical pesticides. Therefore, vigorous research efforts for exploring newer but pragmatic tools for the control of biotic stresses have become imperative under the climate change scenario.

The scenario of climate change drastically fluctuates the pathogenic nature of the microbes. Climate change may activate sleeper pathogens to become more aggressive fungus/bacteria/virus/nematode and attract the attention of scientists due to considerable losses incited by them. However, it is disheartening to observe that the crop grown with the hard labor of the farmers is damaged by various biotic and abiotic stresses. The crop losses estimate range from 15 to 100% due to climate calamities coupled with damage by harmful pests and diseases. Sustain food securities in developing countries under the enormous threat of emerging new strain of virulent pathogens is an uphill task due to more recombination in pathogen and depleting trend of available nutrients in the soil. The global change in climatic and frequent commence of new race/strain of pathogens may lead to influence the infection process, pathogenic behavior, perpetuation habit, and severity, which may directly affect crop production. Reliable and authentic information about the diagnosis and detection of the pathogen is the backbone of management strategies. The early and accurate diagnosis of plant physiological abnormalities is a crucial component of any crop-management strategy preparedness. Plant health abnormalities can be managed most effectively if control measures are introduced at an initial stage of disease development. Visualization of sign or symptoms on the plant parts have only appeared after getting colonization of infection, before that defense mechanism of the plant has defeated by the virulent pathogen. Reliance on symptoms is often not adequate in this regard, since the disease may be well underway when symptoms first appear, and symptom expression can be highly variable. General pathological biological techniques for disease diagnosis and pathogen detection are usually highly accurate but too slow and not amenable to large-scale application. Acceptable accuracy and precision of naked-eye visual disease assessments have often been achieved using traditional disease scales during the past 80 years. The assessment of visual symptoms is essential for the diagnosis of plant diseases. However, these methods are too subjective.

New technologies offer the opportunity to assess disease with greater reliability, precision, and accuracy. Visible light photography and digital image analysis have been increasingly used over the last 30 years, as the software has become more sophisticated and user-friendly for tentative disease diagnosis and suggesting probable management options to the farmers. The inception of polymerase chain reaction (PCR) by Nobel laureate Kary Mullis had a profound impact on nucleic acid-based detection and biodiversity analysis of all types of fungi, bacteria, viruses, viroids, nematode, fastidious bacteria, phytoplasma, and algae. While nucleic acid technology is the only choice for detecting pathogens that have not been cultured or biographic in nature, DNA-based methods and serological techniques have not yet completely replaced classical microbiology and visual inspection. The complementary information generated by those techniques may be utilized for confirmation of individual pathogens by way of conducting pathogenicity. Current and future methodology for detection of plant disease include immunological and DNA-based methods, proximate detection approaches based on the analysis of volatile compounds and genes as biomarkers of disease, sensors based on phage display biophotonics, and remote sensing (RS) technologies in combination with spectroscopy-based methodologies.

This book, *Innovative Approaches in Diagnosis and Management of Crop Diseases,* comprises critical updated reviews and research articles on important diseases of different crops and their recent innovation in the detection of plant pathogens and the most appropriate management strategies with recent technologies. This edited book consists of selected 40 chapters, contributed by renowned and expert scientists working on the different aspects of plant pathology in India. The information on various topics is at advanced as well as comprehensive levels, covering important diseases of crops, incited by bacteria, fungi, viruses, viroids, phytoplasma, and nematode of commercial field and horticultural crops and their management. Chapters cover recent advances in diagnosis and detection of diseases of rice, wheat, pulses, guava, aonla, cruciferous, cucurbits, ginger, sesame, cotton, pigeonpea, field pea, small millets, and maize, in addition of that book thoroughly accommodated the multidimensional ways of extraordinary recent technologies generated and their future prospects are comprising in the chapter of individual disease. This book is a standard reference work, offering available basic facts, re-evaluating and reviewing the past research, and providing the new and current discoveries on the subject and up-to-date information on nanotechnologies, green nanotechnologies, soil-borne diseases, host resistance, bio-intensive management options, biodiversity, the ecology of the seed-borne pathogen, bio-fumigation, versatile biocontrol agent, plant quarantine, plant immunization, and climate change. We are grateful to all authors for contributing their valuable original article to this book. The editorial help of Sandra Jones Sickels, Vice President, Editorial and Marketing, Apple Academic Press Inc., Publishing Department, Taylor and Francis is sincerely appreciated. It is our hope that this book will serve the science fraternity as well, and equally hope that the book will stimulate young scholars to work on biological control, nanotechnology, host RNAi defense, recombinant coat protein, NGS, CRISPR Cas9 genome editing, plant quarantine, biodiversity, plant immunization, and recent detection and diagnosis of plant pathogens.

—**R. K. Singh, PhD**
Gopala, PhD

CHAPTER 1

Paradigm Shift in Detection of Rice Diseases and Their Management

MANAS KUMAR BAG,[1] PRAHLAD MASURKAR,[2] ANUPRITA RAY,[1] and RAKESH KUMAR SINGH[2]

[1]Crop Protection Division, ICAR-National Rice Research Institute, Cuttack, Odisha – 753006, India, E-mail: manas.bag@gmail.com (M. K. Bag)

[2]Department of Mycology and Plant Pathology, Institute of Agricultural Sciences, Banaras Hindu University, Varanasi, Uttar Pradesh – 221005, India

ABSTRACT

Rice is the staple food crop for about 65% of the Indian population, contributing 40% of total food grain production, thus occupies a pivotal role in the food and livelihood security of people. During the last 60 years, productivity has increased by five times, and this growth has come mainly from yield increase. Increasing productivity has now become challenging due to the degradation of natural resources and the problem of changing climate. Thus, future production has to be included in a more efficient and environment friendly production system. An efficient production system includes efficient management of rice diseases, cause losses both quantitatively and qualitatively, due to pathogens such as fungi, bacteria, and viruses. Major fungal diseases which affect the crop health are blast, sheath blight, brown spot, false smut, bacterial blight, bacterial streak, bakanae, etc. The impacts of these diseases are very devastating and widespread in different rice-growing areas. In order to minimize the disease-induced damage during the rice-growing period and for maximizing productivity it is necessary to monitoring health of plant and detecting pathogen in early stage for avoiding further spread of disease and initiating management practices at proper time. In this book chapter, we will focus on the conventional detection method such as visual observation to paradigm shift to advanced detection methodology including use of modern tools along with use of portable monitoring sensors and judiciary use of disease management practices such as cultural practices, source of planting material, epidemiology, forecasting models, biological control and use of new generation chemicals including recent focused research on development of nano-formulation to bio-intensive approaches.

1.1   INTRODUCTION

Increasing population and changing climate scenario became challenge to the food growers to surplus the demand of food in the world. Staple food like wheat, rice, potato, etc., play a major role in fulfilling the global hunger. Among these, rice is one of the central pillars of food security on the Indian perspective that contributes about 77% of total area and food grain production (Singh, 2011). Continuous reduction of landholding is the main cause behind the less production of rice. However, increase in incidence of pest and diseases are major constraints in achieving the targeted food production. On an average pest and diseases cause loss up to 37% of rice yield (Sparks et al., 2012). Improved detection and integrated management practices can minimize the losses due to rice diseases.

Detection and management of diseases are the most important aspect of any cropping program. It is extremely important to identify the disease by various detection methods so that better management strategy can be applied at an early stage of disease initiation, which resulted in lesser economic losses to rice production. In conventional method, visual observation is the most important way of detection but that needs substantial amount of time to establish the disease which is not acceptable. Therefore, in today's context of complexities of disease, rapid diagnostic methods are required which can give accurate and timely solution of proper identification of disease at the earliest time. In the current perspective, advanced, and rapid methods like PCR, LAMP, FISH (fluorescence *in situ* hybridization), ELISA, immunofluorescence (IF), flow cyclometry, thermography, fluorescence imaging (FI), remote sensing and hyperspectral technique and portable sensors are used for accurate and specific detection of pathogen (Cai et al., 2014; Kempf et al., 2000; Ward et al., 2004; Chitarra and Van den Bulk, 2003; Bravo et al., 2004; Charle et al., 2007; Burling et al., 2011; Delalieux et al., 2007; Shipway et al., 2000). Focusing on the appropriate detection method for disease of rice may help in decision making for proper management of the crop.

1.2 DISEASE DETECTION ON THE BASIS OF VISUAL SYMPTOMS

1.2.1 RICE BLAST

This disease is caused by *Magnaporthe oryzae*, a hemibiotroph estimated to destroy 10–30% of rice crop which is enough to feed 210–740 million people (Boddy, 2016). In the beginning, white to green lesions or spots with a border of dark green appears on the leaves. When the lesions become older it became elliptical or spindle-shaped with whitish to gray center along with red to brownish or necrotic border. The spot resembles with diamond shape, wide in the center and pointed towards either end. In the later stage of disease formation lesions coalesce; enlarges growth which results to kill the entire leaves (Asibi et al., 2019).

The symptoms initially appear at the junction of leaf and stem sheath necrosis is observed. Collar infection result in entire leaf killing and it extended around the leaf sheath sometime fungus also produces spores on the lesions.

Rice neck and panicles: neck of the plant is a very important part because it holds the seed or panicle. When blast fungus infects the neck leads to a condition called rotten neck. Infection at neck causes more economic loss because it causing seed failure (blanking) or the entire panicle would be chaffy (Zhu et al., 2005). The fungus also infects the pedicels. Infection on pedicels made panicles brown and seedless which is again a major concern.

1.2.2 BROWN SPOT OF RICE

This disease caused by the fungus *Bipolaris oryzae* Subr and Jain (= *Helminthosporium oryzae*). The visual symptoms produced by fungus from seedling to dough stage. At the seedling stage, infected nurseries shows brownish scorched appearance. Disease usually called sesame leaf spot due to peculiar spot produced by pathogen on the leaves. The spots are initially appeared as minute brown dots, later became cylindrical or oval to circular (resemble sesame seed) (Sunder et al., 2014). The size of spots is around 0.5 to 2.0 mm in breadth, which again coalesces to form large patches. In later stages, several spots coalesce, and whole leaf dries up. The fungus infection also occurs on the panicles and neck with brown color appearance. Some symptoms on seeds look like black or brown spots are covered with olivaceous velvety growth. Dark brown or black spots also appear on the glumes. At severe stage discoloration of seeds observed and these affected grains sometimes get shriveled (Padmanabhan, 1973; Barnwal et al., 2013).

1.2.3   SHEATH BLIGHT OF RICE

This disease caused due to fungus *Rhizoctonia solani* having anastomosis group AG1-1A. In this disease, symptom appears as green or gray ellipsoid lesions on the sheath of rice in acropetal succession. The size of ellipsoidal lesions varies from 0.5 to 3 cm. In the later stage of infection, lesions appear on other parts of leaves, coalesce, and covering the entire stem and sheath of the plant leading to stem lodging from the waterline. Due to lodging, water transport blocks which resulted disturbance in canopy architecture and also affect the photosynthetic activity of the plant (Yellareddygari et al., 2014; Agrios, 2005; Ceresini, 1999).

1.2.4   FALSE SMUT OF RICE

The causal organism of this disease is *Ustilaginoidea virens* telomporph *Villosiclava virens* (Tanaka et al., 2008). The correct information regarding this disease is an enigma. The visual symptoms appear at the grain filling stage, after which management is difficult. The primary appearance is recognized through presence of white to dull white mycelium covered in spikelet, and when crop mature the color of the mycelium changes which may due to production of conidia and sclerotia (Bag et al., 2017). The size of the infected individual grain become more than normal and formed a ball-like structure. The ball changes its color from white to yellow, then olive green and later black, which was filled with a mass of sclerotia (Yong et al., 2018).

1.2.5   BAKANE OF RICE

Balance disease is also known as foolish seedling of rice caused due to infection of fungus, *Fusarium fujikuroi* (Wulff et al., 2010). This disease easily distinguishable in the field due to hyper or hypo elongation of stem, pale green flag leaf, yellowish-green leaves unlike the normal plant. Later development of rigid or wiry adventitious roots on the first or second nodes and death of infected plant may also occur before maturity. Generally white mycelium and pink (occasionally) sporodochia are produced on stem just above water level when severe infestation observed (Zaunudin et al., 2008; Agarwal et al., 1989). Browning of stalk, leaf, and ear of rice at the flowering period observed and this is due to toxin fumonisins produced by the fungus *Fusarium fujikuroi* (Sultana et al., 2019).

1.2.6   BACTERIAL BLIGHT OF RICE

Xanthomonas oryzae pv. *oryzae* is the causal organisms of bacterial blight of rice and cause severe yield loss (Arshad et al., 2015). Bacterial blight appears at the nursery stage as well as in the later stages of plant growth. Younger plants less than 21 days old are the most susceptible and favor the disease. Bacterial spots appear as water-soaked lesions started from leaf tips and progress downwards along the margins. These water-soaked lesions enlarging and turn yellow and ultimately leading to the death of the plants. Bacterial blight manifested either by leaf blight or "Kresek" (acute wilting of young plants) phase (Nino-Liu et al., 2006).

1.3   DETECTION TECHNIQUES

1.3.1   DIRECT DETECTION TECHNIQUES

1.3.1.1   POLYMERASE CHAIN REACTION (PCR)

The polymerase chain reaction (PCR) technique has been used on the principle of DNA hybridization and replication. Initially, PCR technique used for the detection of highly specific diseases caused by bacteria and viruses in mammals (Cai et al., 2014). Now, this technique also has been used for plant pathogen detection. Besides basic PCR methods, nowadays, new and improved advanced methods such as RT-PCR (highly sensitive), multiplex PCR (simultaneously detect different DNA or RNA by running a single reaction) (James, 1999; Nassuth et al., 2000). The PCR-based diagnostic assay was developed to detect *Bipolaris oryzae*, causal agent of brown spot of rice. Universal primers ITS1 and ITS4, used to obtain the rDNA sequence of *Bipolaris oryzae* specific primers, were designed based on *B. oryzae* ITS sequence data. The species-specific primers BoVf and BoVr are used in the detection of *B. Oryza* that amplify a 275 bp fragment from the DNA of *B. Oryza* but not from other species. This technique is also used for knowing the amount of *Magnaporthe oryzae* pathogen proliferation in blast-infected plants. *M. oryzae* proliferation was recognized by the real-time PCR. In real-time PCR, a combination of primer pair and Taqman probe specific to MHP1, a unigene encoding hydrophobin is used which is indispensable for normal rice plants for virulence expression. The reliability of this method is also checked by DNA samples from the rice leaf infected with compatible or incompatible strains of *Magnaporthe oryzae* (Su'udi et al., 2013).

1.3.1.2   *LOOP-MEDIATED ISOTHERMAL AMPLIFICATION (LAMP)*

Identification of pathogen based on visual symptom of the diseases is gradually reducing its importance because similar type of symptoms produce by different pathogens makes confusion in decision-making. In this era of molecular-based detection method, available techniques like Nested PCR, qPCR, and loop-mediated isothermal amplification (LAMP) assay, etc., gives more accurate and efficient diagnosis. Amplification of very small amount of targeted DNA sequence with highly efficient polymerase enzyme, in a static temperature (usually 63–67°C) and loop structure formation of take place when the LAMP primers amplify their target DNA sequences. The LAMP primers are a set of six number primers, i.e., Forward Inner Primer (FIP), backward inner primer (BIP), forward loop primer (FLP), backward loop primer (BLP), F3 and B3 (Notomi et al., 2000).

This assay constructed for manage the raw plant material without any precedent pervasive extraction process or incubation method; furthermore, it does require neither full-fledged laboratory nor advanced analytical knowledge. For efficient control of plant diseases, it is essential to detect the disease at an early stage with a reliable precision (Sankaran et al., 2010). This technique has been used in the detection of plant pathogenic bacteria with greater success (Ruinelli et al., 2017). Reason is isothermal nature of reaction and protocol can be followed in the field using a simple water bath or a heating block with fluorescence (FL) reader. The advantage of this technique is its less time requirement between sampling and sample analysis and therefore convenient for regulatory laboratories and quarantine officials (Hodgetts et al., 2015). Also suitable for nonspecialized laboratories having portable detection device which is simple to operate. Now primers have been developed for rice leaf blight (*Xanthomonas oryzae* pv. *oryzae*) and leaf streak (*Xanthomonas oryzae pv. oryzicola*) to use in reliable and sensitive LAMP assay, is so specific that even it can detect the pathovar along with distinguish between African and Asian lineage of pathogen (Lang et al., 2014).

In LAMP assay, the most suitable target region of 'DNA sequence of interest' identified this can be achieved by the use of comparative genomics, where all the related strains of the pathogen were BLAST and against the database derived from the related strains of the same pathogen. In BLAST the nt/nr also check of the sequences which was segmented earlier as approximately 500 bp and then these probes were screened which allow us to select the large contigs and non-transposable element which is highly specific to the pathogen DNA sequence. Now on the basis of these conserve region of DNA of pathogen we are able to synthesize the primers for amplification of the region of pathogen.

With the current increase in the design of LAMP detection methods, agencies like phytosanitary are able to detect quarantine organisms rapidly and prevent their spread. LAMP is easy to follow, require nominal training, and thus accessible to field workers. LAMP assays now used at entry points like airports for screening and limiting quarantine pests import and a valuable addition for screening planting material to avert further spread of bacteria. Moreover, long incubation step increases the processing time between sampling and results, and requires extra laboratory work. Thus, long incubation time is not suitable for on-site use either at airports or phytosanitary lab at borders as it would require commodities on analysis to be on hold till results are attained (Notomi et al., 2015; Wang and Turechek, 2016).

1.3.1.3   *FLUORESCENCE IN SITU HYBRIDIZATION (FISH)*

Fluorescence *in situ* hybridization (FISH) is a highly sensitive technique that allows visualization, identification, enumeration, and localization of individual microbial cells simultaneously. Ideally, this method is used to spot the presence or absence of specific DNA sequence in the genome. FL probes are used to bind the specific sequence and are detected through FL or confocal microscope. In this method, mechanism of identification is based upon identification of pathogen-specific ribosomal RNA (rRNA) sequence in plant, probe will bind in each of the sample in the ribosome which provides high affinity and specificity towards the pathogen, but auto inflorescence and photobleaching by pathogens are also a major challenge to this technique. Sole fungal isolate of *Rhizoctonia solani* may possess as many as five different dsRNA, and the presence of specific dsRNA influence pathogenicity in host plants. This method is commonly used to detect dsRNA and thus can be adopted for the identification of different anastomosis group of *R. solani* (Charlton et al., 2009) (Figure 1.1).

1.3.1.4   *ENZYME-LINKED IMMUNOSORBENT ASSAY (ELISA)*

The principle of enzyme-linked immunosorbent assay (ELISA) is based upon an antibody, antigen reaction, and changing of color of reaction (Clark and Adams, 1977). This method is particularly used for viral pathogens.

The viral pathogen secretes antigen are specifically bind with antibodies and conjugated to an enzyme. The detection can be visible as color changes according to the enzyme. Thus, performance could be improved significantly through the application of specific monoclonal as well as recombinant antibodies (Rowhani and Falk, 1995).

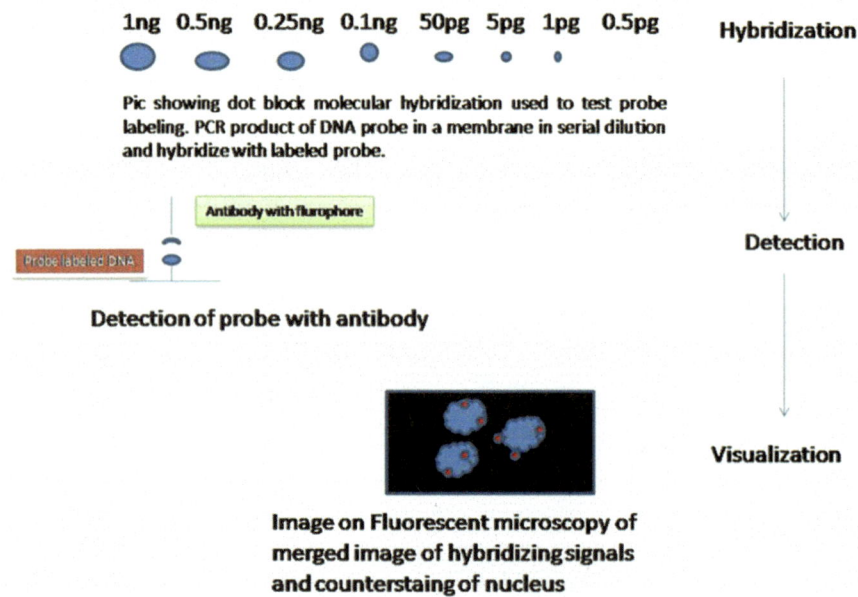

FIGURE 1.1 Steps in fluorescent *in situ* hybridization (FISH). (Source: Adapted from Burns, 2009).

1.3.1.5   IMMUNOFLUORESCENCE (IF)

Immunofluorescence (IF) technique is highly sensitive towards the targeted microbial samples. In this technique, IF microscopy-based detection used. Thin tissue of infected plant sample is fixed to microscopic slide, and detection is achieved by conjugating a specific dye to specific antibody to visualize the distribution of targeted molecule throughout the sample. Bourett and Howard (1992) used fluorophore-labeled anti-actions in electron microscopy and IF microscopy for detection of penetration peg produced by appressoria of blast pathogen *M. oryzae*.

1.3.1.6 FLOW CYTOMETRY (FCM)

Flow cytometry (FCM) is a method commonly used for both routine detection and research in food microbiology and veterinary research (Figure 1.2). It is comparatively new in plant pathology research. Use of FCM for detection of plant diseases has increased because of large number of plant tissue can be analyzed rapidly. FCM used for indexing of plant tissue for any pathogens' genetic sequence for which a specific antibody can be devised and a fluorescent probe attached (D'Hondt et al., 2011). Flow cytometers also used for monitoring of viable but nonculturable plant pathogens.

FCM also used for knowing the sexual mating type of the pathogen. The laser flow cytometer is used to detect the nuclear condition of the somatic strain of the pathogen *Phytophthora infestans*, i.e., a single population of nuclei in non-replicated diplophase (Catal et al., 2010). In *Puccinia* species relative DNA content of pycniospores observed through the FCM (Eliam et al., 1994). Specific detection of *Xanthomonas campestris* var. *campestris* was observed in the mixed population of *Pseudomonas fluorescence* or non-pathogenic *Xanthomonas campestris* strain in crude seed extract by Chittara et al. (2002) with the use of FCM. Variation and instability in the DNA content of the single oospore progeny of *Phytophthora ramorum* was confirmed through this technique (Vercauteren et al., 2011) (Figure 1.2).

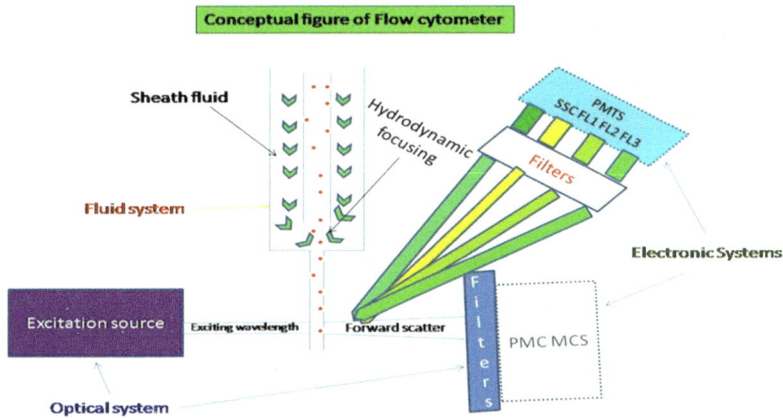

FIGURE 1.2 Flow cytometry: A set of filters and mirrors deflects and passes certain wavelengths of the emitted fluorescence (FL1-3) and scattered laser light (SSC and FSC). The key parts of the electronics system are the photomultiplier tubes (PMTs) that detect the incoming photons, multiply the current they produce and send this electric signal to the computer where it is displayed as a single-parameter histogram or two-parameter dot. (Source: Adapted from D'Hondt et al., 2011).

1.3.2   INDIRECT DETECTION TECHNIQUES

1.3.2.1   THERMOGRAPHY

When plants infected by a pathogen, the individual leaf or whole canopy changes its temperature. The change in canopy temperature changes some wavelength of light which can be detected by thermographic instrument. This early detection of problems can be managed in real-time, and further damage will not be caused by the pathogen. However, this technique also has its limitations as changes in environment can confound the use of thermographic instrument for outdoor use and the technique is also nonspecific (Fang and Ramasamy, 2015) (Figure 1.3).

Leaf temperature of cucurbit crops was monitored to know the difference between diseased plant [powdery mildew pathogen (*Pseudoperenospora cubensis*) infected] and healthy plants. The diseased leaves show 0.8°C less temperature than normal leaves due to abnormal stomata opening (Lindenthal et al., 2005). In another findings apple leaves infected by *Venturia inaequalis* (apple scab), thermographic measurement not only used as a parameter to differ diseased and non-diseased (healthy) leaves but also quantify the disease and is applied in screening system as maximum temperature difference increases with the increasing size of the lesions (Oerke et al., 2011). Another application of thermography is detection through thermal imaging in the pre-symptomatic diagnosis (early disease detection) from remote. In leaves infected with *Plasmopara viticola* amount of water transpired increases, and early detection of irregularity in temperature monitored, and this help to advise early spraying of fungicide (Stoll et al., 2008).

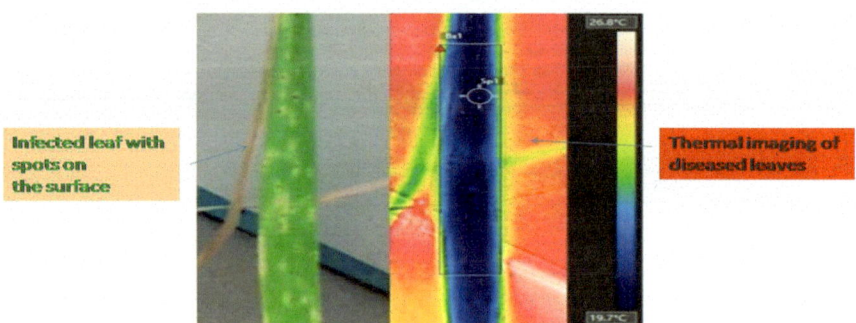

FIGURE 1.3   The area temperature of the diseased sample was significantly higher than other areas of the leaf surface. Dark blue area represents more temperature and other than this area showing less disease so less dark area of leaf tissue.
(Source: Adapted from Zhu et al., 2018).

1.3.2.2   FLUORESCENCE IMAGING (FI)

In fluorescence imaging (FI) technique, chlorophyll is measured on the leaves as a function of light incidence and change in FL can be used to analyze pathogen infection, based on changes in reading of photosynthetic apparatus reading then normal and the photosynthetic electron transport reaction. FI is also a non-invasive technique in which autofluorescence is collected from plants or leaves when ultraviolet excited on them. Chlorophyll A generally give red and far-red region when emitted while blue and green regions is emitted by secondary metabolites and lots of this secondary megabyte act as phenolics rated to plant defense (Buschmann and Lichtenthaler, 1998).

The bandwidth of light 370 nm, 620 nm, and chlorophyll fluorescence spectral 600–800 nm was used for the validation of nitrogen deficiency and imaging leaf rust of wheat (*Puccinia graminis* f. sp. *tritici*). Slightly higher amplitude ratio of R/FR (Red/FarRed) in healthy leaf samples than diseased. While B/G (Blue/Green) amplitude ratio decreases in diseased leaves than healthy leaves of powdery mildew of wheat when observed in fiber-optic fluorescence spectrophotometer (Burling et al., 2010). Early detection of powdery mildew of grapevine (*Plasmopara viticola*) was possible due to maximum quantum yield of photosystem II photochemistry (Fv/Fm) restricted to infected leaf area and effective quantum yield of photosystem II photochemistry (φPSII) lower with larger area of localized spot of leaf infected with pathogen observed at chlorophyll FI with high resolution (Cséfalvay et al., 2009).

1.3.2.3   HYPERSPECTRAL TECHNIQUE

In this technique, the hyperspectral image covers more bands instead of conventional three bands of colored light. In hyperspectral imaging high consistency of color reflectance used over a large range of light spectrum that is beyond human vision. Thus, used for identifying potential minor changes in plant growth and development and further applied in the field of plant diseases. The implementation of the hyperspectral technology (HT) applied in the study of sheath blight and blast of rice. Through color features and color space transformation of images, clear-cut difference between diseased and healthy plants can be observed (Kobayashi et al., 2001).

Rice reflectance measured for determining the spectral region most sensitive to panicle blast infection, i.e., 475–670 nm. After the yellow ripe growth stage near-infrared (NIR) light found most suitable (Kobayashi et al., 2001). Wavelength of 650–700 nm (visible wavelength) was found best for early detection, i.e., three weeks after infection of *Venturia inaequalis* (apple scab) when hyperspectral approaches applied and it also helped in determination of deteriorate leaf tissue vs. good one (Delalieux et al., 2007).

In HT development of different neural networks (NN) for processing hyperspectral data developed. Major type of NNs is single layer perception (SLP), multilayer perception (MLP), radial based functions (RBF), Kohen's self-organizing map (SOP), probabilistic neural networks (PNN) and convolutions neural network (CNN). The hybridization of NNs and hyperspectral approached has emerged as a powerful tool for plant disease detection and diagnosis (Golhani et al., 2018). Hyperspectral reflectance principle used for the determination of different fungal infection level on the rice panicles. Different infection stage of rice panicles which were evaluated was no infection, light, and moderate infection due to glume blast, severe infection of panicles due to false smut disease under the hyperspectral range of 350–2500 nm in the portable spectroradiometer. Various spectra processing method were used for analyzing the reflectance of levels of infection. A learning vector quantization (LVQ) method with some more neural network classifier was employed which were able to discriminate different fungal infection levels of rice panicles (Liu et al., 2010). Hyperspectral sensors (ASD field spectroradiometer) were also available for rice brown spot disease (Liu et al., 2008). Application of HT on bacterial blight of rice also used for the early detection of disease with inclusion of different vegetation indices, i.e., NDVI (Normalized difference vegetation index), SR (Simple Ratio), red edge, MCRI (modified chlorophyll absorption ratio index), NPCI (normalized pigments chlorophyll ratio index), TCARI (transformed chlorophyll absorption ratio index), OSAVI (optimized soil-adjusted vegetation index) observed (Singh et al., 2010). The step-wise discriminate analysis (SDA) in hyperspectral reflectance coupled with satellite data obtained by LISS (linear imaging self-scanner) revealed only four-band, i.e., 760, 990, 680, 540 nm and geospatial maps could analyze for bacterial blight of rice (Das et al., 2015). Another application of hyperspectral imaging is identification of toxicogenic agents, i.e., fungi in the food materials which are highly useful in the food industry. From non-destructive, hyperspectral imaging-based method on maize kernels identifies the toxicogenic fungi where traditional method was not effective (Fiore et al., 2010).

1.3.3 PORTABLE SENSORS

Different types of sensors developed are available for various processes, including environment monitoring and medicinal diagnosis. The principle for detection through portable sensors is based on electrical, chemical, optical, magnetic, or vibration signals. The sensitivity of sensors can be enhanced by the use of bio-recognition elements. There are some biosensors platforms are available on nanomaterials, antibody, DNA/RNA, and bacteriophage.

In the nano materials-based biosensors, metal and metal oxide nanoparticles are used. These nanoparticles work with other foreign biological materials such as antibodies of pathogen, which can be easily detectable by portable sensors. Generally, in metals gold-based nanoparticles are used because of its high sensitivity, specificity, photothermal (i.e., transfer light in to thermal energy), high electric conductivity, electroactivity, and single probe sensors. In addition, nanochips are made in microarrays which contain fluorescent oligo probes and helpful in detecting single nucleotide change and which make these sensors more reliable for detection.

In the portable sensors, quantum dots also used which helps in disease detection. Quantum dots have their advantages as unique optical properties, with fluorescent resonance energy transfer (FRET). The mechanism behind the quantum dots sensitivity is energy transfer between two light reactive molecules. The sensitivity of these sensors increases its applicability for detection of vector also.

Some affinity biosensors are also now applied, which work on the principle of biorecognition elements. Antibody-based biosensors allow rapid and sensitive detection of a wide range of pathogens such as seed, soil to airborne diseases. Antibody-based biosensors worked on the principle of immunogens lies in the coupling of specific antibody with a transducer which converts the binding event to a signal that can be analyzed. Other biosensors which were developed on the basis of DNA/RNA affinity. These biosensors analyze the DNA/RNA probe which attached to the membrane of pathogens DNA/RNA.

Recently Bacteriophage-based biosensors are developed. The interaction between the bacteriophage and the pathogen, impedance of charge transfer reaction at the interface changes which used as signal for detection. The detection by bacteriophages depends upon the CFU/ml of pathogen, if CFU/ml is more than the detection will be on less duration of time and vice-versa.

1.3.4   UNMANED AERIAL VEHICLE (UAV)

A drone or UAV (unmanned aerial vehicle) typically refers to pilotless small aircraft that operate through the combination of telecommunication, including computer vision, object avoidance technique, artificial intelligence, and image analysis. Drones application increases in each and every field because broad coverage applicability from war to peace. Drones depended easy aerial monitoring process, high spatial resolution images and information technology made this technique greater importance in the field of agricultural monitoring (Manjunath et al., 2015; Abu Sari et al., 2018; Norasama et al., 2018).

Detection of plant diseases through some automatic techniques like drones reduces the time of monitoring in a big farm and early detection give more reliability for these UAV (Singh and Mishra, 2015). Disease management in agriculture often speculates that pathogens were distributed in the whole field, but they spread in patches. At present, disease detection predominantly done by human is slow and expensive. Drone approaches holds promise for this. For any disease management, detection of its sign is the primary step easily done by these UAVs. When UAVs were equipped with multispectral digital cameras, capturing high-resolution images of the field crop is possible. On processing and analysis of images, characterization of particular disease and quantification of different levels of the same disease in that field can be done. On experiment basis this practices followed in brown leaf spot of rice (Cai et al., 2019), sheath blight of rice (Zhang et al., 2017), sugarbeet leaf spot (Ziya et al., 2018), peanut leaf spot (Balota and Oakes, 2016), olive verticillium wilt (Calderón et al., 2013), radish *Fusarium* wilt (Dang et al., 2018), grapevine leaf stripe (Di Gennaro et al., 2016), potato blight (Nebiker et al., 2016) and wheat yellow rust (Su et al., 2018) (Figure 1.4).

When the detection of disease confirmed sprayer, system mounted on the UAV used for pesticide spray. Through this accurate site-specific application for a large area of agricultural fields prove a potential platform for disease and pest management (Huang et al., 2009; Zhu et al., 2010). The efficacy of UAV increases with PWM controller (Costa et al., 2015). In Asia pacific region Yamaha RMAX (Giles and Billing, 2015) petrol-powered is in trend. UAV spraying mechanism was regulated WSNs (Wireless Sensor Network) deployed in field, these WSNs gave the feedback which were analyzed and sent to drones through control loops on the basis of that UAV sprays the fungicides (Facial et al., 2014). Recently a GPS coordinates-based quadcopter introduced as an aerial automated pesticide sprayer (AAPS) in the lower altitude environment (Vardhan et al., 2014). Work on the effect of spray volume on deposition of pesticide and by which control of wheat powdery mildew and aphid were assessed (Wang et al., 2019).

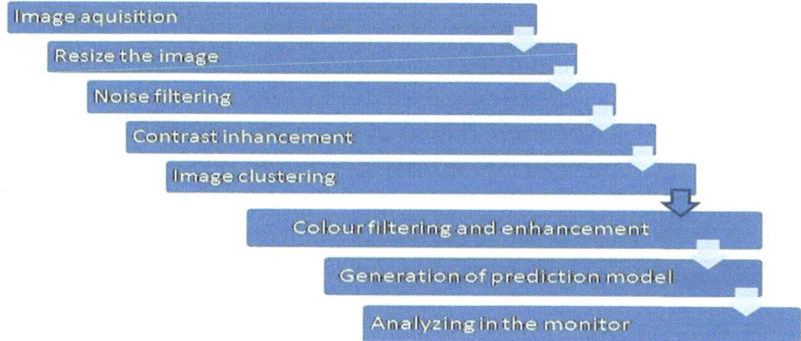

FIGURE 1.4   Algorithm generation for disease and pest damage model.
(Source: Adapted from Hari Shankar et al., 2018).

1.4   MANAGEMENT

Management aspect is the most important aspect for any disease control. However, detection is the most important part of management strategy. Once detection at proper time be possible management process will run smoothly with the help of latest application tools. Now a day instead of control, management of any disease are followed by using all the appropriate tools viz., cultural, biological, breeding approach and in last the chemical uses to maintain the balance between nature and microorganism.

1.4.1   BLAST

The disease occurs at all growth stages of rice and damages the crop. Outbreak rules of rice blast have been developed by Murlidharan and Venkat Rao during 1980 and 1987 by which accurate warning can be issued to farmers. Outbreak rules: moderate to high incidence of leaf blast must be observed in trap nurseries at weekly intervals with a locally susceptible cultivar with high plant population density, high application of fertilizers. Night temperature less than 20°C coupled with high relative humidity that is more than 90% for 7 to 10 days and cloudiness, dew deposition, rainfall or drizzle. For neck blast at tillering stage, there must be some neck blast symptoms present. Low night temperature less than 20°C during panicle initiation to flowering stage. If the above-mentioned rule not satisfied, then the other supporting conditions compensate the rules and lead to an outbreak of leaf or neck blast. For example, if high temperature was prevalent at night so the intense dewfall, extended due point for frequent drizzle compensate it (Pasalu et al., 2006). Kaundal et al. (2006) introduced a machine learning techniques model for forecasting rice blast in India. Six weather variables were selected viz. temperature (max and min), RH (morning and evening), rainfall, and rainy days per week. Blast is mostly preferred by particular air and soil temperatures, relative humidity (RH), hours of continuous leaf wetness (LW), degree of light intensity, duration, and timing of dark periods. All of these have been considered as very crucial for development of the disease.

Cultural practices exert a deep influence on blast disease development as intensive planting of rice at the field; high fertilizer dose, overwintering of conidia on grass host *viz.; Panichum repens, Brachia riamutica, Digitaria sanquinalis,* and *Leersia hexandra* are the weeds which serves as source of primary inoculums. In biological control, seed treatment with *Trichoderma, Bacillus subtilis, Pseudomonas fluorescens* or soil application consider effective and promising for blast control (Nanda and Gangopadhyay, 1983; Chen et al., 2019). Transgenic rice having two chitinase genes (ech42 and nag 70) and one beta-1,3-glucanase (gluc78) of *T. atroviride* resulted in resistance to *Rhizoctonia solani* and *Magnaporthe grisea* in rice (Liu et al., 2004). Neem coated urea 60 and 90 kg per ha was effective in controlling blast.

Padmanabhan (1975) suggested that rice blast disease was governed by the dominant or recessive or by digenic/polygenic genes. Recently in four sterile lines of rice cultivars, i.e., Y58S, Guangzhang63S, C815S and HD9802S, introgression of nine resistant (Pi37, Pit, Pid3, Pigm, Pi36, Pi5, Pi54, Pikm, and Pb1) broad-spectrum cloned blast resistance genes were used (Jiang et al., 2019). So far blast resistance genes pig and pid3 shows significant enhancement in resistance for whole rise growing period. CAR14 variety found resistant to blast disease (Chou, 2019). Among the 1314 germplasm accessions (ICAR-IIRR, NBPGR) evaluated for leaf blast resistance at Hazaribagh, 19 accessions (IC No. 245865, 246277, 246403, 246274, 454167, 121865, 199562, 218270, 245927, 246012, 246228, 246273, and 246659) were highly resistant (SES scores 0, 1, 2) (Jena et al., 2018). Yadav et al. (2019) found some resistance lines to blast, i.e., CM76, Chanpe, Jyoti, Jaya, Birun, etc.

Need-based management of nitrogen application are advised. It is advised to avoid topdressing nitrogenous fertilizer when conditions are conducive for blast development and better to apply fungicides and wait for manageable condition. On appearance of 1–2 leaf spot per plant fungicides like Isoprothiolane 40EC @1.5 ml or Propiconazole 75WP@ 1 ml per liter of water should be applied for blast control. Application of Tebuconazole 50% + Trifloxystrobin 25% WG @ 0.4 g per liter of water give satisfactory result in blast management (Dutta et al., 2012).

1.4.2 BROWN SPOT

Earlier it was considered as nutritional deficiency but later recognized as disease. On analyzing of resistance sources, three QTLs *viz.*, QTLs, qBS2, qBS9 and qBS 11 are found on the chromosomes 2, 9, and 11 have been found associated with brown spot disease in rice cultivar Taduka with latest qBS 11 having major effect (Sata et al., 2008). Out of 573 Assam rice collections, 22 accessions were found moderately resistant (MR) viz., ARC-5846, 5918, 5956, 5550, 6017, 6058, 6101, 6110, 6170, 6622, 7080, 7335, 10618, 10670, 10922, 10934, 11206, 11434, 11566, 11641, 11679, and 12006 (Jena et al., 2018).

For chemical management Propiconazole 25EC @ 1 ml or Mancozeb 75WP or Carbendazim 50WP @ 2 g of water or Carbendazim 64%+Mancozeb 8% 75WP @ 1.5 g per liter of water was found most effective in checking this disease.

1.4.3 BACTERIAL BLIGHT

Bacterial blight most commonly occurs at the places where more cyclonic Storms and intense monsoon rains happens the above condition with the field of high dose of nitrogen fertilizer prone to epidemics of this disease. There are lots of reason for disease dissemination and further development, such as rainstorm strong wind SAP sucking insect and intercellular operations. Heavy rains coupled with high temperature presence of deep irrigation water and severe winds favors' disease, and severe summer and drought suppress the disease. Soil moisture at and above saturation favors the development of Kresek phase and symptoms appear within 10 days at about 30°C.

For controlling of BB cultural practices such as field sanitation aiming at removing weeds rice straw found helpful. Recently *Pseudomonas aeruginosa* strain BRP 3 was found suppressing BB pathogens (Yasmine et al., 2017). Benzylpenicillin, ampicillin, and kanamycin found best for controlling of BB pathogen (Khan, 2012).

Resistance variety as improved Samba masuri developed by introgression of bacterial blight resistance gene *Xa 21 Xa 13* and *Xa 5* by Sundaram et al. (2014). *Oryza barthii* possessed resistance gene and shown resistance against most of the Indian races of *Xoo* especially in Eastern India. Works at NRRI, reported BR-4-39-51-2, BR-51-49-6, IR3796-14-2, ARC-5925, and ARC 5943 as highly resistant and another 50 lines as resistant to BLB kresek phase. A total of 5000 lines were screened for bacterial blight resistance and 50 were resistant. Some of them are AC 36797, 35799, 36370, 36362, 35720, 36357, 36253, 35734, 36369, 35719, 35740, 36283, 35714, and 36294 (Jena et al., 2018). Recently work on integration of Xa 38 a new BB resistance gene of rice was developed (Yugander et al., 2018).

1.4.4   SHEATH BLIGHT

Late and thin planting is advised to avoid the disease and application of recommended dose of fertilizer for reducing the disease. Sheath blight occurs at with an average temperature of 30°C. The pathogen is ubiquitous and has a wide host range affecting all grasses and broad-leaved weeds hence keeping bund clean will help in checking the disease from further spread from primary sources. Removal of infected stubbles crop residue from the field also advised. Seed treatment with *Pseudomonas aurofaciens* also found effective for controlling the disease. *Trichoderma harzianum* produces trichodermin that can inhibit *R. solani*. Formulations of *T. viride* (1%W/P, 1.15% WP, and 6% WP) and *T. harzianum* 1% WP are being marketed in India by many registered companies (Danger et al., 2019).

Very few varieties showing stable resistance to sheath blight, i.e., YSBR1, Tetep, Teqing, and Jasmine 85 (Chen et al., 2000; Meena et al., 2000; Pinson et al., 2008; Jia et al., 2012). Recently 2 QTLs qSB-3 and qSB-6 has been identified, which give the pathway for the development of new resistance variety for this disease (Chen, 2019). A tolerant donor CR 1014 has been identified during the screening work at ICAR-NRRI, India for disease, which has been utilized for developing mapping populations and transferring the tolerance gene to mega variety 'Swarna' (Jena et al., 2018).

On appearance of initial symptom spray Propiconazole 75 WP @ 1 ml or Validamycin 3 L @ 2 ml or Hexaconazole 5EC @ 2 ml or Thifluzamide

24SC @ 1 ml or Carbendazin 12% + Mancozeb 63% WP @ 2 g per liter of water. Biswas and Bag (2010) reported new QoI fungicides Kresoxim methyl, azoxystrobin, metaminostrobin, and trifloxystrobin and combinations with other groups were highly effective against sheath blight of rice. Tebuconazole 50% + Trifloxystrobin 25% WG @ 0.4 g per liter water gave good result in management of sheath blight disease (Bag et al., 2016).

1.4.5 FALSE SMUT OF RICE

Continuous rice cropping and moderate nitrogen fertilizer rates and conservation tillage serve useful for disease control (Brooks et al., 2009). Till date, there is no definitive pattern of infection process, dissemination method and the influence of weather factor vis-à-vis combination factors responsible for severe infection of false smut pathogen to rice. Nessa et al. (2015) provided a broad but relatively clear picture of the epidemiology of rice false smut disease under natural environment and reported that soil is the source of initiation of epidemic but did not recognize any long or short distance primary or secondary source of infection. At Temperature 22–25C with no less than 48 h of wetness duration considered necessary for successful infection of sexual stage of FS pathogen *Villosiclava virens* and the highest level of disease (92.9%) was obtained at 25°C and 95% RH with 120 h wetness. High level of nitrogen fertilization increases rice foliar growth which allowed for higher humidity below the canopy and created an environment favorable for the development of RFS. Additionally, irrigation has been found to be a major factor which affects the development of RFS. Lower minimum and maximum temperature, high atmospheric humidity (92% and above) before and during early part and less during later part of flowering favored the disease (Mohapatra et al., 2018). Jingganmycin and *Bacillus* mixture have been used in large production areas for control of rice false smut (Chen et al., 2011).

Various workers reported resistance rice genotypes to false smut under natural disease incidence. Out of the seven rice genotypes screened, IRAT 170 was highly resistant, Ex-China was resistant (disease severity score < 1% to false smut), ITA 316 was MR (disease severity score <5%). The resistance level of Japonica type ranged from 20.37 to 92.90%, whereas, resistance level of Indica rice ranged from 68.15 to 83.21%. Works at NRRI revealed that, Ranjit, and Luna Suvarna, were free from infection. Whereas CR Dhan 907, CR Dhan 303, NuaKalajeera, Ketakijoha, NuaDhusara, NuaChinikamini have exhibited moderate resistance against false smut pathogen (Jena et al., 2018). Rice cultivars IR 28 have relatively high in resistance to this disease. Resistance was controlled by two major and several minor genes. Eight QTLs found in Lemont Rice variety which shown resistance reaction against disease. Chemical control Nativo 75 WG, Diazole advised for spray twice at panicle initiation and early flowering stage can help to control the disease incidence (Ansari et al., 2016). Copper hydroxide fungicides reduced false smut balls in harvested rice by 80%, but yield was also often reduced significantly while Bag et al. (2010) reported the effectiveness of a new formulation of copper hydroxide.

1.5   CONCLUSION

Proper monitoring and timely detection are the keys to successful management of plant diseases and management of rice is no exception. Therefore, all modern tools and techniques are equally applicable for detecting rice disease-causing pathogens. Sensitivity and specificity of plant-pathogen detection has been enhanced with the advent of these modern tools and techniques. However, more works needed for precision and applicability at the highest level is the needed. With easy, cheap, and convenient available method of detection at the right time can reduce management cost and increase production and productivity of rice leads to reach the future goal of sustainability.

CONFLICT OF INTEREST STATEMENT

The authors declare that they have no conflict of interest.

KEYWORDS

- **backward inner primer**
- **convolutions neural network**
- **enzyme-linked immunosorbent assay**
- **flow cytometry**
- **fluorescence imaging**
- **modern tools**

REFERENCES

Abu, S. N., Abu, S. M. Y., Ahmad, A., Sahib, S., & Othman, F., (2018). Using LAPER quadcopter imagery for precision oil palm geospatial intelligence (OP GeoInt). *Journal of Telecommunication, Electronic and Computer Engineering, 10*(1), 25–33.

Agarwal, P. C., Mortensen, C. N., & Mathur, S. B., (1989). Seed-borne diseases and seed health testing of rice. *Phytopathology, 3,* 31–35.

Agrios, G. N., (2005). *Plant Pathology* (5th edn.). Elsevier Academic Press.

Ansari, T. H., Khatun, M. T., Ahmed, M., Nessa, B., Khan, M. A. I., Monsur, M. A., & Salam, M. U., (2016). Evaluation of fungicides for the control of rice false smut (*Ustilagi noidea virens*). *Bangladesh Rice Journal, 20*(2), 61–66.

Arshad, H. M. I., Hussain, N., Ali, S., Khan, J. A., Saleem, K., & Babar, M. M., (2013). Behavior of *Bipolaris oryzae* at different temperatures, culture media, fungicides and rice germplasm for resistance. *Pak. J. Phytopathol, 25,* 84–90.

Arshad, H. M. I., Naureen, S., Saleem, K., Ali, S., Jabeen, T., & Babar, M. M., (2015). Morphological and biochemical characterization of *Xanthomonas oryzae pv. Oryzae* isolates collected from Punjab during 2013. *Adv. Life Sci., 3,* 125–130.

Asibi, A. E., Chai, Q., & Coulter, J. A., (2019). Rice blast: A disease with implications for global food security. *Agronomy, 9*(8), 451.

Bag, M. K., (2009). Efficacy of a new fungicide 'trifloxystrobin 25% + tebuconazole 50%' 75WG against sheath blight (*Rhizoctonia solani* Kühn) of rice. *Journal of Crop and Weed, 5*(1), 224–226.

Bag, M. K., Saha, S., & Rai, R. K., (2010). Fungitoxicity of a new formulation of copper hydroxide (Kocide 2000 54 DF) against false smut disease of rice in West Bengal. *Pestology, 34*(1), 2628.

Bag, M. K., Yadav, M. K., & Mukherjee, A. K., (2017). Changing disease scenario with special emphasis on false smut of rice. *SATSA Mukhapatra-Annual Technical Issue, 21,* 219–224.

Bag, M. K., Yadav, M., & Mukherjee, A. K., (2016). Bioefficacy of strobilurin based fungicides against rice sheath blight disease. *Transcriptomics, 4,* 128.

Balota, M., & Oakes, J., (2016). Exploratory use of a UAV platform for variety selection in peanut. In: *Autonomous Air and Ground Sensing Systems for Agricultural Optimization and Phenotyping* (Vol. 9866, p. 98660F).

Baranwal, M. K., Kotasthane, A., Magculia, N., Mukherjee, P. K., Savary, S., Sharma, A. K., Singh, H. B., et al., (2013). review on crop losses, epidemiology and disease management of rice brown spot to identify research priorities and knowledge gaps. *Eur. J. Plant Pathol., 136,* 443–457.

Biswas, A., & Bag, M. K., (2010). Strobilurins in management of sheath blight disease of rice: A review. *Pestology, 34*(4), 23–26.

Biswas, C., Srivastava, S. S. L., & Biswas, S. K., (2010). Effect of biotic, abiotic, and botanical inducers on crop growth and severity of brown spot in rice. *Indian Phytopath., 63,* 187–191.

Boddy, L., (2016). Pathogens of autotrophs. In: *The Fungi* (pp. 245–292).

Bourett, T. M., & Howard, R. J., (1992). Actin in penetration pegs of the fungal rice blast pathogen, *Magnaporthe grisea*. *Protoplasma, 168*(1/2), 20–26.

Bravo, C., Moshou, D., Oberti, R., West, J., McCartney, A., Bodria, L., & Ramon, H., (2004). *Foliar Disease Detection in the Field Using Optical Sensor Fusion* (pp. 1–14).

Brooks, S. A., Anders, M. M., & Yeater, K. M., (2009). Effect of cultural management practices on the severity of false smut and kernel smut of rice. *Plant Disease, 93*(11), 1202–1208.

Bürling, K., Hunsche, M., & Noga, G., (2011). Use of blue-green and chlorophyll fluorescence measurements for differentiation between nitrogen deficiency and pathogen infection in winter wheat. *Journal of Plant Physiology, 168*(14), 1641–1648.

Burns, R., (2009). *Methods in Molecular Biology, Plant Pathology*. Springer Science.

Buschmann, C., & Lichtenthaler, H. K., (1998). Principles and characteristics of multi-color fluorescence imaging of plants. *Journal of Plant Physiology, 152*(2/3), 297–314.

Cai, H., Caswell, J., & Prescott, J., (2014). Nonculture molecular techniques for diagnosis of bacterial disease in animals a diagnostic laboratory perspective. *Vet. Pathol. Online, 51*, 341–350.

Cai, N., Zhou, X., Yang, Y., Wang, J., Zhang, D., & Hu, R., (2019). Use of UAV images to assess narrow brown leaf spot severity in rice. *International Journal of Precision Agricultural Aviation, 2*, 2.

Calderón, R., Navas-Cortés, J. A., Lucena, C., & Zarco-Tejada, P. J., (2013). High-resolution airborne hyperspectral and thermal imagery for early detection of verticillium wilt of olive using fluorescence, temperature and narrow-band spectral indices. *Remote Sensing of Environment, 139*, 231–245.

Catal, M., King, L., Tumbalam, P., Wiriyajitsomboon, P., Kirk, W. W., & Adams, G. C., (2010). Heterokaryotic nuclear conditions and a heterogeneous nuclear population are observed by flow cytometry in *Phytophthora infestans. Cytometry A, 77*, 769–775.

Ceresini, P., (1999). *Rhizoctonia solani. Soilborne Plant Pathogens*, 728.

Chaerle, L., Leinonen, I., Jones, H. G., & Van, D. S. D., (2007). Monitoring and screening plant populations with combined thermal and chlorophyll fluorescence imaging. *J. Exp. Bot., 58*, 773–784.

Charlton, N. D., Tavantzis, S. M., & Cubeta, M. A., (2009). Detection of double-stranded RNA elements in the plant pathogenic fungus *Rhizoctonia solani*. In: Burns, R., (ed.), *Methods in Molecular Biology, Plant Pathology* (Vol. 508, p. 177). Springer Science.

Chen, F., Deng, F., Liu, J., Deng, W., & Yang, J., (2011). Experiments of rice false smut and sheath blight management by using a mixture of *Bacillus subtilis* and *Jinggangmycin. Jiangxi Plant Protection, 34*(4), 177–178.

Chen, W. C., Chiou, T. Y., Delgado, A. L., & Liao, C. S., (2019). The control of rice blast disease by the novel bio fungicide formulations. *Sustainability, 11*(12), 3449.

Chen, Z. X., Zou, J. H., Xu, J. Y., Tong, Y. H., Tang, S. Z., Wang, Z. B., Jiang, R. M., et al., (2000). A preliminary study on resources of resistance to rice sheath blight. *Chin. J. Rice. Sci., 14*(1), 15–18.

Chen, Z., Feng, Z., Kang, H., Zhao, J., Chen, T., Li, Q., & Liu, W., (2019). Identification of new resistance loci against sheath blight disease in rice through genome-wide association study. *Rice Science, 26*(1), 21–31.

Chitarra, L. G., & Van, D. B. R. W., (2003). The application of flow cytometry and fluorescent probe technology for detection and assessment of viability of plant pathogenic bacteria. *Eur. J. Plant Pathol., 109*, 407–417.

Chitarra, L. G., Langerak, C. J., Bergervoet, J. H. W., & Van, D. B. R. W., (2002). Detection of the plant pathogenic bacterium *Xanthomonas campestris* pv. *campestris* in seed extracts of *Brassica* sp. applying fluorescent antibodies and flow cytometry. *Cytometry, 47*, 118–126.

Chou, C., Castilla, N., Hadi, B., Tanaka, T., Chiba, S., & Sato, I., (2019). Rice blast management in Cambodian rice fields using *Trichoderma harzianum* and a resistant variety. *Crop Protection*, 104864.

Clark, M. F., & Adams, A., (1977). Characteristics of the microplate method of enzyme-linked immunosorbent assay for the detection of plant viruses. *J. Gen. Virol., 34*, 475–483.

Costa, F., Ueyama, J., Braun, T., Pessin, G., Osorio, F., & Vargas, P., (2012). The use of unmanned aerial vehicles and wireless sensor network in agricultural applications. In: *IEEE Conference on Geoscience and Remote Sensing Symposium* (pp. 5045–5048).

Cséfalvay, L., Di Gaspero, G., Matouš, K., Bellin, D., Ruperti, B., & Olejníčková, J., (2009). Pre-symptomatic detection of *Plasmopara viticola* infection in grapevine leaves using chlorophyll fluorescence imaging. *European Journal of Plant Pathology, 125*(2), 291–302.

D'hondt, L., Höfte, M., Van, B. E., & Leus, L., (2011). Applications of flow cytometry in plant pathology for genome size determination, detection and physiological status. *Molecular Plant Pathology, 12*(8), 815–828.

Dang, L. M., Hassan, S. I., Suhyeon, I., kumarSangaiah, A., Mehmood, I., Rho, S., Seo, S., & Moon, H., (2018). UAV based wilt detection system via convolutional neural networks. *Sustainable Computing: Informatics and Systems.*

Danger, T. K., Bag, M. K., Kumar, U., & Anand, P., (2019). Microbial pesticides. In: Anand, P., et al., (eds.), *Biopesticides in Indian Agriculture.* Applied Zoologists Research Association, Bhubaneswar, Odisha, India.

Das, P. K., Laxman, B., Rao, S. K., Seshasai, M. V. R., & Dadhwal, V. K., (2015). Monitoring of bacterial leaf blight in rice using ground-based hyperspectral and LISS IV satellite data in Kurnool, Andhra Pradesh, India. *International Journal of Pest Management, 61*(4), 359–368.

Del, F. A., Reverberi, M., Ricelli, A., Pinzari, F., Serranti, S., Fabbri, A. A., Bonifazi, G., & Fanelli, C., (2010). Early detection of toxigenic fungi on maize by hyperspectral imaging analysis. *International Journal of Food Microbiology, 144*(1), 64–71.

Delalieux, S., Van, A. J. A. N., Keulemans, W., Schrevens, E., & Coppin, P., (2007). Detection of biotic stress (*Venturia inaequalis*) in apple trees using hyperspectral data: Non-parametric statistical approaches and physiological implications. *European Journal of Agronomy, 27*(1), 130–143.

Di Gennaro, S. F., Battiston, E., Di Marco, S., Facini, O., Matese, A., Nocentini, M., Palliotti, A., & Mugnai, L., (2016). Unmanned aerial vehicle (UAV)-based remote sensing to monitor grapevine leaf stripe disease within a vineyard affected by esca complex. *Phytopathology Mediterranea*, 262–275.

Dutta, D., Saha, S., Ray, D. P., & Bag, M. K., (2012). Effect of different active fungicides molecules on the management of rice blast disease. *International Journal of Agriculture, Environment and Biotechnology, 5*(3), 247–251.

Eilam, T., Bushnell, W. R., & Anikster, Y., (1994). Relative nuclear-DNA content of rust fungi estimated by flow cytometry of propidium iodide stained pycniospores. *Phytopathology, 84*, 728–735.

Faiçal, B. S., Costa, F. G., Pessin, G., Ueyama, J., Freitas, H., Colombo, A., Fini, P. H., et al., (2014). The use of unmanned aerial vehicles and wireless sensor networks for spraying pesticides. *Journal of Systems Architecture, 60*, 4, 393–404.

Fang, Y., & Ramasamy, R. P., (2015). Current and prospective methods for plant disease detection. *Biosensors, 5*(3), 537–561.

Giles, D. K., & Billing, R. C., (2015). Deployment and performance of a UAV for crop spraying. *Chemical Engineering Transactions, 44*, 307–322.

Golhani, K., Balasundram, S. K., Vadamalai, G., & Pradhan, B., (2018). A review of neural networks in plant disease detection using hyperspectral data. *Information Processing in Agriculture, 5*(3), 354–371.

Hayward, A. C., (1993). The hosts of *Xanthomonas*. In: *Xanthomonas* (pp. 1–119). Springer, Dordrecht.

Huang, Y., Hoffmann, W. C., Lan, Y., Wu, W., & Fritz, B. K., (2009). Development of a spray system for an unmanned aerial vehicle platform. *Applied Engineering in Agriculture, 25*(6), 803–809.

James, D. A., (1999). simple and reliable protocol for the detection of apple stem grooving virus by RT-PCR and in a multiplex PCR assay. *J. Virol. Methods, 83,* 1–9.

Jena, M., Rath, P. C., Mukherjee, A. K., Raghu, S., Guru, P. P. G., Basana, G. G., Prasanthi, G., et. al., (2018). Exploring new sources of resistance for insect pest and diseases of rice. In: Pathak, H., et al., (eds.), *Rice Research for Enhancing Productivity, Profitability and Climate Resilience* (p. 542). ICAR-National Rice Research Institute, Cuttack, Odisha, India.

Jia, L. M., Yan, W. G., Zhu, C. S., Agrama, H. A., Jackson, A., Yeater, K., Li, X., et al., (2012). Allelic analysis of sheath blight resistance with association mapping in rice. *PLoS One, 7*(3), e32703.

Jiang, H., Li, Z., Liu, J., Shen, Z., Gao, G., Zhang, Q., & He, Y., (2019). Development and evaluation of improved lines with broad-spectrum resistance to rice blast using nine resistance genes. *Rice, 12*(1), 29.

Kaundal, R., Kapoor, A. S., & Raghava, G. P. S., (2006). Machine learning techniques in disease forecasting: A case study on rice blast prediction. *BMC Bioinformatics, 7,* 485.

Kempf, V. A., Trebesius, K., & Autenrieth, I. B., (2000). Fluorescent in situ hybridization allows rapid identification of microorganisms in blood cultures. *J. Clin. Microbiol., 38,* 830–838.

Khalili, E., Sadravi, M., Naeimi, S., & Khosravi, V., (2012). Biological control of rice brown spot with native isolates of three *Trichoderma* species. *Brazilian Journal of Microbiology, 43*(1), 297–305.

Khan, J. A., Siddiq, R., Arshad, H. M. I., Anwar, H. S., Saleem, K., & Jamil, F. F., (2012). Chemical control of bacterial leaf blight of rice caused by *Xanthomonas oryzae* pv. *oryzae*. *Pakistan Journal of Phytopathology, 24*(2), 97–100.

Kobayashi, T., Kanda, E., Kitada, K., Ishiguro, K., & Torigoe, Y., (2001). Detection of rice panicle blast with multispectral radiometer and the potential of using airborne multispectral scanners. *Phytopathology, 91*(33), 316–323.

Li, Y., Zhu, Z., Zhang, Y., Zhao, L., & Wang, C., (2008). Genetic analysis of rice false smut resistance using mixed major genes and polygenes inheritance model. *Acta Agron. Sin., 34*(10), 1728–1733.

Lindenthal, M., Steiner, U., Dehne, H. W., & Oerke, E. C., (2005). Effect of downy mildew development on transpiration of cucumber leaves visualized by digital infrared thermography. *Phytopathology, 95,* 233–240.

Liu, Z. Y., Huang, J. F., Shi, J. J., Tao, R. X., & Zhang, H. Z., (2008). Estimating rice brown spot disease severity based on principal component analysis and radial basis function neural network. *Spectroscopy and Spectral Analysis., 28*(9), 2156–2160.

Liu, Z. Y., Wu, H. F., & Huang, J. F., (2010). Application of neural networks to discriminate fungal infection levels in rice panicles using hyperspectral reflectance and principal components analysis. *Computers and Electronics in Agriculture, 72*(2), 99–106.

Liu, Z., Huang, J., Tao, R., & Zhang, H., (2008). Estimating the severity of rice brown spot disease based on principal component analysis and radial basis function neural network. *Spectrosc. Spectr. Anal., 28*(9), 2156–2160.

Liu, M., Sun., Z. X., Zhu, J., Xu, T., Harman, G. E., & Loritto, M., (2004). Enhancing rice resistance to fungal pathogens by transformation with cell wall degrading enzyme genes from *Trichoderma atroviride. J. of Zheijiang University of Science, 5*, 133–136.

Manjunath, K., More, R. S., Jain, N., Panigrahy, S., & Parihar, J., (2015). Mapping of rice cropping pattern and cultural type using remote-sensing and ancillary data: A case study for South and Southeast Asian countries. *International Journal of Remote Sensing*, 6008–6030.

Meena, B., Ramamoorthy, V., Banu, J. G., Thangavelu, R., & Muthusamy, M., (2000). Screening of rice genotypes against sheath blight disease. *J. Ecobiol., 12*(2)103–109.

Mohapatra, S. D., Rath, P. C., Mukherjee, A. K., Raghu, S., Guru, P. P. G., Basana, G. G., Prasanthi, G., et al., (2018). Bio-ecology of rice insect pests and diseases: Paving the way to climate-smart rice protection technologies. In: Pathak, H., et al., (eds.), *Rice Research for Enhancing Productivity, Profitability and Climate Resilience* (p. 542). ICAR-National Rice Research Institute, Cuttack, Odisha, India.

Nassuth, A., Pollari, E., Helmeczy, K., Stewart, S., & Kofalvi, S. A., (2000). Improved RNA extraction and one-tube RT-PCR assay for simultaneous detection of control plant RNA plus several viruses in plant extracts. *J. Virol. Methods, 90*, 37–49.

Nebiker, S., Lack, N., Abächerli, M., & Läderach, S., (2016). *Light-Weight Multispectral UAV Sensors and Their Capabilities for Predicting Grain Yield and Detecting Plant Diseases* (Vol. 41). International Archives of the Photogrammetry, Remote Sensing and Spatial Information Sciences.

Niño-Liu, D. O., Ronald, P. C., & Bogdanove, A. J., (2006). *Xanthomonas oryzae* pathovars: model pathogens of a model crop. *Molecular Plant Pathology, 7*(5), 303–324.

Norasma, C. Y. N., Sari, M. A., Fadzilah, M. A., Ismail, M. R., Omar, M. H., Zulkarami, B., Hassim, Y. M. M., & Tarmidi, Z., (2018). Rice crop monitoring using multirotor UAV and RGB digital camera at early stage of growth. In: *IOP Conference Series: Earth and Environmental Science* (Vol. 169, No. 1, p. 012095). IOP Publishing.

Oerke, E. C., Fröhling, P., & Steiner, U., (2011). Thermographic assessment of scab disease on apple leaves. *Precis. Agric., 12*, 699–715.

Padmanabhan, S. Y., (1973). The great Bengal famine. *Annual Review of Phytopathology, 11*(1), 11–24.

Pinson, S. R. M., Oard, J. H., Groth, D., Miller, R., Marchetti, M. A., Shank, A. R., Jia, M. H., et al., (2008). Registration of TIL:455, TIL:514, and TIL:642, three rice germplasm lines containing introgressed sheath blight resistance alleles. *J. Plant Reg., 2*(3), 251–254.

Rowhani, A., & Falk, B. W., (1995). Enzyme-linked immunosorbent assay (ELISA) methods to certify pathogen (virus)-free plants. In: *Plant Cell, Tissue, and Organ Culture* (pp. 267–280).

Sato, H., Ando, I., Hirabayashi, H., Takeuchi, Y., Arase, S., Kihara, J., Kato, H., Imbe, T., & Nemoto, H., (2008). QTL analysis of brown spot resistance in rice (*Oryza sativa* L.). *Breed. Sci., 58*, 93–96.

Shankar, R. H., Veeraraghavan, A. K., Sivaraman, K., & Ramachandran, S. S., (2018). Application of UAV for pest, weeds, and disease detection using open computer vision. In: *2018 International Conference on Smart Systems and Inventive Technology* (pp. 287–292).

Shipway, A. N., Katz, E., & Willner, I., (2000). Nanoparticle arrays on surfaces for electronic, optical, and sensor applications. *Chem. Phys. Chem., 1*, 18–52.

Singh, B., Singh, M., Singh, G., Suri, K., Pannu, P. P. S., & Bal, S. K., (2012). Hyperspectral data for the detection of rice bacterial leaf blight (BLB) disease. *Proceedings of AIPA.*

Singh, M., (2011). Yield gap and production constraints in rice-wheat system: Scenario from eastern Uttar Pradesh. *Bangladesh. J. Agric. Res., 36*, 623–632.

Singh, V., & Misra, A. K., (2015). Detection of unhealthy region of plant leaves using image processing and genetic algorithm. In: *2015 International Conference on Advances in Computer Engineering and Applications* (pp. 1028–1032).

Sparks, A., Nelson, A., & Castilla, N., (2012). Where rice pests and diseases do the most damage. *Rice Today, 11*(4), 26–27.

Su, J., Liu, C., Coombes, M., Hu, X., Wang, C., Xu, X., Li, Q., Guo, L., & Chen, W. H., (2018). Wheat yellow rust monitoring by learning from multispectral UAV aerial imagery. *Computers and Electronics in Agriculture, 155*, 157–166.

Su'udi, M., Kim, J., Park, J. M., Bae, S. C., Kim, D., Kim, Y. H., & Ahn, I. P., (2013). Quantification of rice blast disease progressions through TaqMan real-time PCR. *Molecular Biotechnology, 55*(1), 43–48.

Sultana, S., Kitajima, M., Kobayashi, H., Nakagawa, H., Shimizu, M., Kageyama, K., & Suga, H., (2019). A natural variation of fumonisin gene cluster associated with fumonisin production difference in *Fusarium fujikuroi*. *Toxins, 11*(4), 200.

Sundaram, R. M., Chatterjee, S., Oliva, R., Laha, G. S., Leach, J. E., & Sonti, R. V., (2014). Update on bacterial blight of rice: Fourth international conference on bacterial blight. *Rice, 7*(12). pmid:26055994.

Sunder, S., Singh, R. A. M., & Agarwal, R., (2014). Brown spot of rice: An overview. *Indian Phytopathology, 67*(3), 201–215.

Vardhan, P. H., Dheepak, S., Aditya, P. T., & Arul, S., (2014). Development of automated aerial pesticide sprayer. *International Journal of Engineering Science and Research Technology, 3*(4).

Vercauteren, A., Boutet, X., D'hondt, L., Van, B. E., Maes, M., Leus, L., Chandelier, A., & Heungens, K., (2011). Aberrant genome size and instability of *Phytophthora ramorum* oospore progenies. *Fungal Genet. Biol., 48*, 537–543. doi: 10.1016/j.fgb.2011.01.008.

Wang, G., Lan, Y., Qi, H., Chen, P., Hewitt, A., & Han, Y., (2019). Field evaluation of an unmanned aerial vehicle (UAV) sprayer: Effect of spray volume on deposition and the control of pests and disease in wheat. *Pest Management Science, 75*(6), 1546–1555.

Ward, E., Foster, S. J., Fraaije, B. A., & Mccartney, H. A., (2004). Plant pathogen diagnostics: Immunological and nucleic acid-based approaches. *Ann. Appl. Biol., 145*, 1–16.

Wulff, E. G., Sørensen, J. L., Lübeck, M., Nielsen, K. F., Thrane, U., & Torp, J., (2010). *Fusarium* spp. associated with rice bakanae: Ecology, genetic diversity, pathogenicity and toxigenicity. *Environmental Microbiology, 12*(3), 649–657.

Xu, J., Xue, Q., Luo, L., & Li, Z., (2001). Preliminary report on quantitative trait loci mapping of false smut resistance using near-isogenic introgression lines in rice. *Acta Agric Zhejiang, 14*(1), 14–19.

Yadav, M. K., Aravindan, S., Ngangkham, U., Raghu, S., Prabhukarthikeyan, S. R., Keerthana, U., & Pramesh, D., (2019). Blast resistance in Indian rice landraces: Genetic dissection by gene-specific markers. *PloS One, 14*(1), e0211061.

Yasmin, S., Hafeez, F. Y., Mirza, M. S., Rasul, M., Arshad, H. M., Zubair, M., & Iqbal, M., (2017). Biocontrol of bacterial leaf blight of rice and profiling of secondary metabolites produced by rhizospheric *Pseudomonas aeruginosa* BRp3. *Frontiers in Microbiology, 8*, 1895.

Yellareddygari, S. K. R., Reddy, M. S., Kloepper, J. W., Lawrence, K. S., & Fadamiro, H., (2014). Rice sheath blight: A review of disease and pathogen management approaches. *J. Plant Pathol. Microb, 5*, 241.

Yong, M., Deng, Q., Fan, L., Miao, J., Lai, C., Chen, H., & Yang, B., (2018). The role of *Ustilagi noidea virens* sclerotia in increasing incidence of rice false smut disease in the subtropical zone in China. *European Journal of Plant Pathology, 150*(3), 669–677.

Yugander, A., Sundaram, R. M., Singh, K., Ladhalakshmi, D., Rao, L. V. S., Madhav, M. S., & Laha, G. S., (2018). Incorporation of the novel bacterial blight resistance gene Xa38 into the genetic background of elite rice variety improved *Samba mahsuri. PloS One, 13*(5), e0198260.

Zainudin, N. A. I. M., Razak, A. A., & Salleh, B., (2008). Bakanae disease of rice in Malaysia and Indonesia: Etiology of the causal agent based on morphological, physiological and pathogenicity characteristics. *J. Plant Protec. Res., 4*, 48.

Zhang, D., Zhou, X., Zhang, J., Huang, L., & Zhao, J., (2017). Developing a small UAV platform to detect sheath blight of rice. *IEEE International Geoscience and Remote Sensing Symposium*, 3190–3193.

Zhou, Y. L., Xie, X. W., Zhang, F., Wang, S., Liu, X. Z., & Zhu, L. H., (2014). Detection of quantitative resistance loci associated with resistance to rice false smut (*Ustilagi noidea virens*) using introgression lines. *Plant Pathol., 63*(2), 365–372.

Zhu, H., Lan, Y., Wu, W., Hoffmann, W. C., Huang, Y., Xue, X., & Fritz, B., (2010). Development of a PWM precision spraying controller for unmanned aerial vehicles. *Journal of Bionic Engineering, 7*(3), 276–283.

Zhu, W., Chen, H., Ciechanowska, I., & Spaner, D., (2018). Application of infrared thermal imaging for the rapid diagnosis of crop disease. *IFAC-Papers on Line, 51*(17), 424–430.

Zhu, Y. Y., Fang, H., Wang, Y. Y., Fan, J. X., Yang, S. S., & Mew, T. W., (2005). Panicle blast and canopy moisture in rice cultivar mixtures. *Phytopathology, 95*, 433–438.

Ziya, A., Mehmet, M. O., & Yusuf, Y., (2018). Determination of sugar beet leaf spot disease level (*Cercospora beticola* Sacc.) with image processing technique by using drone. *Curr. Inves. Agri. Curr. Res., 5*, 3.

CHAPTER 2

Diagnosis and Management of Fungal Diseases of Rice Prevalent in Telangana State, India

S. KIRAN,[1] M. SUREKHA,[2] and S. M. REDDY[2]

[1]Department of Botany, Satavahana University, Karimnagar, Telangana, India

[2]Department of Botany, Kakatiya University, Warangal, Telangana, India

ABSTRACT

Rice is an essential basic food for the majority of the people in the world. Throughout the world, 90 millions of people are either rice producers or consumers. In Telangana State, rice covers 25 lakh hectors and yields annually only 77.50 lakh tones which are far below the Nations average production. In spite of its wide acceptance as principle, food production of rice has not increased proportionately, which may be attributed to various reasons, including fungal diseases. Thus the salient features of our investigations on the prevalence of fungal diseases of rice and their management in the Telangana region are summarized along with updated knowledge from various sources. Significant progress has been witnessed in the diagnosis of plant diseases, especially with reference to fungal diseases. Variety of methods, including ES for RPD2, convolution of neural network (CNN), local binary patterns histograms (LBPH), etc., are being developed to identify few rice diseases and also in turn, helpful to manage yield loss during and post-harvest stages of rice production. Therefore, various diagnostic methods with reference to rice diseases along with their merits and demerits are included in this chapter. A comprehensive list of fungal diseases of rice plant all over the globe, along with the most common and significant diseases in this region is described. A detailed account of some important fungal diseases such as sheath blight, blast, stem rot, sheath rot, kernel smut, and false smut are discussed comprehensively. Various predisposing factors contributing to major damage and contributing to yield loss at the same time increase production cost are analyzed. A comprehensive view of various methods adapted to protect the rice crop and minimize yield loss is discussed comprehensively. Some of the innovative methods which are eco-friendly and at the same time contribute to increased production are also given. Further, rice is prone to mycotoxin contamination in the field and during storage by various mycotoxigenic fungi due to unseasonal heavy rains at the time of harvest and high moisture content during storage. Mycotoxins, when ingested, cause a variety of adverse effects in both humans and animals. Therefore, the need was felt to include this aspect in the review.

2.1 INTRODUCTION

Rice is one of the ancient crops cultivated by man, mostly related to the wild grass which was domesticated. According to pre-historic studies, the first variety of rice "Indica" was cultivated in the regions of the Eastern Himalayas, including Burma, Thailand, Laos, Vietnam, and Southern China. The variety "japonica" from Southern China was introduced to India earlier to the Greeks. The rice cultivation in china goes back to 4000 years. Another school of thought believes that in 2000 B.C., rice plant originated in south has spread to China, Korea, and the Philippines and then to Japan and Indonesia. Subsequently, Greek and Arab travelers introduced rice to Europe, Portugal, Netherlands, colonies in West Africa and America through the Columbian Exchange of natural resources. Thereafter, Moors taken it to Spain in 700 A.D. and then to South America at the beginning of the 17th century. Presently it is a major agriculture and economic product of the entire world. China, India, Indonesia, Bangladesh, the Philippines, and Pakistan are the main rice producers.

The culture, diets, and economy of millions of people around the world is rice dependent, and for more than 50% of the world population, "rice is life." In the view of above facts, the United Nations has declared the year 2004 as the "International Year of Rice."

India is the homeland for rice production. The rice is cultivated in almost all parts of India. The wild perennial rice introduced to India in the northern plains around 1400 B.C., is still cultivated in Assam and Nepal. Further, in course of time, it has been cultivated in other regions of the country. Rice contributes 4.49% to the foreign exchange and 1.3% GDP (Anonymous,

2013). According to the statistics, the annual production of rice for the year 2015–2016 was about 104.32 million tons.

Rice crop suffers from about 50 diseases of which fungal diseases form a major component (Mew and Gonzales, 2002). However, some of the important fungal diseases of rice of economic significance are rice blast, brown spot, sheath blight, sheath rot, false smut, stem rot, Bakanae, and post-harvest diseases (Sharma and Bambawale, 2008). According to Mew and Gonzales (2002), these diseases are responsible for about 14–18% yield losses worldwide. Ribot et al. (2008) feel that rice production is facing a crisis because of diseases, the decrease of cultivated land, water resources, fertilizers, and chemicals. FAO has assessed that diseases, insects, and weeds account for 25% failure of rice crop. Pattanayak et al. (2006) has precisely mentioned the various steps to be taken during cultivation of rice in northeast India to secure an optimum yield.

2.2 ECONOMIC IMPORTANCE OF RICE

Rice is a staple food as its most important component is starch with 8% protein and 1% fat content. It is also used by brewers to make alcoholic malt. Rice straw and rice bran are used as animal feed. Rice straw is also used in ceramic, glass, and paper industries. Brown rice is rich in fiber and vitamins like B1 (thiamine), B2 (riboflavin), niacin or nicotinic acid. On the other hand, the white rice is poor in vitamins and minerals, due to the removal of bran and seed coat during the process of polishing. The nutritional value of rice depends on various factors such as variety, soil, environment, and methods of cooking. Rice because of its easily digestible quality is used as a diet for infants, old people, and for people with stomach and intestinal diseases.

Agriculture has undergone a significant change due to climate change with regard to input use, varietal diversity, and cropping pattern during the last five decades. The intensity and profile of rice diseases has been significantly influenced by changes in climatic parameters, intensity, and cropping pattern, increased use of inputs, varietal diversity composition in a particular region, and extensive cultivation of few selected high-yielding varieties. It is reported that a rise in CO_2 concentration increased leaf blast and sheath blight development, indicating that potential risks of rice blast and sheath blight when rice is grown under elevated CO_2 (Kobayashi et al., 2006).

2.3 DISEASE DIAGNOSIS

With the increase of population growth *geometrically* and food production arithmetically, there is an enormous amount of pressure on food-producing resources. The production of rice has to be increased proportionately as it the principal food for many people in the world. Such a situation appears to be remotely possible due to decreased land availability and acute storage of water resources. Diseases, pests, and weeds accentuate the situation. Dean et al. (2005) feel that the amount of rice destroyed by blast disease every year is alone significant to feed 60 million people. Hence, evolving extremely sensitive and accurate methods for early diagnosis of disease is of prime importance.

Traditional and main technique used for identification of plant-pathogen is culture-based morphological observation which usually costs a long time (Ward et al., 2004; McCartney et al., 2003). Hence, it will not be much useful to farmer to diagnose disease in the initial infection stage for timely control of the disease.

Rani et al. (2013) have made an attempt to develop an expert system to diagnose rice diseases. Al-Ahmar (2009); Lopez-Morales et al. (2008); Babu et al. (2010); and Boys and Sun (1994) have developed expert systems for early diagnosis of disease in date palm, tomato, and potato, respectively. Similarly, diseases of olive crops (Gonzales and Andujar, 2008), Jamaican coffee (Mansingh et al., 2007), rice (Sarma et al., 2010) and micronutrient deficiency in crops (Patil et al., 2009) could be effectively detected. Several such diagnostic methods are also available which can predict precisely possible epidemics and protect the crop from yield losses.

Amurtha Devi and Muthukannan (2014) have evolved the image segmentation algorithm to detect the diseased portion of the rice leaf. Kurniawati et al. (2009) have proposed image processing techniques for diagnosis of paddy disease. In this process, the crops are photographed, and the captured images are processed in Raspberry pi3 module, which sends the data to the cloud. The cloud account is synchronized with Android mobile, so that the farmers or end users can monitor the field using this app from anywhere and at anytime. IoT technology will be of great help in the crop production and thus leads to improved productivity. An automated disease diagnostic system will help to prevent and control of plant disease in the shortest time, minimize the economic loss, reduce the pesticide use and boost the quality and quantity of agriculture products (Liang et al., 2019).

Liang et al. (2019) have developed deep convolutional neural network method to recognize rice blast disease. Ramesh and Vydeki (2018) proposed IoT technology for assessing rice blast disease. Sreekantha and Kavya (2017) proposed an IoT system by which the farmer can collect information on soil moisture, pest incidence, weather, including crop growth and animal invasion which will help to increase the crop production.

Ward et al. (2004); and Mondal and Shanmugam (2013) have developed nucleic acid-based methods and immunoassay for diagnosis of plant pathogens. Polymerase chain reaction (PCR) methods are very specific, sensitive, and accurate for detection of blast-disease (Harman et al., 2003; Lieven, 2003; Sun et al., 2010). Qi and Yang (2002) could detect *M. oryzae* in rice plant with PCR and RNA-based northern blot/phospho-imaging analysis. Suudi et al. (2013) have developed a method for the specific detection of *M. oryzae* by detecting as low as 1 pg MHP, DNA from *M. oryzae.* Yang et al. (2014) have developed an ultrasensitive electrochemical biosensor for early diagnosis of blast fungus before the disease symptoms appear in rice plant which will facilitate the farmers to take necessary steps to manage the disease. Besides this ease of fabrication, operational convenience, sensitivity, and specificity, fast analysis, robust resistibility to complex matrix and cheap, etc., has made this method a promising candidate for the early diagnosis and fast for detection of *M. oryzae* in rice. However, PCR-based methods of detection of pathogens has some disadvantages such as complicated and long nucleic acid extraction procedure and relating high assay cost, etc., (De Boer and Lopez, 2012). However, the PCR-based methods target only a single pathogen which make comprehensive screening of complex samples non-economical.

Through enzyme-linked immunoassay (ELISA) is another sensitive and specific method for detection of plant pathogenic fungi, its use is limited by the utilization of specific antibody which is very difficult to obtain.

Electrochemical methods hold potential as a next-generation molecular detection strategy for the diagnostic purpose due to their high sensitivity, rapid response, low cost, and excellent capability with miniaturization technologies (Wang et al., 2013).

2.4 FUNGAL DISEASES

Agronomic practices have globally changed the scenario of rice disease, particularly in Asia (Mew et al., 2004). Many diseases hitherto considered as minor have become very serious in many rice-growing areas. For example, false smut of rice, which was once considered as the sign of bumper harvest, has now become a serious problem in many countries (Ladhalakshmi et al., 2012). The neck blast and brown spot diseases have spread widely in Asia (Khan et al., 2014). Strange and Scott (2005); and Talbot (2003) feel that rice blast has significant negative effect on rice production. Bakanae disease, which was considered a minor problem, has become a major threat to rice production in Northwestern India.

Diseases in different agroclimatic zones of Telangana state are listed in Table 2.1. With increasing area of rice cultivation under intensive system in Telangana resulted in the incidence of different fungal diseases. Some of the important diseases in different zones of Telangana of economic significance are listed in Table 2.1.

TABLE 2.1   Scenario of Rice Diseases in Different Agro-Climatic Zones of Telangana

Name of the Zone	Diseases
Northern Telangana Zone	Blast, BLB, BPH, gall midge, leaf folder, WBPH, panicle mite, sheath rot, stem borer.
Central Telangana Zone	Gall midge, stem borer, leaf folder, BPH, WBPH, mealybug, BLB, blast, sheath rot.
Southern Telangana Zone	Stem borer, gall midge, BPH, WBPH, gundhy bug, cutworm, mealybug, blast, BLB, sheath rot

Naturally infected *Cynodon dactylon, Dactyloctenium aegyptium, Dichanthium annulatum, Echinochloa crus-Galli, E. colona, Euphorbia microphylla, Leptochloa chinensis,* and *Paspalum distichum* like weeds growing in and around paddy fields have been found to be naturally infected by the pathogen and can be an important source of primary infection of rice (Singh et al., 2012). In this chapter, the discussion will be confined to some major and emerging diseases of rice (Table 2.2). The diseases of rice of limited economic significance are précised in Table 2.3.

TABLE 2.2   Fungal Diseases of Rice of Major Importance

Name of the Disease	Name of the Fungus
Blast	*Pyricularia grisea* and *Dactylaria oryza*
Brown spot	*Drechslera oryzae*
False smut	*Ustilaginoidea virens*
Stem rot	*Sclerotium oryzae*
Sheath blight	*Rhizoctonia solani*
Sheath rot	*Sarocladium oryzae*

TABLE 2.3 Fungal Diseases of Rice of Limited Importance

Name of the Disease	Name of the Pathogen	Favorable Conditions	Symptoms	Management	References
Aggregate sheath spot	*Rhizoctonia oryzae-sativae*	Potassium deficient soils	Infects lower leaf sheaths at the waterline during the tillering stage. Lesions are circular to elliptical with gray-green to straw-colored center surrounded by distinct brown margin. Frequently, additional margins from around the initial lesion, producing concentric bands.	Apply Nitrogen fertilizer and spray Quadriss fungicide	Bruce et al. (2008)
Leaf smut	*Entoloma oryzae*	Acidic soils with low moisture and temperature of 10–20°C	Infects the endosperm of seed and partially or completely replaces with masses of dark spores that ooze out of the seed in moist weather. Infected plant may be pale-green or yellow with stunted growth with erect and stiff blades, infected stem rots.	soil pH below 6.0, frequent watering and excess nitrogen fertilizers should be avoided	Mukherjee and Uma Maheswari (2018)
Downy mildew of rice	*Sclerophthora macrospora*	Cool temperature (15°C to 25°C), high humidity and diffusive sunlight	Angular yellow spots on the upper leaf surface which turn to brilliant-yellow then turn brown with yellow margins. Infected seedlings will get dwarfed and twisted with chlorotic, yellow to whitish spots	Sowing pathogen-free seeds. Seed surface sterilization with ethanol	Jin Jee et al. (2002)

TABLE 2.3   *(Continued)*

Name of the Disease	Name of the Pathogen	Favorable Conditions	Symptoms	Management	References
Eyespot	*Drechslera gigantea*	–	Infected regions develop into minute, longitudinally elongated, oval lesions with white to straw-colored necrotic centers surrounded by narrow, dark-brown margin	As loss is negligible. Control measures are not warranted.	Sang-Won (1980)
Kernel Smut	*Neovossia barclayan (Tilletia horrida)*	Acidic soils with excess thatch and low moisture and high nitrogen rates	Primarily appears as dark-brown	Use of semi-dwarf varieties with a history of smut resistance, reduce nitrogen fertilizer rates and floodwater depths in fields and spraying propiconazole	Elshafey (2018)
Leaf scald	*Microdochium oryzae (Monographella albescens)*	Prolonged foliar dew, temperatures ranging from 24°C to 28°C, close spacing, and excess nitrogen	Zonate lesions of alternating oblong light-tan and dark-brown halos starting from leaf tips or edges. Lesions with light brown. Coalescing of lesions results in blighting of a large part of the leaf blade.	Avoid excess nitrogenous fertilizer, using cultivars with high levels of basal resistance and seed treatments using fungicides or their foliar spray.	Quintana et al. (2018); Leonardo et al. (2016)
Narrow-Brown Leaf Spot	*Cercospora janseana*	Potassium deficient soils, and temperature ranging from 25°C–28°C	Light to dark-brown, linear, lesions on leaves and upper leaf sheath which progress parallel to the vein. Lesions may enlarge and connect together forming brown linear necrotic regions. On glumes, lesions are usually shorter but wider than those on the leaves.	Carbendazim 500 g/ha or Mancozeb (2 kg/ha)	Mukherjee and Uma Maheswari (2018)

TABLE 2.3 *(Continued)*

Name of the Disease	Name of the Pathogen	Favorable Conditions	Symptoms	Management	References
Pecky Rice disease	*Cochliobolus miyabeanus, Curvularia spp., Fusarium spp. Microdochium oryzae* and *Sarocladium oryzae*	High wind	Roughly circular lesion which in some cases appears as a shrunken area. In general, pecky rice kernels are accompanied by brown to black discoloration, the whole kernel or portion of the kernel may get affected.	Proper insect control	Godwin Premi et al. (2019)
Stack burn disease	*Trichoconis padwicii* (*Alternaria padwickii*)	Humid atmosphere and high temperatures (26–28°C)	Stack burn spots typically large, oval, or circular, with a dark-brown margin. The center of the spot is initially tan and eventually becomes white or nearly white. Mature spots have small, dark, or black dots in the center.	Seed treatment with Thiram or Captan or Mancozeb at 2 g/kg, hot water treatment at 54°C for 15 minutes	Dharam Singh and Maheshwari (2001)
Udbatta disease	*Ephelis oryzae*	Warm temperature and high humidity	The entire ear head is converted into a straight compact cylindrical black spike-like structure since the infected panicle is matted together by the fungal mycelium, the spikelets are cemented to the central rachis and the size is remarkably reduced	Seed treatment with Captan or Thiram or Granosan MDB + Vitavax, hot water treatment of the seeds (50–54°C) for 10 minutes	Lee and Gunnell (1992)

TABLE 2.3 *(Continued)*

Name of the Disease	Name of the Pathogen	Favorable Conditions	Symptoms	Management	References
Water-Mold	*Achlya* and *Pythium*	It is more pronounced when water temperatures are either low or unusually high.	The endosperm becomes liquefied and oozes out as a white, thick liquid when the seed is mashed. The embryo turns yellow-brown and finally dark-brown. If affected seeds germinate, it results in stunted seedling.	Seed treatment with fungicide will check the growth of water molds	Rush (1985)
Bakanae disease	*Fusarium fujikuroi* F. *andiyazi*, *F. fujikuroi*, *F. proliferatum* and *F. verticillioides*	High humidity and cloudy weather during heading stage temperature ranging from 30 to 35°C. High nitrogen application.	Infected seedlings in nursery are lean and lanky, much taller, and die after some time. In the main field, the affected plants have tall lanky tillers with longer internodes and aerial adventitious roots from the nodes. The root system is fibrous and bushy. The plants are killed before ear head formation or they produce only sterile spikelets	Treating seeds with Thiram or Captan or Carbendazim will check the disease effectively	Hossain et al. (2015); Bashyal et al. (2016)
Black smut of rice	*Tilletia barclayana*	Dew and high humidity	Infected glumes become dark and black, pustules burst. In a severe infection a short beak-like or spur-like outgrowth is produced by the rupturing glumes	crop rotation, limiting nitrogen fertilizer	Ou (1985)

TABLE 2.3 *(Continued)*

Name of the Disease	Name of the Pathogen	Favorable Conditions	Symptoms	Management	References
Sheath blotch	*Pyrenochaeta oryzae*	Disease develops near the water line in lowland fields	The blotches are oblong, about one inch long and brownish at first and gradually the center becomes gray or grayish-brown and is dotted with the black fruiting bodies of the fungus	Protect the crop from insect damage and to destroy/burn the infected straws	Akanda et al. (2003)
Black Sheath rot	*F. moniliforme, F. avenacea, F. graminearum* and *F. proliferatum, Gaemannomyces graminis var. graminis*	High amount of nitrogen and high relative humidity	Dark-brown to black discoloration of the leaf sheaths and reduced tillering, poor grain fill and lodging.	Application of a systemic pesticide, Tridemorph, and phosphamidon. Seed treatment with Mancozeb and Benomyl effectively eliminates seed-borne inoculum.	Datnoff et al. (1993)
Scab	*Fusarium graminearum*	High humidity	Produces bright-orange spore masses on the dead grains	–	Nyvall et al. (1995)
Root rot	*Fusarium* spp., *P. dissotocum,* and *P. spinosum*	Cool temperatures and is less may severe in drill-seeded rice.	Brown to black discoloration. necrosis, and root decay. Under severe conditions young seedlings can die, and mature plants lack support from the roots and lodge or even float causing harvest problems	Planting vigorous, healthy rice seedlings treated with a seed protectant fungicide	Saremi and Okhovat (2004)
Black kernel	*Curvularia lunta C. tuberculata C. verruculosa, C. cymbopogonis*	–	Glumes are discolored and in severe infection the rice kernel shows black discoloration	Seed treatments with fungicide	Shamsi et al. (2003)

TABLE 2.3   *(Continued)*

Name of the Disease	Name of the Pathogen	Favorable Conditions	Symptoms	Management	References
Grain discoloration of rice	*Drechslera oryzae, Curvularia lunata, Sarocladium oryzae, Phoma* sp., *Microdochium* sp., *Nigrospora* sp., and *Fusarium* sp.	Humid conditions	Grains are infected either after milk stage or after harvest or during storage. Infection may be internal or external, causing discoloration of the glumes or kernels. Dark-brown or black spots appear on grains.	Application of Carbendazim + Thiram + Mancozeb (1:1:1) 0.2% at 50% flowering stage will check the malady to a significant level	Thach (2001)
Seedling blight and seed decay	*Bipolaris oryzae, Pythium* sp., *Rhizoctonia solani, Achlya* sp. and *Sclerotium rolfsii.*	Cold and wet weather	Seedling diseases cause spotty, irregular stands through seed decay, pre-emergence, and post-emergence disease	the use of high quality seed, an approved seed treatment, shallow seeding of early-planted rice and planting in warm soil	Ibrahim AboDabab (2014)

2.5 BLAST OF RICE

Blast is one of the most important diseases of rice and is one of the major hindrances for rice production in different parts of the world (Dar et al., 2011). The disease is caused by *Magnaporthe oryzae* (Anamorph: *Pyricularia oryzae* Cav.) and is reported from more than 85 countries of the world (Gilbert et al., 2004). Several rice blast epidemics have occurred all over the world, resulting in 50 to 90% of yield losses (Chaudhary et al., 1994; Agrios, 2005; Yashaswini et al., 2017). It is estimated to cause loss of enough rice to feed more than 60 million people. The blast disease of rice was first reported from China as rice fever disease, in 1637 and subsequently from Japan (1704) and Italy (1828) as imochi-byo and brusone diseases, respectively. Rice blast was recorded in California in 1996 and since then found on grasses in the mid-western United States. In India blast disease of rice was first recorded in 1913 and in 1919 in the Tanjore delta of Tamil Nadu which resulted in a major epidemic outbreak. Since then seven epidemics of blast disease occurred between 1980 and 1987 in four states resulting a major yield losses (Nagarajan, 1988). During 1978 epidemic out break more than 40% panicles of some cultivars were infected by blast disease. Generally, rice blast disease outbreaks occur in the month of February, due to low temperatures at night (22 to 23°C) and heavy dew during the day and heavy damage of the crop occurs in August due to light drizzling for many days (Kim and Kim, 1993). Agriculturally important crops such as rye, wheat, barley, and pearl millet are also susceptible for blast disease.

Based on the infected part of the plant, the disease is named as leaf blast, collar blast, nodal blast, neck blast, and panicle blasts which are described below:

1. **Leaf Blast:** The lesions on the leaves appear as small bluish-green spots, these spots under moisture conditions enlarge to form the spindle-shaped spots with gray central region and dark-brown margins. Lesions enlarge and coalesce to form big linear spots which eventually kill the leaf.
2. **Collar Rot: In this,** the lesion appears at the junction of the leaf blade and sheath which appear as a brown collar. This rot sometimes may cause the death of the leaf. A significant economic loss occurs when collar rot damages the flag or penultimate leaf.
3. **Nodal Blast:** Infected nodes appear black-brown as the disease progress, different spots coalesce and large necrotic areas appear on the leaf, which results in the death of the leaf. Spots also appear on

sheath. Infection at the nodes may result in the breakup of all the plant parts above the infected node.

4. **Neck Blast:** The fungus infects the peduncle at the flowering stage and the lesion changes to brownish-black color. Symptoms start from the node and reaches panicle. Early infection causes, failure of grain filling, whereas partial grain filling is observed in late infection. The glumes of heavily infected panicles show black to brown spots. The yield loss may be from 30–61% depending upon the stage of infection. The infection at the milking stage leads to the premature death of the entire panicle.

5. **Panicle Blast:** The pathogen causes brown lesions on the branches of the panicles and on the spike-let pedicles, resulting panicle blast. Panicles turn white to gray. Partially infected panicles show gray-brown lesions on the stem and on the panicle branches of florets. Infection of the neck, panicle branches, and spikelet pedicles may occur together or may occur separately.

Blast disease occurs throughout a field rather limiting to a localized area. Late planting, frequent showers, overcast skies, and warm weather favor the development of blast. Spores produced in abundance on blast lesions by pathogen become air-borne, disseminating the fungus to a considerable distance resulting secondary infection.

The major disease of rice is blast disease because of its widely occurrence and high incidence rate. Environment conditions such as drought and soil stress favors pathogen. Because of the potential devastation caused to rice production, rice blast disease finds its place in biological terrorism. Valent (2004) considered the blast disease as the world's major disease of rice. Sesma and Osbourn (2004) reported that the leaf blast fungus also infects roots, and through root invasion, it becomes systemic in the plant and causes characteristic symptoms on the leaves, nodes, rachis, and glumes.

2.5.1   PREDISPOSING FACTORS

Temperature of 22 to 28°C, relative humidity of more than 95%, longer duration of dew deposit, leaf wetness (LW) period for more than >10 hrs, cloudy, and drizzling weather, high nitrogen content of soil, degree of host susceptibility and straw of the previously infected crop heaped nearby, more of rainy days and availability of collateral hosts are ideal conditions for spread of the disease resulting severe epidemics.

2.5.2 DISEASE FORECAST

Forecast of rice blast disease can be made on the basis of low night temperature (20–26°C) associated with a high relative humidity (90%) prevailing for
a week or more at the three susceptible stages of crop growth, i.e., seedling
stage, post transplanting, tillering stage and neck emergence stage. The first
blast disease-forecasting model was developed in Japan and was named as
BLAST, later other models; PYRICULARIA, PYRIVIEW, BLASTAM,
EPIBLA, and PBLAS were developed.

2.5.3 MANAGEMENT

1. Resistant varieties like IR-36, IR-64, IR-20, Gauthami, Parijatha,
   Rasi, Sasyashree, Salivahana, Simhapuri, Srinivas, Tikkana, ASD
   18, ADT36, ADT39, and CO47 are to be cultivated. Cultivation of
   highly susceptible varieties viz., IR50 and TKM6 should be avoided
   in disease favorable seasons. Extensive research has to be undertaken
   to identify and tag genes for neck blast resistance. The variety IR-64
   has at least five genes for blast resistance and showed resistance for
   the last 20 years (Khush and Jena, 2009).
2. The field bunds should be free from weed hosts.
3. Treat seed with Captan or Thiram or Carbendazim or Tricyclazole.
   Spray carbendazim to the nursery.
4. Edifenphos or Carbendazim or Tricyclazole or Iprobenphos are to be
   sprayed on the fields.
5. Affected straw and stubbles are to be burned.
6. Employ disease-free seeds.
7. Use of fertilizers based on soil test-split application of nitrogen.
8. Water stagnation in the field to be avoided.
9. Destroy alternate host such as *Panicum repens, Digitaria marginata,
   Brachiaria mutica,* etc., which are in the vicinity of the field.
10. Biological control can be achieved through by employing seed
    treatment, for seedling root dip and foliar spray after 20–25 days of
    transplanting.

2.6 SHEATH BLIGHT OF RICE

Sheath blight is also known as oriental leaf and sheath blight, *Pellicularia*
sheath blight, sclerotial blight and banded blight of rice. It is commonly called

sheath blight because of blighting of the leaf sheaths. It was first reported by Miyake in 1910 from Japan. The causal organism of the disease was *Sclerotium irregular*. Paracer and Chahal reported this disease from Gurdaspur (Punjab) in 1963. Later this disease was also reported from temperate and tropical rice-growing areas by Singh et al. (2004) and Srinivasachary and Savary (2011). Singh et al. (2016) gave an excellent account of sheath blight. Savary et al. (2000) reported 5–10% of yield loss in Asia. Srinivasachary et al. (2011) reported a wide host range of the pathogen. Plants belonging to more than 32 families and 188 genera are infected this pathogen.

In India, loss due to sheath blight disease has been reported to be 54.3% (Chahal, 2003). Similarly, Chen et al. (2012) reported about 6 million tons of yield loss every year in China. Nitrogen fertilizer promotes yield loss significantly (Tang et al., 2007); probably due to the resistance to sheath blight is governed by several minor genes or QTL each with small effect. Rice sheath blight is the major threat to rice production for the last two decades (Skamnioti and Gurr, 2009).

Plants in the growth stage of late tillering or early internode elongation are infected by sheath blight disease, normally lesions develop below the sheaths of lower leaves as oval-to-elliptical, green-gray, water-soaked spots. The lesions will dry and become grayish-white to tan with a brownish border at senesce stage of plant growth. Older plants are highly susceptible. The white sclerotia produced superficially on or near the lesions turn dark-brown at maturity. At the latter stage of infection, lesions expand and the center of the lesions become white with an irregular black to brown border. Under favorable conditions, infection spreads rapidly to upper parts of the plant, i.e., leaf blades, forming irregular shaped dark-colored lesions with brown borders. In the early heading and grain filling stages of growth, frequent rainfall accelerates the rate of infection; at these stages, heavily infected plants produce poorly filled grain, in the lower portion of the panicle. The infection of the Culm leads to reduced carbohydrate reserves and additional loss is resulted from increased lodging or reduced Ratoon production.

High incidence of sheath blight disease is recorded in places with abundant irrigation facilities. The severity of the disease on depends upon cultivation practices, growth stages of the plant at the time of infection, usage of nitrogen fertilizers (Norman et al., 2003), susceptibility of the host (Tang et al., 2007), high relative humidity (> 90%), temperature (30–32°C) and closer planting. In recent years the susceptible varieties with high yielding potential have gained much prominence in the cultivation of rice, thus, in turn, has led to the rapid increase in sheath blight disease.

Sclerotial bodies left in the field from previous crop and weeds, serve as a source of inoculum. Seed, soil, wind, and water are principle transmitters of the disease.

2.6.1 PRECAUTIONS

Incidence of disease may be reduced by planting resistant varieties. In fields with a previous history of the disease, excessive planting of seedlings and high nitrogen fertilizer application should be avoided. The sheath blight disease incidence may be reduced by long-term crop rotation. Application of foliar fungicides may be economical for reducing sheath blight.

2.6.2 PREDISPOSING FACTORS

The conducive factors for disease development in the field are maximum air, temperature (25–30°C), morning high relative humidity (80–100%) and LW (Biswas et al., 2011).

2.6.3 DISEASE FORECAST

Hashiba (1984) reported that the model curve of vertical development of the lesions calculated from temperature, relative humidity, and susceptibility of leaf sheath almost coincided with vertical development of lesions. In models such as blights and its modified version blight as IRRI (Kobayashi et al., 1995), factors like RH between rice hills have been integrated with quantity of sclerotia per unit area of paddy field and susceptibility of leaf sheath and these estimate disease development as vertical and horizontal.

2.6.4 MANAGEMENT

1. *Rhizoctonia solani* has a broad host range and no rice cultivars with complete immunity to sheath blight are available till date (Zue et al., 2014). Radha, Pankaj Swarnadhan, ARC 18119, etc., are the most cultivated varieties. Gaihre et al. (2015) and Richa et al. (2016)

improved resistance in rice against sheath blight using QTLs or by introducing sheath blight defense genes (Zuo et al., 2008, 2011; Zhu et al., 2014) identified few high breeding potential genes or QTLs for sheath blight.

2. Amend the field with neem cake or Farmyard manure.

3. Flow of irrigation water from infected fields to healthy fields should be avoided.

4. Burning the stubbles and deep plowing during summer will reduce the disease incidence.

5. The strobilurin in general and azoxystrobin in particular have been found to be effective in disease reduction and increase the crop yield (Slaton et al., 2003). The disease can be controlled by spraying Carbendazim spray (500 g/ha).

6. Spraying *P. fluorescens* at boot leaf stage and the other after 10 days will reduce the incidence of disease.

7. The disease can be successfully controlled by avoiding the growth of certain alternative host plants like tobacco, water hyacinth, sugar-cane, bean, soybean, tomato, brinjal hyacinth bean, and green gram nearer to rice filed.

8. Check brown planthopper population.

9. Seed treatment with diniconazole, carbendazim, hexaconazole, and mancozeb resulted in the reduction of the disease (Singh et al., 2010).

10. Treating the seeds with salicylic acid (SA), gamma-aminobutyric acid, and chitosan resulted in the immunity and reduce sheath blight severity (Liu et al., 2012).

11. Another effective method to control sheath blight disease was the biological method. Different strains of bacteria were found to be effective in disease control. Radjacommare et al. (2004); and Karnwal and Mannan (2018) reported the efficiency of *Pseudomonas fluorescens* and different strains of *Bacillus spp.* in the control of sheath blight. Similarly, Li et al. (2003) reported the efficiency of *Bacillus subtilis* and *B. megaterium* in inhibiting sclerotia formation by the pathogen. Seed coating of *Pseudomonas*, GRP3 strain followed by root dipping of germinated seedlings inhibited the sclerotia formation of *R. solani* (Pathak et al., 2004). Incidence of seedling disease decreased significantly when sowing seeds treated with *Trichoderma* or carbendazim (Sivalingam et al., 2006).

2.7 FALSE SMUT

This disease was reported from the Thirunelveli district of Tamil Nadu, India (Cooke, 1878). Subsequently, it was reported from more than 40 countries which include almost all rice-growing regions of the world (Biswas, 2001). False smut of rice is caused by *Ustilaginoidea virens* (Zhou et al., 2008; Sanghera et al., 2012). The loss in production of rice was reported between 0.2 and 49% in different states of India depending on severity of disease and varieties of rice cultivated (Dodan and Singh, 1995). In China, 50–60% of infection with the yield reduction of 5–30% was reported (Sanghera et al., 2012). Over 25% yield reduction due to false smut was reported from Tumbes Valley in Peru (Atia, 2004). In hybrid rice cultivation, incidence of false smut during the rainy season reduced the yield to 3 tons per hectare against 6 tons per hectare (Elazegui et al., 2009). In Egypt, yield losses ranged from 1 to 11% (Atia, 2004). Globally, yield loss due to false smut has been reported to range between 3 and 81% depending on the rice variety and disease intensity (Haiyong et al., 2015).

False smut disease is also known as pseudo-smut or green smut. In recent years this disease has emerged as a major disease. The incidence of the disease is particularly more in hybrid varieties. The pathogen, *Ustilaginoidea virens,* in its life cycle produces both ascospores and chlamydospores with multiple propagules. Tanaka et al. (2008) have proposed a new name *Villosiclava virens* for the sexual stage of false smut fungus. The low temperatures, high humidity with moderate rainfall during the period of flowering are favorable conditions for disease development. The survival of pathogen as spore ball in the soil or contaminated rice grain replaces the developing rice kernel. The chlamydospores germinate lately during growing season and releases spores into the air. The primary infection occurs when the spores reach the seed and grows along with the plant and reaches the flower.

The incidence of false smut was considered as an omen of good harvest, but now it has became a menace. Both the quality and quantity of rice grains is reduced by the disease and it also affects the germination vigor of the seeds (Sanghera et al., 2012). The symptoms appear only after flowering, by the time grains in the panicle get infected by the pathogen (Atia, 2004).

Ascospores, which are the primary source of infection to rice plants, are produced by sclerotia, whereas air-borne chlamydospores are the source of secondary infection (Ashizawa et al., 2010). The individual ovaries/grains are transformed into greenish spore balls of velvety appearance by the pathogen.

At the initial stages of infection, the balls of false smut are small and slowly enclose the floral parts. The early balls are slightly flattened and smooth and covered by a thin membrane. As pathogen grows rapidly, false smut ball bursts and release chlamydospores and becomes orange then later yellowish-green or greenish-black. The false smut balls generate sclerotia in an autumn season where the temperature difference between day and night is large (Fan et al., 2016). False smut balls formed on rice panicle produce a number of metabolites like ustilotoxins (a, b, c, d, and e). The false smut contaminated grains and straw contain ustilotoxins, which are toxic to both human and animals (Koiso et al., 1994).

False smuts balls are formed randomly in panicles of rice are collected during harvest. The disease spread varies within the field, between fields and is more severe near the drainage (Nessa et al., 2015). Rainfall at the time of booting and heading stage of rice crop turns the false smut disease into an epidemic However, epidemics varies with varieties, fields, and seasons.

2.7.1 PREDISPOSING FACTORS

The main source of infection is air-borne ascospores, while chlamydospores are secondary source of infection. Temperature (20°C), high relative humidity (above 92%) moderate rainfall with intermittent clear and drizzling weather during flowering are favorable for disease development. Grasses and wild rice species are alternate hosts and serve as source of inoculum.

2.7.2 PRECAUTIONS

Early transplantation of seedlings had higher incidence of infection compared to late planting (Dodan and Singh, 1995). The false smut disease can be avoided by early sowing (Sanne, 1980). False smut disease in susceptible cultivars can be reduced to some extent by conservation tillage, continuous rice cropping, and moderate nitrogen fertilizer (Brooks et al., 2009). The disease at the initial stages can be reduced by using sclerotia free seeds for sowing and by cleaning the bunds. Cultivation practices also play a major role in reduction of disease incidence. Rice cultivated under furrow irrigation system was less susceptible to disease when compared to flooded fields. The mechanism behind the reduction of disease is still not clear (Brooks et al., 2010).

2.7.3 MANAGEMENT

False smut disease can be controlled through cultural, biological, and chemical methods. Mohiddin et al. (2012) reported that prochloraz and carbendazim combination against false smut. Pannu et al. (2010) also reported a reduction in false smut by spraying copper oxychloride 50 WP (0.25%) and propiconazole 25 EC (0.1%) at booting stage. Similarly, Chlorothalonil 75 WP spraying during flowering also reduces disease (Kaur et al., 2015). Foliar spraying of trifloxystrobin 25% and tebuconazole 50% (Nativo 75WG) at booting or 50% panicle emergence stage will check the disease.

Fujione 40 EC and Carbenadazim 50% WP when sprayed at the booting stage has significantly reduced the false smut disease (Bagga and Kaur, 2006). Application of propiconazole 25EC (0.1%) followed by trifloxystrobin and tebuconazole 75 WG combination at booting or at the time of 50%panicle emergence stage was found to be more effective in controlling the disease than other treatments (Raji et al., 2016).

Sanjeet et al. (2018) suggested that the integrated management method was more effective and desirable for the control of false smut disease. Kavya, IR-64 and IR-30864 and KRH-4 varieties are reported to be resistant to this disease. Raji et al. (2016) reported that the bulb extract of *Allium sativum*, *Curcuma longa*, *Lantana camara,* and *Aegle marmelos* and *Cymbopogon flexuous* reduced the disease. Andargie et al. (2017) reported *Antennariella placitae* as potential endophytic fungi in reducing false smut. Application of simeconazole at submerged condition 3 weeks before heading has been highly effective (Tsuda et al., 2006).

In order to check the disease effectively, several missing links for the disease cycle with the life cycle of the pathogen are need to be investigated. Guo et al. (2012) for the effective management of the false smut disease have stressed the need of extensive work in the following areas. (1) The development of a more rapid and effective system to evaluate rice resistance and susceptibility to the disease. (2) The screening of resistance rice germplasms that can efficiently control false smut disease and be used for breeding of resistant varieties. (3) Resistance mechanism of the resistance germplasms. (4) The host ranges and possible colonizing organs other than florets and finally the mechanism of pathogenesis of *Ustilaginoidea virens* is to be understood.

However, much work has taken on the *Ustilaginoidea virens* life cycle, methods for transformation of *Ustilaginoidea virens* for transgenic analysis, artificial inoculation, and understanding disease mechanism. Such studies are lacking on the host side (Kurauchi et al., 2006; Li et al., 2008). Though the disease is much more severe when the rice-heading stage is located in rainy and high humidity days, convincing evidence is lacking (Guo et al., 2012). Similarly, the role of alternate hosts of *Ustilagi noidea virens* is lacking. Hence, future studies should focus on the development of suitable system to evaluate rice resistance and susceptibility. The present methods of sheath injection method are limited to booting stage only and is time consuming. Screening of resistant rice germplasms that can efficiently control false smut disease. Further, the resistance mechanisms of resistant varieties should be deeply investigated. Further, the host ranges of pathogen and different ranges other than florets are to be investigated. Sanjeet et al. (2018), while discussing the management of false smut disease stressed the need of exploration of resistant genes, use of resistant varieties, chemical, non-chemical, and biological methods. Brooks et al. (2009) feel that combined use of conservation tillage continuous rice cropping and moderate nitrogen fertility nearly eliminates dales smut.

2.8 BROWN SPOT

Brown rot disease which is the most devastating disease in the world is caused by *Drechslera oryzae* (*Helminthosporium oryzae*). In India the first epidemic of this disease was reported from Krishna-Godavari delta in 1918–1919 and the second from Bengal. During 1942–1943 epidemic attack, 50–90% of the rice crops were destroyed. This resulted in a major famine in India and Bangladesh, in which 2 million people died of starvation (Chakrabarti, 2001). However, the relationship between plant disease epidemics and famines is not a simple phenomenon; many other factors also play a vital role for the occurrence of such major social consequences (Chakrabarti, 2001; Hossain et al., 2004).

India and South and South-East Asian countries till date report the prevalence of brown spot disease (Savary et al., 2000; Reddy et al., 2010). It causes an average of 10% yield losses (Savary et al., 2000, 2006). Increased variability in rainfall and severe drought conditions are favoring factors for more frequent attack of pathogen (Savary et al., 2005). *H. oryzae* causes both quantity and quality losses up to 7–45% (Jatoi et al., 2016). *Curvularia tuberculata* is also reported to be associated with this disease in Pakistan (Majeed et al., 2016).

2.8.1 SYMPTOMS

In the main field, the pathogen infects the rice crop right from seedling to milky stage. Symptoms appear as small spots on the leaf blade, leaf sheath, coleoptile, glume, and more prominent on the leaf blade and glumes. In Japan, this disease is associated with a physiological disorder known as akiochi.

The spots appear as cylindrical or oval, dark-brown with yellow halo later becoming circular. At later stage of infection several spots coalesce and the leaf dries up (Biswas et al., 2008). The severity of the disease leads to seedlings death due to which the seedbeds are often recognized by burnt appearance from far away. The discoloration of grains is caused due to dark-brown or black spots on glumes; this in turn may affect seed germination, seedling mortality quality, and quantity of grains. The fungus produces certain phytotoxins which breaks the protein fragment of cell wall resulting in partial disruption of integrity of cell.

2.8.2 FAVORABLE CONDITIONS

Pathogen requires high relative humidity above 80%, temperature (25–30°C) and excess of nitrogen fertilizers favor for the disease development.

2.8.3 MANAGEMENT

1. Collateral hosts and infected debris should be removed from the field.
2. Slow-release nitrogenous fertilizers use is advisable.
3. Use of slow-release nitrogenous fertilizers is advisable.
4. Grow tolerant varieties *viz.,* Co44 and Bhavani.
5. Use of disease-free seeds.
6. Seeds to be treated with Thiram or Captan at 4 g/kg. The 20% of nursery should be sprayed with Edifenphos 40 ml or Mancozeb 80 g, if needed it should be sprayed after every 15 days. Edifenphos 500 ml or Mancozeb 2 kg/ha should sprayed if the infection reaches three-stage and the same can be repeated after 15 days.
7. *Drechslera oryzae* can be controlled by fungicides effectively spectacular but this is a relatively short-term measure.
8. Brown spot disease can be controlled eco-friendly by treating seeds with *Trichoderma viride* @ 6 g/kg and Propiconazole (Biswas et al., 2008; Vishal Gupta et al., 2018). Similarly, antagonistic *Pseudomonas* can suppress brown spot disease.

9.   By process of induced resistance can be improved (Shoresh et al., 2010).

2.9   STEM ROT

Sclerotium oryzae (Teleomorph: *Leptosphaeria salvinii*) is a causal organism of stem rot disease. A heavy loss was reported from Japan in 1910. IRRI studies indicated 18–56% of loss due to stem rot fungus and it is a wound parasite. Disease is noticed in rice fields only during the later stages of maturity. In the primary stage of infection, small black lesions are formed on the outer leaf sheath small black lesions are formed, and at the later stage of infection, these spots enlarge and infect the inner leaf sheath. At an advanced stage of infection, a large number of sclerotia appear in the rotting tissues. The culm collapses and plants lodge. After harvest the sclerotia are carried in stubbles.

In fields, the disease appear in circular to irregular areas The pathogen infects the rice plant either during late tillering or at early reproductive stages near the waterline, this accelerates the death percentage of rice plants and cropping becomes difficult. At initial stage of infection rectangular black lesions with angular borders appear on the leaf sheath and at later stage the lesions become larger and diffuse to an irregular shape lesions which deeply penetrate into the culm. The injury to the stem increases with the maturity of the rice plant and reaches to its peak at harvest stage. Weak stalks break at this stage, and harvesting of plants become difficult. Early infection leads to poor yield.

In stubbles and straw, the pathogen survives in the form of sclerotia and is carried through irrigation water. The sclerotia survive in the upper layers (2–3 inches) of the soil for long periods. In the field half-life of sclerotia is about 2 years. The survival of viable sclerotia in fields is up to 6 years after the harvest of a rice crop. The sclerotia are easily carried away by floodwater thus; they come in contact with rice tillers near the waterline, germinate, and infect rice tillers.

2.9.1   *FAVORABLE CONDITIONS*

An infestation of leafhoppers and stem borer and high doses of nitrogen fertilizers favors the diseases.

2.9.2 MANAGEMENT

To eliminate sclerotia plowing deeply in summer and burning stubbles is one of the ways to check the disease. Flow of irrigation water from infected to healthy fields to be avoided. Soil should be maintained dry by effectively draining irrigation water. Crop rotations, use of early maturing varieties, controlling irrigation and destroying rice stubble are some of other control measures. Resistant or non-lodging varieties proved to be more desirable.

Konthoujam et al. (2007) identified low land rice cultivars susceptible and resistant to stem rot of rice from Manipur valley. Fungicides, bavistin, and topsin-M were effective in controlling the stem rot disease (Sharma and Verma, 1985). Similarly, Gangopadhyay and Padhi (1984) recommended Hinosan and lihocin combination as the best fungicidal combination for control of stem rot of rice. Cintas and Webster (2001) suggested other alternative methods such as adding of the straw residue, rolling of the straw to enrich soil contact, removal of residue fall, and burning the straw.

2.10 SHEATH ROT

Sheath rot caused by *Sarocladium oryzae* was one of the major diseases of rice first described by Sawada (1922) from Taiwan. Miah et al. (1985) first reported sheath rot from Bangladesh. Shamsi (1999), after examining 794 sheath rot affected rice samples from 317 varieties/lines collected from different parts of Bangladesh, concluded that along with *Sarocladium oryzae*, *Curvularia lunata*, *Drechslera oryzae* and *Nigrospora oryzae* are also associated with same disease symptoms. Sheath rot of rice proved to be potential threat to rice cultivation as losses caused by this disease varies with the Nation such as Taiwan (3–20%), Philippines (53%), total destruction Thailand (85%), USA (11%) and India (3–90%) (Srinivasan, 1980). It is also causes significant grain yield loss (15%), which further increases with the tungro virus infection (Singh and Dodan, 1986).

Symptoms of sheath rot infection are the appearance of dark reddish-brown irregular lesions with gray center on the leaf sheath. Disease advancement leads to enlarge and fusion of lesions covering the entire lamina and also panicle fails to develop and remains within the sheath and rot. Even if the panicle, even if the emerges no seed setting fails resulting either shriveled partially filled seeds or no seed setting (IRRI, 2003; Sakthivel, 2001).

2.10.1 FAVORABLE CONDITIONS

Factors like insect injury, presence of entry points; large amount of nitrogen fertilizers, high relative humidity, dense crop growth, and leaf canopy and prevalence of temperature from 20 to 28°C at heading to maturity of the rice crop promotes sheath rot. Two phytotoxins helvolic acid and cerulenin produced by the pathogen in culture, applied to sheath develops similar symptoms (Gnanamanickam and Mew, 1991).

2.10.2 MANAGEMENT

Cultural practices like removal of infected stubbles after harvest and optimum plant spacing can check the disease to some extent. Calcium sulphate and zinc sulphate foliar sprays were found to be more effective in controlling sheath rot disease. Seed treatment and foliar spraying with carbendazim, benomyl, and copper oxychloride, edifenphos or mancozeb were found to reduce sheath rot. Soil application of gypsum, neem seed kernel extract (5%) or neem oil (3%) or Ipomoea or Prosopis leaf powder extract (25 Kg/ha) in two spits were effective in the control sheath rot of rice.

2.11 MYCOTOXIN PROBLEM

Mycotoxin contamination of paddy is another problem of great concern. A large number of molds were reported to be associated with paddy either in the field on the standing crop or during its storage (Reddy et al., 2004; Butt et al., 2011; Kiran et al., 2012). Many molds which are associated with the grains in the field in an inactive state, molds in the storage where hostile conditions prevail for grains, become active causing either seedling disease or deterioration of seed and elaborate variety of mycotoxins which are health hazardous to animals and man, when consumed. Among these molds species of *Fusarium, Aspergillus, and Penicillium* are dominant and produce mycotoxins like, trichothecenes, zearalenone, fumonisins, afla-toxins, etc. These toxic metabolites are either carcinogenic, hepatotoxic, tremorgenic, teratogenic, cardiotoxic, nephrotoxic, and cause a variety of health hazards in man which may lead to death. The mysterious deaths of man reported earlier could be attributed to mycotoxins. Many of these toxins are organ-specific, systemic, in their action, thermostable, and cannot be metabolized as a result, they accumulate and reach threshold level and become fatal. In view of the exposed nature of paddy, high humidity in the fields, adverse conditions prevailing during storage, many molds colonize paddy. The problem is more aggravated in places where harvest season coincides with the wet months (Bilgram et al., 1981). Further, floods or unseasonal rains during harvest favor the mycotoxigenic molds growth and elaboration of mycotoxins in rice was reported from different parts of Andhra Pradesh (Laxama et al., 1999; Reddy et al., 2004; Kiran et al., 2015). Konishi et al. (2006) reported the natural contamination of ochratoxin A in retailed rice. Megalla et al. (2007) recorded the zearalenone production in rice by *Fusarium graminearum*. Tanaka et al. (2007) reported the incidence of DON, ZEA, sterigmatocystin, ochratoxin A, nivalenol, and citrinin in different rice samples. Reddy et al. (2008); and Toteja et al. (2006) reported infestation of paddy with different fungi at different stages from harvest to consumption. Reddy (2008) also reported a variety of molds in paddy stored in closed tin. Incidence of mycotoxigenic fungi on food grains varied significantly with the storage structure (Hemanth Raj et al., 2007; Kiran et al., 2013). Chary and Reddy (1987) reported the heavy infestation dehusked paddy in the flooded area of Warangal. Chary et al. (1986) conducted investigations on the factors affecting the contamination of paddy with citrinin. Begum and Samajapathi (2000); Abd-Allah and Ezzar (2005); and Vinod Kumar et al. (2008) have also reported the incidence of different mycotoxins in paddy from different regions of India. Similarly, Maheshwar et al. (2010); Qais et al. (2011); Ok et al. (2014); and Rofiat et al. (2015) also reported natural incidence of mycotoxins on paddy from other parts of the world.

From the above facts it is clear that the incidence of mycotoxigenic fungi and mycotoxins in paddy is not simple to brush aside. Under unpredicted environmental conditions, unscientified cultural, harvesting, transport, and storage conditions, the mycotoxin problem assumes significance in terms of valuable food grain loss and health problems arising in human and livestock due to consumption of infested paddy. Mycotoxin menace needs to be tackled through a multiphase approach involving farmers, warehouse managers, merchants, and consumers. Hence, coordinated efforts are needed to protect the paddy from the mycotoxigenic molds:

1. Selection of good paddy cultivar exhibiting resistance to the growth of mycotoxigenic fungi and mycotoxin elaboration.
2. Proper management to alleviate environmental stress conditions during the crop growth and grain storage.

3. Harvesting the crop at maturity and drying for optimal seed moisture and ventilation. Periodical drying of seed meant for long time storage is desirable.
4. Avoid insect infestation by maintaining optimal conditions.
5. Proper education of different stakeholders about the biological significance of mycotoxins in the health of man is a prudent approach in protecting the health of man.

2.12 CONCLUSIONS AND PERSPECTIVES

Complexity of some of the disease of rice needs a deeper understanding of interaction of different factors like soil, water, physiology plant-water relation, host-plant interaction with the pathogen nature of pathogen and biological environmental conditions. Further, whole-plant approach is desirable in breeding for disease resistance and biochemical conditions (Carvalho et al., 2010). Understanding of global change and socio-economic condition in relation to rice production needs to be investigated. The rice plant system may be taken as a model for unfolding complex pathosystem responding to environmental changes all over the globe. Further, the critical analysis of data available information regarding epidemiological processes, source of resistance, and biocontrol methods needs to be understanding the complexity of pathogen needs to be analyzed.

Research to understand the interaction of Biocontrol agent and host plant resistance and its growth-related genes at the molecular level may reveal the physiology of environmental stress of the disease and their interaction. Further, an integrated approach involving the development of resistant varieties, cultural methods, use of efficient biocontrol methods and biopesticides, perfection, and implementation of prediction system and need-based application of fungicides is the need of the hour.

ACKNOWLEDGMENT

The authors duly acknowledge the Head, Department of Botany, Kakatiya University, Warangal, and Satavahana University, Karimnagar for providing laboratory facilities.

CONFLICT OF INTEREST STATEMENT

The authors declare that they have no conflict of interest with an organization or individual.

KEYWORDS

- **enzyme-linked immunoassay**
- **fungal diseases**
- **local binary patterns histograms**
- **mycotoxins**
- **neural network**
- **yield loses**

REFERENCES

Abd-Allah, E. F., & Ezzat, S. M., (2005). Natural occurrence of citrinin in rice grains and its biocontrol by *Trichoderma hamatum*. *Phytoparasitica, 33*, 73–84.

Agrios, G. N., (2005). *Plant Pathology* (5th edn., p. 922). Elsevier-Academic Press, San Diego, CA.

Ahonsi, M. O., & Adeoti, A. Y. A., (2002). False smut on upland rice in eight rice-producing locations of Edo state, Nigeria. *J. Sustain. Agri., 20*, 81–94.

Al-Ahmar, A. M., (2009). An object-oriented expert system for diagnosis of fungal diseases of date palm. *International Journal of Soft Computing, 4*, 201–207.

Akanda, S. I., Wagih, M. E., Tomda, Y., & Maino, M. K., (2003). Sheath blotch of rice: A new report in Papua New Guinea. *Papua New Guinea Journal of Agriculture, Forestry and Fisheries, 46*, 47–48.

Amutha, D. D., & Muthukannan, K., (2014). Analysis of segmentation scheme for diseased rice leaves. In: *Advanced Communication Control and Computing Technologies (ICACCCT), International Conference* (pp. 1374–1378). IEEE.

Andargie, M., Congyi, Z., Yun, Y., & Li, J., (2017). Identification and evaluation of potential biocontrol fungal endophytes against *Ustilaginoidea virens* on rice plants. *World. J. Microbiol. Biotechnol., 33*, 120–125.

Anonymous, (2013). *Agriculture Statistics of Pakistan* (p. 29). Govt. of Pak. Ministry of Food, Agriculture and Livestock Division (Economic Wing), Islamabad.

Ashizawa, T., Takahashi, M., Moriwaki, J., & Hirayae, K., (2010). Quantification of the rice false smut pathogen *U. virens* from soil in Japan using real-time PCR. *European. J. Plant. Pathol., 128*, 221, 222.

Atia, M. M. M., (2004). Rice false smut (*Ustilaginoidea virens*) in Egypt. *J. Plant. Dis. Protect., 111*, 71–82.

Babu, M. S. P., Murty, N. V. R., & Narayana, S. V. N. L., (2010). A web-based tomato crop expert information system based on artificial intelligence and machine learning algorithms. *International Journal of Computer Science and Information. Technologies, 1*, 6–15,

Bagga, P. S., & Kaur, S., (2006). Evaluation of fungicides for controlling false smut (*Ustilaginoidea virens*) of rice. *J. Indian Phytopath., 59*, 115–117.

Bashyal, B. M., Aggarwal, R., Sharma, S., Gupta, S., Rawat, K., Singh, D., & Krishnan, S. G., (2016). Occurrence, identification and pathogenicity of *Fusarium* spp. associated with bakanae disease of basmati rice in India. *Euro. J. of Pl. Patho, 144*, 457–466.

Begum, F., & Samajpathi, N., (2000). Mycotoxin production on rice, pulses and oil seeds. *Naturwissenschaffen, 87*, 275–277.

Bilgrami, K. S., Prasad, T., Misra, R. S., & Sinha, K. K., (1981). Aflatoxin contamination in maize under field conditions. *Indian Phytopath., 34*, 67–68.

Biswas, A., (2001). False smut disease of rice: A review. *Environ. Biol., 19*, 67–83.

Biswas, B., Dhaliwal, L. K., Chahal, S. K., & Pannu, P. P. S., (2011). Effect of meteorological factors on rice sheath blight and exploratory development of a predictive model. *Indian J. Agric. Sci., 81*, 256–260.

Biswas, S. K., Rattan, V., Srivastava, S. S. L., & Singh, R., (2008). Influence of seed treatment with biocides and foliar spray with fungicides for management of brown leaf spot and sheath blight of paddy. *Indian. Phytopath., 61*, 55–59.

Boys, D. W., & Sun, M. K., (1994). Prototyping an expert system for diagnosis of potato diseases. *Comput. Electronics in Agric, 10*, 259–267.

Brooks, S. A., Anders, M. M., & Yeater, K. M., (2009). Effect of cultural management practices on the severity of false smut and kernel smut of rice. *Plant. Dis., 93*, 1202–1208.

Brooks, S. A., Anders, M. M., & Yeater, K. M., (2010). Effect of furrow irrigation on the severity of false smut in susceptible rice varieties. *Plant. Dis., 94*, 570–574.

Bruce, A. L., Eric, B., Grace, J., John, F. W., Johan, S., William, H., & Chris, V. K., (2008). Nitrogen and potassium fertility impacts on aggregate sheath spot disease and yields of Rice. *Plant Production Science, 11*, 260–267.

Butt, A. R., Yaseen, S. I., & Javaid, A., (2011). Seed-borne mycoflora of stored rice grains and its chemical control. *J. Anim. Plant. Sci., 21*, 193–196.

Carvalho, M. P., Rodrigues, F. A., Silveira, P. R., Andrade, C. C. L., Baroni, J. C. P., Paye, H. S., Jose, E., & Loureiro, J., (2010). Rice resistance to brown spot mediated by nitrogen and potassium. *Journal of Phytopathology, 158*, 160–166.

Chahal, K. S., Sokhi, S. S., & Rattan, G. S., (2003). Investigations on sheath blight of rice in Punjab. *Indian Phytopath., 56*, 22–26.

Chakrabarti, N. K., (2001). Epidemiology and disease management of brown spot of rice in India. In: *Major Fungal Disease of Rice: Recent Advances* (pp. 293–306). Kluwer Academic Publishers.

Chary, M. P., & Reddy, S. M., (1987). Mycotoxins contamination of dehusked rice in the flooded area of Warangal. *Natl. Acad. Sci. Lett., 10*, 129–132.

Chary, M. P., Girisham, S., & Reddy, S. M., (1986). Influence of different seed-borne fungi of rice on citrinin production by *P. citrinum. J. Food Sci. Technol., 23*, 160–162.

Chaudhary, B., Rampur, P. B., & Lal, K. K., (1994). *Neck blast-resistant Lines of Radha-17 Isolated* (Vol. 19, p. 11). IRRN (Philippines).

Chen, Y., Zhangm, A. F., Wang, W. X., Zhang, Y., & Gao, T. C., (2012). Baseline sensitivity and efficacy of thifluzamide in *Rhizoctonia solani. Ann. Appl. Biol., 161*, 247–254.

Cintas, N. A., & Webster, R. K., (2001). Effects of rice straw management on *Sclerotium oryzae* inoculum, stem rot severity, and yield of rice in California. *Plant Disease, 85,* 1140–1144.

Cooke, M. C., (1878). Some extra European fungi. *Grevillea, 7,* 13–15.

Dar, M. S., Hussain, S., Darzi, A. B., & Bhat, S. H., (2011). Morphological variability among various isolates of *Magnaporthe grisea* collected from paddy growing areas of Kashmir. *Int. J. Pharma. Sci., 8,* 90–92.

Datnoff, L. E., Elliott, M. L., & Jones, D. B., (1993). Black sheath rot caused by *Gaeumannomyces graminis* var. *graminis* on rice in Florida. *Plant Disease, 77,* 210.

De Boer, S. H., & López, M. M., (2012). New grower-friendly methods for plant pathogen monitoring. *Annual Review of Phytopathology, 50,* 197–218.

Dean, R. A., Talbot, N. J., Ebbole, D. J., Farman, M. L., Mitchell, T. K., Orbach, M. J., Thon, M., et al., (2005). The genome sequence of the rice blast fungus *Magnaporthe grisea. Nature, 434,* 980–986.

Dharam, S., & Maheshwari, V. K., (2001). The influence of stack burns disease of paddy on seed health status. *Seed Research, 29,* 205–209.

Dodan, D. S., & Ram, S., (1995). Effect of planting time on the incidence of blast and false smut of rice in Haryana. *Indian Phytopathol., 48,* 185, 186.

Elazegui, F. A., Castilla, N. P., Nieva, L. P., & Vera, C. C. M., (2009). *Notes on Rice Diseases.* http://www.knowledgebank.irri.org (accessed on 11 February 2021).

Elshafey, R. A. S., (2018). Biology of rice kernel smut disease causal organism *Tilletia barclayana* and its molecular identification. *J. Phytopathol Pest Manage, 5,* 108–128.

Fan, L. L., Yong, M. L., Li, D. Y., Liu, Y. J., Lai, C. H., Chen, H. M., Cheng, F. M., & Hu, D. W., (2016). Effect of temperature on the development of sclerotia in *Villosiclava virens. Journal of Integrative Agriculture, 15,* 2550–2555.

Gaihre, Y. R., Yamagata, Y., Yoshimura, A., & Nose, A., (2015). Identification of QTLs involved in resistance to sheath blight disease in rice line 32R derived from *Tetep. Tropical. Agr. Dev., 59,* 154–160.

Gangopadhyay, S., & Padhi, B., (1984). *Stem Rot: Epidemiology and Control* (pp. 66–67). Annual Report of CRRI.

Gilbert, M. J., Soanes, D. M., & Talbot, N. J., (2004). Functional genomic analysis of the rice blast fungus *Magnaporthe grisea. Applied Mycology and Biotechnology, 4,* 331–352.

Gnanamanickam, S. S., & Mew, T. W., (1992). Biological control of blast disease of rice (*Oryza sativa* L.) with antagonistic bacterial and its mediation by a *Pseudomonas* antibiotic. *Ann. Phytopathol. Soc. Japan, 58,* 380–385.

Godwin, P. M. S., Narmadha, R., & Bernatim, T., (2019). A brief survey on diseases of paddy plant. *J. Pharm. Sci. Res., 11,* 2739–2743.

Gonzalez-Andujar. J. L., (2008). Expert system for pests, diseases and weeds identification in olive crops. *Expert Syst. Appl., 36,* 3278–3283.

Guo, X., Yan, L., Jing, F., Liang, L., Fu, H., & Wenming, W., (2012). Progress in the study of false smut disease in rice. *Journal of Agricultural Science and Technology, 2,* 1211–1217.

Haiyong, H., Wongkaew, S., Jie, Y., Xuehui, Y., Xiaojun, C., Shiping, W., Qigqun, T., et al., (2015). Biology and artificial inoculation of *Ustilaginoidea virens* (Cooke) Takahashi in rice. *Afr. J. Microbiol. Res., 9,* 821–830.

Harmon, P. F., Dunkle, L. D., & Latin, R., (2003). A rapid PCR-based method for the detection of *Magnaporthe oryzae* from infected perennial ryegrass. *Plant Dis., 87,* 1072.

Hashiba, T., (1984). Forecasting and estimation of yield losses by rice sheath blight disease. *Japan Agric. Res. Quarterly, 18,* 92–98.

Hemanth, R. M., Niranjana, S. R., Chandra, N. S., & Shekar, S. H., (2007). Health status of farmers saved paddy, sorghum, sunflower and cowpea seeds in Karnataka, India. *World. J. of Agri. Sci., 2,* 167–177.

Hossain, M. S., Ayub, A. M., Mollah, M. I. U., Khan, M. A. I., & Sajjadul, I. A. K. M., (2015). Evaluation of fungicides for the control of bakanae disease of rice caused by *Fusarium moniliforme* (Sheldon). *Bangladesh. Rice J., 19,* 49–55.

Hossain, M., Khalequzzaman, K. M., Mollah, M. R. A., Hussain, M. A., & Rahim, M. A., (2004). Reaction of breeding lines/cultivars of rice against brown spot and blast under field condition. *Asian Journal of Plant Sciences, 3,* 614–617.

Ibrahim, E. A. M., & AboDabab, M. S., (2014). Seed discoloration and their effect on seedlings growth of Egyptian hybrid rice. *Res. J. Seed. Sci., 7,* 63–74.

IRRI, (2003). *Rice Doctor.* Philippines: The International Rice Research Institute.

Jatoi, G. H., Manzoor, A. A., Javad, A. T., Shabana, M., Naimatullah, M., Sultan, A., Azhar-ul-Din, Shahid, H., & Abdul, S. M., (2016). Efficacy of selected fungicides on the linear colony growth of the *Helminthosporium oryzae* caused by brown spot disease of rice. *Pak. J. Biotechnol., 13,* 13–17.

Jin, J. H., Seong-Sook, H., & Jin-Hyeuk, K., (2002). A simple method for sporangial formation of the rice downy mildew pathogen *Sclerophthora macrospora. Plant. Pathol. J., 18,* 77–80.

Karnwal, A., & Mannan, M., (2018). Application of *Zea mays* L. rhizospheric bacteria as promising biocontrol solution for rice sheath blight. *Pertanika. J. Trop. Agric. Sci., 41,* 1613–1626.

Kaur, Y., Lore, J. S., & Pannu, P. P. S., (2015). Evaluation of rice genotypes for resistance against false smut. *Plant Dis. Res., 30,* 46–49.

Khan, M. A. I., Bhuiyan, M. R., Hossain, M. S., Sen, P. P., Ara, A., Siddique, M. A., & Ali, M. A., (2014). Neck blast disease influences grain yield and quality traits of aromatic rice. *C R Biol., 337,* 635–641.

Khush, G. S., & Jena, K. K., (2009). Current status and future prospects for research on blast resistance in rice (*Oryza sativa* L.). In: Wang, G. L., & Valent, B., (eds.), *Advances in Genetics, Genomics and Control of Rice Blast Disease.*

Kim, C. K., & Kim, C. H., (1993). The rice leaf blast simulation model EPIBLAST. In: Frits, P. D. V., Paul, T., & Klaas, M., (eds.), *Systems Approaches for Agricultural Development* (pp. 309–321).

Kiran, S., Naresh, A., Surekha, M., Ram, R. S. A., & Reddy, S. M., (2012). Incidence of mycotoxin producing fungi on stored paddy Warangal District of A.P., India. *International Journal of Recent Scientific Research, 11,* 897–900.

Kiran, S., Surekha, M., & Reddy, S. M., (2013). Succession of mycotoxigenic fungi in relation to storage and mycotoxin production on stored paddy (*Oryza sativa* L.). *Indian Phytopathology, 66,* 258–262.

Kiran, S., Surekha, M., & Reddy, S. M., (2015). Fungal infestation and mycotoxins contamination of paddy of flooded areas of Godavari belt region of Telangana State, India. *Research Science Reporter, 5,* 30–35.

Kobayashi, T., Ijiri, T., Mew, T. W., Maringas, G., & Hashiba, T., (1995). Computerized forecasting system (blight IRRI) for rice sheath blight disease in the Philippines. *Annals. Phytopathol. Soc. Japan, 61,* 562–568.

Kobayashi, T., Ishiguro, K., Nakajima, T., Kim, H., Okada, M., & Kobayashi, K., (2006). Effects of elevated atmospheric CO_2 concentration on the infection of rice blast and sheath blight. *Phytopathology, 96,* 425–431.

Koiso, H. C., Li, Y., Iwasaki, S., Hanaoka, K., Kobayashi, T., Sonoda, R., Fujita, Y., et al., (1994). Ustiloxin, antimitotic cyclic peptides from false smut balls on rice panicles caused by *Ustilaginoidea virens. Journal of Antibiotics, 47,* 765–773.

Konishi, S. Y., Nakajima, M., Tabata, S., Ishikuro, H., Tanaka, T., & Norizuki, H., (2006). Occurrence of aflatoxins, ochratoxin-A and fumonisins in retailed foods in Japan. *J. Food Prot., 69,* 1365–1370.

Konthoujam, J., Chhetry, G. K. N., & Ranjit, S., (2007). Symptomatological significance and characterization of susceptibility/resistance group among low land rice cultivars towards stem rot of rice in Manipur valley. *Indian Phytopathology, 60,* 478–481.

Kurauchi, K., Kudo, Y., Kimura, T., & Uemura, T., (2006). Difference in resistance to false smut disease between rice cultivars in Aomori prefecture. *Ann. Rept. Plant Prot. North Jpn., 57,* 17–21.

Kurniawati, N. N., Abdullah, S. N. H. S., Abdullah, S., & Abdullah, S., (2009). Investigation on image processing techniques for diagnosing paddy diseases. *Soft Computing and Pattern Recognition,* 272–277.

Ladhalakshmi, D., Laha, G. S., Singh, R., Krishnaveni, D., Srinivas, P. M., Mangrauthia, S. K., Prakasam, V., et al., (2012). *False Smut: A Threatening Disease of Rice* (p. 32). Technical Bulletin No. 63, Indian Institute of Rice Research (ICAR), Rajendra Nagar, Hyderabad.

Laxma, R. G., Srinivas, M., & Reddy, S. M., (1999). Incidence of mycotoxins in foods of tribals of Godavari belt. *Proc. Nat. Symposium. Fungi in Diversified Habitats,* 68–73.

Lee, F. N., & Gunnell, P. S., (1992). Udbatta. In: Webster, R. K., & Gunnell, P. S., (eds.), *Compendium of Rice Diseases* (p. 29). St. Paul, Minnesota, USA: APS Press.

Leonardo, A., Rayane, S. P., & Fabricio, A. R., (2016). Microscopic aspects of silicon-mediated rice resistance to leaf scald. *Phytopathology, 106,* 133–141.

Li, X. M., Hu, B. S., Xu, Z. G., & Mew, T. W., (2003). Threshold population sizes of *Bacillus subtilis* B5423-R to suppress the occurrence of rice sheath blight. *Chin. J. Rice Sci., 17,* 360–364.

Li, Y. S., Zhu, Z., Zhang, Y. D., Zhao, L., & Wang, C. L., (2008). Genetic analysis of rice false smut resistance using mixed major genes and polygenes inheritance model, *Acta Agron. Sin., 34,* 1728–1733.

Liang, W. J., Hong, Z., Gu-Feng, Z., & Hong-Xin, C., (2019). Rice blast disease recognition using a deep convolutional neural network. *Sci. Rep., 9,* 2869.

Lieven, B., Margreet, B., Alfons, C. R. C., Vanachter, C., André, L., Bruno, P. A., Cammue, B., & Thomma, P. H. J., (2003). Design and development of a DNA array for rapid detection and identification of multiple tomato vascular wilt pathogens. FEMS *Microbiol. Lett., 223,* 113.

Liu, H., Tian, W., Li, B., Wu, G., Ibrahim, M., Tao, Z., Wang, Y., Xie, G., Li, H., & Sun, G., (2012). Antifungal effect and mechanism of chitosan against the rice sheath blight pathogen, *Rhizoctonia solani. Biotechnol. Lett., 34,* 2291–2298.

Lopez-Morales, V., Lopez-Ortega, J., Ramos, F., & Munoz, L. B., (2008). JAPIEST: An integral intelligent system for the diagnosis and control of tomato diseases and pests in hydroponic greenhouses. *Expert Syst. Appl., 35,* 1506–1512.

Maheshwar, P. K., & Janardhana, G. R., (2010). Natural occurrence of toxigenic *Fusarium proliferatum* on paddy (*Oryza sativa* L.) in Karnataka, India. *Tropi. Life. Sci. Res., 21,* 1–10.

Majeed, R. A., Shahid, A. A., Saleem, M. Z., Asif, M., Zahid, M. A., & Haider, M. S., (2016). First report of *Curvularia tuberculata* causing brown leaf spot of rice in Punjab, Pakistan. *The American Phytopathological Society, 100,* 1791.

Mansingh, G., Reichgelt, H., & Bryson, K. M., (2007). CPEST: An expert system for the management of pests and diseases in the Jamaican coffee industry. *Expert Syst. Applic., 32,* 184–192.

McCartney, H. A., Foster, S. J., Fraaije, B. A., & Ward, E., (2003). Molecular diagnostics for fungal plant pathogens. *Pest Manage. Sci., 59,* 129–142.

Megalla, S. E., Bennett, G. A., Ellis, J. J., & Shotwell, O. I., (2007). Production of deoxynivalenol and zearalenone by isolates of *Fusarium graminearum* Schw. *J. Basic Microbiol., 26,* 415–419.

Mew, T. W., & Gonzales, P. A., (2002). *Handbook of Rice Seed-Borne Fungi* (p. 83). International Rice Research Institute, Los Banós, Philippines.

Mew, T. W., Leung, H., Savary, S., Vera, C. C. M., & Leach, J., (2004). Looking ahead in rice disease research and management. *Crit. Rev. Plant Sci., 23,* 103–127.

Miah, S. A., Shahjahan, A. K. M., Hossain, M. A., & Sharma, M. M., (1985). Survey of rice diseases in Bangladesh. *Tropical Pest Management, 31,* 204–213.

Miyake, I., (1910). Studies on the fungi of rice plants in Japan. *J. College Agric. Imperial Univ. Tokyo, 2,* 237–276.

Mohiddin, F. A., Bhat, F. A., Gupta, V., Gupta, D., & Kalha, C. S., (2012). Integrated disease management of false smut of rice caused by *Ustilaginoidea virens*. *Trends in Biosciences, 5,* 301, 302.

Mondal, K. K., & Shanmugam, V., (2013). Advancements in the diagnosis of bacterial plant pathogens: An overview. *Biotechnol. Mol. Biol. Rev., 8,* 1–11.

Mukherjee, B., & Uma, M. N., (2018). Biological control of narrow brown leaf spot and leaf smut disease in paddy crops by some antagonistic fungi. *Global Journal of Medical Research: K Interdisciplinary, 18,* 29–39.

Nagarajan, S., (1988). Epidemiology and loss of rice, wheat and pearl millet crops due to diseases. In: *International Symposium on Crop Losses and Diseases Outbreaks in Tropics and Control Measures* (p. 209). Tropical Agriculture Research Centre, Japan.

Nessa, B., Salam, M. U., Haque, A. H. M. M., Biswas, J. K., MacLeod, W. J., et al., (2015). Flyer: A simple yet robust model for estimating yield loss from rice false smut disease (*Ustilaginoidea virens*). *Am. J. Agric. Biol. Sci., 10,* 41–54.

Norman, R. J., Wilson, Jr. C. E., & Slaton, N., (2003). Soil fertilizers and mineral nutritional in US mechanized rice culture. In: Smith, C. W., & Dilday, R. H., (eds.), *Rice: Origin, History and Production* (pp. 31–412). John Wiley: NJ. USA.

Nyvall, R. F., Porter, R. A., & Percich, J. A., (1995). First report of scab on cultivated wild rice in Minnesota. *Plant Dis., 79,* 82.

Ok, H. E., Kim, D. M., Kim, D., Chung, S. H., Chung, M. S., Park, K. H., & Chun, H. S., (2014). Mycobiota and natural occurrence of aflatoxin, deoxynivalenol, nivalenol and zearalenone in rice freshly harvested in South Korea. *Food Control, 37,* 284–291.

Ou, S. H., (1985). *Rice Diseases* (p. 380). Commonwealth Mycological Institute, England.

Pannu, P. P. S., Thind, T. S., & Goswami, S., (2010). Standardization technique for artificial creation of false smut of rice and its management. *Indian Phytopathol., 63,* 234, 235.

Pathak, A., Sharma, A., Johri, B. N., & Sharma, A. K., (2004). Pseudomonas strain GRP3 induces systemic resistance to sheath blight in rice. *Int. Rice Res. Notes, 29,* 35, 36.

Patil, S. S., Dhandra, B. V., Angadi, U. B., Shankar, A. G., & Joshi, N., (2009). Web-based expert system for diagnosis of micronutrients deficiencies in crops. *Proceedings of the World Congress on Engineering and Computer Science I* (pp. 20–22). San Francisco, USA.

Pattanayak, A., Bujarbaruah, K. M., Sharma, Y. P., Ngachan, S. V., Dhiman, K. R., Munda, G. C., Azad, T. N. S., et al., (2006). Technology for increased production of upland rice and lowland waterlogged rice. In: *Proceedings of Annual Rice Workshop* (pp. 9–13). Hyderabad.

Qais, A. L. N., Amra, H. A., & Ali. A. B., (2011). Natural occurrence of mycotoxins in corn grains and some corn products. *Pak. J. Life Soc. Sci., 9*, 1–6.

Qi, M., & Yang, Y., (2002). Quantification of *Magnaporthe grisea* during infection of rice plants using real-time polymerase chain reaction and northern blot/phosphoimaging analyses. *Phytopathology, 92*, 1–8.

Quintana, L., Susana, G., & Aldo, O., (2018). *Microdochium oryzae* associated with rice leaf scald disease in Paraguay. *Australian Journal of Basic and Applied Sciences, 12*, 56–58.

Radjacommare, R., Kandan, A., Nandakumar, R., & Samiyappan, R., (2004). Association of the hydrolytic enzyme chitinase against *Rhizoctonia solani* in rhizobacteria-treated rice plants. *J. Phytopathol., 152*, 365–370.

Raji, P., Sumiya, K. V., Renjisha, K., Dhanya, S., & Narayanankutty, M. C., (2016). Evaluation of fungicides against false smut of rice caused by *Ustilaginoidea virens*. *International Journal of Applied and Natural Sciences, 5*, 77–82.

Ramesh, S., & Vydeki, D., (2018). Rice blast disease monitoring using mobile app. *International Journal of Engineering and Technology, 7*, 400–402.

Rani, P. M. N., Rajesh, T., & Saravanan, R., (2013). Development of expert system to diagnose rice diseases in Meghalaya State. In: *Fifth International Conference on Advanced Computing* (ICoAC) (pp. 8–14).

Reddy, C. S., Laha, G. S., Prasad, M. S., Krishnaveni, D., Castilla, N. P., & Nelson, A., (2010). Characterizing multiple linkages between individual diseases, crop health syndromes, germplasm deployment and rice production situations in India. *Field Crops Research, 120*, 241–253.

Reddy, C. S., Reddy, K. R. N., Kumar, R. N., Laha, G. S., & Muralidharan, K., (2004). Exploration of aflatoxin contamination and its management in rice. *J. Mycol. Pl. Pathol., 34*, 816–820.

Reddy, K. R. N., (2008). *Estimation and Prevention of Aflatoxin Contamination in Rice*. PhD. dissertation, Osmania University. Hyderabad, India.

Reddy, K. R. N., Reddy, C. S., Abbas, H. K., Abel, C. A., & Muralidharan, K., (2008). Mycotoxigenic fungi, mycotoxins and management of rice grains. *Toxin Reviews, 27*, 287–317.

Ribot, C., Hirsch, J., Balzergue, S., Tharreau, D., Notteghem, J. L., Lebrun, M. H., & Morel, H. B., (2008). Susceptibility of rice to the blast fungus, *Magnaporthe grisea*. *Journal of Plant Physiology, 165*, 114–124.

Richa, K., Tiwari, I. M., Kumari, M., Devanna, B. N., Sonah, H., Kumari, A., Nagar, R., et al., (2016). Functional characterization of novel chitinase genes present in the sheath blight resistance QTL: QSBR11-1 in rice line. *Tetep. Front Plant Sci., 7*, 244.

Rofiat, A. S., Fanelli, F., Atanda, O., Sulyok, M., Gozzi, G., Bavaro, S., Krska, R., Logrieco, A. F., & Ezekiel, C. N., (2015). Fungal and bacterial metabolites associated with natural contamination of locally processed rice (*Oryza sativa* L.) in Nigeria. *Food Addit. Contam. A, 32*, 950–959.

Rush, M. C., (1985) Controlling seed-rot and seedling diseases in rice. *Phytopathology, 175,* 1329.

Sakthivel, N., (2001). Sheath rot disease of rice: Current status and control strategies. In: Sreenivasapraasad, S., Johnson, R., & Manibhushanrao, K., (eds.), *Major Fungal Diseases of Rice: Recent Advances* (pp. 271–283). Dordrecht: Springer.

Sanghera, G. S., Ahanger, M. A., Kashyap, S. C., Bhat, Z. A., Rather, A. G., & Parray, G. A., (2012). False smut of rice (*Ustilaginoidea virens*) under temperate agro-climatic conditions of Kashmir, India. *Elixir Bio Technol., 49,* 9827–9831.

Sang-Won, A., (1980). Eyespot of rice in Colombia, Panama and Peru. *Plant Dis., 64,* 878–880.

Sanjeet, K. S., & Shankar, S. B., (2018). Integrated management of false smut disease in rice: A review. *International Journal of Chemical Studies, 4,* 48–51.

Sanne, G., (1980). Studies on false smut disease caused by *Ustilaginoidea virens* on paddy in Karnataka, India. *Intern. Rice Res. Newsl., 5,* 4, 5.

Saremi, H., & Okhovat, S. M., (2004). The occurrence of root rot and crown rot of rice in Gilan and Zanjan provinces, Iran. *Commun. Agric. Appl. Biol. Sci., 69,* 525–529.

Sarma, S. K., Singh, K. R., & Singh, A., (2010). An expert system for diagnosis of diseases in rice plant. *International Journal of Artificial Intelligence, 1,* 1–6.

Savary, S., Castilla, N. P., Elazegui, F. A., & Teng, P. S., (2005). Multiple effects of two drivers of agricultural change, labor shortage and water scarcity, on rice pest profiles in tropical Asia. *Field Crops Research, 91,* 263–271.

Savary, S., Teng, P. S., Willocquet, L., & Nutter, F. W. Jr., (2006). Quantification and modeling of crop losses: A review of purposes. *Annual Review of Phytopathology, 44,* 89–112.

Savary, S., Willocquet, L., Elazegui, F. A., Castilla, N., & Teng, P. S., (2000). Rice pest constraints in tropical Asia: Quantification of yield losses due to rice pests in a range of production situations. *Plant Disease, 84,* 357–369.

Savary, S., Willocquet, L., Elazegui, F. A., Teng, P. S., Du, P. V., Zhu, D., Qiyi, T., et al., (2000). Rice pest constraints in tropical Asia: Characterization of injury profiles in relation to production situations. *Plant Disease, 84,* 341–356.

Sawada, K., (1922). *Descriptive Catalogue of Formosan Fungi II* (Vol. 2, p. 135). Report. Govt. Res. Inst. Dep. Agric. Formosa.

Sesma, A., & Osbourn, A. E., (2004). The rice blast pathogen undergoes developmental processes typical of root-infecting fungi. *Nature, 431,* 582–586.

Shamsi, S., (1999). *Investigations into the Sheath Rot Disease of Rice (Oryza sativa L.) in Bangladesh* (pp. 1–127). PhD Thesis. Department of Botany, University of Dhaka.

Shamsi, S., Khan, A., Shahzahan, A., & Miah, S., (2003). Fungal species associated with sheaths and grains of sheath rot affected rice varieties from Bangladesh. *Bangladesh J. Bot., 32,* 17–22.

Sharma, J. P., & Verma, R. N., (1985). Effect of various concentration of fungicides *in vitro* on Sclerotium state of *Corticium rolfsii. Indian Phytopathology, 38,* 358–360.

Sharma, O. P., & Bambawale, O. M., (2008). Integrated management of key diseases of cotton and rice. *Integrated Management of Plant Pest and Diseases, 4,* 271–302.

Shoresh, M., Harman, G. E., & Mastouri, F., (2010). Induced systemic resistance and plant responses to fungal biocontrol agents. *Annual Review of Phytopathology, 48,* 21–43.

Singh, R., & Dodan, H., (1986). Sheath rot of rice. *Int. J. Trop. Plant. Dis., 13,* 139–152.

Singh, R., Sunder, S., & Dodan, D. S., (2010). Standardization of inoculation method in nursery beds and management of sheath blight of rice through host resistance, chemicals and botanicals. *Indian Phytopathol., 63,* 286–291.

Singh, R., Sunder, S., & Dodan, D. S., (2012). Status and weed hosts of *Rhizoctonia solani* Kuhn incitant of sheath blight of rice in Haryana. *Plant Dis Res., 27,* 225–228.

Singh, R., Sunder, S., & Kumar, P., (2016). Sheath blight of rice: Current status and perspectives. *Phytopathol., 64,* 340–351.

Singh, S. K., Shukla, V., Singh, H., & Sinha, A. P., (2004). Current status and impact of sheath blight in rice (*Oryza sativa* L.): A review. *Agric. Rev., 25*(4), 289–297.

Sivalingam, P. N., Vishwakarma, S. N., & Singh, U. S., (2006). Role of seed-borne inoculum of *Rhizoctonia solani* in sheath blight of rice. *Indian Phytopathol., 59,* 445–452.

Skamnioti, P., & Gurr, S. J., (2009). Against the grain: safeguarding rice from rice blast disease. *Trends in Biotechnol., 27,* 141–150.

Slaton, N. A., Cartwright, R. D., Meng, J., Gbur, E. E. Jr., & Norman, R. J., (2003). Sheath blight severity and rice yield as affected by nitrogen fertilizer rate application method and fungicide. *Agron. J., 95,* 1489–1496.

Sreekantha, D. K., & Kavya, A. M., (2017). Agricultural crop monitoring using IOT-A study In: *11*[th] *International Conference on Intelligent Systems and Control (ISCO)* (pp. 134–139).

Srinivasachary, S., Willocquet, L., & Savary, S., (2011). Resistance to rice sheath blight (*Rhizoctonia solani* Kuhn) (teleomorph: *Thanatephorus cucumeris* (A.B. Frank) Donk) disease: Current status and perspectives. *Euphytica, 178,* 1–22.

Srinivasan, S., (1980). *Yield Losses Caused by Sheath Rot* (Vol. 5, p. 4). Int. Rice Res Newsletter.

Strange, R. N., & Scott, P. R., (2005). Plant disease: A threat to global food security. *Annu. Rev. Phytopathol., 43,* 83–116.

Sun, J. F., Najafzadeh, M. J., Vicente, V., Xi, L. Y., & De Hoog, G. S., (2010). Rapid detection of pathogenic fungi using loop-mediated isothermal amplification, exemplified by *Fonsecaea* agents of chromoblastomycosis. *J. Microbiol. Methods, 80,* 19–24.

Suudi, M., Jinyeong, K., Jong-Mi, P., Shin-Chul, B., Donghern, K., Yong-Hwan, K., & Pyung, A., (2013). Quantification of rice blast disease progressions through Taqman real-time PCR. *Molecular Biotechnology, 55,* 43–48.

Talbot, N. J., (2003). On the trail of a cereal killer: Exploring the biology of *Magnaporthe grisea. Annu. Rev. Microbiol., 57,* 177–202.

Tanaka, K., Sago, Y., Zheng, Y., Nakagawa, H., & Kushiro, M., (2007). Mycotoxins in rice. *Int. J. Food. Microbiol., 119,* 59–66.

Tanaka, T., Ashizawa, T., Sonoda, R., & Tanaka, C., (2008). *Villosiclava virens* gen. nov., com. nov., teleomorph of *Ustilaginoidea virens*, the causal agent of rice false smut. *Mycotaxon., 106,* 491–501.

Tang, Q., Peng, S., Buresh, R., Zou, Y., Castilla, N. P., Mew, T. W., & Zhong, X., (2007). Rice varietal difference in sheath blight development and its association with yield loss at different levels of N fertilization. *Field Crops Res., 102,* 219–227.

Thach, (2001). Survey on seed-borne fungi and its effects on grain quality of common rice cultivars in the Mekong Delta. *Omonrice, 9,* 107–113.

Toteja, G. S., Mukherjee, A., Diwakar, S., Singh, P., Saxena, B. N., Sinha, K. K., Sinha, A. K., et al., (2006). Aflatoxin B$_1$ contamination of parboiled rice samples collected from different states of India: A multi-center study. *Food. Addit. Contam., 23,* 411–414.

Tsuda, M., Sasahara, M., Ohara, T., & Kato, S., (2006). Optimal application timing of simeconazole granules for control of rice kernel smut and false smut. *J. Gen. Plant Pathol., 72*, 301–304.

Valent, B., (2004). Plant disease: Underground life for rice foe. *Nature, 431*, 516, 517.

Vinod, K. M., Basu, S., & Rajendran, T. P., (2008). Mycotoxin research and mycoflora in some commercially important agricultural commodities. *Crop. Prot., 27*, 891–905.

Vishal, G., Naveed, S., Razdan, V. K., Seethiya, M., Kausar, F., Sharm, S., & Rai, P. K., (2018). Management of brown spot of rice (*Oryza sativa* L.) caused by *Bipolaris oryzae* by bio-control agents. *Int. J. Curr. Microbiol. App. Sci., 7*, 3472–3477.

Wang, Z., Li, L., Xuejun, Y., Hailin, G., Aigui, G., & Jianxiu, L., (2013). Genetic diversity analysis of *Cynodon dactylon* (bermudagrass) accessions and cultivars from different countries based on ISSR and SSR markers. *Biochemical Systematics and Ecology, 46*, 108–115.

Ward, E., Foster, S. J., Fraaije, B. A., & McCartney, H. A., (2004). Plant pathogen diagnostics: Immunological and nucleic acid-based approaches. *Ann. Appl. Biol., 145*, 1–16.

Yang, W., Hongyan, Z., Mengxue, L., Zonghua, W., Jie, Z., Shihua, W., Guodong, L., & Feng, F., (2014). Early diagnosis of blast fungus, *Magnaporthe oryzae*, in rice plant by using an ultra-sensitive electrically magnetic-controllable electrochemical biosensor. *Analytica Chimica Acta, 850*, 85–91.

Yashaswini, C., Reddy, P.N., Pushpavati, B., Rao, S., Madhav, S. (2017). Prevalence of rice blast (*Magnaporthe oryzae*) incidence in South India. *Bul. Env. Pharmacol. Life Sci., 6*, 370–373.

Zhou, Y. L., Pan, Y. J., Xie, X. W., Zhu, L. H., Wang, S., & Li, Z. K., (2008). Genetic diversity of rice false smut fungus *Ustilaginoidea virens* and its pronounced differentiation of populations in North China. *Journal of Phytopathology, 156*, 559–554.

Zhu, Y. J., Zuo, S. M., Chen, Z. X., Chen, X. G., Li, G., Zhang, Y. F., Guiquan, Z., & Xuebiao, P., (2014). Identification of two major rice sheath blight resistance QTLs, qSB11 HJX74 and qSB11 HJX74 in field trials using chromosome segment substitution lines. *Plant Dis., 98*, 1112–1121.

CHAPTER 3

New Insights into the Identification and Management of Wheat Diseases

T. L. PRAKASHA,[1] A. N. MISHRA,[1] J. B. SINGH,[1] S. V. SAI PRASAD,[1] and SURESH CHAND[2]

[1]ICAR-Indian Agricultural Research Institute, Regional Station, Indore – 452 001, Madhya Pradesh, India, E-mail: prakash7385@gmail.com (T. L. Prakasha)

[2]School of Life Sciences, Devi Ahilya University, Indore – 452 001, Madhya Pradesh, India

ABSTRACT

Wheat is one of the most important cereal crops of the world, constituting a major source of food to a vast population. Many biotic factors like pests and diseases, as well as abiotic factors like drought, heat, salinity, etc., affect global wheat production to a great extent. The important diseases affecting wheat are rusts (leaf, stripe, and stem rusts), powdery mildew, smuts, bunts, *Fusarium* head blight, tan spot, and spot blotch. Soil bore root rots, viruses, bacterial pathogens, and nematodes also cause substantial yield reduction in wheat in some areas. The early identification of diseases, especially seed-borne diseases, diseases caused by viruses and bacteria, has proved vital in minimizing losses by taking appropriate management measures. Many conventional methods are often employed to identify wheat diseases which may be non-accurate and time-consuming. Many modern methods like immunoassays, nucleic acid-based techniques, optical sensor-based methods like thermography, fluorescence imaging (FI), hyperspectral techniques, biosensors are being increasingly used in accurate detection of wheat pathogens. Internet-based applications have enabled the farmers to identify the pest and disease problems in their fields. The use of host resistance and chemicals are the main methods of managing wheat diseases. The marker-assisted selection improved the chances of stacking more than one resistance gene into a single cultivar. The plant-mediated RNAi and CRISPR/Cas9 technologies are being utilized to develop plants with disease resistance.

3.1 INTRODUCTION

Wheat is one of the most important food crops of the world, providing staple food for billions of people around the world. Wheat, along with rice and maize, provides the majority of the food requirements of the global population. Wheat is planted in an area of ~220 million ha with annual production of 749 million tons (FAOSTAT, 2016). Providing food forever increasing global population, which is now at 7.6 billion and projected to reach 9.8 billion by 2050 is a major challenge and nearly 880 mt of wheat will be required to feed this population (International Grains Council, 2019).

However, wheat production is plagued by many pest and disease attacks, which results in heavy losses. The major diseases affecting wheat production are stripe rust, leaf rust, stem rust, powdery mildew, head blight, Karnal bunt, flag smut, and spot blotch. Along with these fungal diseases, many viral, bacterial pathogens and nematodes cause localized yield reduction. The early identification and managing these diseases is highly important to minimize the losses caused by these pathogens. Therefore, knowing different fungal, bacterial, viral pathogens, their distribution, existing variability, methods of dissemination, etc., are important in minimizing crop losses.

Annual global wheat production losses in both developing and developed countries are reported to be around 12.4% (Oerke et al., 1994). Therefore, early identification and management strategies are important in order to reduce the losses caused by wheat pathogens.

3.2 WHEAT DISEASES

3.2.1 WHEAT RUSTS

The rusts of wheat belong to genus *Puccinia*, family Pucciniaceae, order Uredinales, and class Basidiomycotina. Stem rust (=Black rust) caused by *P. graminis* f. sp. *tritici,* leaf rust (=Brown rust) caused by *P. triticina,* yellow rust (=stripe rust) of wheat is caused by *P. striiformis* f. sp. *tritici* are the three rust diseases affecting global wheat production. These are obligate parasites capable of growing only on living host tissue, and are heterocious requiring two different hosts to complete their life cycles.

Stem rust causes enormous losses to wheat production all over the world. The pathogen attacks all the above-ground parts like stem, ears, leaves, leaf sheath, and awns. The black/brown colored pustules appear on both upper and lower leaf surfaces, stems, leaf sheaths, ear heads and awns. The epidermal layer will rupture, releasing a powdery mass of brick-red colored uredospores. Later these pustules turn black due to teliospore production which is resting spores. The day temperature of 25–30°C with mild nights (15–20°C), and free moisture from rain or dew favor the stem rust development (Schumann and Leonard, 2000). The pathogen completes its life cycle on wheat and barberry. The pycnial and aecial stages were found to be on barberry and uredial and telial stages on wheat.

Leaf rust occurs throughout the world due to its wide range of temperature adoptability. Leaf rust epidemics have been reported in Northwestern India during 1971–1972 and 1972–1973 leading to wheat production losses of eight and ten lakh tones, respectively. Minute, round, orange colored pustules appear irregularly on the leaves, and they may appear occasionally on stem, leaf sheath, and spikes. Gray to black colored telia appears on lower leaf surface towards the end of the season. This fungus completes its sexual cycle mostly on *Thalictrum speciosissimum* where the fungus produces pycniospores and receptive hyphae. Temperature of around 20–25°C with free moisture (rain or dew) causes epidemics.

Yellow rust of wheat has become a devastating disease, especially in a very cool climate. Orange to orange-yellow pustules containing uredospores appear like narrow stripes in between veins on leaves. In severe conditions, symptoms appear on leaf sheaths, glumes, and stem. Severe infection may lead to coalescing of pustules, drying of leaves. At the end of the season, dark colored teliospores replace uredospores. Shriveled grains are formed in the severely affected plants. Temperature of 10–20°C with high humidity favor the disease development. *Berberis* spp. and *Mahonia aquifolium* have recently been reported as the alternate hosts of *P. striiformis* f. sp. *tritici*.

3.2.2   LOOSE SMUT

It is a seed borne disease caused by *Ustilago tritici* (Pers.) Rostr. (= *Ustilago segetum var. tritici*) leading to an annual loss of 3–4%. The disease is favored whenever a cool, humid climate accompanied by light showers occurs especially during flowering time. The disease incidence is more in

Northern India. The disease could be identified only during ear head emergence time when the black smutted spikes emerge from the flag leaf. All the grains in-ear heads will be replaced with black powdery mass. Diseased ears usually appear a few days earlier than the healthy ear heads. The plant growth will be stunted with few tillers. Primary infection comes from the seed as mycelium remains dormant in seed embryo and causes infection during the next crop season.

3.2.3   KARNAL BUNT

It was first reported from Karnal in Haryana, India by Mitra (1931). This disease is of quarantine importance and listed in around 21 countries as quarantine disease. It is caused by fungus *Tilletia indica* Mitra (=*Neovossia indica* (Mitra) Mundkur). Along with yield losses, its major effect is on quality of the wheat flour. The flour is fishy in odor due to production of trimethylamine chemical if more than 3% bunted grains are mixed with healthy grains and will be unfit for consumption. Karnal bunt affected grains are either partially or wholly bunted, and the latter get converted into black powder of bunt spores. In an ear head, a few grains get converted into black powdery mass. This fungus is both seed and soil-borne and disease symptoms could be observed only during examination of grains.

3.2.4   POWDERY MILDEW

It is caused by *Erysiphe graminis* f. sp. *tritici*. The disease appears initially as superficial grayish small patches of white cottony growth which later cover entire leaves and above-ground parts. The white cottony growth turns into brownish due to cleistothecia formation later in the season. The leaves will dry prematurely. The cleistothecia remain in the infected wheat straw and serve as the primary source of inoculum in the next crop season. Warm temperatures of 15–22°C with short periods of high humidity favors the fungal infection and development.

3.2.5   SPOT BLOTCH

It is caused by *Bipolaris sorokiniana*, and is one of the most serious foliar diseases, especially in warmer areas. In India, this disease causes extensive yield losses in Eastern Gangetic Plains which has warmer temperatures along with high humidity (Duveiller et al., 2005). It also causes serious crop losses in other southeast Asian countries like Nepal and Bangladesh. The disease symptoms appear on coleoptiles, leaves, stems, roots, and awns. Initially, it appears as dark brown necrotic spots without yellow halo (Chand et al., 2002) which later coalesce leading to large blotches eventually killing entire leaf. High disease pressure may lead to stunted growth and reduced tillering producing shriveled grains. In Bangladesh and China, it was reported to cause a yield loss of 15% (Alam et al., 1998; Xiao et al., 1998). The disease is seed transmitted and conidia survive in soil. The warm temperature of 18–32°C along with high humidity favor the disease development (Duveiller et al., 2005).

3.2.6   HEAD BLIGHT/HEAD SCAB

It is caused by many species of the *Fusarium*, and is one of the most devastating diseases of wheat (O'Donnell et al., 2004). The most important species is *F. graminearum* (teleomorph *Gibberella zeae*) (Dean et al., 2012) which infects many grasses and cereal crops like wheat, barley, maize, rye, rice, and triticale. Warm weather with high humidity during flowering and maturity are the congenial factors for disease occurrence. The fungus enters through floral organs and spike becomes pale colored and later turns pink, especially at the base and edges of the glumes. Under favorable conditions, fungi infect entire spike and adjacent spikelets and grain will become shriveled, low in weight (Gilchrist and Dubin, 2002).

3.2.7   ALTERNARIA BLIGHT

It is caused by *Alternaria triticina*, leading to significant yield losses in wheat. Its epidemic was reported in eastern Uttar Pradesh during 1964–1965 and 1965–66. Usually, 45–50 days old wheat seedlings are prone to this disease. The symptoms manifest as discolored, oval lesions on the lower leaf surface, which later progresses upwards, lesions enlarge and coalesce to form irregular, dark blotches, often with chlorotic margins. Disease affects older leaves first and youngest leaves are not affected. The grains get discolored and shriveled, and under severe conditions, the affected leaves dry up. Temperature around 25°C, coupled with high humidity, favors the disease. Wheat is the only host of *Alternaria triticina,* and the disease is both seed and soil-borne. The fungus survives as conidia and mycelium on seed surface and within seeds, respectively.

3.2.8   YELLOW EAR ROT OR TUNDU

It was first reported from Punjab state of India. This is a complex disease caused by Ear cockle nematode (*Anguina tritici*) and a bacterium, *Rathayibacter tritici* (*Corynebacterium michiganensis tritici*). Affected plants show twisting and crinkling of lower and middle leaves. Slime covers the entire ear, glumes, and stems. The plants remain stunted and stems become deformed. Under humid conditions, slime oozing is visible from the affected parts, but slime becomes hard and dark yellow to brown under dry conditions. The grains will be replaced by nematode galls. The nematode serves as the vector of bacteria. The bacteria alone cannot multiply and cause the disease. Bacterial multiplication is favored by low temperature of 10–15°C with high humidity (70–100%). The nematodes survive as nematode galls and start the infection when nematode gall mixed seeds is used for sowing.

3.2.9   EAR COCKLE

This disease caused by a nematode *Anguina tritici* is commonly reported from Eastern Europe, parts of Asia and Africa. In India, it is reported to be present in the states of Bihar, Jharkhand, eastern U.P., and Chhattisgarh. The initial symptom is the basal swelling of the stem at the ground level and affected plants remain stunted in growth. Affected plants show profuse tillering without any increase in productive tillers. Grains get replaced with nematode galls which may vary in number (1–5/ear). The nematodes survive as galls and are transferred from one area to another through infected seeds.

3.2.10   VIRUS AND VIRUS-LIKE DISEASES

Many viral diseases reported on wheat causing yield losses in many countries. Among viruses (Wiese, 1987; Lapierre and Signoret, 2004) barley yellow dwarf virus (BYDV), soil-borne wheat mosaic virus (SWMV) and wheat yellow mosaic virus (WYMV) are reported to be most important. Aphid transmitted BYDV is the most well-known viral disease (Henry and Plumb, 2002). The infection by BYDV leads to yellowing of leaves, stunting of plants and symptoms appear in scattered patches initially. Later on circular patches can be observed in the field due to secondary spread. Tiller number is reduced in case of early infection. Soil-borne cereal mosaic virus (SBCMV) and Soil-borne wheat mosaic virus (SBWMV) are transmitted by the fungus *Polymyxa graminis,* which is also soil-borne and remain viable in soil for at least 15 years. The viruses get transmitted from one place to another place by the movement of soil particles with fungus. The infected plants show pale green to yellow streaks on leaves and leaf sheaths, plants will be stunted. These symptoms appear in distinct patches. This fungus also reported to be a vector for other viruses like Indian peanut clump and Chinese wheat mosaic virus. Due to obligate nature of the vector, identification, and management of these viruses is difficult (Kanyuka et al., 2003).

On durum and bread wheat, chlorosis, streaking, yellowing of leaves, ears with plant stunting was reported and through leaf dip electron micros-copy, PCR, and rolling circle amplification (RCA) association of any of the wheat viruses were ruled out. It was proved that the symptoms were due to the presence of 16SrXI-B group phytoplasma by amplification of 16S rRNA through PCR and RFLP technique. The disease incidence was up to 20% in different genotypes and was severe, especially on durum wheat (Rao et al., 2017).

The diseases cause heavy losses in wheat crop production. Nearly 10–22% yield losses were reported from 18 diseases studied over 34 years in Kansas State (Bockus et al., 2001). An estimate arrived on 22 developing countries' data indicated 3.7% yield loss over 10 years due to leaf rust disease alone. Yield losses of up to 20% were reported due to spot blotch in Indo Gangetic plains of India (Duveiller, 2004). Powdery mildew reported to affect at least 12 million hectares in China and South America (Costamilan, 2005). The early identification and management of these diseases is necessary to mini-mize the crop losses due to these dreaded diseases.

3.3   RECENT ADVANCES IN IDENTIFICATION OF WHEAT DISEASES

Effective crop disease management always depends on early identification of plant diseases which enables to contain its further spread. The plant diseases can be identified using both direct as well as indirect methods. The direct methods involve symptomatology studies, serology, and molecular methods; indirect methods involve the measure of various parameters like temperature difference, volatiles identification, transpiration rate change, etc. DNA-based molecular methods as well as many serological methods are the most accurate methods for identifying plant diseases and are highly reliable. Many new technologies developed for plant disease identification are being utilized in wheat crop disease identification. These are highly reliable, precise, and accurate (Martinelli et al., 2015). Many of the wheat diseases (especially those appearing at ear head emergence stage) are difficult to detect in the early stages. Therefore, utilizing new techniques may aid in the management of these diseases.

3.3.1 VISUAL METHODS

The traditional method of identifying a disease is by observing and interpreting different characteristic symptoms produced in wheat viz., pustules, leaf spots, blight, fungal structures like sclerotia, encrusted mycelium, seed discoloration (De Tempe and Binnerts, 1979), seed galls, smutted seeds, etc. These could be augmented by examination of leaf sections, seeds, etc., under a microscope to identify the disease. Karnal bunt-affected seed can be identified through stereomicroscope (Begum and Mathur, 1989). Some seed-borne pathogens (*Tilletia indica* and *T. barclayana*) can also be detected by observing water used to wash infected wheat seeds (Begum and Mathur, 1989). By soaking bunted wheat seeds (Agarwal and Verma, 1983; Agarwal and Srivastava, 1985) in 2% NaOH solution for 20 hr at 20°C, the disease can be identified as infected seed appears shiny jet black colored in contrast to pale yellow color of healthy seed. Incubation and embryo count methods were also employed for effective identification of seed-borne diseases like ear cockle and bunt (De Tempe and Binnerts, 1979; Rennie, 1982). The visual observations may be followed by isolation and culturing of pathogen in artificial media followed by microscopic observation of the morphological characteristics of the pathogens viz., spore morphology, sporulation patterns, production, and characteristics of sporulating structures producing asexual and sexual spore forms (Narayanasamy, 2011). However, culturing method is time-consuming, and it depends on the individual skills and experience of the person doing the diagnosis (McCartney et al., 2003).

3.3.2 SEROLOGICAL METHODS

These methods are based on specific antigen-antibody bindings and are widely used for the detection of viral, fungal, phytoplasma, and bacterial pathogens. The polyclonal, monoclonal, and phage-displayed recombinant antibodies have been produced for different pathogens. The various methods of immunological techniques include immunodiffusion test, enzyme-linked immunosorbent assay (ELISA), radio-immunosorbent assay (RISA), dipstick immunoassay, dot immunobinding assay (DIBA), tissue blot immunoassay (TBIA), western blot analysis, immunosorbent Electron Microscopy and flow cytometry (FCM). Of these, ELISA is most prominently used in plant disease diagnosis. The antibody which is bound on a solid support like microtiter plates will react with antigen and color is produced by adding another antibody conjugated with an enzyme (like alkaline phosphatase) along with a substrate (p-nitrophenyl phosphate). The intensity of the colored product is measured spectroscopically, which is proportional to the amount of antigen/pathogen present. Polyclonal antisera were produced in rabbits against *Tilletia indica* and *T. barclayana* and differences were observed between the two in their reactivity patterns (Holland et al., 1990). Micro-titer ELISA technique is used for Karnal bunt detection at an early stage (Varshney, 1999). Seed immunoblot binding assay (SIBA) was developed by Kumar et al. (1998) for early detection of this disease. When Karnal bunt infected seeds are kept on nitrocellulose membrane, the teliospores (antigens) will be adsorbed onto this membrane; the color developed is having a direct correlation with the severity of KB infection. These serological methods are rapid, robust, cost-effective, and simpler, however, they are not highly specific, and sensitive when compared with nucleic acid-based methods.

3.3.3   NUCLEIC ACID-BASED TECHNIQUES

The nucleic acid-based techniques are the most rapid, useful, reliable, and efficient ones for disease diagnosis. The genome of every living organism has a set of unique, conserved nucleotide sequences. Highly conserved genomic regions are present in diverse organisms which are distantly related. These techniques are not dependent on the gene expression and thus are independent of environmental factors.

3.3.3.1   POLYMERASE CHAIN REACTION (PCR)

Polymerase chain reaction (PCR) technique has been in use to detect plant pathogens due to its rapidity, sensitivity, reliability, and accuracy. This method involves the synthesis of specific DNA fragment which will be amplified using enzymes to detect sequence which is specific to pathogens. The specific DNA fragment is amplified using forward and reverse oligo-nucleotide primers through repeated cycles of DNA denaturation, primer annealing with complementary sequences and extension of the annealed primers with a thermostable DNA polymerase. The several copies of specific DNA fragment are synthesized with these repetitive cycles (Mullis, 1987).

Several PCR variants have been developed with enhanced accuracy and sensitivity. These are reverse transcription-polymerase chain reaction (RT-PCR), nested PCR, multiplex real-time PCR, quantitative real-time PCR, co-operational PCR (Co-PCR), Bio-PCR, fluorescence *in situ* hybridization (FISH), loop-mediated isothermal amplification (LAMP).

The choice of PCR primers is important in PCR reactions. For detection of pathogen variations when DNA sequence is not known, random amplified polymorphic DNA (RAPD) is a primer of choice. The arbitrary sequences of about 10 nucleotides are selected and utilized in the PCR and products are separated on a gel. However, these are not suitable for pathogen diagnosis as the results are difficult to reproduce and DNA from another source may interfere. Therefore, primers which are specific to the pathogen of interest are essential for diagnosis. In fungal pathogens, ribosomal genes (internal transcribed spacer (ITS), transcribed external spacer (ETS) intergenic spacer (IGS) regions) are targeted for primer designing. *Tilletia indica* was identified using two sets of oligonucleotide primers targeting DraI fragments of mitochondrial DNA (Smith et al., 1996). The primers designed based on mitochondrial DNA were successfully used to differentiate *T. indica* from *T. walker* (Frederick et al., 2000). Primers developed to identify RAPD amplified fragments successfully identified *Fusarium poae* which causes head blight (Parry and Nicholson, 1996). Similar successful attempts were made to identify *F. avenaceum* by Turner et al. (1998) who developed primers specific to *F. avenaceum*, with no cross-reaction with *F. tricinctum.*

The multiple DNA or RNA can be detected using Multiplex PCR in a single reaction which reduces the need of multiple PCR assays (James et al., 2006). In this method, two or more target sequences are amplified simultaneously within a single reaction. *Septorium tritici* (leaf blotch) and *S. nodorum* (leaf and glume blotch) were detected by multiplex PCR by designing the primers targeting the β-tubulin gene (Fraaije et al., 2001). This technique was utilized for simultaneous detection of *Puccinia graminis* f. sp. *tritici, P. triticina,* and *Blumeria graminis* f. sp. *tritici* (Chen et al., 2015). Multiplex reverse transcription PCR protocol was developed to simultaneously identify five strains of Barley and Cereal yellow dwarf virus (B/CYDVs), and single strains of Wheat spindle streak mosaic (WSSMV), SBWMV and Wheat streak mosaic virus (WSMV) (Deb and Anderson, 2008).

Real-time polymerase chain reaction (RT-PCR or quantitative PCR (qPCR), a technique which is based on PCR, can be used to monitor the amplification of a targeted DNA in real-time. This technique measures the intensity of fluorescent signals, which are produced by intercalating non-specific fluorescent dyes or sequence-specific fluorescent dye-labeled DNA probes which enable its detection only after its hybridization with complementary sequence. This method was used to quantify *Fusarium spp.* in wheat grains (Waalwijk et al., 2004). This method was also used to detect low titer of B/CYDVs in wheat samples (Martin et al., 2000).

3.3.3.2 LOOP-MEDIATED ISOTHERMAL AMPLIFICATION (LAMP) METHOD

It was first reported by Notomy and is increasingly being used for pathogen detection. This method is simple, cheap, accurate, and highly specific, does not require any laboratory set up and can be performed in field conditions. The amplification of nucleic acid is performed under isothermal conditions using three or more primers that are highly specific to target DNA. This method uses auto-cycling strand displacement DNA synthesis using *Bst* DNA polymerase, dNTPs, specific primers and the target DNA template. The method produces many stem-loop DNA structures with inverted repeats of the target sequence. Turbidity is observed after the completion of amplification due to the production of magnesium pyrophosphate, a by-product of the reaction which is measured by fluorescent dyes such as Syber Green which changes its color from orange to green as the LAMP amplicons are produced. Several LAMP methodologies have been developed to detect wheat pathogens. LAMP-based detection method was developed to detect *Puccinia striiformis* f. sp. *tritici* DNA in uredospores and wheat seedlings with latent infections. It was successful in detecting fungus after 24 hr of inoculation with the sample DNA as low as 2 pg/μl (Huang et al., 2011). Using two LAMP primers, *Ustilago tritici* (loose smut fungus) was detected in wheat seed samples and the detection limit was observed to be 100 fg μL^{-1} (Yan et al., 2019).

3.3.4   INDIRECT DETECTION METHODS

These methods will be able to detect pathogens indirectly based on stress profiling, gaseous metabolites of plants and plant volatiles produced during infection. These methods identify the diseases based on their effect on plant cell physiology. Thermography, fluorescence imaging (FI), and hyperspectral techniques are some indirect methods used for plant disease detection.

3.3.4.1   SPECTROSCOPIC AND IMAGING TECHNIQUES

Plant health will be evaluated by these methods by exposing plant tissue surface with specific wavelength of light and intensity of reflected light is measured. Various methods viz., fluorescence spectroscopy, infrared spectroscopy, nuclear magnetic resonance (NMR) spectroscopy, RBG imaging, FI, hyperspectral imaging, and thermography (Ray et al., 2017) are being utilized for plant disease detection. These spectroscopic and imaging techniques may be automated and with automated agricultural vehicle these methods can help to prevent early spread of the disease:

1. **Reflectance/Transmittance Spectroscopy:** This technique was used in identifying wheat grain characteristics in visible or the near-infrared (NIR) range. Internal seed infestation was assessed using NIR spectroscopy in wheat seeds (Dowell et al., 1998) which detected head scab, vomitoxin, and ergosterol in single wheat grain sample.
2. **Thermography Technique:** It enables imaging surface temperature changes of leaves and crop canopies. Thermographic cameras will be used to capture these infrared radiations emitted from crop canopies and difference is analyzed. The water loss through stomata which is affected by various plant pathogens can be measured by thermography and water loss due to transpiration could be measured without any external temperature impact (Fang and Ramasamy, 2015).
3. **Hyperspectral Imaging (Remote Sensing):** It is a highly useful method to scan large geographical area in small time and can obtain plant health conditions over a wide range of 350 and 2500 nm spectrum. It is robust, non-destructive, and can be used for real-time rapid analysis of the imaging data. Presently most of the remote sensing studies are conducted to detect and monitor a single plant disease, and a few are conducted for multiple diseases. Using the

leaf level spectral analysis, a model for discriminating wheat leaf infected with stripe rust and powdery mildew was developed with more than 80% accuracy (Yuan et al., 2013). Based on the canopy hyperspectral remote sensing data, powdery mildew, stripe rust, and wheat aphid effects were identified with more than 90% accuracy by using stepwise discriminate analysis and hierarchical clustering (Qiao et al., 2010). Using near-infrared reflectance spectroscopy data of individual wheat leaves (Li et al., 2013), stripe rust and leaf rust were identified with identification accuracy of 96%. Using canopy hyperspectral data, stripe rust and leaf rust were identified with overall identification accuracy 82% (Wang et al., 2015).

4. **Fluorescence Imaging Technique:** In this method, chlorophyll fluorescence on the leaves is measured. The changes in fluorescence due to changes in the photosynthetic apparatus and photosynthetic electron transport reactions can be used to detect plant pathogens. Using this technique, the temporal and spatial changes in fast chlorophyll fluorescence kinetics parameters and red/NIR reflectance were assessed for powdery mildew and leaf rust infection on wheat seedlings. The technique was able to identify these pathogen infection 2–3 days before visual symptoms or significant changes in normalized-differenced-vegetation index (NDVI) (Kuckenberg et al., 2009). By using this technique, the delayed chlorophyll fluorescence induction phase in photosystem II was observed in wheat plants infected with *Bipolaris sorokiniana* causing common root rot disease. The chlorophyll *a* fluorescence induction Parameters could be used for early identification of *B. sorokiniana* infection in wheat plants (Matorin et al., 2018).

3.3.4.2 BIOSENSORS FOR DISEASE DETECTION

Different biosensors have been commercially developed for detection, diagnosis, and monitoring of different pathogens. TheseThese sensors are able to detect different signals viz., electrical, chemical, electrochemical, optical, magnetic, or vibrational emitted by the analytes. The limit of detection and specificity could be enhanced by nanomaterial matrices as transducers and bio-recognition components like DNA, RNA, enzymes, antibodies, etc., respectively (Fang and Ramasamy, 2015). These biosensors are accurate and rapid, thus facilitating effective disease management. These biosensors can be broadly categorized as optical biosensors, volatile biosensors, electrochemical biosensors, mass sensitive biosensors, and nanoparticle-based biosensors. In the case of the optical biosensor, the changes in the amplitude, polarization, or frequency of the input light produced due to changes in physical/chemical changes induced by the pathogen are measured. The recognition element may be enzyme, antibodies, or microbes. These sensors are broadly categorized as fluorescence-based biosensors, chemiluminescence-based biosensors and surface plasmon resonance (SPR)-based biosensors (Ray et al., 2017). Out of these sensors, SPR-based biosensor had been used to detect *Fusarium culmorum* in wheat (Zezza et al., 2006). This technique measures change in effective refractive index on the surface without the need of labeling particles and the interaction of the biomolecules on the surface can also be obtained in real-time (Skottrup et al., 2007). An oligonucleotide probe was developed using a specific primer for *F. culmorum,* which is biotinylated and was used for specific detection of the wheat pathogen in both culture material and naturally infected wheat plants successfully (Zezza et al., 2006).

3.4   WHEAT DISEASE MANAGEMENT

The major wheat diseases are stem, leaf, and stripe rusts, loose smut, flag smut, Karnal bunt, powdery mildew, leaf blotch/blight and head blight. For the effective disease management, knowing, and identifying the symptoms is highly essential. Early, rapid, and accurate identification of wheat diseases is possible through several modern direct and indirect methods. Traditionally wheat diseases are being managed by several cultural, biological, and chemical methods.

Cultural management methods include choice of healthy and certified seeds, crop rotation, fertilizer management, and irrigation management. Choosing healthy and certified seeds will help in managing many of the seed-borne diseases of wheat like Karnal bunt, loose smut, ear cockle, tundu disease, etc. Wheat followed by maize crop system reported to be having higher disease incidence, and replacing it with other pulses or non-cereal crop was reported to be beneficial (Teich and Nelson, 1984). As *Fusarium* spp. survives saprophytically on wheat crop debris, burying the debris leading to faster decomposition was beneficial in disease management (Dill-Macky et al., 1998). The increased dose of nitrogen fertilizers was conducive for head blight occurrence in wheat, barley, and triticale (Heier et al., 2005).

Many biological control agents are effectively utilized in integrated disease management (IDM). Many fungal (*Trichoderma* sp., *Clonostachys rosea, Cladosporium cladosporioides,* and *Cryptococcus flavescens*) and bacterial (*Bacillus* sp., *Streptomyces* sp., *Brevibacillus* sp., *Paenibacillu spolymixa, Lysobacter enzymogenes,* and *Pseudomonas fluorescens*) bioagents are used for the management of Karnal bunt, head blight, etc.

Several chemicals have been effectively utilized in the management of wheat diseases. Many of the triazole group fungicides like propiconazole, tebuconazole, metconazole, prothioconazole are being used to manage various wheat diseases. Seed treatment of wheat seeds with mancozeb, thiabendazole, difenoconazole, carboxin, carbendazim are recommended for the management of many of the seed-borne diseases (Dill-Macky, 1997). However, the ill effects of chemicals on the environment is the major hindering factor when chemicals are used for the disease management, and hence use of host resistance is the most commonly used approach to manage several wheat diseases.

3.4.1   UTILIZING HOST RESISTANCE IN WHEAT DISEASE MANAGEMENT

The most economical and eco-friendly method of managing wheat diseases is to utilize host resistance in breeding and deploying effective resistant wheat varieties to multiple diseases, and this is relatively inexpensive in the long term and environmentally safe compared to use of chemicals. To date, there are 78 leaf rust, 78 stripe rust, 59 stem rust, 58 powdery mildew, 10 bunt resistance genes in wheat have been cataloged along with many more temporarily designated genes, QTLs (McIntosh et al., 2017; Kolmer et al., 2018).

The rust resistance genes are of two types-seedlings (all-stage) and adult plant resistance (APR) genes. Seedling resistance genes are race specific, major gene controlled, have no or little environmental effect and are often overcome by new virulent rust pathotypes. Generally, wheat varieties carrying major genes often become susceptible in a relatively short span of time (5–8 years) due to the emergence of new pathogen races. Race non-specific resistance genes are race non-specific, minor gene controlled, quantitative, and expressed at adult plant stage and are generally durable in nature (Lindhout, 2002; Bariana et al., 2001). The durability of APR genes has led to the worldwide utilization of APR genes in developing new cultivars

(Rajaram et al., 1997). The durability of these genes led to the discovery of multiple APR genes or quantitative trait loci (QTLs) which provide partial resistance which is often addictive in nature. Most of the QTLs identified for resistance to different diseases have small effect.

3.4.1.1   ANTICIPATORY BREEDING

It is the development of improved wheat varieties which are resistant to future emerging pathogen race population. The knowledge on existing race flora and genetic diversity among wheat germplasm is essential to breed the future-ready wheat crop. Maintaining genetic diversity is highly essential for breeding wheat against future challenges posed by newer pathogen race flora. In the case of wheat rusts, mutation, recombination, and migration play an important role in evolving new pathotypes.

3.4.1.2   GENE PYRAMIDING

It is the simultaneous introgression of multiple resistance genes into a single genotype. Many resistance gene pyramids were made through conventional breeding approaches by introgressing newer R genes into released cultivars. It is considered that multiple resistance genes in a single cultivar should be more durable as the evolution of a pathotype virulent to multiple genes is very difficult. In the case of obligate foliar diseases like rusts and mildews, pyramiding combinations of defeated R genes have no effect or limited effectiveness on pathogen (Pederson and Leath, 1988). In the case of conventional breeding, it was difficult to identify and pyramid several undefeated R genes in a cultivar. Due to the development of marker-assisted selection, now it is possible to identify different genes through different linked molecular markers. These markers allow breeders to select the plants with target genes of interest.

3.4.1.3   MARKER ASSISTED SELECTION (MAS)

Marker-assisted selection (MAS) helps in accelerating conventional breeding. Conventional breeding involves the selection of phenotype, which is a result of complex genotype x environment interaction. The fixing of target genes could be achieved in early generation using MAS as markers are not affected by environmental conditions and can be used to detect gene of interest (Dubcovsky, 2015). Introgression of minor genes has been a challenge due to scarce genotype carrying multiple minor genes, masking effect of major genes associated and environmental effects leading to challenges in their selection. The molecular markers reduced dependence on laborious screening methods and thus aiding in minor genes introgression (Ragimekula et al., 2013). MAS is successfully utilized as a faster, reliable method for the identification and transfer of resistance genes. Many of the resistance genes are being transferred into common background. Many of the varieties were developed with multiple resistance genes through MAS. Wheat variety, '*Unnat* PBW 343,' was developed by pyramiding leaf rust resistance genes *Lr37* and *Lr17* and stripe rust resistance genes *Yr70* and *Yr76* in the background of wheat variety PBW343 (Sharma et al., 2019). Leaf rust and stem rust resistance genes like-*Lr19*, *Sr31*, and *Sr24* (Wessels and Botes, 2014) have been transferred into single background. Efforts were made to pyramid *Fusarium* head blight resistance genes *Fhb1*, *Fhb2,* and *Fhb4* into single breeding lines (Randhawa et al., 2013). An elite bread wheat cultivar Yang158 was successfully utilized as background parent for transferring three powdery mildew gene combinations *Pm2* + *Pm4a*, *Pm2* + *Pm21*, *Pm4a* + *Pm21* using markers. The near-isogenic lines were developed having these genes and combinations (Liu et al., 2000). Some of the rust resistance gene which is most effective under Indian conditions and their molecular markers are listed in Table 3.1.

TABLE 3.1   Rust Resistance Genes and Their Molecular Markers

SL. No.	Resistance Genes	Primer	Primer Sequence	References
1.	*Sr26*	Sr26#43	5'-AAT CGT CCA CAT TGG CTT CT-3' 5'-CGC AAC AAA ATC ATG CAC TA-3'	Mago et al. (2005)
2.	*Sr32*	csSr32#1	5'-GGT TTG GTG GCA ACT CAG GT-3' 5'-CAT AAG CCA AAG AGG CAC CA-3'	Mago et al. (2013)
3.	*Sr33*	BARC152	5'-CTTCCTAAAATCGGGCAACCGCTT-GTTG-3' 5'-GCGTAATGATGGGAGTGGCTATA-GGGCAGTT-3'	Sambasivam et al. (2008)
		CFD15	5'-CTC CCG TAT TGA GCA GGA AG-3' 5'-GGC AGG TGT GGT GAT GAT CT-3'	
4.	*Sr39-Lr35*	Sr39#22r	5'-AGA GAA GAT AAG CAG TAA ACA TG-3' 5'-TGC TGT CAT GAG AGG AAC TCT G-3'	Mago et al. (2009)

TABLE 3.1   *(Continued)*

SL. No.	Resistance Genes	Primer	Primer Sequence	References
5.	*Sr43*	Xcfa2040	5'-TCA AAT GAT TTC AGG TAA CCA CTA-3'	Niu et al. (2014)
			5'-TTC CTG ATC CCA CCA AAC AT-3'	
		Xrwgs30	5'-CTC TTG GTG CCA CAC TCT GA-3'	
			5'-TCA GTT CCC TCC CAT TCA TC-3'	
6.	*Yr5*	Yr5_ insertion	5'-CTC ACG CAT TTG ACC ATA TAC AAC T-3'	Marchal et al. (2018)
			5'-TAT TGC ATA ACA TGG CCT CCA GT-3'	
7.	*Yr15*	Xgwm413	5' TGCTTGTCTAGATTGCTTGGG 3'	Peng et al. (2000)
			5' GATCGTCTCGTCCTTGGCA 3'	
8.	*Lr32*	WMC43	5'-TAG CTC AAC CAC CAC CCT ACT G-3'	Thomas et al. (2010)
			5'-ACT TCA ACA TCC AAA CTG ACC G-3'	
		BARC135	5'-ATC GCC ATC TCC TCT ACC A-3'	
			5'-GCG AAC CCA TGT GCT AAG T-3'	
9.	*Lr39*	GDM35	5'-CCT GCT CTG CCC TAG ATA CG-3'	Pestsova et al. (2000)
			5'-ATG TGA ATG TGA TGC ATG CA-3'	
10.	*Lr47*	PS10	5' GCT GAT GAC CCT GAC CGG T 3'	Helguera et al. (2000)
			5' GGG CAG GCG TTT ATT CCA G 3'	

3.4.2   GENE EDITING

This technique uses different tools that alter target DNA sequence inside the genome of an organism. These techniques are based on production of specific DNA double-strand brakes (DSBs), which trigger the repair of DNA in the cell. The repairing is by either error-prone non-homologous end joining (NHEJ) or through homologous recombination (HR) which is error-free. The NHEJ leads to insertions or deletions which results in knockdown of gene function and HR leads to precise modification in the genome/gene. So this technique can be used to edit specific target DNA (Huassain et al., 2019). The technique of wheat genome editing using nucleases (zinc finger nucleases-ZFNs, Transcription activator-like effector nucleases-TALENs) and CRISPR/Cas9 were developed to create specific mutations at target site (Puchta and Fauser, 2013). ZFNs are the synthetically prepared restriction enzymes which are able to cleave long double-stranded DNA sequences. The enzyme is prepared by fusion of FokI DNA restriction enzyme's non-specific

DNA cleavage domain and a Cys2-His2 zinc finger domain. The TALENs are the transcription factors produced and transferred to plant cell by *Xanthomonas* bacteria. When these are fused with a nuclease, it causes DNA breaks at target locations. The DNA-recognition protein domain should be tailor specific to each target DNA and therefore due to their complicated construction, CRISPR/Cas9 system has been increasingly popular as gene-editing tool as CRISPR/Cas9 does not require any protein engineering steps and can be used to target new sequences by simply changing sequence of single-guide RNA. It has more success rate, versatile, universal, and comparatively less expensive (Mushtaq et al., 2019).

3.4.2.1 THE CRISPR/CAS SYSTEM

CRISPR/Cas9 (Clustered Regularly Interspaced Short Palindromic Repeats/ CRISPR Associated Gene9) system performs gene editing by Cas9 nuclease which will be guided by small RNAs (sgRNAs) to the target gene through base pairing. For designing sgRNA constructs, conserved sequences having 2–6 DNA base pair sequence adjacent to DNA sequence targeted by the Cas9 nuclease is selected and utilized (Mushtaq et al., 2019). The CRISPR/ Cas9 tool can be introduced through *Agrobacterium*-mediated, biolistic transformation of explants, or by direct transformation of protoplasts. Using CRISPR/Cas9 system, multiple genes can be targeted with a single construct (Borrelli et al., 2018). Cas9 nuclease will introduce cuts in target DNA which is complementary to the variable region of sgRNA with an adjacent proto-spacer adjacent motif (PAM). Therefore, the variable region of sgRNA can be designed so that the Cas9 nuclease will generate double-stranded breaks at required targets (Jinek et al., 2012). This system of gene editing was first successfully utilized to develop powdery mildew resistance in wheat. Wang et al. (2014) used TALEN and CRISPR/Cas9 techniques for simultaneous modification of multiple homeoalleles encoding mildew resistance locus (MLO). They showed that all three TaMLO homologs were resistant to powdery mildew. CRISPR-Cas9 was used to generate transgenic plants carrying mutations in the TaMLO-A1 allele. Recently this technique was used to simultaneous knockdown of three homeologs of wheat EDR1 (enhanced disease resistance 1) gene, which is a negative regulator of powdery mildew resistance. The TaEDR1 wheat plants generated were resistant to powdery mildew disease (Zhang et al., 2017).

3.4.2.2   *RNA INTERFERENCE (RNAI)*

RNAi is one of the common regulatory mechanisms for eukaryotic gene expression, which is now a better tool for functional gene analysis. This process utilizes a sequence of RNA for inactivation of specific mRNA sequence resulting in the inhibition of a gene function. In this process, the Dicer enzyme cleaves long double-stranded RNA (dsRNA) molecules into small interfering RNA (siRNA) having nearly 20 nucleotides. The siRNA unwinds to form two single-stranded RNA (ssRNA) called passenger RNA and guide RNA. RNA-inducing silencing complex (RISC) will be formed with guide RNA and the passenger strand is degraded in the cell. These RISC will cleave target mRNA after attaching onto mRNA (Agrawal et al., 2003). The major application of RNAi technology in wheat is for disease and pest control through virus and host-induced gene silencing (HIGS) platforms (Nowara et al., 2012). Recently, HIGS was developed through plant-mediated RNAi for the management of insect pests and diseases. The short interfering molecules complementary to the target pathogen produced by the host plants are transferred to the target pathogen resulting in silencing of the targeted gene (Nowara et al., 2010). In wheat, host induced gene silencing was reported by silencing virulence effector gene *Avra10,* resulting in reduced development of powdery mildew fungus *Blumeria graminis* (Nowara et al., 2010). The coexpression of three hairpin RNAi constructs corresponding to chitin synthase gene (Chs3b) of *Fusarium graminearum* causing head scab disease in wheat cultivars led to high stable resistance to head blight and seedling blight disease (Cheng et al., 2015). Chen et al. (2015) reported a similar approach wherein wheat plants with RNAi hairpins against *F. culmorum* genes β-1,3-glucansynthase gene FcGls1, or FcGls1 + mitogen-activated protein kinase (FcFmk1)+ chitin synthase V myosin-motor domain (FcChsV), had high degree of resistance against *Fusarium* disease.

3.4.3   *ROLE OF INFORMATION AND COMMUNICATION TECHNOLOGY (ICT)*

The impact of information and communication technology (ICT) is huge in detection of insect pests and diseases and their management in modern agriculture. The agriculture production information, market facilitation, and financial intermediation services are important, and ICT can play an important role in addressing these challenges. Mobile telephony is also being utilized as a choice for the delivery of ICT services and solutions. Most mobile applications are being developed for improving agriculture and provide a lot of information on production practices, pest, and disease information, market intelligence, etc. Mobile applications for pest and disease management are the latest digital tools which will allow the farmers to identify different pests and diseases affecting their crops by using their mobile phones and they can have management measures for them (http://www.icrisat.org/mobile-app-for-pest-and-disease-management-of-crops). A mobile android application, m-WHEAT, which was developed in Ethiopia, provides information on pest and disease diagnostics of wheat crops along with their management practices. A database was made which consists of symptoms of different pests and diseases along with their photographs and their management practices in English and other local languages. By selecting a pest/disease, the farmers could get information on their symptoms and management practices (Paul Mansingh et al., 2017).

AgriApp, an android application, provides information on crop production and crop protection measures, and it is available in many languages viz., Kannada, English, Hindi, Telugu, Tamil, and Marathi. MyCrop Wheat application was developed to include pest/disease diagnosis, integrated management measures, different varieties, economic risk analysis, and crop monitoring tips (Paul Mansingh et al., 2017). Tata Consultancy Services Ltd. has developed the mobile-based application mKRISHI, which allows farmers to submit their queries to experts and get expert advice. It also provides information on local weather conditions, soil conditions, pests, and diseases, and market prices (https://www.tcs.com/enabling-digital-farming-with-pride).

KEYWORDS

- **adult plant resistance**
- **CRISPR/Cas9**
- **nuclear magnetic resonance**
- **nucleic acid-based techniques**
- **radio-immunosorbent assay**
- **reverse transcription-polymerase chain reaction**

REFERENCES

Agarwal, V. K., & Srivastava, K. D., (1985). NaOH seed soak method for routine examination of rice seed lots for rice bunt. *Seed Res., 13*, 159–161.

Agarwal, V. K., & Verma, H. S., (1983). A simple technique for detection of Karnal bunt infection in wheat seed samples. *Seed Res., 11,* 110–112.

Alam, K. B., Banu, S. P., & Shaheeda, M. A., (1997). The occurrence and significance of spot blotch disease in Bangladesh. In: Duveiller, E., Dubin, H. J., Reeves, J., & McNab, A., (eds.), *Int. Workshop on Helminthosporium Disease of Wheat: Spot Blotch and Tan Spot* (pp. 63–66). CIMMYT, El Batan, Mexico.

Bariana, H. S., Hayden, M. J., Ahmad, N. U., Bell, J. A., Sharp, P. J., & McIntosh, R. A., (2001). Mapping of durable adult plant and seedling resistance to stripe and stem rust disease in wheat. *Aust. J. Agric. Res., 52,* 1247–1255.

Begum, S., & Mathur, S. B., (1989). Karnal bunt and loose smut in wheat seed lots of Pakistan. *FAO Plant Protect Bull., 37,* 165–173.

Bockus, W. W., Appel, J. A., Bowden, R. L., Fritz, A. K., Gill, B. S., Martin, J., Sears, R., et al., (2001). Success stories: Breeding for wheat disease resistance in Kansas. *Plant Dis., 85,* 453–461.

Chand, R., Singh, H. V., Joshi, A. K., & Duveiller, E., (2002). Physiological and morphological aspects of *Bipolaris sorokiniana* conidia surviving on wheat straw. *J. Plant. Path., 18,* 328–332.

Chen, Y., Gao, Q., & Huang, M., (2015). Characterization of RNA silencing components in the plant pathogenic fungus *Fusarium graminearum. Sci. Rep., 5,* 12500.

Cheng, W., Song, X., & Li, H., (2015). Host-induced gene silencing of an essential chitin synthase gene confers durable resistance to *Fusarium* head blight and seedling blight in wheat. *Plant Biotech. J., 13*(9), 1335–1345.

Costamilan, L. M., (2005). Variability of the wheat powdery mildew pathogen *Blumeria graminis* f. sp. *tritici* in the 2003 crop season. *Fitopatol. Brasileira, 30,* 420–422.

De Tempe, J., & Binnerts, J., (1979). Introduction to methods of seed health testing. *Seed. Sci. Tech., 7,* 601–636.

Dean, R., Van, K. J. A., Pretorius, Z. A., Hammond-Kosack, K. E., Di Pietro, A., Spanu, P. D., Rudd, J. J., et al., (2012). The Top 10 fungal pathogens in molecular plant pathology. *Mol. Plant Pathol., 13,* 414–430.

Deb, M., & Anderson, J. M., (2008). Development of a multiplexed PCR detection method for barley and cereal yellow dwarf viruses, wheat spindle streak virus, wheat streak mosaic virus and soil-borne wheat mosaic virus. *J. Virol. Methods, 148,* 17–24.

Dill-Macky, R., (1997). *Fusarium* head blight: Recent epidemics and research efforts in the upper Midwest of United States. In: Dubin, H. J., Gilchrist, L., Reeves, J., & McNab, A., (eds.), *Fusarium Head Scab: Global Status and Future Prospects.* Mexico, DF, CIMMYT.

Dowell, F. E., (1998). Automated color classification of single wheat kernels using visible and near-infrared reflectance. *Cereal Chem., 75,* 142–144.

Dubcovsky, J., (2017). *Marker-Assisted Selection in Wheat.* http://maswheat.ucdavis.edu/ (accessed on 11 February 2021).

Duveiller, E., (2004). Controlling foliar blights of wheat in the rice-wheat systems of Asia. *Plant Dis., 88,* 552–556.

Duveiller, E., Kandel, Y. R., Sharma, R. C., & Shrestha, S. M., (2005). Epidemiology of foliar blights (spot blotch and tan spot) of wheat in the plains bordering the Himalayas. *Phytopath., 95,* 248–256.

Fang, Y., & Ramasamy, R. P., (2015). Current and prospective methods for plant disease detection. *Biosensors, 4,* 537–561.

FAOSTAT, (2016). http://www.fao.org/faostat/en/#data/QC (accessed on 11 February 2021).

Fraaije, B. A., Lovell, D. J., Coelho, J. M., Baldwin, S., & Hollomon, D. W., (2001). PCR-based assays to assess wheat varietal resistance to blotch (*Septoria tritici* and *Stagonospora nodorum*) and rust (*Puccinia striiformis* and *Puccinia recondita*) diseases. *Euro. J. Plant. Pathol., 107,* 905–917.

Frederick, R. D., Synder, K. E., Tooley, P. W., Berthier-Schaad, Y., Peterson, G. L., Bonde, M. R., Schaad, N. W., & Knorr, D. A., (2000). Identification and differentiation of *Tilletia indica* and *T. walkeri* using the polymerase chain reaction. *Phytopathol., 90,* 951–960.

Gilchrist, L., & Dubinm, H. J., (2002). *Fusarium* head blight. In: Curtis, B. C., Rajaram, S., & Gómez, M. H., (eds.), *Bread Wheat Improvement and Production.* FAO Plant Production and Protection Series No. 30.

Helguera, M., Khan, I. A., & Dubcovsky, J., (2000). Development of PCR markers for wheat leaf rust resistance gene *Lr47. Theor. Appl. Genet., 101,* 625–631.

Henry, M., & Plumb, R., (2002). *Barley yellow dwarf luteoviruses* and other virus diseases. In: Curtis, B. C., Rajaram, S., & Gomez, M. H., (eds.), *Bread Wheat: Improvement and Production, Plant Production and Protection Series No. 30* (pp. 331–344). FAO, Rome.

Huang, C., Sun, Z., Yan, J., Luo, Y., Wang, H., & Ma, Z., (2011). Rapid and precise detection of latent infections of wheat stripe rust in wheat leaves using loop-mediated isothermal amplification. *J. Phytopath., 159*(78), 582–584.

Hussain, A., Imran, Q. M., & Yun, B., (2019). *CRISPR/Cas9-Mediated Gene Editing in Grain Crops.* Recent Advances in Grain Crops Research. doi: http://dx.doi.org/10.5772/intechopen.88115.

International Grains Council. (2021). https://www.igc.int/downloads/gmrsummary/gmrsumme.pdf (accessed on 11 February 2021).

James, D., Varga, A., Pallas, V., & Candresse, T., (2006). Strategies for simultaneous detection of multiple plant viruses. *Can. J. Pl. Pathol., 28,* 16–29.

Jinek, M., Chylinski, K., Fonfara, I., Hauer, M., Doudna, J. A., & Charpentier, E., (2012). A programmable dual-RNA-guided DNA endonuclease in adaptive bacterial immunity. *Science, 337,* 816–821.

Kanyuka, K., Ward, E., & Adams, M., (2003). *Polymixa graminis* and the cereal viruses it transmits. *Mol. Plant. Pathol., 4,* 393–406.

Kolmer, J. A., Bernardo, A., Bai, G., Hayden, M. J., & Chao, S., (2018). Adult plant leaf rust resistance derived from toropi wheat is conditioned by *Lr78* and three minor QTL. *Phytopath., 108,* 246–253.

Kuckenberg, J., Tartachnyk, I., & Noga, G., (2009). Temporal and spatial changes of chlorophyll fluorescence as a basis for early and precise detection of leaf rust and powdery mildew infections in wheat leaves. *Precision Agriculture, 10*(1), 34–44.

Kumar, A., Singh, A., & Garg, G. K., (1998). Development of seed immunoblot binding assay for detection of Karnal bunt (*Tilletia indica*) of wheat. *J. Plant Biochem. Biotech., 7,* 119–120.

Lapierre, H., & Signoret, P. A., (2004). Virus diseases of *Triticum spp.* In: *Viruses and Virus Diseases of Poaceae (Gramineae)* (pp. 554–609). INRA, Paris.

Li, X. L., Ma, Z. H., Zhao, L. L., Li, J. H., & Wang, H. G., (2013). Early diagnosis of wheat stripe rust and wheat leaf rust using near-infrared spectroscopy. *Spectrosc. Spect. Anal., 33*(10), 2661–2665.

Liang, Z., Chen, K., & Li, T., (2017). Efficient DNA free genome editing of bread wheat using CRISPR/Cas9 ribonucleoprotein complexes. *Nat. Commun., 8,* 14261.

Lindhout, P., (2002). The perspectives of polygenic resistance in breeding for durable disease resistance. *Euphytica, 124*(2), 217–226.

Liu, J., Liu, D., Tao, W., Li, W., Wang, S., Chen, P., Cheng, S., & Gao, D., (2000). Molecular marker-facilitated pyramiding of different genes for powdery mildew resistance in wheat. *Plant Breed, 119*(1), 21–24.

Mago, R., Bariana, H. S., Dundas, I. S., Spielmeyer, W., Lawrence, G. J., Pryor, A. J., & Ellis, J. G., (2005). Development of PCR markers for the selection of wheat stem rust resistance genes *Sr24* and *Sr26* in diverse wheat germplasm. *Theor. Appl. Genet., 11,* 496–504.

Mago, R., Verlin, D., Zhang, P., Bansal, U., Bariana, H., Jin, Y., Ellis, J., Hoxha, S., & Dundas, I., (2013). Development of wheat-*Aegilops speltoides* recombinants and simple PCR-based markers for *Sr32* and a new stem rust resistance gene on the 2S#1 chromosome. *Theor. Appl. Genet., 126,* 2943–2955.

Mago, R., Zhang, P., Bariana, H. S., Verlin, D. C., Bansal, U. K., Ellis, J. G., & Dundas, I. S., (2009). Development of wheat lines carrying stem rust resistance gene *Sr39* with reduced *Aegilops speltoides* chromatin and simple PCR markers for marker-assisted selection. *Theor. Appl. Genet., 124,* 65–70.

Marchal, C., Zhang, J., Zhang, P., Fenwick, P., Steuernagel, B., Adamski, N. M., Boyd, L., et al., (2018). BED-domain containing immune 1 receptor confer diverse resistance spectra to yellow rust. *Nature Plants, 4,* 662–668.

Martin, F. N., & Kistler, H. C., (1990). Species-specific banding pattern of restriction endonuclease-digested mitochondrial DNA from the genus Pythium. *Exp. Mycol., 14,* 32–46.

Martinelli, F., Scalenghe, R., Davino, S., Panno, S., Scuderi, G., Ruisi, P., Villa, P., et al., (2015). Advanced methods of plant disease detection: A review. *Agron. Sustain. Dev., 35,* 1–25.

Matorin, D. N., Timofeev, N. P., Glinushkin, A. P., Bratkovskaja, L. B., & Zayadan, B. K., (2018). Effect of fungal infection with *Bipolaris sorokiniana* on photosynthetic light reactions in wheat analyzed by fluorescence spectroscopy, *Moscow Univ. Biol. Sci. Bull., 73*(4), 203–208.

McCartney, H. A., Foster, S. J., Fraaije, B. A., & Ward, E., (2003). Molecular diagnostics for fungal plant pathogens. *Pest Manag. Sci., 59*(2), 129–142.

McIntosh, R. A., Dubcovsky, J., Rogers, W. J., Morris, C., & Xia, X. C., (2017). Catalogue of gene symbols for wheat: 2017 supplement. https://shigen.nig.ac.jp/wheat/komugi/genes/macgene/supplement2017.pdf (accessed on 11 February 2021).

Mitra, M., (1931). A new bunt on wheat in India. *Ann. Bot., 18*(2), 178–179.

Mullis, K., (1987). Specific synthesis of DNA in vitro via a polymerase-catalyzed chain reaction. *Methods Enzymol., 155,* 335–350.

Mushtaq, M., Sakina, A., Wani, S. H., Shikari, A. B., Tripathi, P., Zaid, A., Galla, A., et al., (2019). Harnessing genome editing techniques to engineer disease resistance in plants. *Front. Plant Sci., 10,* 550.

Narayanasamy, P., (2011). *Microbial Plant Pathogens-Detection and Disease Diagnosis* (pp. 5–199). Springer Netherlands.

Niu, Z., Klindworth, D. L., Yu, G., Friesen, T., Chao, S., Jin, Y., Cai, X., et al., (2014). Development and characterization of wheat lines carrying stem rust resistance gene *Sr43* derived from *Thinopyrum ponticum*. *Theor. Appl. Genet., 127,* 969–980.

Nowara, D., Gay, A., & Lacomme, C., (2012). HIGS: Host-induced gene silencing in the obligate biotrophic fungal pathogen *Blumeria graminis. Plant Cell, 22*(9), 3130–3141.

O'Donnell, K., Ward, T. J., Geiser, D. M., Kistler, H. C., & Aoki, T., (2004). Genealogical concordance between the mating-type locus and seven other nuclear genes supports formal recognition of nine phylogenetically distinct species within the *Fusarium graminearum* clade. *Fungal Genet. Biol., 41,* 600–623.

Oerke, E. C., Dehne, H. W., Schonbeck, F., & Weber, A., (1994). *Crop Production and Crop Protection: Estimated Losses in Major Food and Cash Crops.* Elsevier, Amsterdam.

Paul, M. J., Velmurugan, L., Shenkuta, D. T., & Bayissa, D. D., (2017). Pests and disease diagnostic mobile tool "m-wheat" for wheat crop in Ethiopia. *Journal of Agricultural Informatics, 8*(2), 44–54.

Peng, J. H., Fahima, T., Roeder, M. S., Huang, Q. Y., Dahan, A., Li, Y. C., Grama, A., & Nevo, E., (2000). High-density molecular map of chromosome region harboring stripe-rust resistance genes *YrH52* and *Yr15* derived from wild emmer wheat, *Triticum dicoccoides. Genetica, 109,* 199–210.

Pestsova, E., Ganal, M. W., & Röder, M. S., (2000). Isolation and mapping of microsatellite markers specific for the D genome of bread wheat. *Genome, 43,* 689–697.

Puchta, H., & Fauser, F., (2013). Gene targeting in plants: 25 years later, *The International Journal of Developmental Biology, 57*(6, 7, 8), 629–637.

Qiao, H. B., Xia, B., Ma, X. M., Cheng, D. F., & Zhou, Y. L., (2010). Identification of damage by diseases and insect pests in winter wheat. *J. Triticeae Crops, 30*(4), 770–774.

Ragimekula, N., Varadarajula, N. N., Mallapuram, S. P., Gangimeni, G., Reddy, R. K., & Kondreddy, H. R., (2013). Marker-assisted selection in disease resistance breeding. *J. Plant Breed. Genet., 1*(2), 90–109.

Rajaram, S., Singh, R. P., & Van, G. M., (1997). Breeding wheat for wide adaptation, rust resistance and drought tolerance. In: Manjit, S. K., (ed.), *Crop Improvement for the 21*st *Century* (pp. 139–163). Kerala, India.

Randhawa, H. S., Asif, M., Pozniak, C., Clarke, J. M., Graf, R. J., & Fox, S. L., (2013). Application of molecular markers to wheat breeding in Canada. *Plant Breed., 132,* 458–471.

Rao, G. P., Prakasha, T. L., Priya, M., Thorat, V., Kumar, M., Baranwal, V. K., Mishra, A. N., & Yadav, A., (2017). First report of association of *Candidatus phytoplasma* cynodontis (16SrXI-B Group) with streak, yellowing, and stunting disease in durum and bread wheat genotypes from central India. *Plant Dis., 101*(7), 1314.

Ray, M., Ray, A., Dash, S., Mishra, A., Achary, K. G., Nayak, S., & Singh, S., (2017). Fungal disease detection in plants: Traditional assays, novel diagnostic techniques and biosensors. *Biosensors and Bioelectronics, 87,* 708–723.

Rennie, W. J., (1982). Wheat loose smut. *ISTA Handbook on Seed Health Testing: Section 2.* Working Sheet No. 48, International Seed Testing Association, Zürich, Switzerland.

Sambasivam, P. K., Bansal, U. K., Hayden, M. J., Dvorak, J., Lagudah, E. S., & Bariana, H. S., (2008). Identification of markers linked with stem rust resistance genes *Sr33* and *Sr45.* In: *The 11*th *International Wheat Genetics Symposium.* https://ses.library.usyd.edu. au/bitstream/handle/2123/3267/P066.pdf?sequence=1&isAllowed=y (accessed on 11 February 2021).

Schumann, G. L., & Leonard, K. J., (2000). *Stem Rust of Wheat (Black Rust)*. The plant health instructor. doi: 10.1094/PHI-I-2000-0721-01 Updated 2011.

Sharma, A., Sohu, V. S., & Bhardwaj, S. C., (2019). Pre-emptive breeding for stripe rust resistance to checkmate the rapidly evolving pathogen. In: Sai, P. S. V., Mishra, A. N., & Singh, G. P., (eds.), *Current Trends in Wheat and Barley Research and Development* (pp. 121–136). ICAR-IARI Regional Station, Indore, India.

Skottrup, P., Frøkiær, H., Hearty, S., O'Kennedy, R., Hejgaard, J., Nicolaisen, M., & Justesen, A. F., (2007). Monoclonal antibodies for the detection of *Puccinia striiformis* urediniospores. *Mycol. Res., 111*(3), 332–338.

Teich, A. H., & Nelson, K., (1984). Survey of *Fusarium* head blight and possible effects of cultural practices in wheat fields in Lambton county in 1983. *Can. Plant Dis. Surv., 64,* 11–13.

Thomas, J., Nilmalgoda, S., Hiebert, C., McCallum, B., Humphreys, G., & DePauw, R., (2010). Genetic markers and leaf rust resistance of the wheat gene *Lr32*. *Crop Science, 50,* 2310–2317.

Turner, A. S., Leesb, A. K., Rezanoora, H. N., & Nicholsona, P., (1998). Refinement of PCR-detection of *Fusarium avenaceum* and evidence from DNA marker studies for phonetic relatedness to *Fusarium tricinctum*. *Plant Pathol., 47,* 278–288.

Varshney, G. K., (1999). In: *Characterization of Genetic Races of Karnal Bunt (Tilletia indica) of Wheat by Immunobinding Assay Procedure*. MSc Thesis, G.B. Pant Univ. of Agric. and Tech., Pantnagar.

Waalwijk, C., Van, D. H. R., De Vries, I., Van, D. L. T., Schoen, C., Costrel-Decorainville, Haeuser-Hahn, I., et al., (2004). Quantitative detection of *Fusarium* species in wheat using TaqMan. *Euro. J. Plant Pathol., 110*(5, 6), 481–494.

Wang, H., Qin, F., Liu, Q., Ruan, L., Wang, R., Ma, Z., Li, X., et al., (2015). Identification and disease index inversion of wheat stripe rust and wheat leaf rust based on hyperspectral data at canopy level. *J. Spectr., 2015*(1), 1–10.

Wang, Y., Cheng, X., Shan, Q., Zhang, Y., Liu, J., Gao, C., & Qiu, J., (2014). Simultaneous editing of three homoeoalleles in hexaploid bread wheat confers heritable resistance to powdery mildew. *Nat. Biotechnol., 32,* 947–951.

Wessels, E., & Botes, W. C., (2014). Accelerating resistance breeding in wheat by integrating marker-assisted selection and doubled haploid technology. *S. Afr. J. Plant Soil, 31*(1), 35–43.

Wiese, M. S., (1987). *Compendium of Wheat Diseases* (2nd edn.). American Phytopathological Society, St. Paul, Minnesota.

Xiao, Z., Sun, L., & Xin, W., (1998). Breeding for resistance in *Heilongjiang province*. In: Duveiller, E., Dubin, H. J., Reeves, J., & McNab, A., (eds.), *Helminthosporium Blight of Wheat: Spot Blotch and Tan Spot* (pp. 162–169). Mexico, DF: CIMMYT.

Yan, H., Zhang, J., Ma, D., & Yin, J., (2019). qPCR and loop-mediated isothermal amplification for rapid detection of *Ustilago tritici*. *PeerJ., 7,* 7766.

Yuan, L., Zhang, J. C., Zhao, J. L., Huang, W. J., & Wang, J. H., (2013). Differentiation of yellow rust and powdery mildew in winter wheat and retrieving of disease severity based on leaf-level spectral analysis, *Spectrosc. Spect. Anal., 33*(6), 1608–1614.

Zezza, F., Pascale, M., Mulè, G., & Visconti, A., (2006). Detection of *Fusarium culmorum* in wheat by a surface plasmon resonance-based DNA sensor. *J. Microbiol. Methods, 66*(3), 529–537.

Zhang, X., Wang, C., Zhang, Y., Sun, Y., & Mou, Z., (2012). The Arabidopsis mediator complex subunit16 positively regulates salicylate-mediated systemic acquired resistance and jasmonate/ethylene-induced defense pathways. *The Plant Cell, 24,* 4294–309.

Zhang, Y., Bai, Y., Wu, G., Zou, S., Chen, Y., Gao, C., & Tang, D., (2017). Simultaneous modification of three homoeologs of TaEDR1 by genome editing enhances powdery mildew resistance in wheat. *Plant J., 91*(4), 714–724.

CHAPTER 4

Detection and Management Approaches for Bakanae (Foot Rot) Disease in Rice

SACHIN KUMAR JAIN,[1] KAMAL KHILARI,[2] and MUKESH DONGRE[3]

[1]*Department of Plant Pathology, Amar Singh College, Lakhaoti, Bulandshahr, Uttar Pradesh, India, E-mail: sachinjain1115@gmail.com*

[2]*Department of Plant Pathology, S.V.P. University of Agriculture and Technology, Modipuram, Meerut, Uttar Pradesh, India*

[3]*Department of Plant Pathology, RVSKVV, College of Agriculture, Indore, Madhya Pradesh, India*

ABSTRACT

Bakanae (foot rot) disease is an important disease of rice caused by the fungus *Fusarium moniliforme* Sheldon. It is a cosmopolitan disease distributed in all the rice-growing areas of the world. The typical symptoms of this disease are yellowish-green thin leaves, abnormal stem elongation, and rooting of root stem joint as well as also from stem node. The yield loss by this disease is ranging from 3.0–95.4% and the incidence of this disease varies with regions and cultivars grown. Different practices were applied for the management of this disease and to reduce the losses. It has been reported that genotypes *viz.,* MAUB 2009-1, PAU 3456-46-6-1-1, HKR 96-561, HKR 96-565, HKR 07-40, HKR 08-13, HKR 08-21, HKR 08-22 and PNR 600 are highly resistant against this disease. Some practices such as late transplanting (July, 15) of rice seedling have produces lowest symptoms of foot rot disease. Bakanae disease reduced drastically when paddy nursery was uprooted in standing water as compared to uprooting of nursery in moist condition. Application of *T. viride* at 5 g/kg seed reduced 60% disease in Pusa Basmati-1121. Seed soaking in carbendazim (Bavistin) solution at 1 and 2 g/kg seed help in disease suppressed considerably. Seedling dip in 0.1% carbendazim solution for 6 hours before transplanting reduced the bakanae incidence in transplanted crop. Foliar application of carbendazim at flowering stage also reduced the grain infection.

4.1   INTRODUCTION

Bakanae (foot rot) disease is an important and emerging disease of rice which causing severe losses in many basmati varieties of rice. It occurs in both upland and lowland rice fields. This disease is widely distributed in almost all the rice-growing areas. Previously this disease was as a minor disease, but now it is becoming a major disease particularly in basmati rice variety PB-1121 which is most popular in farmers (Khilari et al., 2011). Bakanae disease was first time reported during 1828 in Japan. In India, this disease was reported as foot rot disease of rice in the year 1931 by Thomas. Bakanae disease can be observed throughout the all-growing stages of rice. The disease is also known as foolish plant, foot rot, *Fusarium* blight, white stalk, and elongation disease in different countries. 'Bakanae' is a Japanese word; the mean of this word are bad or naughty or stupid seedlings.

4.2   GEOGRAPHICAL DISTRIBUTION

Bakanae (foot rot) disease is caused by *Fusarium moniliforme* (*G. fujikuroi*) fungus which is widely distributed in temperate as well as tropical environments throughout the world. In Thailand, Japan, China, Bangladesh, Pakistan, and Nepal, foot rot disease has been gaining a major concern since the last decade. It has never been observed in the rice-growing area of South-Eastern Australia (Lanoiselet, 2008). However, in 1965, Heaton and Morschel reported *Fusarium* foot rot disease in Northern Territory, Australia. In India, it has been observed in all states, viz., Uttarakhand, Uttar Pradesh, Haryana, Punjab, Bihar, and Rajasthan (Gupta et al., 2014). In Haryana state, foot rot disease was first recorded during Kharif in 1988 in both group's scented tall varieties and high-yielding dwarf varieties (Sunder et al., 1997). This disease has been reported from different districts of Western Uttar Pradesh in basmati rice varieties (Jain et al., 2019).

4.3 YIELD LOSSES

The yield loss due to bakanae disease varied from location to location and season to season. Under favorable disease development conditions, it is known to cause substantial losses in grain yield (Singh and Sunder, 1997). Average 70% Losses due to this disease have been reported in many parts of the world (Singh et al., 1996). This disease is responsible to caused significant losses of approximately 20% in Asia (Cumagun et al., 2015). However, the disease has been reported to cause severe losses but information on yield loss and the relationship between incidence and loss in grain is limited. In Pakistan, loss by this disease was estimated 10–50% since last decade (Bhalli et al., 2001; Khokhar and Jaffrey, 2002). Pavgi and Singh (1964) reported 15% and 40–50% losses due to bakanae disease in India and Japan, respectively.

4.4 SYMPTOMOLOGY

Symptom of this disease appears on whole plant. In the seedbed, infected seedling can die before transplanting or immediately after transplanting. The critical symptoms of this disease appear on infected plants. Infected plants become thin with yellowish-green and pale green leaves and also several inches taller than normal plants. Root initiation from nodes and white powdery growths of fungus appear on the lower portion of the infected plants. In later infection, few infected tillers survive to maturity stage; if tillers survive, they develop partially filled grains, sterile or empty grains.

4.5 CAUSAL ORGANISM

The causal organism of this disease is *Fusarium moniliforme* Sheldon. The teleomorph stage of this fungus is *Gibberella fujikuroi* Sawada, which produce gibberellins hormone that is a growth-regulating hormones. Gibberellin hormone is responsible to increase the height of rice plant. Based on morphological characters, many plant pathologists have reported that *F. moniliforme* is the only species involved in causing bakanae disease of rice. In recent years, many researchers gave contradictory views in regard to the possibility of the involvement of other *Fusarium* species in the production of foot rot symptoms. Gupta et al. (2015) reported that three mating populations viz., *F. fujikuroi*, *F. verticilloides* and *F. proliferatum* of the *G. fujikuroi* complex have been associate with foot rot disease of rice but *F. fujikuroi* Nirenberg were reported to be more virulent than other species. The teleomorph stage of fungus, i.e., *Gibberella fujikuroi* has been reported in China, Japan, and Taiwan (Sun, 1975).

4.5.1   MORPHOLOGICAL CHARACTERS OF CAUSAL ORGANISM

Hyphae of *F. moniliforme* are hyaline, branched, and septate. The pathogen produces two types of spores: micro-conidia and macro-conidia. Micro-conidia are club or oval-shaped with a flattened base and 0 to 1-septate and measured 4–10 × 1–2 µm in size (Ilija et al., 2009; Pradeep et al., 2013). Conidiogenous cells may be monophialides and polyphialides (Leslie and Summerell, 2006) and are produced abundantly. The micro-conidia are more or less arranging in chains and remain joined or cut off in false heads (Ilija et al., 2009). Macro-conidia are slightly sickle-shaped and delicate, relatively slender or almost straight with no significant curvature and measured 6.5–12.5 µm in diameter (Leslie and Summerell, 2006). These are narrow at both ends and are occasionally somewhat bent into a hook at the apex and the basal cell is seven distinctly or barely notched, and a number of septa varies from 3 to 5. Morphological characterization of *Fusarium* species has been also done by many workers based on the observed morphological features like shapes, sizes, and formation of macroconidia and micro-conidia and formation of conidiophores (Yadav et al., 2014).

4.5.2   HOST RANGE OF THE CAUSAL ORGANISM

The causal organism of this disease, *F. moniliforme* have the wide host range. Rice, barley, sorghum, maize, wheat, sugarcane, pine, rye, and asparagus from South East Asia, United States and Africa have been reported as primary hosts of bakanae pathogen (Puhalla and Spieth, 1985; Petrovic et al., 2013). Several alternate hosts *viz.*, proso millet, cowpea, banana, tomato, subabool, barnyard grass, and early water grass may also serve as reservoir of inoculum in the field. *F. fujikuroi* survived in early water grasses and barnyard scored positive for Koch's Postulates. This indicated the possibility of the grasses to be source of inoculums for the disease (Carter et al., 2008).

4.6 DISEASE CYCLE

The causal organism of bakanae disease is seed-borne, soil borne, and having the capacity to produce large amount of conidia on diseased plants (Ou, 1985). Conidia and ascospores are adhering to the seed coat work as the primary source of inoculums. Conidia and ascospores germinate and infect seedlings through the roots and crown (Sun, 1975). Dissemination of the conidia by wind and water caused secondary infections in the rice field. Under favorable conditions, numerous conidia produce on infected plants that subsequently infect neighboring plants at the flowering stage. Infection may also through spores and mycelium that are left in the water used for soaking seeds (Karov et al., 2009). The plant becomes systemically infected. The fungus can infect the seeds without producing visibly distinguishable symptoms and can even be isolated from healthy looking seeds. Symptoms development may influenced by the amount of inoculums occurred and the strain of the pathogen (Cumagun, 2015) and the temperature (Lanoiselet, 2008). Symptoms do not appear below 20°C temperatures. Soil temperature of 35°C is most favorable for infection and application of nitrogen also favors the development of this disease.

4.7 PRACTICES FOR MANAGEMENT OF BAKANAE (FOOT ROT) DISEASE

4.7.1 VARIETAL RESISTANCE

Varietal resistance is the most cost-effective and reliable method of disease management. Cultivation of resistant varieties has provided effective and durable disease control. It was observed that basmati/scented germplasm and cultivars are more susceptible to bakanae disease as compared to non-scented rice cultivars (Pannu et al., 2012; Gupta et al., 2014). Basmati rice cultivars viz., Pusa Basmati 1121, Pusa Basmati 1176 and Pusa Basmati 1509 are severely infected by this disease than compared to rice varieties viz., Pusa Basmati 1401, Pusa 2511, CSR 30, Dehradun basmati and Pakistani basmati which were also found to be infected by the disease with 2.0–22.8% incidence in Northern part of India (Bashyal et al., 2012; Gupta et al., 2014). Khan et al. (2000) found that most basmati rice varieties cultivated in Pakistan are susceptible to bakanae disease. It was observed that two rice varieties (IR-6 and DR-82) were found highly resistant, four resistant (IR-8, DR-83,

KS-282 and DM-15-1-95) and one (52616) moderately resistant (MR) to bakanae disease. Sunder et al. (2014) reported that rice genotypes like HKR 96-561, HKR 96-565, HKR 07-40, HKR 07-53, HKR 08-13, HKR 08-21, HKR 08-22 are highly resistant to bakanae disease.

4.7.2   CULTURAL METHOD

Uses of cultural practices are the non-chemical method of plant disease management. All cultural practices are the preventive methods. Many practices reduce the density and activity of the pathogen. Cultural practices include clean cultivation, nutrient management, time of planting, etc.

4.7.2.1   SOWING OF CLEAN SEED

The pathogen of this disease is mainly seed borne therefore; sowing of clean seeds help to minimize the disease occurrence. The effective method for the disease management is the use of clean non infested seeds.

4.7.2.2   USE OF BRINE SOLUTION

Dipping the seed before sowing in Brine solution (saltwater) may be an effective method to separate the lightweight seed. In saltwater, lightweight seeds can be separate from seed lots which help in reduce seed borne inoculums (Cother and Lanoiselet, 2002).

4.7.2.3   BURNED THE CROP RESIDUES

Crop residues of previously infected crop in the field may help the survival of the pathogen. Destruction of crop residues having infection of pathogen may provide some benefits by limiting the amount of inoculum that may carry over to the next crop (Gupta et al., 2004).

4.7.2.4   UPROOTING METHOD

Uprooting of nursery seedlings from water logging field and dry field may be influenced disease incidence. Sunder et al. (2014) reported that disease incidence reduced considerably in plots where nursery was uprooted in standing water. On mean basis, the bakanae incidence was observed 9.4 and 23.0% in CSR 30 and PB-1121 in plots planted with nursery uprooted in standing water as against 28.8 and 58.1% in plots planted with nursery uprooted under moist soil condition, respectively. The higher disease incidence under moist condition facilitation of pathogen entry through injured roots.

4.7.2.5   SOWING TIME

Adjustment of sowing time of rice may help to reduce the disease incidence. Bagga et al. (2007) observed minimum disease incidence in late-planted rice crop by the end of July, which may be due to low temperature. Bal and Biswas (2018) observed the lowest symptoms of bakanae disease under late transplanting after the second week of July.

4.7.2.6   FERTILIZER MANAGEMENT

Application of fertilizers in standing rice crop can be influenced the disease incidence. Nutrients may affect the survival of pathogens either through a direct stimulatory or toxic effect. High dose of nitrogenous fertilizers may be increased the disease incidence because higher levels of nitrogen affect the survival of fungus. It was observed that a combination of NPK, $ZnSO_4$, and $FeSO_4$ reduced the survival of pathogen significantly (Sunder and Satyavir, 1998a).

4.7.2.7   ROUGING

It is controversial that Rouging of infected rice plants can be reduces or increases disease incidence. Anon (1977) reported that from field where infected tillers were rouged from time to time, disease incidence and yield loss was higher (16.5 and 33.75%) than that of infected tillers not rouged (8.8 and 21.25%). Whereas, Hossain et al. (2007) observed that Rouging of infected tillers at 30 DAT, increased yield by 2.02–7.5%.

4.7.3   PHYSICAL CONTROL

Physical control is efficient and effective control method available except the treatment of seeds with chemical fungicides. Pathogen of this disease is mainly seed borne. Hot-cold water treatment is good practice for disinfection of infected rice seed. Hot water treatment of seeds to be an efficient method for eliminate of seed-borne pathogens observed by Jensen (1888). Miyasaka et al. (2000) reported that soaking of seeds into hot water at 60°C for 10 minutes before sowing, reduced bakanae disease in nursery and fields.

4.7.4   BIOLOGICAL CONTROL

Biological control is a method to control of disease by the application of biological agents. Some bacterial agents like *Pseudomonas* and *Bacillus* species have been used for controlling bakanae disease of rice (Rosales and Mew, 1997). Luo et al. (2005) reported that *Bacillus* spp. viz., *B. subtilis* and *B. megaterium* are effective to control of bakanae disease. Root drenching with *B. oryzicola* suspension reduced the incidence and severity of bakanae disease significantly reported by Hossain et al. (2016). The most common bio-control agent like *Trichoderma strictipilis* and *T. atroviride* reduces the vegetative growth of bakanae pathogen (Bhramaramba and Nagamani, 2013). Seed treatment with *T. viride* at 5 g/kg seed provided about 60% disease control in rice variety Pusa Basmati-1121 (Kumar et al., 2016). Lower disease incidence was reported in fields treated with FYM 10 t/ha + *Trichoderma*+ *Pseudomonas* (Wyawahare et al., 2012).

4.7.5   CHEMICAL CONTROL

Uses of fungicides have a great impact on suppressing the seed-borne pathogens and enhance the overall germination percentage. Impact of different fungicides is influenced by where the fungicides are used. For control of bakanae disease, fungicides can be applied as according to the following methods.

4.7.5.1   SEED TREATMENT

This disease is mainly seed-borne. Conidia and ascospores are adhering to the seed coat work as primary source of inoculums. Seed treatment with fungicides can completely eliminate to fungus spores from seed coat. Ahangar et al. (2012) reported that Carbendazim + Mancozeb up on seed dressing are most efficient method to controlling the bakanae disease. Hossain et al. (2015) found that fungicides Bavistin, Sunphanate, Carzeb, and Nativo completely eradicated the pathogen from the infected seeds at low dose (2.5 gm/L) dose. Seed treatment with sodium hypochlorite is also effective in reducing the disease incidence. Seeds are soaked in a premixed solution of 5 gallons of bleach to 100 gallons of water for 2 hours and then drained and soaked in freshwater.

4.7.5.2   SEEDLING TREATMENT

Dipping of seedling in fungicides solution may be an effective method for disease management. Some scientists were recorded the good effect of seedling treatment on reducing disease incidence. Bagga and Sharma (2006) reported that seedling treatment with Bavistin or benomyl at 0.1% for 6 and 8 hours is reduced disease effectively. Tilt 25 EC at 0.05% was also found most effective in controlling bakanae but showed signs of phytotoxicity. Bal and Biswas (2018) reported that seed treatment with bavistin 50 WP @ 0.2% solution + seedling dip treatment in bavistin 50 WP @ 0.2% + uprooting the infected seedlings in the nursery was found to be the best for controlling foot rot disease in basmati rice.

4.7.5.3   SOIL DRENCHING

Soil drenching with fungicides before seed sowing in nursery also found effective in reducing disease occurrence. Pannu et al. (2009) reported that seed treatment followed by seedling dip treatment with carbendazim at 0.2% and soil drenching with carbendazim (Aurangzeb et al., 1998) is found effective to management of this disease.

4.7.5.4   FOLIAR SPRAY

Foliar application of benzimidazoles (benomyl and carbendazim) at 0.1% significantly reduced the disease incidence and increased the grain yield (Biswas and Das, 2002). Benomyl and carbendazim (Hajra et al., 1994; Sasaki, 1987) and propiconazole (Pannu et al., 2009) protected the grains from infection and checked the spread of the disease.

4.8 CONCLUSION

Due to the increase in population, there is an urgent need to increase rice production globally and also reduce the losses which cause by biotic and non-biotic stresses. In biotic stresses, bakanae (foot rot) disease is the most harmful threat, particularly in basmati/scented rice. Bakanae disease is caused by *Fusarium moniliforme* Sheldon (*G. fujikuroi* Sawada) which is widely distributed in all over the world. The common symptoms produced in this disease include yellowish-green thin leaves, abnormal stem elongation. High yield losses ranging from 3.0–95.4% was observed by various scientists in different parts of the world. Lots of management practices are available for the disease, but the effectiveness is dependent on their integrated use. Sanitation measures, fertilizers doses, sowing date and other aspects of disease management help to reduces losses. Lowest disease symptoms development was observed under late transplanting, i.e., 15 July. Transplanting of basmati rice before 15 July is not advisable for effective control of the disease. Using of resistant rice cultivars is a powerful tool to reduce the losses and also reduce the harmful effect of destructive pesticides. Some strains of *Trichoderma, Pseudomonas,* and *Bacillus* have been found effective against this disease which can be use for suppressing the pathogen activity and population. Seed treatment with Bavistin 50 WP @ 0.2% solution + seedling dip treatment in Bavistin 50 WP @ 0.2% + uprooting the infected seedlings in the nursery was found to be the best for controlling bakanae disease in basmati rice. An integrated approach involving the development of resistant varieties, use of cultural practices, use of efficient bio-control agents and chemicals may be recommended for the management of bakanae disease.

KEYWORDS

- **bakanae**
- **basmati rice**
- **foot rot**
- ***Fusarium moniliforme***
- **grain yield**
- **micro-conidia**

REFERENCES

Ahangar, M. A., Najeeb, S., Rather, A. G., Bhat, Z. A., Parray, G. A., & Sanghara, G. S., (2012). Evaluation of fungicides and rice genotypes for the management of bakanae. *Oryza, 49*, 121–126.

Anonymous, (1977). *Annual Report* (pp. 64, 65). Mimeographed Paper, BRRI, Joydebpur, Dacca.

Aurangzeb, M., Ahmed, J., & Ilyas, M. B., (1998). Chemical control of bakanae disease of rice caused by *Fusarium moniliforme. Pak. J. Phytopathol., 10*(1), 14–17.

Bagga, P. S., & Sharma, V. K., (2006). Evaluation of fungicides as seedling treatment for controlling bakanae/foot rot (*Fusarium moniliforme*) disease in Basmati rice. *Indian Phytopath., 59*, 305–308.

Bagga, P. S., Sharma, V. K., & Pannu, P. P. S., (2007). Effect of transplanting dates and chemical seed treatments on foot rot disease of basmati rice caused by *F. moniliforme. Pl. Dis. Res., 22*, 60–62.

Bal, R. S., & Biswas, B., (2018). Epidemiology and management of foot rot in basmati rice. *J. Krishi Vigyan, 6*(2), 87–94.

Bashyal, B. M., & Aggarwal, R., (2013). Molecular identification of *Fusarium* species associated with bakanae disease of rice (*Oryza sativa*) in India. *Ind. J. Agri. Sci., 83*, 72–77.

Bashyal, B. M., Aggarwal, R., Gupta, S., & Banerjee, S., (2012). *Ecology and Genetic Diversity of Fusarium spp. Associated with Bakanae Disease of Rice* (p. 577). Biotech Books Publisher, New Delhi (India).

Bhalli, J. A., Aurangzeb, M., & Iiyas, M. B., (2001). Chemical control of bakanae disease of rice caused by *Fusarium moniliforme. J. Bio. Sci., 1*(6), 483–484.

Bhramaramba, S., & Nagamani, A., (2013). Antagonistic *Trichoderma* isolates to control bakanae pathogen of rice. *Agric. Sci. Digest., 33*.104–108.

Biswas, S., & Das, S. N., (2002). Fungicidal spraying for control of bakanae disease of rice in field. *J. Mycopathol. Res., 40*(2), 211–212.

Bonman, J. M., (1992). Root and crown disease, bakanae. In: Webster, R. K., & Gunnell, P. S., (eds.), *Compendium of Rice Diseases* (p. 27). APS Press; St. Paul, MN, USA.

Carter, L. L. A., Leslie, F. J., & Webster, R. K., (2008). Population structure of *Fusarium fujikuroi* from California rice and water grass. *Phytopathol., 98*, 992–998.

Cother, E., & Lanoiselet, V., (2002). Crop nutrition. In: Keaky, L. M., & Clampett, W. S., (eds.), *Production of Quality Rice in South-Eastern Australia* (p. 10). (RIRDC: Kingston, ACT, Australia). Plant Health Australia, Rice Industry Biosecurity Plan, 2009. Contingency Plan: Bakanae.

Cumagun, C. J. R., (2015). *The Rise of the Bakanae Disease of Rice in the Philippines.* FARMD: Forum for Agricultural Risk Management Development in University of the Philippines Los Baños http://www.agriskmanagementforum.org/farmd/content/rise-bakanae-disease-rice-philippines (accessed on 8 March 2021).

Fiyaz, R. A., Krishan, S. G., Rajashekara, H., Yadav, A. K., Bashyal, B. M., Bhowmick, P. K., Singh, N. K., et al., (2014). Development of high throughput screening protocol and identification of novel sources of resistance against bakanae disease in rice (*Oryza sativa* L.). *Ind. J. Genet., 74*(4), 414–422.

Gnanamanickam, S. S., (2009). An overview of progress in biological control. In: Gnanamanickam, S. S., (ed.), *Biological Control of Rice Diseases: Progress in Biological Control Series* (pp. 43–51). Springer; Dordrecht, Netherlands.

Gupta, A. K., Singh, Y., Jain, A. K., & Singh, D., (2014). Prevalence and incidence of bakanae disease of rice in Northern India. *J. Agri. Search, 1*(4), 233–237.

Gupta, A. K., Solanki, I. S., Bashyal, B. M., Singh, Y., & Srivastava, K., (2015). Bakanae of rice: An emerging disease in Asia. *Journal of Animal and Plant Science, 25*, 1499–1514.

Gupta, P., Sahai, S., Singh, N., Dixit, C., Singh, D., Sharma, C., Tiwari, M., Gupta, R., & Garg, S., (2004). Residue burning in rice-wheat cropping system: Causes and implications. *Curr. Sci., 87*, 1713–1717.

Hajra, K. K., Ganguly, L. K., & Khatua, D. C., (1994). Bakanae disease of rice in West Bengal. *J. Mycopathol. Res., 32*(2), 95–99.

Heaton, J. B., & Morschel, J. R., (1965). A foot rot disease of rice, variety blue bonnet, in Northern Territory, Australia, caused by *Fusarium moniliforme* Sheldon. *Tropical Science, 7*, 116–121.

Hossain, K. S., Miah, M. A. T., & Bashar, M. A., (2011). Preferred rice varieties, seed source, disease incidence and loss assessment in bakanae disease. *J. Agro for. Environ., 5*, 125–128.

Hossain, M. A., Latif, M. A., Kabir, M. S., Kamal, M. M., Mian, M. S., Akter, S., & Sharma, N. R., (2007). *Dissemination of Integrated Disease Management Practices Through Farmers' Participatory Field Trial* (p. 1215). A Report on Agricultural Technology Transfer (ATT) Project. Bangladesh Agricultural Research Council, New Airport Road, Dhaka.

Hossain, M. T., Khan, A., Eu Jin, C., Md Harun-Or, R., & Chung, Y. R., (2016). Biological control of rice bakanae by an endophytic *Bacillus oryzicola* YC7007. *Plant Pathol. J., 32*(3), 228–241.

Ilija, K. K. S. K., Mitrw, S. K., & Kostadin, N. E. D., (2009). *Gibberella fujikuroi* Wollenweber, the new parasitical fungus on rice in the Republic of Macedonia. *Prot. Nat. Sci., 116*, 175–182.

Jaehwan, R., Kim, G., Shin, D., Cho, Y., Kim, Y., & Kang, U., (2009). *Control of Bakanae Diseases by Hot Water Treatment in Rice.* National Institute of Crop Science, RDA, Korea.

Jain, S. K., Khilari, K., Dongre, M., & Pal, S., (2019). Occurrence of bakanae disease of rice in Western Uttar Pradesh, India. *Int. J. Curr. Microbiol. App. Sci., 8*(05), 207–212.

Jensen, J. L., (1888). The propagation and prevention of smut in oats and barley. *J. Royal Agric. Soci. Engl. Sec., 2*(24), 397–415.

Karov, I. K., Mitrev, K. S., & Emilija, D. K., (2009). *Gibberella fujikuroi* (Sawada) Wollenweber-the new parasitical fungus on rice in the republic of Macedonia. *Proc. Nat. Sci. Matica. Srpska Novi. Sad., 116*, 175–182.

Khan, J. A., Jamil, F. F., & Gill, M. A., (2000). Screening of rice germplasm against bakanae and bacterial leaf blight. *Pak. J. Phytopathol., 12*, 6–11.

Khilari, K., Bhanu, C., Sharma, R., Gupta, A., & Gangwar, B., (2011). Bakanae disease-a serious threat to basmati rice cultivation. In: *Proceeding Advances in Biotechnology in Agriculture Crops for Sustaining Productivity, Quantity Improvement and Food Security* (p. 12).

Khokhar, L. K., & Jaffrey, A. H., (2002). Identification of sources of resistance against bakanae and foot rot disease of rice. *Pakistan J. Agri. Res., 17*, 176–177.

Lanoiselet, V., (2008). Bakanae. *Rice Industry Biosecurity Plan* (pp. 1–10). Australia.

Leslie, J. F., & Summerell, B. A., (2006). *The Fusarium Laboratory Manual* (p. 388). Blackwell Publishing Ltd, UK.

Luo, J., Guan-lin, X., Bin, L., Luo, Y., Li-han, Z., Xiao, W., Bo, L., & Wen, L., (2005). Gram positive bacteria associated with rice in China and their antagonists against the pathogens of sheath blight and bakanae disease in rice. *Rice Sci., 12*, 213–218.

Mandal, D. N., & Chaudhuri, S., (1988). Survivality of *Fusarium moniliforme* shelder under different moisture regimes and soil conditions. *Int. J. Trop. Pl. Dis., 6*, 201–206.

Miyasaka, A., Ryoichi, S., & Masataka, I., (2000). Control of the bakanae disease of rice by soaking seeds in hot water for the hydroponically raised seedling method in the long-mat type rice cultivation. *Proc. Kanto-Tosan. Pl. Prot. Soc., 47*, 31–33.

Ou, S. H., (1985). *Rice Diseases* (2nd edn., p. 380). Commonwealth Mycological Institute, Kew, Surrey, England, UK.

Pannu, P. P. S., Kaur, J., Singh, G., & Kaur, J., (2012). Survival of *Fusariu mmoniliforme* causing foot rot of rice and its virulence on different genotypes of rice and basmati rice. *Ind. Phytopath.,* 149–209.

Pannu, P. P. S., Singh, N., Rewal, H. S., Sabhiki, H. S., & Raheja, S., (2009). Integrated management of foot rot of basmati rice. *National Symposium of INSOPP on "Plant Pathology in the Challenging Global Scenario" Organized by Indian Society Plant Pathologists* (pp. 13–32). NBPGR, New Delhi.

Pavgi, M. S., & Singh, J., (1964). Bakanae and foot rot of rice in Uttar Pradesh, India. *Pl. Dis. Reptr., 48*, 340–342.

Petrovic, T., Burgess, L., Cowie, I., Warren, R., & Harvey, P., (2013). Diversity and fertility of *Fusarium sacchari* from wild rice (*Oryza australiensis*) in Northern Australia, and pathogenicity tests with wild rice, rice, sorghum and maize. *European J. Pl. Path., 136*, 773–788.

Pradeep, F. S., Shakila, M. B., Palaniswamy, M., & Pradeep, B. V., (2013). Influence of culture media on growth and pigment production by *Fusarium moniliforme* isolated from paddy field soil. *World Appl. Sci. J., 22*(1), 70–77.

Puhalla, J. E., & Spieth, P. T., (1985). A comparison of heterokaryosis and vegetative compatibility among varieties of *Gibberella fujikurai (Fusarium monoliforme)*. *Exp. Mycol., 9*, 39–47.

Rosales, A. M., & Mew, T. W., (1997). Suppression of *Fusarium fujikuroi* in rice by rice-associated antagonistic bacteria. *Pl. Dis., 81*, 49–52.

Rosales, A. M., Nuque, F. L., & Mew, T. W., (1986). Biological control of bakanae diseases of rice with antagonistic bacteria. *Phil Phytopath., 22*, 29–35.

Sasaki, T., (1987). *Epidemiology and Control of Rice Bakanae Disease* (Vol. 74, pp. 1–47). Bull. Tohoku Natn. Agric. Exptl. Stn.

Singh, R., & Sunder, S., (1997). Foot rot and bakanae of rice: Retrospects and prospects. *Int. J. Tropical Plant Dis., 15*, 153–176.

Singh, R., & Sunder, S., (2012). Foot rot and bakanae of rice: An overview. *Review Plant Pathology, 5*, 566–604.

Sun, S. K., (1975). The diseases cycle of rice bakanae disease in Taiwan. *Proc. Natl. Sci. Counc. Repub. China., 8*, 245–256.

Sunder, S., & Satyavir, (1998a). Survival of *Fusarium moniliforme* in soil, grains and stubbles of paddy. *Ind. Phytopath., 51*, 47–50.

Sunder, S., Satyavir, & Singh, A., (1998). Screening of rice genotypes for resistance to bakanae disease. *Ind. Phytopath., 51*, 299, 300.

Sunder, S., Satyavir, & Virk, K. S., (1997). Studies on correlation between bakanae incidence and yield loss in paddy. *Indian Phytopathology, 50*, 99–101.

Sunder, S., Singh, R., & Dodan, D. S., (2014). Management of bakanae disease of rice caused by *Fusarium moniliforme*. *Indian Journal of Agricultural Sciences, 84*(1/2), 48–52, 224–228.

Thomas, K. M., (1931). A new paddy disease in Madras. *Madras Agric., 19,* 34–36.

Tung, L. D., & Serrano, E. P., (2011). Effects of warm water in breaking dormancy for rice seed. *Omon Rice, 18,* 129–136.

Wulff, E. D., Sorensen, J. L., Lubeck, M., Nelson, K. F., Thrane, U., & Torp, J., (2010). *Fusarium spp.* associated with rice bakanae: Ecology, genetic diversity, pathogenicity and toxigenicity. *Envir. Microbiol., 12,* 649–657.

Wyawahare, K. K., Sharma, O. P., & Yaspal, (2012). Field evaluation of *Trichoderma* and *Pseudomonas* against bakanae disease of rice caused by *Fusarium fujikuroi. Ind. Phytopath., 65S,* 56.

Yadav, R. S., Tyagi, S., Javeria, S., & Gangwar, R. K., (2014). Effect of different cultural condition on the growth of *Fusarium moniliforme* causing bakanae disease. *Eur. J. Mol. Biotechnol., 4,* 95–100.

Yamashita, T., Eguchi, N., Akanuma, R., & Saito, Y., (2000). Control of seed borne diseases of rice plants by hot water treatment of rice seeds. *Ann. Report Kanto-Tosan. Pl. Prot. Soc., 47,* 7–11.

CHAPTER 5

Biotic and Abiotic Stresses in Cotton Crop in Punjab, India

RUPESH KUMAR ARORA and PARAMJIT SINGH

Punjab Agricultural University (PAU), Regional Research Station, Bathinda – 151001, Punjab, India, Mobile: 9646687131, E-mail: rkarora@pau.edu (R. K. Arora)

ABSTRACT

Cotton is one of the most important on-perishable crops of our country grown in the *Kharif* season. In Northern India, Haryana, Punjab, and Rajasthan states are the prominent cotton-growing states. In Punjab, the cultivation of cotton crop is done mainly in the South Western Region of Punjab, i.e., Bathinda, Mansa, Fazilka, Mukstar, Moga, Faridkot, Sangrur, and Barnala. The maximum area of the cotton crop in Punjab is under the American cotton (*Gossypium hirsutum* L.). Irrespective of the larger area of the South-Western Region of Punjab state under cotton, biotic, and abiotic stresses lead to cause the setback in the production or productivity of the cotton crop. Among the biotic stresses, during the cropping season in relation to the diseases, e.g., cotton leaf curl disease, bacterial blight, root rot, and leaf spot are the important diseases and are causing economic loss. Cotton Leaf Curl Disease (CLCuD) is an important viral disease, considered as the major threat to production of cotton especially American cotton in Northern India. The bacterial blight caused by the bacterial plant pathogen, i.e., *Xanthomonas* sp. is seed-borne in nature and survives in infected crop debris, can infect any part of the cotton crop at any stage. The root rot and leaf spot is the fungal disease. Incidence of root rot usually noticed at the time of first irrigation or after rain. Incidence of leaf spot usually noticed at later stages in the September-October months around 120–140 days old crop. Among the abiotic stress, i.e., Para wilt, Tirak, yellowing of the cotton crop, drying of the cotton crop due to water stagnation, leaf reddening of the cotton crop, phyo-toxicity in cotton crop due to pesticides, etc., are the important constraints. These all-abiotic stresses are being noticed since the last 4–5 years in the cotton belt of Punjab and causing great yield loss. An incidence of abiotic stresses are becoming increasing in the long duration cotton crop as it leads to more exposure to the change in the environmental conditions, improper cultural practices, or sprays of the pesticides adopted by the farmers in raising the cotton crop.

5.1 INTRODUCTION

Cotton is one of the nature's best gifts to mankind. India ranks first in area and production of cotton crop in the world but still lags behind in productivity to 36th Position (Anonymous, 2017). In India, there are nine major cotton-growing states which fall under three zones viz. the North Zone, the Central Zone and the Southern Zone.

The state of Punjab comes under the Northern Zone and cultivation of cotton is done mainly in the South Western Region of Punjab, i.e., Bathinda, Mansa, Fazilka, Muktsar, Faridkot, Barnala, Ferozpur, and Sangrur.

Cotton, an important non-perishable crop in Punjab, faces many constraints during the cropping season, leads to low production or productivity. The major contributing constraints declining productivity of the crop categorized into two main categories *viz.* biotic and abiotic. The diseases caused by the virus, fungus, and bacteria, e.g., Cotton leaf curl disease, bacterial blight, root rot and leaf spot are considered as the important biotic constraint in addition to the pest's infestation in cotton.

Among the biotic constraints, disease caused by the virus, i.e., cotton leaf curl virus disease (CLCuD) transmitted by whitefly causing enormous losses by breaking the resistance breakdown in cotton varieties and hybrids except *desi* cotton. Farmers are facing economic losses to their crop due to the CLCuD not only in the South-Western region of Punjab, but throughout the cotton belt of Northern India.

5.2 COTTON LEAF CURL VIRUS DISEASE (CLCUD)

Cotton leaf curl virus disease (CLCuD) is one of the most important deadly diseases of cotton in the South Western region of Punjab. Incidence of CLCuD is being noticed in almost all the *Bt* cotton hybrids and varieties grown except *desi* cotton and reached up to the maximum 100% incidence when weather conditions are congenial to the CLCuD. Although, the severity of CLCuD is varied from Grade I to Grade VI as per the scale used for the CLCuD disease severity grade (Monga, 2014).

Another major problem in the cultivation of cotton in most of the area of the South-Western region of Punjab is the poor-quality tube well water which is unfit for irrigation, causing drastic effects on the germination/plant stand of the crop. Farmers have to depend on the canal water, which is usually not available well in time and leads to delayed sowing that also aggravates the problem of CLCuD.

It was observed in the experimental trial that, in delayed sowing, the greater number of plants fall under high disease severity grade of CLCuD, resulting in higher yield loss as compared to the recommended date of sowing.

All the (*Gossypium hirsutum* L.) hybrids and varieties are susceptible to the CLCuD leads to the emergence of the disease complex symptoms, i.e., an increase in the variants of the cotton leaf curl virus. The CLCuD caused by one or more strains or new strains of existing virus caused the CLCuD disease complex, which leads to the frequent resistance breakdown of the *Bt* cotton hybrids or varieties.

With the introduction of the *Bt* cotton hybrids, the farmers are more inclined to grow the *Bt* cotton hybrids in the South Western region of Punjab. Benefits of *Bt* technology are realized in farmer's community well, as it eradicates the need of spray against bollworm pests (American Bollworm, Spotted Bollworm, etc.), and allows uniform picking with considerable increase in yield.

Initially, insecticide sprays were severely decreased against the bollworm pest, but it leads to the resurgence of the secondary pests (whitefly), which acts as the vector of CLCuD in cotton. Moreover, the area of the un-recommended *Bt* cotton hybrids (PAU, Ludhiana) (susceptible to CLCuD) is also increasing gradually with reduction in area under *desi* cotton (resistant to CLCuD).

5.2.1 SYMPTOMS

The initiation of disease is characterized by small vein thickening (SVT) type symptoms on the lower sides of the young upper leaves of plants with netted like appearance. Later on, upward/downward leaf curling occurs. Leaves remain small, thickened, and appears as a cup-shaped. Small leaflets (enations)/leaf-like outgrowths develop on the undersides of leaves on the main and lateral veins.

The damage caused by the CLCuD depends upon the stage of the plant infected. The infection at younger stages leads to more damage, the crop gets severely infected, resulted in severe stunted growth of the plant leads to reduction in the yield and fiber quality is also deteriorated. The diseased plants become infected and have twisted internodes. The number of fruiting bodies is reduced in the diseased plants (Table 5.1).

Arora and Singh (2016) conducted an experiment during the Kharif seasons of two succeeding years 2014–2015 to determine the resistant source in *Bt* and non-*Bt* cotton hybrids. They screened the 65 *Bt* cotton hybrids and 1 non-*Bt* cotton hybrid LHH 144 against the CLCuD. They examined all plants for the incidence and severity of CLCuD. They revealed that out of 65 *Bt* cotton hybrids and non-*Bt* LHH 144 showed highly resistant reaction towards CLCuD.

Zubair et al. (2017) reported that CLCuD is a determinant problem in the Indian subcontinent, and it causes the great economical losses to the cotton production. They found that a major factor Cotton leaf curl Multan beta-satellite (CLCuMuB) is associated with cotton leaf curl causing begomoviruses.

TABLE 5.1   Screening of Varieties Against Cotton Leaf Curl Virus Disease in Bathinda

SL. No.	Hybrids/Varieties	PDI	Disease Reaction
1.	LH 2108	46.67	S
2.	LH 2076	40.00	MS
3.	F 2383	53.33	HS
4.	HS 6	61.00	HS
5.	RST 9	65.50	HS
6.	RS 2013	70.67	HS
7.	F 846	71.67	HS

In another experimental trial on one of the *Bt* cotton hybrid conducted at RRS, Bathinda, it was noticed that if all the plants in the cotton field were at higher disease severity grade (VI Grade) it leads to 80% yield loss. If all the plants of cotton field were in Grade I, yield loss were of 7.8%, in Grade II, yield loss of 23.5%, in Grade III, yield loss of 28.1%, in Grade IV, yield loss of 35.9%, in Grade V, yield loss of 67.4%, respectively (Graph 1).

5.2.2 MANAGEMENT

1. American cotton should not be grown in and around citrus orchards and adjoining (bhindi) crops. As American cotton are prone to CLCuD, emphasis should be given to promote to grow *Desi* cotton as it is resistant to CLCuD.
2. At the initial stages, diseased plants of CLCuD should be uprooted and destroyed from time to time.
3. The site of the field of the cotton crop should be clean. Volunteer plants or the weeds acting as the collateral or alternate host for the source of the CLCuD infection must be eradicated. The weed, i.e., *Kanghibuti*, and *Peelibuti*, acting as collateral hosts must be destroyed.
4. The vector of CLCuD, i.e., whitefly can be managed by spraying of neem-based pesticides and chemical pesticides recommended by the Punjab Agricultural University, Ludhiana (2020).

Nimbecidine or Achook (Neem based biopesticide)-1.0 liter/acre should be sprayed at the initial stages of the cotton crop at the time of the appearance of the whitefly. Later on, the different chemical pesticides are recommended by the PAU, Ludhiana (Table 5.2).

TABLE 5.2   Insecticides Recommended Against Whitefly*

SL. No.	Insecticides	Dosages Per Acre
1.	Ulala 50 WG (Flonicamid)	80 g
2.	Polo/Craze (Diafenthiuron)	200 g
3.	Osheen 20 SG(Dinotefuran)	60 g
4.	Lano/Deta 10 EC(Pyriproxyfen)	500 ml
5.	Oberon/Voltage 22.9 SC(Spiromesifen)	200 ml
6.	Fosmite/Gold Mit 50 EC(Ethion)	800 ml
7.	Dantotsu 50 WG(Clothianidine)	20 g
8.	Applaud 25 SC(Buprofezin)	400 ml

*As per the package and practices for Kharif Crops, PAU, Ludhiana.
Source: Anonymous (2020).

5.3   BACTERIAL BLIGHT

The responsible bacterium was initially called *Pseudomonas malvacearum*, then *Bacterium malvacearum, Xanthomonas malvacearum* and finally *X. campestris* pv. *malvacearum* (Delannoy et al., 2005) The first detailed description of Bacterial blight was made in the United States by Atkinson (1891) with the identification of several symptoms, including angular leaf spots, water-soaked lesions, stem black arm, boll rot and plantlet burning (Innes, 1993; Zomorodian and Rudolph, 1993).

Bacterial blight caused by *Xanthomonas axonopodis* pv. *malvacearum* (Xam) is a worldwide disease of cotton (Yeshwant et al., 2005) and affects the yield and fiber quality.

Depending on the infected plant parts, disease shows different types of symptoms such as seedling blight, angular leaf spot, vein blight, black arm, and boll rot (Hillocks, 1981). As the disease affects the leaves, reduce the photosynthetic activity by destroying the chlorophyll content in leaves. Cotton fields infected with bacterial blight have been found to show as much as 80% yield loss in certain areas (Jalloul et al., 2015).

This pathogen is seed borne and survives in infected crop debris of a previous crop. It may be present either in the field or on infected crop debris from a previous season or maybe introduced at planting with infected seeds. Bacterial blight can affect any part of the plant. It can affect the leaves, boll, etc. The vein blight is the predominates in the Punjab state.

5.3.1   SYMPTOMS

The symptoms initiate as minute, water-soaked, and angular lesions appeared on the leaves on both sides, which later on turn brown and transformed into black angular dead lesions. The bacterial plant pathogen can infect any part of the plant, e.g., their young developing bolls, leaves, bracts, etc. In progressive of the disease, leaves may be shed prematurely resulting in extensive defoliation and severely affected the yield of the cotton crop.

5.3.2   EPIDEMIOLOGY

High relative humidity more than 85% and optimum temperature over and around 25°C lead to development of symptoms of bacterial blight. This type of congenial weather conditions are appeared in the cotton field just after the rains. The rainy season in the Punjab state varied from the July to September months. Usually, the rains occurred late in the September months except in 2019 in which rains occurred early. If the rains occurred late in that condition, the crop is likely to mature, and the incidence of Bacterial blight do not cause any significant yield loss.

5.4   SOOTY MOLD IN COTTON

Sooty mold is the saprophytic fungus, i.e., *Capnodium, Cladosporium,* etc., grows on the honeydew secretions of the sucking pests. It can be seen on the number of the crops. In Bathinda districts, incidence of the Sooty mold reported on the citrus crops and in Cotton.

The incidence of sooty mold was widely noticed in the cotton crop in 2015 at the time of the epidemic of whitefly. Sooty mold interferes with the photosynthesis activity of the crop, and it leads to the great yield loss. At the time of the high severity of the sooty mold, fiber quality of the cotton crop is also deteriorated.

5.4.1   SYMPTOMS

In case of *Myrothecium roridium,* the disease is characterized by circular to semi-circular brown colored spots with broad violets margins appears on leaves, bracts as well as on bolls.

In *Alternaria gossypina,* in the early stages, the spots have a pale green area with irregular margins. As the spots enlarge, irregular concentric zones are formed. Disease-infected leaves were shredded in later stages.

In the case of *Cercospora* sp, small, circular to irregular spots having a whitish center with dark brown margin. In severely infected leaves, necrotic central portion may fall out, giving shot hole appearance in the leaves, and later on, leaves become shredded.

5.4.2   MANAGEMENT

The cotton crop can be managed from the leaf spots infected crop by the spray of Amistar Top 325 SC @ 0.1% immediately on the appearance of the symptoms of the disease. The spray can be repeated at intervals of 15 to 20 days.

5.5   ROOT ROT

The root rot disease is considered as the minor problem in cotton crop, but their incidence or severity are found to be increasing since last 4–5 years at farmer's field. Root rots disease cause losses on almost all vegetables, flowers, and several field crops. During 2014–2018, more the 500 cotton fields of the Bathinda districts and adjoining areas were visited, percent incidence of Root rot lies in the range of (0–20.0%) in patches in the scattered plots in the *Bt* cotton hybrid in the South Western Region of Punjab.

An incidence of root rot usually appeared in June at the vegetative stage during the time of first irrigation and become vigorous during July onwards if coincides with rains. Sometimes these types of symptoms noticed in the later stages in the month of July or even in August. Rotting of the roots of the plants was seen, and the plants can be uprooted easily completely.

The pathogen overwinters usually in the form of resting stage *viz.* sclerotia in the soil and once established in a field remains there indefinitely and spreads with rain splashes, irrigation, and anything else which acts as the carrier for the pathogen.

5.6   PATHOGEN ASSOCIATED WITH THE ROOT ROT DISEASE IN COTTON

The name *Rhizoctonia* commonly referred to as "rhizoc" is derived from the Greek rhiza "root" and Ktonos "killer." It is an important pathogen of ornamental nursery plants, vegetable seedlings, and bedding plants; it also causes a wide range of significant diseases in horticulture and field crops (Pegg and Manners, 2014).

The reports of the plant pathogens, i.e., *Rhizoctonia solani* and *R. bataticola* were found to be associated with the root rot in cotton. Both the pathogens have a diverse host range, e.g., vegetables, flowers, and several field crops. The seedling diseases in cotton-root rot/stem rot generally caused by *Rhizoctonia* sps. or might be some soil borne fungal pathogen were associated.

The no of plant pathogenic soil borne fungi might be associated with the root rot disease complex. Till date, only the fungi associated with the root rot disease complex is the *Rhizoctonia solani* and *Rhizoctonia bataticola* is established in the cotton belt.

5.6.1 SYMPTOMS

Symptoms vary with the stage of the plant and time of infection in accordance with the congenial environmental conditions for its development. Initially plant becomes yellowish in color as compared to healthy plants. The disease generally spreads in round patches in the field and lead to death of the cotton plant. The affected plants can easily be pulled out of the ground. Later on, the bark of the roots is broken into rotten shreds and foul smell appears from the rotten basal portion of stem and roots.

5.6.2 MANAGEMENT

1. The field to raise the cotton drop should be properly drained and wet poorly drained areas should be avoided.
2. There should be wide spaces among plants for good aeration of the soil surface and of plants.
3. Soil solarization and deep plowing during the summer and crop rotation for 3 years is beneficial to some extent only.

5.7 ABIOTIC STRESSES OR DISORDER

Abiotic stress or disorder usually raised mainly due to the change in the environmental conditions or improper cultural practices followed in the cultivation technology of the cotton crop. It means it is non-pathogenic in nature and can be caused by the drastic change in nature, or sometimes man-made factors are also responsible for the abiotic stresses. The abiotic stresses can be seen in any of the field crops.

The cotton variety or hybrid having a long duration is one of the most important factors responsible for the exposure of the cotton crop to the abiotic stresses leads to the yield loss in the cotton crop. Every year, during the cotton cropping season, a new kind of problem arises and leads to a setback in the growth of the cotton crop. A different form of the abiotic stresses appeared in the cotton crop are described below:

5.8 PARAWILT

Parawilt is becoming the most important physiological disorder in cotton, nowadays seen in the *Bt* cotton hybrid sown in more than 90% areas of the cotton field. It occurs in a sporadic manner. The association of the pathogen were not existing it occurred due to the drastic change in the environmental conditions. It was well known as the mysterious wilt; it was called by different names like sudden wilt, new wilt and often termed as parawilt by cotton researchers (Mayee, 1988). In cotton, it was reported in the rainy season of 1978 in an intra hirsutum hybrid JKHY-1 from Adilabad district of Andhra Pradesh (Srinivasan, 1984). Parawilt or Sudden wilt incidence noticed due to the disturbance in the uptake and supply of water to the plant. The conditions in the environment like high temperature and bright sunlight prevails for a time, if followed by heavy rain favors the incidence of parawilt. In Para wilt affected plants, shriveling or drooping down of leaves were noticed immediately from top towards bottom leads to death of the plants within 24–48 hours. Although, the root system was remains intact, and if the fields were dry, then the plant cannot be uprooted easily as the root system is intact. It generally occurs after droughts when the crop is heavily irrigated or there is heavy rain. It may occur in any of the variety/hybrid of cotton.

5.8.1   MANAGEMENT

1.   The drainage system should be proper to avoid the water stagnation in the field especially in heavy soils.
2.   The Para wilt can be managed with by the spray of Cobalt chloride@10 mg per one lire of water (10 ppm) immediately after the appearance of symptoms of Parawilt. If the permanent wilting will be set in the cotton field or due to the delayed in the spray of Cobalt chloride then there will be no recovery.

5.9   TIRAK

Tirak is one of the most important abiotic stresses or physiological disorder. Bad boll opening is popularly known as Tirak. It mainly occurred in the later stages at the time of the maturity of the cotton crop in the border areas of the Punjab adjoining Rajasthan and Haryana state when there is high temperature, dry weather, dry spells, low humidity, etc., during the flowering and fruiting stages. It is characterized by the yellowing and reddening of leaves, followed by the bad/poor opening of the bolls. Tirak can be observed in the cotton crop sowing very early or nutrient deficient or sandy soil. Tirak does not cause any major economic loss during the last 4–5 years in the South Western Region of Punjab.

5.9.1 MANAGEMENT

Timely sowing of the cotton crop, adequate fertilizer or nutrient applications and timely and appropriate irrigations are some of the precautionary measures. All these factors are interlinked and proper cultural practices should be followed right from the time of sowing till the maturity of the cotton crop.

5.10 YELLOWING OF THE COTTON CROP

Cotton crops have the indeterminate growth habit and in need of the regular uptake of the nutrients. Although, the requirement of the nutrients are varied as per the growth of the cotton crop (Mullins and Burmester, 2010). During the reproductive phase, the deficiency of the nutrients leads to the yellowing of the cotton crop. Yellowing of the cotton leaves generally seen in the cotton crop raised in the sandy, highly alkaline or acidic soil or nutrient (e.g., nitrogen) deficient soil. As in the sandy or highly alkaline or acidic soil interferes in the nutrient availability to the cotton plants leads to the yellowing of the cotton crop. Sometimes yellowing occurred due to due to water stress or water stagnation conditions. These types of symptoms observed in the initial stages (30–40 DAS) of the cotton crop or at the later stages at the time of full bloom or boll formation stage. At the later stage, additional nutrient required for the flower initiation, boll formation and its maintenance.

5.10.1 MANAGEMENT

It can be managed by implementing the proper cultural practices. In case of nutrient deficiency, application of fertilizers, especially urea and other macro and micronutrients should be applied as per the package of practices for *Kharif* crops of Punjab. Foliar spray of Urea @1% or 2% can be done for the fast recovery of the cotton crop.

5.11   DRYING OF THE COTTON CROP DUE TO WATER STAGNATION

Drying of the cotton crop observed in the waterlogged condition maintains for the longer times. These types of problems were seen in the Saldurgarh area due to the waterlogging conditions seen 2 year back. In 2019, a larger number of the areas of the cotton crop was dried in the Bathinda and adjoining district due to the water stagnation by heavy rainfall. Prolonged water stagnation during the cropping season of the cotton crop leads to drying of the cotton crop as the cotton crop is very sensitive to the water.

5.12   LEAF REDDENING OF THE COTTON CROP

At later stages, during full boom and boll development stages, reddening of the leaves of the cotton crop were seen. These types of symptoms were generally seen in the mature leaves of the cotton crop or the cotton crop raised in the sandy soil. It might be due to the demand of more nutrients, water stress, or excess or change in the climatic conditions, etc. It is due to nutrient deficiency, and if not properly managed, it can lead to yield loss. Magnesium being associated with maintaining healthy green leaves and higher photosynthesis occurred in the center part of the chlorophyll molecule (Rao and Santosh, 2017). Seed treatment with PSB, foliar, and soil application of humic acid along with fertilizer mixture improved seed cotton yields and delayed leaf reddening by chelating and hormonal action (Raju, 2017).

5.12.1   MANAGEMENT

It can be properly managed by the two sprays of the Magnesium Sulphate (1%) in 100 liters of water at 15 days intervals at the time of the peak demand of the nutrients during full bloom and boll development stages.

5.13   PHYTOTOXICITY IN COTTON CROP DUE TO PESTICIDES

The phytotoxic symptoms were seen in the cotton crop were damaged due to the faulty application of fertilizers or pesticides or tank mixing of insecticides or herbicides.

During morning hours, improper broadcasting of urea at the time of flower initiation and boll formation stages leads to the charring of the leaves.

This is due to the sticking of the granules of the urea to the leaves of the cotton crop leads to the charring of the cotton leaves.

A faulty spray of the herbicides or exposure of the cotton crop to the herbicides 2,4-D (2,4-dichlorophenoxyacetic acid) leads to the deformity of the leaves of the cotton crop, and a setback in the growth of the cotton crop were occurred. Even exposure of the cotton crop to the 2,4-D herbicides leads to the damaged of the floral parts and leaves of the cotton crop. Leaves become malformed and become "monkey palms." Heavy exposure or sprays of 2,4-D leads to the complete damage of the apical buds.

KEYWORDS

- **abiotic stress**
- **biotic stress**
- **cotton**
- **cotton leaf curl Multan beta-satellite**
- ***Gossypium hirsutum* L.**
- **small vein thickening**

REFERENCES

Anonymous, (2017). *Package of Practices for Kharif Crops of Punjab* (p. 186). The Punjab Agricultural University of Ludhiana.

Anonymous, (2020). *Package of Practices for Kharif Crops of Punjab* (p. 197). Punjab Agricultural University of Ludhiana.

Arora, R. K., & Singh, P., (2016). Screening of *Bt*-cotton hybrids for resistance to cotton leaf curl virus disease in the southwestern region of Punjab. *Pl. Dis. Res., 31*(1), 120, 121.

Atkinson, G. F., (1891). The black rust of cotton. *Ala. Agric. Exp. Stn. Bull., 27*, 1–16.

Central Institute for Cotton Research (CICR). (2014). Technical Bulletin No: 2. *Abiotic Stresses in Cotton: A Physiological Approach.* www.cicr.org.in (accessed on 11 February 2021).

Delannoy, E., Lyon, B. R., Marmey, P., Jalloul, A., Daniel, J. F., Montillet, J. L., Essenberg, M., & Nicole, M., (2005). Resistance of cotton towards *Xanthomonas campestris pv. Malvacearum. Annual Review of Phytopathology, 43*, 63–82.

Hillocks, R. J., (1981). Cotton disease research in Tanzania. *Tropical Pest Management, 27*(1), 1–12.

Innes, N. L., (1983). Bacterial blight of cotton. *Biol. Rev., 58*, 157–176.

Jalloul, A., Sayegh, M., Champion, A., & Nicole, M., (2015). Bacterial blight of cotton. *Phytopathol. Mediterr., 54*, 3–20.

Mayee, C. D., & Choular, A. B., (1988). Parawilt of cotton. In: *National Seminar on 'Changing Pest Situation in Current Agricultural Scenario in India.'* Indian Agricultural Research Institute, New Delhi.

Monga, D., (2014). *Technical Bulletin No. 2/2014.* CICR, Sirsa, Haryana.

Mullins, G. L., & Burmester, C. H., (2010). Relation of growth and development to mineral nutrition. In: Stewart, J. M., Oosterhuis, D. M., Heitholt, J. M., & Mauney, J. R., (eds.), *Physiology of Cotton* (pp. 97–105). Springer, New York.

Pegg, K., & Manners, A., (2014). *Plant Health Biosecurity, Risk Management and Capacity Building for the Nursery Industry.* Agri-science Queensland, Department of Agriculture, Fisheries and Forestry (DADD), a part of NY11001.

Raju, A. R., (2017). Leaf reddening in *Bt* hybrid cotton. *Agri. Res. Tech: Open Access, 3,* 1–4.

Rao, S., & Santhosh, U. N., (2017). Management of leaf reddening through soil and foliar nutrition in irrigated *Bt* cotton (*Gossypium hirsutum* L.). *J. Cotton Res. Dev., 31*(2), 244–248.

Sheo, R., (1988). *Grading System for Cotton Diseases* (pp. 1–7). Nagpur, CICR, Technical Bulletin.

Srinavasan, K. V., (1984). The new wilt disease affecting cotton. In: *Proceedings of the Group Discussion on New Wilt of Cotton* (pp. 9–13). CICR, Nagpur.

Yeshwant, R. M., Cleide, B., & Viviani, B., (2005). A semi-selective agar medium to detect the presence of *Xanthomonas axonopodis pv. malvacearum* in naturally infected cotton seed. *Fitopatologia Brasileira [Fitopatologia Brasileira], 30*(5), 489–496.

Zomorodian, A., & Rudolph, K., (1993). *Xanthomonas campestris pv malvacearum*: Cause of bacterial blight of cotton. In: Swings, J. G., & Civerolo, E. L., (eds.), *Xanthomonas* (pp. 25–30). London: Chapman & Hall.

Zubair, M., Zaidi, S. S. A., Shakir, S., Amin, I., & Mansoor, S., (2017). Review: An insight into cotton leaf curl Multan beta satellite, the most important component of cotton leaf curl disease complex. *Viruses, 9,* 280.

CHAPTER 6

Modern Approaches for Management of Sesame Diseases

N. RANSINGH,[1] B. KHAMARI,[2] and N. K. ADHIKARY[3]

[1]*Associate Professor, College of Agriculture, OUAT, Bhawanipatna, Odisha, India, E-mail: nirakar.ranasingh@gmail.com*

[2]*Assistant Professor, Department of Plant Pathology, IAS, SOADU, Bhubaneswar, Odisha, India*

[3]*Junior Pathologist, ICAR-AICRP on Sesame and Niger, Institute of Agriculture Science, University of Calcutta, West Bengal, India*

ABSTRACT

Sesame crop is threatened by several diseases causing huge yield loss. Fungal diseases like *Alternaria* leaf blights, *Cercospora* leaf spot (CLS), *Phytophthora* blight, Powdery Mildew, *Macrophomina* root and stem rot and Bacterial disease like Bacterial leaf spot and blight are important out of them. Apart from these, the crop is also affected by phyllody, a phytoplasmal disease. Intensity of different diseases depends on various factors like climatic conditions, pathogen inoculums, stage of crop, etc. Spraying of Wettable sulfur @ 0.2% during winter from October to 1st week of February is most effective in managing powdery mildew disease. For bacterial diseases, Streptocycline (250 ppm) is used as seed treatment as well as spray application. For reducing fungal diseases such as *Alternaria* blight and CLS disease, seed should be treated with Thiram (0.3%) or Carbendazim 50WP (0.1%) followed by two foliar sprayings of Mancozeb + Carbindazim @0.2%. The vector population can be brought down by seed treatment with Imidacloprid followed by foliar spray with Thiomethaxam, for managing phyllody incidence. *Macrophomina* root and stem rot, a very destructive disease of sesame can be managed effectively by application of biological control agent, *Trichoderma viride* as seed treatment @10 g/kg seed and as soil application @ 2.5 kg/ha.

6.1 GEOGRAPHICAL DISTRIBUTION AND ECONOMIC IMPORTANCE

Sesamum (*Sesamum indicum* L.) is an important oilseed crop domesticated over 3000 years ago and belongs to the family Pedaliaceae. It is one of the most ancient and important oilseed crops with 6.5 million hectares cultivated worldwide, producing more than 3 million tons of seed. A major contribution of world production (68%) was from India, Sudan, Myanmar, and China only (Chattopadhaya et al., 2015). The crop is cultivated worldwide including semi-arid tropics, sub-tropics, and temperate areas (Raikwar and Srivastava, 2013). India leads in production of sesame throughout the world (FAOSTAT, 2016).

India is among the top five countries of the world in oilseed production estimating 25.5 million tons annually. Nine edible oilseeds are cultivated in India and sesamum ranks fifth in production, after groundnut, rapeseed, soybean, and sunflower (Chattopadhaya et al., 2015).

Due to richness in oil content (about 50%), protein (about 23%) and carbohydrate (15%), cultivation of sesame is preferred in India (Ranganatha et al., 2012). Apart from that, it is also a rich source of linoleic acid, Vitamin E, A, B1, and B2 (Brar and Ahuja, 1979). It has also different anti-bacterial, anti-viral, anti-fungal properties. This is the reason why sesame is often called the "Queen of oil seeds" (Weiss, 1971). It is also known as "the seed of immortality" due to the presence of many antioxidants such as sesamin, sesamol, sesamolin are present in sesame which increases its medicinal value manyfold (Bedigian, 1985; Moazzami, 2006; El-Bramway and Mahesh, 2010). A major part of sesame oil is used for edible purposes, whereas a minor part is used for industrial purposes. It is used for the manufacturing of paints, soap, skincare products, cosmetics, pharma products, and insecticides. Sesame oil is also used in cooking and health food industries due to lack of cholesterol. Sesame seed, especially white one is extensively used in bread, breadsticks, cookies, candies, pasta, and curry dishes whereas black seeded sesame has medicinal properties. The oil cake is rich in methionine, cysteine, arginine, and tryptophan and are used for cattle feed, poultry feed. It can also be used as manure because it contains good amount of Nitrogen (6.0–6.2%), phosphorous (2.0–2.2%) and potash (1.0–1.2%).

It is a short day plant that can be grown in the plains and hilly regions up to an elevation of 1,300 m with 400–600 mm of rainfall annually. The crop prefers well-drained, light to medium textured soils with good water holding capacity, moderate fertility and pH ranges from 5–8. Due to its short duration and drought-tolerant nature, it fits well in many cropping systems.

6.2 CONSTRAINTS IN SESAME CROP PRODUCTION

In spite of its high nutritive value and wide use, it is given less importance when it comes to cultivation practices. It is generally grown in a short period, between two major crops following improper management practices, which are the main reason for its low productivity. It may also due to a lack of proper knowledge and biotic and abiotic stresses. Low and unreliable yields, shattering, high production cost leads to lower returns to the farmers (Murthy et al., 1985).

Cultivated sesame encounters a large number of abiotic stresses such as salinity, drought, waterlogging, frost, hailstorm damage, sandy soil, alkalinity, and acidic soils, etc. (Yousif et al., 1972). Early senescence, photosensitivity, and susceptibility to various biotic and abiotic stresses are major productivity constraints in sesame (Sudhakar and Sree Rangaswamy, 1989). Heavy rains during crop growth may attract many fungal pathogens. Similarly, biotic factors like several diseases such as leaf spots, leaf blight, *Fusarium* wilt, stem, and root rot, powdery mildew, stem, and crown rots, phyllody, and incidence of Phytophthora blight affects sesame crop heavily (Verma, 2002). This chapter focuses on important diseases of sesame and modern approaches for their management.

6.3 STEM AND ROOT ROT OF SESAME

It is a destructive disease of sesame causing 5–100% yield loss (Sundara-raman, 1933; Murugesan et al., 1978). It is otherwise known as charcoal rot, dry rot, stem rot or root rot.

6.3.1 GEOGRAPHICAL DISTRIBUTION AND ECONOMIC IMPORTANCE

Stem and root rot is distributed mostly in tropical and subtropical countries having arid to semiarid climatic conditions (Gray et al., 1990; Abawi and Pastor-Corrales, 1990; Diourte et al., 1995; Wrather et al., 2001). In India, it is mostly seen in dry areas of Madhya Pradesh, Rajasthan, Gujrat, Bihar, Maharastra, Chattishgarh, Delhi, Haryana, Uttar Pradesh, Karnataka, Tamil Nadu and Odisha with crop loss ranging from 5% to 100% (Sundararaman, 1931; Khare et al., 1973; Vyas, 1981; Khamari et al., 2018b) depending on favorable environmental conditions, susceptibility of cultivar, stage of crop and severity of disease. If management practices such as field preparation, sowing quality seed, proper fertilization, irrigation, timely sowing, and adequate crop protection are not provided properly then there will be a severe yield loss (Khaleifa, 2003; El-Bramawy, 2006; El-Shakhess and Khalifa, 2007). Apart from yield, other biometric parameters are also directly affected by the disease incidence (Khamari et al., 2016).

6.3.2 SYMPTOMS

It is a devastating disease which affects all the stages of plant. It leads to poor seedling establishment, reduction of vigor, discoloration of plant, dry root rot, and finally wilting of plant which reduces the productivity (Abawi and Corrales, 1990; Khamari et al., 2016). The disease affects the seed at the initial stage, causing seed rot leading to improper germination and rotting of young seedlings, resulting into poor plant population in the field (Yu and Park, 1980; Gonzalez and Subero, 1984; Khare et al., 1973; De Mooy and Burke, 1990; Pun et al., 1998; Das et al., 2015; Mayek-Perez et al., 2001; Su et al., 2001; Khamari et al., 2016). In a later stage, small brownish-colored lesion appears on the stem regions. It gradually spreads in both the directions affecting the whole plant (Khamari et al., 2016). Affected plants turn black color due to the presence of microsclerotia (Dhingra and Sinclair, 1978; Abdou et al., 1980; Khamari et al., 2016). Not only the aerial parts, but also it attacks the below-ground parts especially the root system which results in poorly developed secondary roots and reduced its number (Khamari et al., 2016). In advance stage, it becomes deform and rot. Finally, the whole plant wilts, which leads to poor development of capsule and ultimately a significant reduction in total plant yield. Being systemic in nature, pathogen moves through xylem vessel and totally degrades the vascular tissue. The infected vascular bundle turns to black color. Hence, there is blockage in the flow of nutrients, which results in wilting and death of plant in later stages (Khamari et al., 2016). The disease severity is more in stress condition as defense system of plant become weak and the pathogenic activity may increases several folds. In short, the disease includes a series of symptoms affecting different parts of the plant showing various types of symptoms (Singh et al., 1990). There is also a report of the development of spindle-shaped lesions with dark border and light gray center covered with small pinhead-sized microsclerotia and sometimes pycnidia (Pun et al., 1998; Gulya et al., 2002).

6.3.3   THE PATHOGEN

Macrophomina phaseolina (Tassi.) Goid is the notorious pathogen responsible for this destructive disease which exists in two forms. In pathogenic phase it is known as *Macrophomina phaseolina* which produces pycnidia but in saprophytic phase (*Rhizoctonia bataticola*), it produces microsclerotia (Holliday and Punithalingam, 1970; Dhingra and Sinclair, 1972, 1978; Halsted, 1890; Mihail and Taylor, 1995). The pathogen attacks more than 500 plant species (Dhingra and Sinclair, 1972; Mihail and Taylor, 1995; McCain and Scharpf, 1989; Wyllie, 1993; Songa and Hillocks, 1996; Smith and Carvil, 1997; Su et al., 2001; Mayek-Perez et al., 2001; Saleh et al., 2010; Mahdizadeh et al., 2011). Despite of wide host range, *Macrophomina* is a monotypic genus with only one species, i.e., "*phaseolina*" (Sutton, 1980) due to extremely intraspecific variations (Dhingra and Sinclair, 1972; Echavez-Badel and Perdomo, 1991). In host pycnidia are rarely produced (Knox-Davies, 1966). Production of pycnidia in a host largely depends on the type of host, nature of fungal isolate (Ahmed and Ahmed, 1969). However, it can be induced *in vitro* by providing different incubation methods (Mihail and Taylor, 1995; Gaetan et al., 2006).

Macrophomina phaseolina, the destructive pathogen shows morphological variability (Mihail and Taylor, 1995; Mayek-Pérez, 1997), physiological variability (Manici et al., 1995; Mihail and Taylor, 1995), Pathogenic variability (Manici et al., 1995; Mihail and Taylor, 1995; Miklas et al., 1998; Mayek-Pérez et al., 2001; Su et al., 2001) and genetic variability (Chase et al., 1994; Mihail and Taylor, 1995; Jones et al., 1997; Mayek-Pérez et al., 2001; Su et al., 2001). The virulence of pathogen also varies. These characters help it to adapt in various environmental conditions.

The pathogen produces pale yellow to brown color hyphae which are generally perpendicular to main hyphae (Dhingra and Sinclair, 1977). That hypha are aggregated and coupled with melanin pigments to form microsclerotia. The color of microsclerotia is usually black and the size ranged from

50–150 μm depending upon the type of host and the media used (Short and Wyllie, 1978).

It is primarily a polyphagous, cosmopolitan, soil-borne as well as seed-borne pathogen showing a high degree of variability with heterogeneous host specificity, infecting a large species of monocots as well as dicots (Dhingra and Sinclair, 1973; Su et al., 2001; Kaur et al., 2011).

6.3.4   *DISEASE CYCLE*

Macrophomina phaseolina survives in soil as well as in seed. It exists in the form of microsclerotia in the absence of host. These microsclerotia persist in soil for a very long period of time. It varies from 2–15 years depending upon environmental conditions and host residues (Short et al., 1980; Baird et al., 2003). These are potent resting structure having the potential to germinate quickly from each and every cell and produce a germ tube which finally infects the susceptible host, acting as the source of primary inoculum (Ayanru and Green, 1978; Mihail et al., 1994). These resting structures are adapted to a wide range of environmental conditions, such as moisture, temperature, or nutrient stresses (Short et al., 1980). It requires temperature, ranging from 28–35°C for germination (Mihail, 1989). These microsclerotia produce appressoria after germination, which helps in penetration of the host epidermis by secreting various cell wall degrading enzymes (CWDEs). The pathogen may also penetrate through natural openings or wounds (Bowers et al., 1999; Mayek-Perez et al., 2002). These mycelia colonize in vascular tissue by entering the xylem vessels. It Clogs the vascular system and restricts the conduction process resulting in wilting of plant (Abawi and Pastor-Corrales, 1990; Wyllie, 1988). The pathogen produces various enzymes and toxins which may helps in the development of characteristic symptoms of rotting, wilting, and blight, etc., (Jones and Wang, 1997; Kuti et al., 1997).

The severity of the disease depends on microsclerotial density or pathogen inoculum present in the field (Madden, 1980; Kenerley and Bruck, 1987; McCain and Scharpf, 1989; Dawar and Ghaffar, 1998; Moradia, 2011; Uma Maheswari et al., 2001; Khamari et al., 2019). The pathogen also survives in seed apart from soil (Richardson, 1979). Seeds produced from infected plants becomes rot and do not germinate, when sown in the field. Even if it germinates, after few days, the stems become brownish, thin, wire-like and toppled down from that point though the upper portion remain green (Khamari et al., 2016).

6.3.5 EPIDEMIOLOGY

The rate of disease development and disease severity is directly relates with the pathogen propagules present in the soils, as well as the mode of pathogen invasion (Madden, 1980; Kenerley and Bruck, 1987; Khamari et al., 2019). The disease is more common at the maturity stage of plant that to in dry weather (Almeida et al., 2003). The pathogenic ability increases with water stress and rise in temperature between 28 and 35C (Jain and Kulkarni, 1965; Abawi and Pastor-Corrales, 1990; Manici et al., 1995; Diourte et al., 1995; Cardona, 2006; Akhtar et al., 2011; Khamari et al., 2018c). The disease incidence depends mostly on age of plant, atmospheric temperature, and soil humidity (Rodriguez and Zammlw, 1985). Soil temperature of 80–95°F (27–35°C) for 2 to 3 weeks favors development of disease (Yang and Navi, 2003). The disease incidence is positively related to temperature but negatively correlated with relative humidity (Deepthi et al., 2014).

6.3.6 MANAGEMENT OF DISEASE

Management of this disease is a bit tough as well as not economical as the pathogen is soil-borne, highly variable and affects a wide range of plant species (Gulya et al., 2002). The management strategies should be formulated after proper knowledge on biology as well as survival of the pathogen and stage of the crop. It can be managed in various ways like cultural methods, use of organic products, biocontrol agents, chemicals, use of resistant varieties, or their combinations.

6.3.6.1 CULTURAL CONTROL

It is the safest method to manage the disease by altering cultural practices. It is not only ecofriendly but also pocket friendly. Modifications in cultural practices can save the crop from deadly diseases. It may be field preparation, choosing the right cropping system, collection of healthy seed material, proper sowing time, spacing, timely irrigation, fertilization, timely harvest, and proper storage. There are several methods to manage sesame diseases.

Intercropping of sesame with other crops can reduce disease incidence. Inter or mixed cropping of sesame with moth bean or mung bean may minimize the stem and root rot disease incidence. Mixed cropping of sesame with Urd (*Phaseolus mungo L.*), Cow-pea (*Vigna sinensis*), Moong (*Phaseolus aureus*), Guar (*Gramopsis tetragonoloba*) and Moth (*Phaseolus aconitifolius*) also reduce root and stem rot disease of sesame (Daftari and Verma, 1975). Rotation of sesame crop with non-host crop is effective against the diseases as the pathogen survives in soil. Small grain and corn be used for this. The effectiveness of this method depends on the type of crop and field where it is grown.

As the pathogen is seed-borne in nature, the use of clean and disease-free seeds are mandatory. Treatment of seeds before sowing provides a good plant population in the field by protecting the seedlings from disease.

Soil solarization is also an important component for the management of soil-borne disease. The propagules of *M. phaseolina* effectively reduced by soil solarization up to 20% as compare to unsolarized soil after 30 days (Dubey et al., 2009). Soil solarization solely or mixed with *Trichoderma pseudokoningii* and *Emericella nidulans* solely or in consortium reduces incidence of disease (Ibrahim and Abdel-Azeem, 2015). Combination of soil amended with powder form of eucalyptus leaves and soil solarization has synergistic effect, which increase the number of healthy plants by reducing disease (Ibrahim and Abdel-Azeem, 2015). Transplanting of sesame can protect plants for a long time against soil-borne diseases than seed treatment (Deacon and Berry, 1993; Elewa et al., 1994, 2011; Ziedan, 1998; Sahab et al., 2001; Mostafa et al., 2003; Alasee (Najwa), 2006).

Stem and root rot disease is more severe if it is grown in stress conditions such as high temperatures, drought or poor fertility, etc. Therefore, such cultural practices should be chosen that minimizes plant stress and reduce the risk of charcoal rot. Reduction in plant populations, proper spacing, timely irrigation, and optimization of fertility levels, especially phosphorus can reduce stress in plant. It may not control charcoal rot but can reduce the risk and finally reduce disease impact on yield.

Cultural methods are more effective and economical against the soil-borne pathogen. However, it takes a long time to manage the diseases. Again due to a high degree of variability, competitive saprophytic ability, as well as polyphagous nature of pathogen, cultural methods like crop rotation are inefficient (Almeida et al., 2001; Pearson et al., 1984).

6.3.6.2 BIOLOGICAL CONTROL

It is a method to manage the disease with living organisms. Different species of *Trichoderma* such *T. viride* and *T. harzianum, T. polysporum* are reported not only to reduce stem and root rot disease incidence but also enhance germination and boost plant growth in different crops to an appreciable level both *in vitro* as well as *in vivo* conditions (Ramezani, 2008; Kumari et al., 2012; Arora and Dhurwe, 2013; Senjaliya and Nathawat, 2015). *Trichoderma viride* as seed treatment is effective for management of root rot fungus *M. phaseolina* in various crops (Anis et al., 2010). Application of *T. harzianum* as both seed treatment as well as soil application significantly brought down the disease incidence (Choudhary et al., 2018) and enhance grain yield. Soil application of *T. viride* @ 2.5 kg/ha is effective and economical for the management of root and stem rot disease of sesame (Gupta et al., 2018).

The use of biological agents such as *T. harzianum, T. viride, B. subtilis* and *Pseudomonas flourescens* to manage soil-borne pathogenic fungi is an attractive possibility (Kang and Kim, 1989; Hyun et al., 1999; Harman et al., 2004; Deacon and Berry, 1993; Leclere et al., 2005; Abou Sereih (Neven) et al., 2007). *T. viride* coils around and penetrate into hyphae of another soil-borne pathogen, i.e., *F. oxysporum* f. sp. *sesame* which cause wilt disease of sesame (Chung and Choi, 1992; Harman et al., 2004; Elewa et al., 2011).

There are many bacterial biocontrol agents such as *Bacillus subtilis* and *Pseudomonas fluorescens* which are effective against stem and root rot pathogen (Rajpurohit, 1999; Bhattacharyya et al., 2014; Ashwini et al., 2014; Vyas and Patel, 2015; Singh and Verma, 2015; Savaliya et al., 2016; Anis et al., 2010). Fluorescent pseudomonad produces chitinase and β-1,3-glucanase which helps in inhibiting the growth of *Fusarium oxysporum, M. phaseolina,* and *Sclerotinia sclerotiorum* (Gupta et al., 2002, 2006). *Pseudomonas chlororaphis* subsp. *aurantiaca also* significantly inhibits *M. phaseolina in vitro* and reduce damping-off *in vivo* (Marisa et al., 2014). *Bacillus subtilis* another biological control agent also has the capacity to reduce growth, sporulation, and sclerotial formation of pathogen (Shin et al., 1987; Jacobsen et al., 2004; Leclere et al., 2005; Elewa et al., 2011). Bacterial antagonists are found to be more effective biocontrol over *Trichoderma* spp. (Savaliya et al., 2016).

Combination of both fungal as well as bacterial antagonist as seed treatment as well as soil treatment provides more promising results in terms of seed germination, growth, plant stand, and total harvested produce. Treatment of seed and soil application with *Trichoderma viridae* and *P. Fluorescens* @5 g/kg of seed and 2.5 kg/ha respectively is effective against the disease (Gupta and Ranganatha, 2014).

6.3.6.3 MYCORRHIZA

It is the symbiotic association of fungi with sesame root. The association of this mycorrhiza in roots of sesame plant can reduce the colonization of fungal pathogen in the rhizosphere. At flowering stage of sesame plant root colonization of vesicular-arbuscular mycorrhizal fungi (VAM) is more (Phillips and Hayman, 1970). Application of *Glomus* spp. in sesame not only reduces colonization of fungal pathogens in sesame rhizosphere but also diminishes their virulence and enhances lignin contents in root system of plant (Khalifa, 1997; Sahab et al., 2001; El-Fiki et al., 2004; Linderman, 1992; Ziedan, 1998; Mostafa et al., 2003; Ziedan et al., 2010, 2011). VAM improves resistance of plants by increasing antifungal chitinase enzymes in roots (Dehne et al., 1978). Another mycorrhiza, i.e., *Lums* spp. (VA mycorrhizae) significantly increases biometric parameters of plants such as plant height, number of branches, and number of pods. Treatment with mycorrhiza stimulates colonization of selective bacteria such as bacteria belong to the Bacillus group in the sesame rhizosphere which shows antagonistic potential to fungal pathogens (Ziedan et al., 2011).

Application of mycorrhizae and biocontrol agent such as *Trichoderma viride* or *Bacillus subtilis* in consortium are more effective than the individual, for controlling Macrophomina disease incidences and increase morphological characters and seed yield of sesame (Ziedan et al., 2011). *Trichoderma* spp. along with VA mycorrhizae (*Glomus* spp.) not only protects sesame plant from wilt and root-rot disease but also significantly increased seed yield (Khalifa, 1997; Sahab et al., 2001; Abou Sereih (Neven) et al., 2007). Mycorrhiza along with bacterial biocontrol agent such as *B. subtilis* is also effective (Sahab et al., 2001; Jacobsen et al., 2004; Leclere et al., 2005).

6.3.6.4 ORGANIC AMENDMENTS

Application of organic products such as oil cake, farmyard manure, green manure, and vermicompost (VC) may reduce harmful microorganisms from the rhizosphere by promoting growth of antagonistic microorganisms. Beside this, these organic products are rich in sufficient amount of alkaloids which suppress the pathogens. Mustard cake, neem cake, groundnut cake and sesame cake are few of the oil cakes which are effective against the disease (Dhingani et al., 2013; Meena et al., 2014; Khamari and Patra, 2018a). Among which mustard cake gives better performance (Gemawat and

Verma, 1971; Jha et al., 2000; Dhingani et al., 2013; Meena et al., 2014). Apart from oil cakes, different manures like farmyard manure, VC, and goat manure can also used in the field for reduction of disease incidence (Kumari et al., 2012; Dhingani et al., 2013; Khamari and Patra, 2018a).

6.3.6.5 EFFECTIVENESS OF PLANT EXTRACTS

Extracts of many commonly available plants which have effectiveness against the disease can be a possible alternative to the hazardous chemicals as it is easily available in our localities, environment friendly and suits to the pockets of farmers. It neither disturb to the ecosystem nor leaves residues in the final product. Garlic, onion, acacia, ginger, neem, turmeric, datura, and Karanj are commonly available plants which are effective against the pathogen *in vitro* (Tandel et al., 2010; Dhingani et al., 2013; Meena et al., 2014; Sandipan, 2014; Savaliya, 2015). The aqueous extracts of *Cymbopogon citratus* (Bankole and Adebanjo, 1995) and Powder of *Datura fastulosa* (datura) can be used against *M. phaseolina* in pot (Enteshamul et al., 1996).

6.3.6.6 EFFECT OF OILS

Essential oils isolated from various sources have fungicidal as well as bactericidal properties as it provides a barrier between pathogen and host. Oils are reported to suppress much air-borne, seed-borne as well as a soil-borne pathogen (Murthy and Amonkar, 1973). Actinidine isolated from *Nepeta clarkei* was found to be quite effective *in vitro* against *M. phaseolina* (Saxena and Matela, 1997). Garlic oil, neem oil, palmarosa oil, clove oil are reported effective against *M. phaseolina* (Kazmi et al., 1995; Alice et al., 1996; Kumari et al., 2012; Khamari et al., 2018a). Neem oil is more or equally effective as compared to benomyl and carbendazim (Kazmi et al., 1995: Alice et al., 1996; Ilyas et al., 1997; Sharma and Basandrai, 1997; Lokhande et al., 1998; Dubey and Kumar, 2003). Oils can be utilized as seed dresser for effective management of disease.

6.3.7 INTEGRATED APPROACH FOR MANAGEMENT

A combination of Neem cake (250 kg/ha) and *Trichoderma viride* (2.5 kg/ha) can also give good results against the disease incidence (Rajpurohit,

2008, 2013). The combined application of neem cake and *Trichoderma viride* at the rate of 250 kg/ha and 2.5 kg/ha, respectively, works well against the disease (Rajpurohit, 2008, 2013). The use of *Pseudomonas fluorescens* as seed treatment and soil amendment with mustard cake, VC, and FYM successfully reduce Macrophomina root rot of chickpea (Khan and Gangopadhyay, 2008). Sundaravadana (2002) mentioned that soil application of $ZnSO_4$ followed by combined application of *T. viride* + $ZnSO_4$ significantly reduces root rot incidence.

6.3.7.1   CHEMICAL CONTROL

As resistant germplasm against virulent pathogen is not there, utilization of systemic fungicides has potential approach to reduce the inoculum density of this soil-borne disease (Reznikov et al., 2016). It is relatively cheap and more effective (Azeez and Morakinyo, 2010). It is an easy method to practice such as prepares solutions, drench soil, and application with irrigation, which is quick in action (El-Fiki et al., 2004).

There are many fungicides that were reported to be effective against *Macrophomina phaseolina* both *in vitro* and *in vivo*. It reduces the disease incidence as well as sclerotia production (Ramadoss and Sivapraskasam, 1994; Dubey and Kumar, 2003). It is a virulent pathogen of sesame which can be managed by seed treatments before sowing (Nayyar et al., 2014). Seeds treated with carbendazim, Captan, etc., boost germination of seed by reducing disease incidence (Shukla and Singh, 1974). Seed dressing (Benomyl @5 g/kg seed) combined with or without chlorothalonil or dicloran (5 kg/feddan) as soil treatment reduces sesame diseases (El Deb et al., 1985). Carbendazim, captan, thiram, mancozeb, indofil M-45, Tebuconazole, propiconazole, and vitavax or raxil are effective against the disease and reduce root rot incidence (Sinha and Khare, 1977; Ramadoss and Sivapraskasam, 1994; Dubey and Kumar, 2003; Kumari et al., 2012; Khamari and Patra, 2018d). The new generation chemicals like Tebuconazole 2DS as seed treatment and tebuconazole 25.9 EC as soil drenching are most effective against charcoal rot incidence and increase yield (El-Fiki et al., 2004; Kumar and Jain, 2004; Lokesha and Benagi, 2007; Nagmani et al., 2014; Choudhary et al., 2018; Khamari and Patra, 2018d).

Fungicides in combination enhanced the yield more effectively as compared to application of bio-agents (Elaigwu et al., 2017). Combination of VC with bavistin reduces the root rot incidence in pots conditions (Kumari et al., 2012). A new generation combination fungicide, i.e., Nativo (Trifloxys-trobin 25% + Tebuconazole 50%) disrupts the metabolism of pathogen and hampers their growth and development. It forms covalent bond with scle-rotia and interrupts its ionic concentration (El-Fiki et al., 2004b; Kumar et al., 2016; Khamari and Patra, 2018d). Application of combination fungicide such as Carbendazim 12% + mancozeb 63% WP also found effective against the disease (Khamari and Patra, 2018d). The plants protected with fungicides gave more number of capsules per plant, more number of seeds per capsule, and finally increase the yield (Deepthi et al., 2014).

6.3.7.2 HOST RESISTANCE

Diseases of sesame are mostly managed through fungicides and cultivation techniques at farmers' level. However, some of the diseases can also be handled by resistance breeding as natural genetic variation exists for disease resistance. Since wild sesame species found in India and Nigeria shows resistance to root rot, bacterial diseases, and pests, those can be crossed with *Sesamum indicum* to select resistant breeds to root rot. Mutation breeding can also be practiced. A large numbers of morphological and physiological mutants are cultivated in diseased areas (Ashri, 1981, 1985; Micke et al., 1987). Different management practices such as management of irrigation, fertilization regimes, and application of systemic fungicides are very expen-sive, have temporary effect, and are not eco-friendly. In that case, resistant genotypes are the panacea as it provides a practical, economical, long-term, and environmentally benign means of limiting the damage by disease (Wang et al., 2001). Several varieties of sesame were reported resistant to stem and root rot such as Maporal, Arawaca, and 439 (Rodriguez and Zammlw, 1985), RT-1 and C50 (Mishra et al., 1973). Variety Ansanggae was resistant to seed-ling blight, root rot and wilt diseases of sesame in Korea (Lee et al., 1985). N58-2 is found semi resistant to pod infection by *M. phaseolina* (Mishra et al., 1973). Varieties M-3-1, G-5, ES-89, and TC-66 are tolerant to the disease (Daftari and Verma, 1975.).

Theoretically, no genotype identified resistant to stem and root rot disease so far (Almeida et al., 2001). Wide adaptability and broad host range along with variable nature of pathogen are the major hindrance on development of resistant variety. Apart from that lack of effective selection criteria and evalu-ation methods along with insufficient genetic knowledge of the resistance traits restricted the development of resistant genotypes (El-Bramawy and

Abdul Wahid, 2006; Kavak and Boydak, 2006). High temperature, drought, and the presence of mature plant tissues are often considered necessary for complete symptom development. Hence, these are the major requirements in a resistance-screening program (Odvody and Dunkle, 1979; Wyllie, 1989; Edmunds, 1994; Diourte et al., 1995). The resistant genotypes have different biochemicals which resist pathogen attack. The amounts of total phenols are higher in highly resistant and resistant entries as compared to susceptible and highly susceptible entries (El-Fiki et al., 2004). However, most of the time it is seen that the resistance of a plant to diseases gradually reduces with time (Abd-El-Ghany et al., 1974; Bakheit et al., 1988; El-Shazly et al., 1999).

Mutants may play a good role in the management of charcoal rot disease if a number of management strategies, including cultural practices, biological control, solarization, chemical, etc., will be integrated. Four mutants' viz., NS13P1, NS163-1, NS270P1, and NS26004 can be used to manage the disease following the disease management strategies (Akhtar et al., 2011). Proper management strategies help to reduce the number of microsclerotia that serve as the primary source of inoculum under optimum conditions with dry summer. Generally, control methods for *M. phaseolina* targets a reduction of fungal propagules in the soil as well as within the plant (Singh et al., 1990; Lodha et al., 1997).

Breeding plans mostly include disease resistance and high-yield. Sesame breeds produced by natural variation and selection has partial resistance to diseases like phyllody, bacterial leaf spot and *Fusarium* wilt (Weiss, 1983). Resistance to certain diseases (*Sclerotinia, Fusarium, Rhizoctonia*) do exists but in commercial sesame cultivars, these traits are not included. In the same way, many wild sesame lines are resistant pests but are excluded in commercial varieties. Most of the breeding programs in sesame especially aimed to high yield, but disease resistance are relatively limited. That's why a large gap is seen in seed yield of high-yielding genotypes (El-Shakhess, 1998). Therefore, disease resistance trait should be included in breeding program for better and healthy yield.

6.4   ALTERNARIA LEAF SPOT OF SESAME

6.4.1   *ECONOMIC IMPORTANCE*

Alternaria leaf spot was reported on sesamum for the first time by Kvashina (1928) from North Caucasus as epidemic form, and since then it was reported and described from several parts of the world. A complete loss of sesamum crop in Maryland, United States of America occurred due to the incidence of *Alternaria* blight (Thomas, 1959). In India, it was first time recorded by Dey (1948). A heavy loss occurs due to *Alternaria* blight in sesame growing countries under moist conditions (Kawamura, 1931; Singh, 1974). *Alternaria sesame* is considered as major foliar pathogen, causing significant loss to the crop (Chohan, 1978). *Alternaria alternata* (Kissler) and *Alternaria sesame* are incitant of leaf blight of sesame in India (Siddaramaiah and Hegade, 1984). Many workers estimated that *Alternaria* leaf blight reduced the yield @ 10.73 Kg/ha for every 1% increase in disease index (Dhamu et al., 1981).

The disease is transmitted by seed (Yu et al., 1981). It is distributed worldwide (Leppik and Sowell, 1964). The pathogen can survive between cropping seasons or unfavorable conditions as an infectant in the seeds (Kolte, 1985).

6.4.2 SYMPTOMS

Brown colored spot of round to irregular shape varying from 1 to 8 mm in diameter are observed on leaf blade (Mohanty and Behera, 1958). In the early stages of infection, minute brown spots were found to appear on the leaf blade, which later became darker in color with concentric zonations demarcated with brown lines inside the spots on the upper surface. On the under surface, the spots were grayish-brown in color. These spots coalesce, including leaf blade, and leads to drying and dropping off of the leaves. In advance stage, shot holes appeared and the leaf blade broke in an irregular manner. Spots on the stem and petioles were elongated. Affected floral buds failed to open, shriveled, and dried up. In early stages, the effect of infection by the fungus includes a decline in photosynthetic area due to leaf damage. Damping-off is very common in the seedling stage (Berry, 1960). Seed-borne infection of pathogen was found to result in pre-emergence damping-off (Samuel et al., 1972). Premature defoliation affects growth and yield of plant adversely. Severe infection caused complete defoliation (Kolte, 1985).

Three species of *Alternaria viz., A. longissima, A. Alternate* and *A. sesamicola,* infects sesame-inducing symptoms such as foliage blight, stem necrosis and spots on capsules. All three species also reduced seed germination and seedling stand.

6.4.3   THE PATHOGEN

The causal pathogen of the disease was first recorded in North Caucasus as *Alternaria* sp. by Kvashina (1928). Dey (1948) recorded *Alternaria* sp. on gingerly for the first time in India. It was named as *Macrosporium sesamicola* in Japan (Kawamura, 1931). The fungus was described properly and proposed a new name as *Alternaria sesame* (Mohanty and Behera, 1958). The morphology of the fungus causing leaf blight of sesamum was described properly (Leppik and Sowell, 1964). The conidiophores are simple and sparingly branched, arising directly from hyphae and gradually enlarging near the apex into a clavate conidiogenous cell that would produce a single conidium. Conidia are dilute dull brown in color, narrow, ellipsoid, and tapered to a pointed apex. Transverse and longitudinal distoseptation was evident throughout the enlargement of conidia. Brown discoloration and mycelial growth of *A. Alternate* is observed on the surface of infected sesamum seed. The pathogen is present both internal as well as external surface of seed (Dubey and Singh, 2004). The conidiophores of *A. sesame* is simple and sparingly branched, arising directly from hyphae and enlarging near the apex into a clavate conidiogenous cell producing single conidium. Conidia are dull brown colored, narrow, ellipsoid with a pointed apex, often rostrate (beaked) and muriform (Leppik and Sowell, 1964). It may be catenate (formed in chains) or solitary, typically ovoid or obclavate.

6.4.4   SURVIVAL AND FAVORABLE CONDITIONS

The pathogen survives in seeds and serves as primary inocula for subsequent crops (Neergaard, 1979). Due to a lack of commercial certified seeds, farmers plant their own seeds from previous harvests.

The temperature of 25.9–33.7°C, relative humidity between 89 and 95% and a sufficient intermittent rainfall (1.2–12.8 mm in a meteorological week) is conducive for disease development. Ojiambo et al. (1998) studied prolonged high humidity and frequent rain favored spore's dispersal, infection, and development of leaf blight disease.

6.4.5   INTEGRATED MANAGEMENT

6.4.5.1   BOTANICAL

It is an ecofriendly approach of management of disease. Bulb extract of garlic is a wonder for reducing this foliar disease (Hemalatha, 2006). *Calotropis gigantia* and *Ocimum sanctum* plant extracts also works well in mycelial inhibition of *A. sesame* (Hemalatha, 2006). Neem seed kernel extract (NSKE) is effective (Amaresh et al., 2002). Apart from it, intercropping with maize minimizes the incidence of *Alternaria* leaf spot.

6.4.5.2   BIOLOGICAL

Trichoderma harzanium and *T. viride* works well against the pathogen *in vitro* inhibiting mycelial growth (Savitha et al., 2011; Taware et al., 2014). Application of neem cake and *Trichoderma viride* in soil at the rate of 250 kg/ha and 2.5 kg/ha respectively along with seed treatment of *Trichoderma viride* followed by foliar spray of Azadirachtin @ 3 ml/l on 30 and 45 days after sowing is very effective.

6.4.5.3   CHEMICAL

Spraying of fungicides against this disease is quite common. It is cheap, easy, and well accepted by farmers. Foliar Spraying of Mancozeb @ 0.2% and Propiconazole (0.1%) is highly effective against the disease. Systemic fungicide like carbendazim @ 0.2% or combination fungicide like Carbendazim + mancozeb @ 0.2% also works well against the disease.

6.4.5.4   RESISTANT VARIETY

Resistant variety like TC-325 and SP-70-23 can be used for *Alternaria* blight.

6.5   CERCOSPORA LEAF SPOT (CLS)

Cercospora leaf spot (CLS) is otherwise known as frogeye leaf spot disease. It was first reported by Saccardo in 1876. *Cercospora*, the causal agent of CLS, penetrates the plant directly or through stomata and destroy the leaf tissues forming circular or irregularly shaped brownish spots with a light whitish center resembling 'frogeye.'

6.5.1   ECONOMIC IMPORTANCE

It gained attention due to its high disease severity and the crop loss as high as 70% (Gibbons, 1980; Shane and Teng, 1992).

6.5.2   SYMPTOMS

Initially, the disease appears as minute water-soaked lesions on the leaves, which enlarge to form round to irregular shaped spots of 5–15 mm diameter with whitish center and blackish-purple margin on both the leaf surfaces. The spots, in some cases, are surrounded by a yellow halo. Spots may be light brown, reddish-brown or dark brown in color with a whitish center (Meghvansi et al., 2013a). These spots increase in size and are delimited by veinlets of leaves which later become angular. In case of severe infection, excessive development of spots on leaf leads to drying up and shedding of leaves, causing defoliation of the plant. This disease occurs in Kharif only and not on Rabi crop (Mohanty, 1958). Premature defoliation is severe in humid condition (Kolte, 1985). Defoliation and damage to capsules before maturity results into 20–50% yield loss.

6.5.3   THE PATHOGEN

There are two species of Cercospora such as *Cercospora sesame* and *Cercospora sesamicola* which are associated with the disease. Both the species occurs together and cause 100% loss in seed yield. *Cercospora sesame* causing white spot was first reported by Zimmerman. In India, this pathogen was first reported by Chowdhury (1944). *Cercospora sesamicola* causing severe leaf spot disease was reported by Mohanty (1958). The fungal hypha is irregularly septate, light brown colored and thick-walled. Conidiophores are produced in cluster, 1–3 septate, hyaline at the tip portion and light brown colored at base. Conidia are hyaline to light yellow, elongated, 7–10 septate with broad base and tapering apex.

6.5.4   DISEASE CYCLE

The fungus survives in seed as well as in plant debris. It may survive on seed both externally as well as internally. Thus, it initiated the disease as primary infection, but secondary spread is carried out by wind-borne conidia. When conidia land on the surface of leaves, it germinates in the presence of free moisture. The hyphae emerge from germinated spores infect plant by entering through stomatal openings, wounds or by direct penetration.

After penetrating the leaf surface, fungal hyphae ramify the parenchymous tissue and grow intercellularly (Steinkamp et al., 1979). Cercosporin toxin produced by hyphae damages the cell membranes primarily through lipid peroxidation and leads to necrosis of cells. Pathogen absorbs nutrients that are leached through the damaged leaf cells.

6.5.5   FAVORABLE CONDITIONS

During humid conditions, usually in late spring and summer, the spore germinates. If leaves are frequently damp, the germination process is encouraged. Highly humid (95% and above) conditions are essential for conidial germination. Wide range of temperature (25–30°C) is required for conidia germinate (Kumar et al., 2011). Warm and humid conditions are conducive for the disease development.

6.5.6   DISEASE MANAGEMENT

6.5.6.1   CULTURAL

Alteration of cultural practices is helpful to get rid of this disease to some extent. Maintenance of field sanitation by destroying crop residues is key to disease management strategy. Planting early, i.e., immediately after the onset of the monsoon is helpful to avoid the disease. Intercropping of sesame with pearl millet in the ratio of 3:1 is beneficial for reducing disease incidence.

6.5.6.2   BOTANICALS

Botanicals have effective antifungal activity against many plant pathogens. They act as an ideal source of low cost and eco-friendly management of plant diseases. *Datura stramonium, Tridax procumbense, Azadirachta indica* Juss and NSKE can effectively inhibited conidial germination at 2.5% concentration (Benagi, 1995).

6.5.6.3   CHEMICALS

It is very popularly used method for managing the disease among farmers. Seed should be treated with chemicals like thiram or carbendazim at the rate of 2 g/kg of seed. Spraying of carbendazim @ 0.1% is effective in managing the disease. Three sprays of Mancozeb @ 0.25% at fifteen days interval from the first appearance of disease can also be used for managing the disease.

6.5.6.4   RESISTANT VARIETY

The use of resistant variety, not only reduce the disease incidence but also cut the cost of cultivation by reducing inputs. It is an ecofriendly and ecological approach to reduce the incidence of disease. Resistant variety like RT-127 can be used for this disease.

6.6   PHYTOPHTHORA BLIGHT

6.6.1   SYMPTOMS

The disease symptom initiates with water-soaked spots of chestnut brown color on leaves and stems, which later turns to black. Severity of disease is more in humid weather. The main root of the plant gets affected, and hence it can be easily pulled out. The affected plants shed leaves prematurely. The seeds produced from those plants are shriveled.

6.6.2   THE PATHOGEN

Phytophthora parasitica var. *sesame* is the incitant of this disease. The mycelium of this fungus is non-septate and hyaline. It produces hyaline sporangiophores, which are branched sympodially and bears sporangia at the tip. The sporangia are hyaline, spherical, and papillate. Oospores are the sexual spore or resting spore, which are smooth, spherical, and thick-walled.

6.6.3   SURVIVAL AND SPREAD

It is a soil-borne pathogen that survives in the soil in the form of dormant mycelium and oospores. Dormant mycelium present on the surface of the seed also causes the primary infection. The secondary spread occurs by wind-borne sporangia.

6.6.4 FAVORABLE CONDITIONS

Prolonged rainfall, low temperature (25°C) and high relative humidity (above 90%) favors the disease. Severity of disease is more in heavy soil with high soil moisture.

6.6.5 MANAGEMENT

Proper field sanitation by removal and destruction of infected plant parts is an important step towards the management of disease. Continuous cropping of sesame in the field should be avoided. Follow crop rotations with non-host crop. Intercropping with blackgram in a proportion of 1:3 reduces incidence of Phytophthora blight. Gamma rays have also been used to produce mutants resistant to Phytophthora blight (Pathirana, 1992). Tolerant varieties such as MT-75, TKG-22, and TKG055 can be used. Treat the seed with metalaxyl @4 g/Kg of seed. Foliar spraying of metalaxyl or mancozeb @0.2% is very effective.

6.7 POWDERY MILDEW

6.7.1 SYMPTOMS

It is a foliar disease that affects almost all the above-ground parts of the plant. It initiates with development of grayish-white powdery growth on the upper surface of leaves. It gradually spread to flowers and young capsules with the advancement of the disease. Severely affected parts like leaves may get twisted and malformed. The affected flowers and young capsules shed prematurely. The whitish mycelia turn to dark or black in color in the advanced stage due to the development of cleistothecia.

6.7.2 THE PATHOGEN

The pathogen responsible for this is a fungus. The mycelium of the fungus is hyaline, septate, ectophytic which send its haustoria into the host epidermis.

Primary mycelium produces conidiophores which are short and non-septate bearing long chains of conidia. The conidia are generally hyaline, single-celled, ellipsoid or barrel-shaped. The cleistothecia produced by pathogen are dark in color, globose with hyaline or pale brown myceloid appendages. The asci are ovate with 2–3 ascospores inside each ascus. These ascospores are thin-walled, elliptical, and pale brown in color.

6.7.3   DISEASE CYCLE

The pathogen is an obligate parasite. It perenneates in the infected plant debris present in soil in the form of cleistothecia. The ascospores from these cleistothecia initiate the infection and spread through wind-borne conidia.

6.7.4   FAVORABLE CONDITIONS

Dry humid weather along with low relative humidity aggravates the disease.

6.7.5   MANAGEMENT

6.7.5.1   CULTURAL

Clean cultivation is the key for management of this disease as the pathogen survives in infected plant debris. The infected plant debris and stubbles should be removed from the field and destroy properly. The weed plants and wild host are also removed from in and around the field.

6.7.5.2   CHEMICAL

Applications of Wettable sulfur @ 0.2% or dust Sulphur at 25 kg/ha is effective for the disease. Repeat it after 15 days. Other fungicides such as Tridemorph, Dinocap can also be used for effective management of disease.

6.7.5.3   RESISTANT VARIETY

Varieties like Rajeshwari, SI-1926, KRR-2, etc., are resistant to this disease. Hence, it can be an economical option for low disease incidence.

6.8   BACTERIAL LEAF SPOT

It is a very destructive disease-causing complete death of the plant with 100% yield loss to the crop.

6.8.1   SYMPTOMS

A water-soaked spot develops on the undersurface of the leaf initially which gradually spread to the upper surface. These spots increase in size to become angular due to restrictions by the veins. These dark brown spots coalesce together to form irregular brown patches causing dry leaves. The reddish-brown lesions may also appear on petioles and stem.

6.8.2   THE PATHOGEN

Pseudomonas syringae pv. *sesami,* a bacterium is the incitant of the disease. It is a gram-negative, aerobic, rod-shapedrod-shaped bacterium with one or more polar flagella.

6.8.3   DISEASE CYCLE

The bacterium survives in infected plant tissues. The pathogen is internally seed-borne in nature. The disease spreads through rain splash.

6.8.4   MANAGEMENT

6.8.4.1   CULTURAL

As the pathogen survives in soil, keeping the field free from infected plant debris is the first step towards disease management. Seeds should be collected from healthy source. Use of resistant variety can be a better option.

6.8.4.2   PHYSICAL

Due to internal seed-borne nature pathogen, seed should be treated with hot water at 52°C for 10 minutes.

6.8.4.3   CHEMICAL

Seed can also be treated with various chemicals before sowing. Seed should be dipped in Agrimycin 100 (250 ppm) or streptocycline suspension (250 ppm) for 30 minutes before sowing. Spraying of Streptomycin sulfate or Oxytetracycline hydrochloride at the rate of 100 g/ha twice at 15 days interval effectively manage the disease.

6.9   PHYLLODY

It is a very serious disease which drastically reduces the yield of sesame crop, especially in warm climatic conditions (Salehi and Izadpanah, 1992; Manjunatha, 2012). When the crop is about two months old, disease generally appears and cause loss as high as 80% (Pramod and Mishra, 1992). The yield loss may go up to 100% in severe disease incidence (Sarwar and Haq, 2006). According to one estimation, 1% increase in disease incidence affect the crop badly and reduces yield by 8.36 kg/ha (Maiti et al., 2008).

Sesamum phyllody was first reported in Burma (Myanmar) and was designated as "Green flowering disease/Pothe" (McGibbon, 1924). Later, it was reported in India and known as sepaloid and stenosis (Kashiram, 1930). Phyllody disease is caused by phytoplasma. These are present in the phloem tissue of the infected plants (Doi et al., 1967).

6.9.1   SYMPTOMS

The inflorescence is converted into vegetative part. Except stamens, the whole flower is transformed into leaf-like structures or showed a marked tendency to become leafy. The stamens seldom contained functional pollen and plant becomes completely sterile. In the flowering stage, floral parts get transform; produce abundant vegetative growth and become short stature (Vasudeva and Sahambi, 1955). Evidence suggests that the phytoplasma deregulates a gene involved in flower formation in tomato plants (Pracros et al., 2006).

Shoot apex fasciations, flattening of shoot apex, shortened internodes, intense proliferation of leaf and flower buds in sesame plant are associated with mycoplasma like organism (Tamimi et al., 1989).

Akhtar et al. (2009a) studied symptomatology of sesame phyllody and reported that, different type of phyllody disease symptoms such as floral virescence, phyllody, proliferation; seed capsule cracking, formation of dark exudates on foliage, yellowing, shoot apex fasciation, internode elongation, reduced leaf size and stunting. Disease also show other symptoms like pale green and bushy plant due to reduction in leaf size, reduced internode length, excessive axillary proliferation, floral malformation, abnormal green structure developed in place of normal flower (Manjunatha et al., 2012). The pod of infected plant after flowering is very small.

6.9.2   THE PATHOGEN

The disease was incited by phytoplasma, earlier known as mycoplasma like organisms (MLO) (Das and Mitra, 1998). Being an obligate parasite, these are found in sieve elements of plants and some insect vector. Yellow disease that was presumed to be caused by virus earlier could not be visualized in affected plants (Lee and Davis, 1992). Japanese scientists Doi et al. (1967) were the first to describe about phytoplasma that is responsible for yellow disease.

Phytoplasma are prokaryotes that lack a rigid cell wall, surrounded by a single-unit membrane and are pleomorphic in shape with an average diameter of 0.2–0.8 µm (Doi et al., 1967; McCoy et al., 1989).

6.9.3   TRANSMISSION

Sesame phyllody is not transmitted mechanically or by seeds rather by grafting, dodder, and the leafhopper. The pathogen can produce disease symptoms within 25–35 days in all the grafts.

The disease is transmitted from one plant to another by phloem-feeding insect, leafhopper (*Orosius albicinctus*) belonging to the families Cicadellidae, vegetative propagation through grafts and cuttings (Lee and Davis, 1992). Sastry and Rao (1989) studied phytoplasma diseases of oilseed crops and reported that, leafhopper (*O. albicinctus*) is the vector for transmission of phyllody disease and virus-vector relation has been found to be persistent type. The optimum acquisition period of vector is 3–4 days with inoculation feeding period of 30 minutes. The incubation period of the pathogen in the vector, i.e., leafhoppers maybe 15–63 days and 13–61 days in sesame. Nymphs of insect are incapable of transmitting the phytoplasma. Vector population is generally high during summer and low during winter months.

Leafhopper (*O. albicinctus*) can successfully transmit the phytoplasma phyllody disease and phytoplasma bodies in indicator plants and insect vector (Hosseini et al., 2007). Akhtar et al. (2009a) studied transmission of sesame phyllody and reported that, seed, and sap transmission could not achieve. Disease is transmitted successfully from infected to healthy plant via grafting, dodder, and leafhopper (*O. albicinctus*).

6.9.4   HOST RANGE

Sesame phyllody phytoplasma has a wide host range. It attacks 91 plant species belonging to 36 genera distributed in 12 families. It includes important crop plants like Egyptian clover, carrot, Indian mustard, lucerne, radish, sunhemp, and Indian rape (Sahambi, 1970). Phyllody disease also seen in chickpea (Salib et al., 2005; Akhtar et al., 2009b), brinjal (Das and Mitra, 1998), parthenium (Tessema et al., 2010; Kirdat et al., 2018), wild niger and periwinkle (Omidi et al., 2010).

6.9.5   MANAGEMENT

It can be managed by various ways. It is most important to bring down the vector population.

6.9.5.1   CULTURAL

An appropriate sowing date may be useful in avoiding the severe occurrence of the disease. The incidence of the disease reduces considerably by sowing the crop in early August under Indian conditions. The population of the vector should be low during the growth period of the sesame crop to keep the disease under check (Mathur and Verma, 1973). Avoid growing of sesame near cotton, groundnut, and grain legumes. Phyllody incidence can be reduced by intercropping of sesame with pigeon pea (3:1 row proportion).

6.9.5.2   CHEMICAL

At the time of sowing, soil treatment with thimet 10G at the rate of 10 kg/ ha (Muheet and Chauhan, 1975) or with phorate 10 G at the rate of 11 kg/

ha (Tandon and Banerjee, 1988) or with Temik (Aldicarb) 10G@ 25 kg/ha (Brar and Sandhu, 1976) can be an effective option. Spraying of the crop with methyl-o-demeton (0.1%) or with any other effective systemic insecticide (Muheet and Chauhan, 1975; Tandon and Banerjee, 1988) also works well for managing the disease.

Spraying of Tetracycline @ 500 ppm concentration at the initiation of flowering is effective against phyllody but recovery is temporary. Spraying of manganese chloride can be an option for biochemical control (Purohit and Arya, 1980). Manganese chloride oxidizes the phenol and inhibits the enzymes, bringing the auxin level to normal. Once hyper auxin is oxidized, the plant can grow to its normal condition. Two spraying of endosulfan is very effective against vector (Ali, 2003). Misra (2003) reported ethion and cypermethrin, cypermethrin, and quinalphos, deltamethrin, and triazophos combinations resulted in the lowest phyllody incidence (8.7%).

Seed treatment with insecticide, i.e., imidacloprid, along with spraying of antibiotic tetracycline can be a better option for disease management. Seed treatment with imidacloprid and spraying of thiamethoxam effectively reduces vector population and thereby reduce the disease incidence.

6.9.5.3   RESISTANT VARIETY

Varieties like TBS-05, TBS-09, TBS-02, Sweta, KMR-69, TKG-22, Pragati were resistant to phyllody disease. Hence, it can be used to avoid phyllody disease.

Modern approach of integrated disease management (IDM) strategies should be formulated in such a way that disease incidence should be bring down without hampering the environment. It should be ecofriendly as well as pocket friendly.

KEYWORDS

- Alternaria leaf blight
- Cercospora leaf spot
- macrophomina stem rot
- phyllody
- powdery mildew
- sesame

REFERENCES

Abawi, G. S., & Pastor-Corrales, M. A., (1990). *Root Rot of Beans in Latin America and Africa: Diagnosis, Research Methodologies and Management Strategies.* Cali: Centro International de Agricultura Tropical.

Abd-El-Ghany, A. K., Seoud, M. B., Azab, M. W., Mahmoud, B. K., El-Alfy, K. A. A., & Abd-El-Gwad, M. A., (1974). Tests with different varieties and strains of sesame for resistance to root rot and wilt diseases. *Agric. Res. Rev., 52*, 75–83.

Abdou, Y. A., El Hassan, S. A., & Abbas, H. K., (1980). Effects of exudation from sesame seeds and seedlings on sclerotial germination and mycelium behavior of *Macrophomina phaseolina*, the cause of sclerotial wilt in the soil. *Agric. Res. Rev., 57*(2), 167–174.

Abdou, Y. A., El-Hassan, S. A., & Abbas, H. K., (1980). Seed transmission and pycnidial formation in sesame wilt disease caused by *Macrophomina phaseolina* (Maubl) Ashby. *Agric. Res. Rev. 57*(2), 63–69.

Abou, S. A., El-Aal, S. K. H. A., & Sahab, A. F., (2007). The mutagenic activity of chitosan and its effect on the growth of *Trichoderma harzianum* and *Fusarium oxysporum* f. sp. *sesami. J. Appl. Sci. Res., 5*(6), 450–455.

Ahmed, N., & Ahmed, Q. A., (1969). Physiological specialization in *Macrophomina phaseoli* (Maubl.) Ashby, causing stem rot of jute, *Corchorus* species. *Mycopathol., 39*, 129–138.

Akhtar, K. P., Sarwar, G., & Arshad, H. M. I., (2011). Temperature response, pathogenicity, seed infection and mutant evaluation against *Macrophomina phaseolina* causing charcoal rot disease of sesame. *Arch. Phytopath. Plant Protect, 44*(4), 320–330.

Akhtar, K. P., Sarwar, G., Dickinson, M., Ahmad, M., Haq, M. A., & Iabal, M. J., (2009a). Sesame phyllody disease: Its symptomatology, etiology and transmission in Pakistan. *Turk J. Agric., 33*, 477–486.

Akhtar, K. P., Shah, T. M., Atta, B. M., Dickinson, M., Hodgetts, J., Khan, R. A., Haq, M. A., & Hameed, S., (2009b). Symptomatology, etiology and transmission of chickpea phyllody disease in Pakistan. *J. Pl. Path., 91*(3), 649–653.

Alasee, B., (2006). The use of transplanting as a method for control root rot of sesame with other control methods under greenhouse conditions. *9th Arab Congress of Plant Protection* (p. 196). Damascus, Syria.

Ali, S., & Singh, R. B., (2003). Management of insect pests and diseases of sesamum through integrated application of insecticides and fungicides. *Crop Res., 26*(2), 275–279.

Alice, D., Ebenezar, E. G., & Siraprakasan, K., (1996). Biocontrol of *Macrophomina phaseolina* causing root rot of jasmine. *J. Ecobiol., 8*, 17–20.

Almeida, A. M. R., Torres, E., Farias, J. R. B., Benato, L. C., Pinto, M. C., & Martin, S. R. R., (2001). *Macrophomina phaseolina* in soybean: Effect of tillage system, survival on crop residues and genetic diversity. *Londrina PR Embrapa Soja Circular Tecnica No., 34*, 47.

Amaresh, Y. S., Neruda, V. B., & Somasekhar, B., (2002). Use of botanicals and fungitoxicants against *A. helianthi* (Hansf.) Tsubaki and Nishihara, the causal agent of sunflower leaf blight. *Indian J. Plant Prot., 30*(1), 55–58.

Anis, M., Abbasi, W. M., & Zaki, M. J., (2010). Bioefficacy of microbial antagonists against *Macrophomina phaseolina* on sunflower. *Pak. J. Bot., 42*(4), 2935–2940.

Arora, M., & Dhurwe, U., (2013). Biochemical control of charcoal rot of *Sorghum bicolor* (L.) Moench. *Int. J. Curr. Microbiol. App. Sci., 2*(11), 19–23.

Ashri, A., (1981). Increased genetic variability for sesame improvement by hybridization and induced mutations. In: Ashri, A., (ed.), *Sesame: Status and Improvement* (pp. 141–145). FAO Plant Production and Protection, Paper 29, Rome.

Ashri, A., (1985). *Sesame and Safflower: Status and Potentials.* Proceedings of expert consultations. Food and Agriculture Organization of the United Nations, Rome, Italy.

Ashwini, C., Giri, G. K., & Halgekar, N. Y., (2014). Efficacy of bioagents against seed-borne fungi of black gram. *Int. J. Appl. Biol. Pharm. Technol., 5,* 56–57.

Ayanru, D. K. G., & Green, R. G., (1978). Germinability of sclerotia of *Macrophomina phaseolina. Can. J. Bot., 56,* 1107–1112.

Azeez, M. A., & Morakinyo, J. A., (2010). Genetic diversity of fatty acids in sesame and its relatives in Nigeria. *Eur. J. Lipid Sci. Tech., 113,* 238–244.

Baird, R. E., Watson, C. E., & Scruggs, M., (2003). Relative longevity of *Macrophomina phaseolina* and associated mycobiota on residual soybean roots in soil. *P. Dis., 87,* 563–566.

Bakheit, B. R., El-Hifny, M. Z., Mahdy, E. E., Gurguis, N. R., & El-Shimy, A., (1988). Evaluation of sesame genotypes for relative tolerance to root rot disease. *Assuit. J. Agri. Sci., 19,* 255–264.

Bankole, S. A., & Adebanjo, A., (1995). Inhibition of growth of some plant pathogenic fungi using from some Nigerian plants. *Int. J. Trop. Pl. Dis., 13,* 91–95.

Bashir, M. R., Mahmood, A., Sajid, M., Zeshan, M. A., Mohsan, M., Khan, Q. A. T., & Tahir, F. A., (2017). Exploitation of new chemistry fungicides against charcoal rot of sesame caused by *Macrophomina phaseolina* in Pakistan. *Pak. J. Phytopathol., 29*(2), 257–263.

Bedigian, T., (1985). Sesamin, sesamoline and the original of sesame. *Biochem. Systematics Ecology, 13,* 9–133.

Benagi, V. I., (1995). *Epidemiology and Management of Late Leaf Spot of Groundnut Caused by Phaeosariopsis personata (Berk. and Curt.).* PhD Thesis, Univ. Agric. Sci., Dharwad, Karnataka (India).

Berry, S. Z. (1960). Comparison of cultural variants of *Alternaria sesame. Phytopathology, 50,* 298–304.

Bhattacharyya, S. K., Sengupta, C., Adhikary, N. K., & Tarafdar, J., (2014). *Bacillus amyloliquefaciens*-A Novel PGPR strain isolated from jute based cropping system. *The Bioscan., 9*(3), 1263–1268.

Bowers, G. R., & Russin, J. S., (1999). Soybean disease management. In: Heatherly, L. G., & Hodges, H. F., (eds.), *Soybean Production in the Mid-South.* Boca Raton, FL: CRC Press.

Cardona, R., (2006). Distribución vertical de esclerocios de *Macrophomin phaseolina* en um suelo infestado naturalmente en el estado Portuguesa. *Revista Facultad de Agronomía, Maracaibo, 23*(3), 284–291.

Chase, T. E., Jiang, Y., & Mihail, J. D., (1994). Molecular variability in *Macrophomina phaseolina. Phytopathol., 84,* 1149.

Chattopadhay, C., Kolte, S. J., & Waliyar, F., (2015). Sesame diseases. *Diseases of Edible Oilseed Crops,* 293–328.

Chohan, J. S., (1978). Diseases of oilseed crops, future plans and strategy for control under smallholdings. *Indian Phytopath., 31,* 1–15.

Choudhary, K., Meena, A. K., Dhakar, H., & Kumar, M., (2018). Management of charcoal rot of sesame [*Macrophomina phaseolina* (Tassi.) Goid.] by different bio-agent and fungicide. *Int. J. Chem. Stud., 6*(6), 1556–1559.

Chowdhury, S., (1944). Physiology of *Cercospora sesami* Zimm. *J. Indian. Bot. Soci., 23,* 91.

Daftari, L. N., & Verma, O. P., (1975). An integrated approach to control the root and stem rot of sesamum caused by *Macrophomina phaseoli*. *Presented in Symposium on Plant Disease Problems Organized by Society of Mycology and Plant Pathology*. Udaipur, India.

Das, A. K., & Mitra, D. K., (1998). Hormonal imbalance in brinjal tissue infected with little leaf phytoplasma. *Indian Phytopath., 51*(1), 17–20.

Das, I. K., Kishore, B. B., Srivastava, A. K., Kumar, S., & Arora, D. K., (2015). Ecology, biology and management of *Macrophomina phaseolina*: An overview. In: *Agriculturally Important Microorganisms* (Vol. 1, pp. 193–206).

Dawar, S., & Ghaffar, A., (1998). Effect of sclerotium inoculum density of *Macrophomina phaseolina* on charcoal rot of sunflower. *Pak. J. Bot., 30*, 287–297.

Deacon, J. W., & Berry, L. A., (1993). Biocontrol of soil-borne plant pathogens concepts and their application. *Pestic. Sci., 37*, 417–426.

Deepthi, P., Shukla, C. S., Verma, K. P., & Reddy, S. S., (2014). Yield loss assessment and influence of temperature and relative humidity on charcoal rot development in sesame (*Sesamum indicum*). *The Bioscan, 9*(1), 193–195.

Dehne, H. W., Schonebeck, F., & Baltruschat, H., (1978). Untersuchungen zum Einfluss de endotrophen Mycorrhiza auf Pflanzenkrankheiten: 111. Chitinase-aktivitat und ornithinzyklus. (The influence of endotrophic mycorrhiza on plant disease: Chitinase activity and ornithine-cycle). *Zpflanzenkrankh Pflanzenschutz, 85*, 666–678.

DeMooy, C. J., & Burke, D. W., (1990). External infection mechanism of hypocotyls and cotyledons of cowpea seedlings by *Macrophomina phaseolina*. *Plant Dis., 74*(9).

Dey, P. K., (1948). *Plant Pathology Administrative Report of Agriculture Department* (pp. 43–46). Uttar Pradesh, India.

Dhamu, K. P., Pothiraj, P., Jeyarajan, R., & Balakumar, M., (1981). Estimation of yield loss in sesamum due to diseases. *Madras Agric. J., 68*(10), 648–652.

Dhingani, J. C., Solanky, K. U., & Kansara, S. S., (2013). Management of root rot disease [*Macrophomina Phaseolina* (tassi.) Goid] of chickpea through botanicals and oil cakes. *The Bioscan, 8*(3), 739–742.

Dhingra, O. D., & Sinclair, J. B., (1972). Variation among isolates of *Macrophomina phaseolina* (*Rhizoctonia bataticola*) from the same soybean plant. *Phytopathol., 62*, 511–518.

Dhingra, O. D., & Sinclair, J. B., (1973). Location of *Macrophomina phaseoli* on soybean plants related to culture characteristics and virulence. *Phytopathol., 63*(7), 934–936.

Dhingra, O. D., & Sinclair, J. B., (1978). *Biology and Pathology of Macrophomina Phaseolina* (pp. 141–166). Universidade Federal de Vicosa, Vicosa, Brazil. Imprensa Universitaria, Universidade federal de Vicosa.

Diourte, M., Starr, J. L., Jeger, M. J., Stack, J. P., & Rosenow, D. T., (1995). Charcoal rot (*Macrophomina phaseolina*) resistance and the effects of stress on disease development in sorghum. *Pl. Pathol., 44*, 196–202.

Doi, Y., Terenaka, M., Yora, K., & Asuyama, H., (1967). Mycoplasma or PLT group-like microorganisms found in the phloem elements of plants infected with mulberry dwarf, potato witches' broom, aster yellows. *Annals of the Phytopath. Society of Japan, 33*, 259–266.

Dubey, A. K., & Singh, T., (2004). Seed-borne infection of *Alternaria alternata* and its role in disease development in sesame. *J. Mycol. Pl. Path., 34*(2), 169–171.

Dubey, R. C., & Kumar, R., (2003). Efficacy of Azadirachtin and fungicides on growth and survival of sclerotia of *Macrophomina phaseolina* causing charcoal rot in soyabean. *Indian Phytopathol., 56*, 216, 217.

Dubey, S. C., Bhavani, R., & Singh, B., (2009). Development of Pusa 5 SD for seed dressing and Pusa bio pellet 10 G for soil application formulations of *Trichoderma harzianum* and their evaluation for integrated management of dry root of mung bean (*Vigna radiate*). *Bio. Control, 50*, 231–242.

Echavez-Badel, R., & Perdomo, A., (1991). Characterization and comparative pathogenicity of two *Macrophomina phaseolina* isolates from Puerto Rico. *J. Agri. Univ. Puerto. Rico, 75*, 419–421.

Edmunds, L. K., (1964). Combined relation of plant maturity, temperature, and soil moisture to charcoal stalk rot development in grain sorghum. *Phytopathol., 54*, 514–517.

Ehteshamul-Haque, S., Abid, M., Sultana, V., Ara, J., & Ghaffar, A., (1996). Use of organic amendments on the efficacy of biocontrol agents in the control of root rot and root-knot disease complex of okra. *Nematologia Mediterranea, 24*(1), 13–16.

Elaigwu, M., Oluma, H. O. A., & Ochokwunu, D. I., (2017). *In vivo* and *in vitro* activities of some plant extracts on *M. phaseolina* (tassi) Goid. The causal agent of charcoal rot of sesame in Benau State, Nigeria. *Int. J. Sci. Res. Methodology, 5*(4), 23–39.

El-Bramawy, M. A. S., (2006a). Inheritance of resistance to *Fusarium* wilt in some crosses under field conditions. *Plant Protection Sci., 42*(2), 99–105.

El-Bramawy, M. A. S., (2006b). *Inheritance of Fusarium wilt Disease Resistance Caused by Fusarium oxysporum f sp. Sesami in Some Crosses Under Field Conditions* (Vol. 21, pp. 1–8). Sesame and Safflower Newsletters.

El-Bramway, M., (2010). Genetic analysis of yield component and disease resistance in sesame (*Sesame indicum* L.) using two progenies of diallel crosses. *Res. J. Agr., 43*, 44–56.

El-Deb, A. A., Hual, A. A., Radwan, I. A., Ali, A. A., & Mmmed, H. A., (1985). Varietal reaction and wilt diseases of sesame. *Ann. Agric. Sci. Moshtohor., 23*(2), 713–721.

Elewa, I. S., Sahab, A. F., Mostafa, M. H., & Ziedan, E. H., (2011). Direct effect of biocontrol agents on wilt and root-rot diseases of sesame. *Arch. Phytopathol. Plant Protect, 44*(5), 493–504.

Elewa, L. S., Sahab, A. F., Ziedan, EI-H. E., & Mostafa, M. H., (1994). Transplanting of sesame plants as effective method of cultivation in the control program of soil home disease. In: *5th Conf. Agric. Dev. Res.* (Vol. 1, pp. 159–171). Fac. Agric., Ain Shams Univ., Cairo, Egypt.

El-Fiki, A. I. I., Mohamed, F. G., El-Deeb, A. A., & Khalifa, M. M. A., (2004). Some applicable methods for controlling sesame charcoal rot disease (*Macrophomina phaseolina*) under greenhouse conditions. *Egypt. J. Phytopathol., 32*(1/2), 87–101.

El-Shakhess, A. M., (1998). *Inheritance of Some Economic Characters and Disease Reaction in Some Sesame (Sesamum indicum L.).* PhD Thesis, Cairo University, Cairo.

El-Shakhess, S. A. M., & Khalifa, M. M. A., (2007). Combining ability and heterosis for yield, yield components, charcoal-rot and *Fusarium* wilt diseases in sesame. *Egypt J. Plant Breed, 11*(1), 351–371.

El-Shazly, M. S., Abdul, W. O. A., El-Ashry, M. A., Ammar, S. M., & El-Bramawy, M. A., (1999). Evaluation of resistance to *Fusarium* wilt disease in sesame germplasm. *International Journal of Pest Management, 45*, 207–210.

FAOSTAT, (2016). Food and Agriculture Organization of the United Nations Statistics division. www.faostat.com food/outlook/2016 (accessed on 11 February 2021).

Gaetán, S. A., Fernandez, L., & Madia, M., (2006). Occurrence of charcoal rot caused by *Macrophomina phaseolina* on canola in Argentina. *Plant Dis., 90*(4), 524–524.

Gemawat, F. D., & Verma, O. P., (1971). Diseases of sesamum (*Sesamum indicum* L.) in Rajasthan, a note on the control of charcoal rot. *Madras Agric. J., 58*, 321–323.

Gonzalez, M. C., & Subero, M. L., (1984). *Influencia de M. phaseolina (Tassi) goid enla germinacion de le semilla y desarrollo delas plantulas deajonjoli (Sesamum indicum L.) ucv. Agronimia.* Maracay Iera. Me Morias de Trabajos de Grado (Resumenes).

Gray, F. A., Kolp, B. J., & Mohamed, M. A., (1990). A disease survey of crops grown in the bay region of Somalia, East Africa. *FAO Plant Prot. Bult., 38*, 39–47.

Gulya, T. J., Krupinsky, J., Draper, M., & Charlet, L. D., (2002). First report of charcoal rot (*Macrophomina phaseolina*) on sunflower in North and South Dakota. *Plant Dis., 86*(8), 923.

Gupta, C. P., Dubey, R. C., & Maheshwari, D. K., (2002). Plant growth enhancement and suppression of *Macrophomina phaseolina* causing charcoal rot of peanut by fluorescent *Pseudomonas. Biol. Fertil. Soils, 35*, 399–405.

Gupta, C. P., Kumar, B., Dubey, R. C., & Maheshwari, D. K., (2006). Chitinase-mediated destructive antagonistic potential of *Pseudomonas aeruginosa* GRC1 against *Sclerotinia sclerotiorum* causing stem rot of peanut. *BioControl, 51*, 821–835.

Gupta, K. N., & Ranganatha, A. R. G., (2014). Biological control for charcoal rot (*Macrophomina phaseolina*) of sesame. *Agrotechnol., 2*(4), 142.

Gupta, K. N., Naik, K. R., & Bisen, R., (2018). Status of sesame diseases and their integrated management using indigenous practices. *Int. J. Chem. Stud., 6*(2), 1945–1952.

Halsted, B. D., (1890). *Some Fungous Diseases of the Sweet Potato* (p. 76). New Jersey Agricultural Exp. Station Bull.

Hemalatha, (2006). *Studies on Alternaria Leaf Spot of Sesame (Sesamum indicum).* MSc (Ag.) Thesis submitted to Acharya N. G. Ranga Agricultural University, Hyderabad, Andhra Pradesh.

Holliday, P., & Punithalingam, E., (1970). *CMI Descriptions of Pathogenic Fungi and Bacteria No. 275.* Kew, Surrey, UK: Commonwealth Mycological Institute.

Hosseini, S. A. E., Mirazaie, A., Jafari-Nodooshan, A., & Rahiman, H., (2007). The first report of transmission of phytoplasma associated with sesame phyllody by *Orosius albicinctus. Aus. Pl. Dis. Notes, 2*, 33–34.

http://eagri.org/eagri50/PATH272/lecture12/007.html (accessed on 11 February 2021).

https://en.wikipedia.org/wiki/Powdery_mildew (accessed on 11 February 2021).

https://link.springer.com/chapter/10.1007/978-94-011-5472-7_112 (accessed on 11 February 2021).

https://vikaspedia.in/agriculture/crop-production/integrated-pest-managment/ipm-for-oilseeds/ipm-strategies-for-sesame/sesame-diseases-and-symptoms (accessed on 11 February 2021).

Ibrahim, M. E., & Abdel-Azeem, A. M., (2015). Management of sesame (*Sesamum Indicum* L.) charcoal rot caused by *Macrophomina phaseolina* (Tassi) Goid. through the application of different control measure. *J. Pure Appl. Microbiol., 9*(1), 1–9.

Ilyas, M. B., Iftikar, K., Anwar, W., & Haq, M., (1997). Effect of different neem products on the vegetative growth and sclerotial production of *Macrophomina phaseolina. Pak. J. Phytopathol., 9*, 77–79.

Jacobsen, B. J., Ziedack, N. K., & Larson, B. J., (2004). The role of *Bacillus* based biological control agents in integrated pest management system: Plant diseases. *Phytopathol., 94*, 1272–1275.

Jain, A. C., & Kulkarni, S. N., (1965). Root and stem rot of sesamum. *Indian Oilseeds J., 9*(3), 201–203.

Jha, A. K., Dubey, S. C., & Jha, D. K., (2000). Evaluation of different leaf extracts and oilcakes against *Macrophomina phaseolina* causing collar rot of okra. *J. Research, 12*(2), 225–228.

Jones, R. W., & Wang, H., (1997). Immunolocalization of a β-1,4-endoglucanase from *Macrophomina phaseolina* expressed in planta. *Can. J. Microbiol., 43*(5), 491–495.

Kashiram, S., (1930). Studies in oilseeds. The type Sesamum indicum DC. *Indian Bot. Sci., 18*, 144–146.

Kaur, S., Dhillon, G. S., Brar, S. K., & Chauhan, V. B., (2011). Carbohydrate degrading enzyme production by plant pathogenic mycelia and microsclerotia isolates of *Macrophomina phaseolina* through koji fermentation. *Indus Crop Pro., 36*, 140–148.

Kavak, H., & Boyda, K. E., (2006). Screening of the resistance levels of 26 sesame breeding lines to *Fusarium* wilt disease. *Plant Pathol. J., 5*(2), 157–160.

Kawamura, E., (1931). *New Fungi on Sesamum Indicum D.C* (Vol. 2, pp. 26–29). Fungi (Nippon fungological society).

Kazmi, S., Saleem, S., Ishrat, N., Shahzad, S., & Niaz, I., (1995). Effect of neem oil and benomyl on the growth of the root infecting fungi. *Pak. J. Bot., 27*, 217–220.

Kenerley, C. M., & Bruck, R. L., (1987). Distribution and disease progress of phytophthora root rot of Fraser fir seedlings. *Phytopathol., 77*, 520–526.

Khalifa, M. A., (2003). *Pathological Studies on Charcoal rot Disease of Sesame.* PhD Thesis, Faculty of Agriculture, Moshtohor, Zagazig University, Egypt.

Khalifa, M. M. A., (1997). *Studies on Root-Rot and Wilt Diseases of Sesame (Sesamum indicum L.)* (p. 158). MSc Thesis, Fac. Agric., Zagazig Univ.

Khamari, B., & Patra, C., (2018a). Evaluation of antifungal potency of natural products against stem and root rot of sesame. *J. Pharmacogn. Phytochem., 7*(6), 156–158.

Khamari, B., & Patra, C., (2018d). Evaluation of potentiality of different fungi toxicants against *M. phaseolina,* causing stem and root rot disease of sesame. *International Journal of Chemical Studies, 6*(6), 424–426.

Khamari, B., Bureau, S. K., & Ranasinghe, N., (2018b). Status of sesame diseases grown in different agroclimatic zones of Odisha. *Int. J. Curr. Microbiol. Appl. Sci., 7*(11), 945–948.

Khamari, B., Beura, S. K., Ranasingh, N., & Dhal, A., (2016). Symptomatological study of stem and root rot of sesame. *J. Mycopatholo. Res., 54*(3), 443–445.

Khamari, B., Satapathy, S. N., & Patra, C., (2019). Effect of inoculum load and duration of exposure to *Macrophomina phaseolina* on disease incidence of sesame. *Int. J. Chem. Stud., 7*(1), 1728–1730.

Khamari, B., Satapathy, S. N., Roy, S., & Patra, C., (2018c). Adaptability of *Macrophomina phaseolina,* the incident of stem and root rot disease of sesame to different regime of temperature, pH and light period. *Int. J. Curr. Microbiol. App. Sci., 7*(12), 761–766.

Khan, M. A., & Gangopadhyay, S., (2008). Efficacy of *Pseudomonas fluorescens* in controlling root rot of chickpea caused by *Macrophomina phaseolina. J. Mycol. Plant Pathol., 38*, 580–587.

Khare, M. N., Sharma, H. C., Kumar, S. M., & Chaurasia, R. K., (1973). Current plant pathological problems of soybean and their control. *Proceeding of the Fourth All Indian Workshop on Soybean* (pp. 55–67). GBPUAT, Pantnagar.

Kirdat, K., Thorat, V., & Yadav, A., (2018). The MLSA of group 16SrII phytoplasma associated with phyllody disease of pulses crops, weeds and insect vectors in India. In: *70th Ann. Meeting IPS, Nat. Symp. On Plant Disease Management* (p. 214).

Knox-Davies, P. S., (1966). Further studies on pycnidium production by *Macrophomina phaseoli*. *South Afri. J. Agri. Sci., 9*, 595–600.

Kolte, S. J., (1985). *Disease of Annual Edible Oilseed Crop: Rapeseed, Mustard, Safflower and Sesame Diseases* (Vol. II, pp. 90, 91, 135). CRC Press Inc, Boca Raton, Florida, USA.

Kumar, G. D., Natarajan, N., & Nakkeeran, S., (2016). Antifungal activity of nanofungicide Trifloxystrobin 25% + Tebuconazole 50% against *Macrophomina phaseolina. Afr. J. Microbiol. Res., 10*(4), 100–105.

Kumar, R. K., & Jain, S. C., (2004). *Macrophomina phaseolina* in cluster bean (*Cyamopsis tetragonoloba*) seed and its control. *J. Mycol. Pl. Pathol., 34*(3), 833–835.

Kumar, R., Pandey, M., & Chandra, R., (2011). Effect of relative humidity, temperature and fungicide on germination of conidia of *Cercospora canescens* caused the Cercospora leaf spot disease in mung bean. *Arch. Phytopath. Plant Protect, 44*, 1635–1645.

Kumari, R., Shekhawat, K. S., Gupta, R., & Khokhar, M. K., (2012). Integrated Management against root-rot of mung bean [*Vigna radiata* (L.) Wilczek] incited by *Macrophomina phaseolina. J. Plant Pathol. Microb., 3*, 136.

Kuti, J. O., Schading, R. L., Latigo, G. V., & Bradford, J. M., (1997). Differential responses of guayule (*Parthenium argentatum* Gray) genotypes to culture filtrate and toxin from *Macrophomina phaseolina* (Tassi) Goidanich. *J. Phytopathol., 145*(7), 305–311.

Kvashina, M. E. S., (1928). Preliminary report of the survey of diseases of medicinal and industrial plants in North Caucasus. Bulletin North Caucasian. *Plant Protection Stat.,* 30–46.

Leclere, V., Bechet, M., Adam, A. L., Guez, J. S., Wathelet, B., Ongena, M., Thonart, P., Gancel, F., Ghollet, I. M., & Jacques, P., (2005). Mycosubtilin over production by *B. subtilis* BBG 100 enhances the organism's antagonistic and biocontrol activities. *Appl. Environ. Microbiol., 71*, 4577–4584.

Lee, I. M., & Davis, R. E., (1992). Mycoplasmas which infect insects and plants. In: *Mycoplasmas: Molecular Biology and Pathogenesis: American Soc. Microbiol.* (p. 609).

Leppik, E. E., & Sowell, G., (1964). *Alternaria sesamia* serious seed born pathogen of worldwide distribution. *FAO Pl. Prot. Bull., 12*(1), 13–16.

Linderman, R. G., (1992). Vesicular-arbuscular mycorrhizae and soil microbial interactions. In: Bethlenfalony, G. J., (ed.), *"Mycorrhizae in Sustainable Agriculture"* (pp. 45–70, 124). ASA Special Publ. No. 54, Madison, Wisconsin.

Lodha, S., (1997). Influence of moisture conservation techniques on *Macrophomina phaseolina* population, dry root rot and yield of cluster bean. *Indian Phytopathol., 49*, 342–349.

Lokesha, N. M., & Benagi, V. I., (2004). Studies on cultural variability of isolates of *Macrophomina phaseolina* (Tassi) goid Karnataka. *J. Agric. Sci., 17*(4), 721–724.

Lokhande, N. M., Lanjewar, R. D., & Newaskar, V. B., (1998). Effect of different fungicides and neem products for control of leaf spot of groundnut. *J. Soils Crops, 8*, 44–46.

Madden, L. V., (1980). Quantification of disease progression. *Prot. Ecol., 2*, 159–176.

Mahdizadeh, V., Safaie, N., & Aghajani, M. A., (2011). New hosts of *Macrophomina phaseolina* in Iran. *J. Plant Pathol., 93*(4), 63–89.

Maiti, S., Hedge, M. R., & Chattopadhyay, C., (2008). *Hand Book of Annual Oilseed Crops* (p. 325). Oxford and IBH publishing company, New Delhi.

Manici, L. M., Caputo, F., & Cerato, C., (1995). Temperature responses of isolates of *Macrophomina phaseolina* from different climate regions of sunflower production in Italy. *Plant Dis., 79*, 834–838.

Manjunatha, N., Prameela, H. A., Rangaswami, K. T., Palanna, K. B., & Wickrama, A., (2012). Phyllody phytoplasma infecting sesame (*Sesamum indicum* L.) in South India. *Phytopath. Mollicutes., 2*(1), 29–32.

Mathur, Y. K., & Verma, J. P., (1973). Relation between date of sowing and incidence of sesamum phyllody and abundance of its vector. *Indian J. Entomol., 34*(1), 74–75.

Mayek-Pérez, N., García-Espinosa, R., LÓpez-Castañeda, C., Acosta-Gallegos, J. A., & Simpson, J., (2002). Water relations, histopathology and growth of common bean (*Phaseolus vulgaris* L.) during pathogenesis of *Macrophomina phaseolina* under drought stress. *Physiol. Mol. Plant Pathol., 60*(4), 185–195.

Mayek-Perez, N., Lopez-Castaneda, C., & Acosta-Gallegos, J. A., (1997). Variacion en caracteristicas culturales *in vitro* de aislamientos de *Macrophomina phaseolina* y su virulencia en frijol. *Agrociencia, 31,* 187–195.

Mayek-Perez, N., Lopez-Castaneda, C., Gonzalez-Chavira, M., Garch-Espinosa, R., Acosta-Gallegos, J., De La Vega, O. M., & Simpson, J., (2001). Variability of Mexican isolates of *Macrophomina phaseolina* based on pathogenesis and AFLP genotype. *Physiol. Mol. Plant Pathol., 59,* 257–264.

McCain, A. H., & Scharpf, R. F., (1989). Effect of inoculum density of *Macrophomina phaseolina* on seedling susceptibility of six conifer species. *Eur. J. Forest Pathol., 19,* 119–123.

McCoy, R. E., Caudwell, A., Chang, C. J., Chen, T. A., Dale, J. L., Sinha, R. C., Whitcomb, R. F., et al., (1989). *Plant Disease Associated with Mycoplasma Like Organisms* (Vol. 5, pp. 546–640). Academic Press.

McGibbon, T. D., (1924). *Annual Report of the Economic Botanist* (p. 5). Burma for the year ending 30th June.

Meena, P. N., Tripathi, A. N., Gotyal, B. S., & Satpathy, S., (2014). Bio-efficacy of phytoextracts and oil cakes on *Macrophomina phaseolina* (Tassi) causing stem rot disease of jute, *Corchorus spp. J. Appl. Nat. Sci., 6*(2), 530–533.

Meghvansi, M. K., Khan, M. H., & Veer, V., (2013a). *Biology of Cercospora Leaf Spot Disease.* CreateSpace (an Amazon Company), Charleston, SC, USA (ISBN13: 9781482630619).

Micke, A., Donini, B., & Maluszynski, M., (1987). Induced mutations for crop improvement: A review. *Trop. Agric., 64,* 259–278.

Mihail, J. D., & Taylor, S. J., (1995). Interpreting variability among isolates of *Macrophomina phaseolina* in pathogenicity, pycnidium production and chlorate utilization. *Can. J. Bot., 73,* 1596–1603.

Mihail, J. D., (1989). *Macrophomina phaseolina*: Spatio-temporal dynamics of inoculum and of disease in a highly susceptible crop. *Phytopathol., 79*(8), 848–855.

Mihail, J. D., Obert, M., Taylor, S. J., & Bruhn, J. N., (1994). The fractal dimension of young colonies of *Macrophomina phaseolina* produced from microsclerotia. *Mycologia., 86,* 350–356.

Miklas, P. N., Johnson, E., Stone, V., & Beaver, J. S., (1998). Inheritance and QTL analysis of field resistance to ashy stem blight in common bean. *Crop Sci., 38,* 916–921.

Misra, H. P., (2003). Efficacy of combination insecticides against phyllody, *Ann. Pl. Protec. Sci., 11*(2), 277–280.

Moazzami, A., (2006). *Sesame Seed Lignans.* PhD Thesis, Department of Food Science, SLU Acta University Agricultural Scientiae.

Mohanty, N. N., & Behera, B. C., (1958). Blight of sesame (*Sesame Orientale* L.) caused by *Alternaria sesami* (Kawamura) n. comb. *Current Science, 27,* 492–493.

Mohanty, N. N., (1958). *Cercospora* leaf spot of sesame. *Indian Phytopathol., 11,* 186, 187.

Moradia, A. M., (2011). Effect of inoculum levels of *Macrophomina phaseolina* on groundnut causing dry root rot. *Int. J. Plant Prot., 4,* 199, 200.

Mostafa, M. H., Sahab, A. F., Elewa, I. S., & Ziedan, E. H., (2003). Sesame transplanting strategies for controlling soil-borne diseases. *Egypt. J. Agric. Res., 1*(2), 387–401.

Muheet, A., & Chauhan, L. S., (1975). Control of phyllody of sesamum *(Sesamum Orientale* L.), *Madras Agric. J., 62*(4), 219–220.

Murthy, G. S. S., Joshua, D. C., Rao, N. S., & Bhatia, C. R., (1985). *Induced Mutation in Sesame and Sunflower: Status and Potential, Plant Production and Protection* (pp. 188–190). Food and Agriculture Organization.

Murthy, N. B. K., & Amonkar, S. V., (1973). Effect of natural insecticide from garlic (*Allium sativum* L.) and its synthetic form (diallyl disulphide) on plant pathogenic fungi, *Indian J. Exp. Biol., 12,* 208–209.

Murugesan, M., Shanmugam, M. M., Menon, P. P., Arokiaraj, A., Dhamu, K. P., & Kochubabu, M., (1978). Statistical assessment of yield loss of sesamum due to insect pests and diseases. *Madras Agric. J., 65,* 290–295.

Nagamani, P., Viswanath, K., & Kiran, B. T., (2014). Management of dry root rot caused by *Rhizoctonia bataticola* (Taub.) Butler in chickpea. *Curr. Biotica., 5*(3), 364–369.

Nayyar, B. G., Akram, A., Arshad, M., Akhund, S., & Rafiq, M., (2014). Seed viability test and pathogenicity assessment of most prevalent fungi infecting *Sesamum indicum* L. *IOSR. J. Pharm. Biol. Sci., 9*(5), 21–23.

Neergaard, P., (1979). *Seed Pathology* (Vol. I & II, p. 739). London: Macmillan.

Odvody, G. N., & Dunkle, L. D., (1979). Charcoal stalk rot of sorghum: Effect of environment on host-parasite relations. *Phytopathol., 69,* 250–254.

Ojiambo, P. S., Narla, R. D., Ayiecho, P. O., & Nyabundi, J. O., (1998). Effect of infection level of sesame (*Sesamum indicum* L.) seed by *Alternaria sesame* on severity of *Alternaria* leaf spot. *Trop. Agr. Res. Ext., 1*(2), 125–130.

Omidi, M., Pour, H., & Massumi, R., (2010). Investigations on transmittance status of *Orosius albicinctus* (*Hemiptera*: *Cicadellidae*) as a natural vector of phytoplasmas in south-eastern Iran. *J. Pl. Path., 92*(2), 531–535.

Pathirana, (1992). Gamma ray-induced field tolerance to *Phytophthora* blight in sesame. *Plant Breed, 108,* 314–319.

Pearson, C. A. S., Schwenk, F. W., Crowe, F. J., & Kelley, K., (1984). Colonization of soybean roots by *Macrophomina phaseolina*. *Plant Dis., 68,* 1086–1088.

Phillips, J. M., & Hayman, D. S., (1970). Improved procedures clearing roots and staining parasitic and VA mycorrhizal fungi for rapid assessment of infection. *Trans. Brit. Mycol. Soc., 55,* 158–161.

Pracros, P., Renaudin, J., Eveillard, S., Mouras, A., & Hernould, M., (2006). Tomato flower abnormalities induced by stolbur phytoplasma infection are associated with changes of expression of floral development genes. *Mol. Plant-Microbe Interact., 19*(1), 62–68.

Pramod, K., & Mishra, U. S., (1992). Diseases of *Sesamum indicum* in Rohilkhand: Intensity and yield loss. *Indian Phytopath., 45,* 121, 122.

Pun, K. B., Sabitha, D., & Valluvaparidasan, V., (1998). Studies on seed-borne nature of *Macrophomina phaseolina* in okra. *Plant Dis. Res., 13,* 249–290.

Purohit, S. D., & Arya, H. C., (1980). Phyllody: An alarming problem for sesamum growers. *Agril. Digest., 4*(5), 12.

Raikwar, R. S., & Srivastava, P., (2013). Productivity enhancement of sesame (*Sesamum indicum* L.) through improved production technologies. *Afr. J. Agril. Res., 8*(47), 6071–6078.

Rajpurohit, T. S., & Nema, S., (2013). Studies efficacy of soil amendments with neem cake and biocontrol agent on the incidence of *Macrophomina* stem and root rot of sesame (*Sesamum indicum* L.). *J. Oilseeds Res., 29*(2), 178–179.

Rajpurohit, T. S., (1999). *Trichoderma viride*, Bio-control agent effective against *Macrophomina* stem and root rot of sesame. In: *4th Agricultural Science Congress* (p. 297). Jaipur, India, Abstract.

Rajpurohit, T. S., (2008). Studies on the efficacy of soil amendments with oil cakes on the incidence of *Macrophomina* stem and root rot of sesame. *Geobios, 35*, 225, 226.

Ramadoss, S., & Sivaparakasam, K., (1994). Effect of seed treatment with fungicides and insecticides on the control of root rot and stem fly on cowpea. *Madras Agric. J., 80*, 618–620.

Ramezani, H., (2008). Biological control of root-rot of eggplant caused by *Macrophomina phaseolina*. *American-Eurasian J. Agric. Environ. Sci., 4*(2), 218–220.

Ranganatha, A. R. G., Lokesha, R., Tripnthi, A., Asafa, T., Paroha, S., & Srivastava, M. K., (2012). Sesame improvement-present status and future strategies. *J. Oilseeds Res., 29*(1), 1–26.

Reznikov, S., Vellicce, G. R., González, V., De Lisi, V., Castagnaro, A. P., & Ploper, L. D., (2016). Evaluation of chemical and biological seed treatments to control charcoal rot of soybean. *J. Gen. Plant Pathol., 82*, 273–280.

Richardson, M. J., (1979). *An Annotated List of Seed Borne Diseases* (p. 320). C.M.I. Knew, England.

Rodriguez, M., & Zammlw, C., (1985). Studies on relationships between *Macrophanina phaseoli* (Maubl.) (Ashby) and sesame (*Sesamum indicum* L.) in Venezuela. *Sesame and Safflower Newsletter, 1*, 36.

Saccardo, P. A., (1876). Fungi veneti novi vel citici. *Ser. V. Nuovo Giorn. Bot. Ital., 8*, 161.

Sahab, A. F., Elewa, I. S., Mostafa, M. H., & Ziedan, E. H., (2001). Integrated control of wilt and root-rot diseases of sesame in Egypt. *Egypt. J. Appl. Sci., 16*(7), 448–462.

Sahambi, H. S., (1970). Studies on sesamum phyllody virus, virus-vector relationship and host range. *Pro. First International Symposium Plant Pathology* (pp. 340–351). New Delhi.

Saleh, A. A., Ahmed, H. U., Todd, T. C., Travers, S. E., Zeller, K. A., Leslie, J. F., & Garrett, K. A., (2010). Relatedness of *Macrophomina phaseolina* isolates from tallgrass prairie, maize, soybean and sorghum. *Mol. Eco., 19*, 79–91.

Salehi, M., & Izadpanah, K., (1992). Etiology and transmission of sesame phyllody. *Iran. J. Phytopathol., 135*, 37–47.

Salib, M., Bayliss, B., Dell, B., Hardy, G. E., & Jones, M. G. K., (2005). First report of phytoplasma associated disease of chickpea (*Cicer arietinum*) in Australia. *Australian Pl. Path., 34*, 425, 426.

Samuel, S. G., Govindaswamy, C. V., & Vidhyasekaran, P., (1972). Studies on *Alternaria* blight disease of gingelly. *Madras Agric. J.*, 882–886.

Sandipan, P. B., (2014). Bioefficacy of water-based phytoextracts against dry rot and charcoal rot pathogens of potato caused by *Fusarium* sp. and *Macrophomina Phaseolina* under *in vitro* test condition. *Int. J. Life Sci. Bt. Pharm. Res., 3*(4), 69–73.

Sarwar, G., & Haq, M. A., (2006). Evaluation of sesame germplasm for genetic parameters and disease resistance. *J. Agric. Res., 44*(2), 89–95.

Sastry, K. S., & Rao, R., (1989). *An appraisal of Mycoplasma Diseases of Oil Seed Crops: Problems and Prospects* (pp. 84–94). Directorate of Oilseeds Research, Rajendra Nagar, Hyderabad.

Savaliya, V. A., Bhaliya, C. M., Marviyaand, P. B., & Akbari, L. F., (2015). Evaluation of phytoextracts against *Macrophomina phaseolina* (Tassi) Goid causing root rot of sesame. *J. Bio pest., 8*(2), 116–119.

Savaliya, V. V., Bhaliya, C. M., Akbari, L. F., Amipara, J. D., & Marviya, P. B., (2016). Biological Control of *Macrophomina phaseolina* causing Sesame (*Sesamum indicum*) root rot. *The Bioscan, 11*(2), 769–772.

Savitha, A. S., Naik, M. K., & Ajitkumar, K., (2011). Ecofriendly management of *A. sesame*, inciting blight of sesame. *J. Plant Dis. Sci., 6*(2), 150–152.

Saxena, J., & Mathela, C. S., (1996). Antifungal activity of new compounds from *Nepeta leucophylla* and *Nepeta clarkei*. *Appl. Environ. Microbiol., 62*(2), 702–704.

Senjaliya, B. D., & Nathawat, B. D. S., (2015). Evaluation of bioagents, phyto-extracts and organic amendments on stem rot of groundnut. *Ann. Pl. Protec. Sci., 23*, 98–101.

Shane, W. W., & Teng, P. S., (1992). Impact of Cercospora leaf spot on root weight, sugar yield and purity of *Beta vulgaris*. *Plant Dis., 76*, 812–820.

Sharma, B. K., & Basandrai, A. K., (1997). Effect of biocontrol agents, fungicides and plant extracts on sclerotial viability of *Sclerotinia sclerotiorum*. *Indian J. Agric. Sci., 67*, 132, 133.

Short, G. E., & Wyllie, T. D., (1978). Inoculum potential of *Macrophomina phaseolina*. *Phytopathol., 68*, 742–746.

Short, G. E., Wyllie, T. D., & Bristow, P. R., (1980). Survival of *Macrophomina phaseolina* in soil and residue of soybean. *Phytopathol., 70*, 13–17.

Shukla, B. N., & Singh, B. P., (1974). Effect of fungicidal seed treatment on *Macrophamina* root rot of sesame (*Sesamum indicum*). *Indian J. Mycol. and Plant Path., 3*(2), 208, 209.

Siddaramaiah, A. L., & Hegade, R. K., (1984). Occurrence of *Alternaria alternata* (Kissler) on safflower and sesamum. India. *J. Oilseeds Res., 1*, 83, 84.

Singh, D., (1974). *Location of Alternaria Sesame in Seed of Sesame and its Transmission* (pp. 1–22). A report submitted by the candidate for the candidate for the examination of seed pathology held by the Danish government. Institute of Seed Pathology for Developing Countries Copenhagen, Denmark.

Singh, G., & Verma, R. K., (2015). Compatibility of bioagents and neem products against root rot of soybean. *J. Mycopatho. Res., 43*(2), 211–214.

Singh, S. K., Nene, Y. L., & Reddy, M. V., (1990). Influence of cropping systems on *Macrophomina phaseolina* populations in soil. *Plant Dis., 74*, 812–814.

Sinha, O. K., & Khare, M. N., (1977). Control of seed-borne *Macrophomina phaseolina* and *Fusarium equisiti* of cowpea seeds. *Seed Res., 5*, 20–22.

Smith, G. S., & Carvil, O. N., (1997). Field screening of commercial and experimental soybean cultivars for their reaction to *Macrophomina phaseolina*. *Plant Dis., 81*, 363–368.

Songa, W., & Hillocks, R. J., (1996). Legume hosts of *Macrophomina phaseolina* in Kenya and effect of crop species on soil inoculums level. *J. Phythopathol., 144*, 387–391.

Steinkamp, M. P., Martin, S. S., Hoefert, L. L., & Ruppel, E. G., (1979). Ultrastructure of lesions produced by *Cercospora betlicola* in leaves of *Beta vulgaris*. *Physiol Plant Pathol., 15*, 13–26.

Su, G., Suh, S. O., Schneider, R. W., & Russin, J. S., (2001). Host specialization in the charcoal rot fungus *Macrophomina phaseolina*. *Phytopathol., 91*, 120–126.

Sudhakar, & Rangaswamy, S., (1989). Sesamum biotechnology: Embryo culture in *indicum*. *Oil Crops News Letter., 6*, 48–50.

Sundararaman, S., (1931). *Administration Report of the Mycologist for the Year, 30,* 1929, 1930.

Sundaravadana, S., (2002). *Management of Black Gram (Vigna mungo (L.) Hepper) Root Rot Macrophomina phaseolina (Tassi) Goid with Bioagents and Nutrients.* MSc (Ag.) Thesis, Tamil Nadu Agriculture University, Coimbatore.

Sutton, B., (1980). *The Coelomycetes: Fungi Imperfecti with Pycnidia Acervuli and Stromata.* Kew, England: Commonw. Mycol. Inst. Assoc. Appl. Biol.

Tamimi, K. M., Fattah, F. A., & Al-Hamdani, M. A., (1989). Shoot apex fasciation in Sesamum indicum associated with mycoplasma like organisms. *Pl. Path., 38*, 300–304.

Tandel, D. H., Sabalpara, A. N., & Pandya, J. R., (2010). Efficacy of phytoextracts on *Macrophomina phaseolina* (Tassi) Goid causing leaf blight of green gram. *Int. J. Parma. Biosci., 2*, 1–5.

Tandon, I. N., & Banerjee, A. K., (1988). Control of phyllody and leaf curl of *Sesamum Orientale. Pl. Dis. Rep., 52*(5), 367–369.

Taware, M. R., Gholve, V. M., & Dey, U., (2014). Bio-efficacy of fungicides, bioagents and plant extracts/botanicals against *Alternaria carthami*, the causal agent of *Alternaria* blight of safflower (*Carthamus tinctorius* L.). *Afr. J. Microbiol. Res., 8*(13), 1400–1412.

Tessema, T., Hoppe, B., Janke, J., Henniger, T., Gossmann, M., Bargen, S. V., & Buttner, C., (2010). Parthenium weed (*Parthenium hysterrophorus* L.) research in Ethiopia: Investigation of pathogens as biocontrol as biological control. *Ethiop. J. Agric. Sci., 20*, 107–127.

Thomas, C. A., (1959). Control of pre-emergence damping-off and two leaf spot diseases of sesame by seed treatment. *Phytopathol., 49*, 461.

Uma, M. C., Ramakrishanan, G., & Nallthambi, P., (2001). Role of inoculum level on diseases incidence of dry root rot caused by *Macrophomina phaseolina* in groundnut. *Madras Agric. J., 87*, 71–73.

Vasudeva, R. S., & Sahambi, H. S., (1955). Phyllody in sesamum (*Sesamum Orientale* L.), *Indian Phytopathol., 8*, 124–129.

Verma, M. L., (2002). *Fungal and Bacterial Diseases of Sesame and Their Management Challenges for the Millennium* (pp. 161–192). Jyothi Publication, New Delhi.

Vyas, S. C., & Patel, M. C., (2015). Integrated biological and chemical control of dry rot of chickpea. *Indian J. Mycol. Pl. Pathol., 24*(2), 132–134.

Vyas, S. C., (1981). Diseases of sesame and Niger in India and their control. *Pesticides, 15*, 10–15.

Wang, M., Farnham, M. W., & Thomas, C. E., (2001). Inheritance of true leaf stage downy mildew resistance in broccoli. *J. Amer. Soc. Hort. Sci., 126*(6), 727–729.

Weiss, E. A., (1971). *Castor, Sesame and Safflower, Leonard Hill Books* (pp. 311–355). London.

Weiss, E. A., (1983). *Sesame. Oilseed Crops* (pp. 282–340). London, Longman.

Wrather, J. A., & Koenning, S. R., (2006). Estimates of disease effects on soybean yields in the United States 2003–2005. *J. Nematol., 38*, 173–180.

Wyllie, T. D., (1988). Charcoal rot of soybean-current status. In: Wyllie, T. D., & Scott, D. H., (eds.), *Soybean Diseases of the North Central Region* (pp. 106–113). St. Paul, MN: APS Press.

Yang, X. B., & Navi, S. S., (2005). First report of charcoal rot epidemics caused by *Macrophomina phaseolina* in soybean in Iowa. *Plant Dis., 89*(5), 526.

Yousif, H. Y., Bingham, F. T., & Yermason, D. M., (1972). Growth, mineral composition, and seed oil of sesame (*Sesamum indicum* L.) as affected by NaCl. *Soil Sci. Soc. Am. Proc., 36,* 450–453.

Yu, S. H., Mathur, S. B., & Neergaard, P., (1981). Taxonomy and pathogenicity of four seed-borne *Alternaria* from sesame (*S. indicum*L.). *Trans. British Mycol. Soc., 78,* 447–458.

Yu, S., & Park, J. S., (1980). *M. phaseolina* detected in seeds of *Sesamum indicum* and its pathogenicity. *Korean J. Plant Prot., 19,* 135–140.

Ziedan, E. H. E., (1998). *Integrated Control of Wilt and Root-Rot Diseases of Sesame in A. R. E* (p. 169). PhD thesis, Faculty of Agric. Ain Shams Univ.

Ziedan, E. H., Elewa, I. S., Mostafa, M. H., & Sahab, A. F., (2010). Application of mycorrhizae for controlling root diseases of sesame. *First International Congress MCOMED* (p. 97). Morocco.

Ziedan, E. H., Elewa, I. S., Mostafa, M. H., & Sahab, A. F., (2011). Application of mycorrhizae for controlling root diseases of sesame. *J. Pl. Prot. Res., 51*(4), 355–361.

CHAPTER 7

Approaches for Diagnosis and Management of Banded Blight in Small Millets

A. K. JAIN,[1] S. K. TRIPATHI,[1] and R. P. JOSHI[2]

[1]Department of Plant Pathology, JNKVV, College of Agriculture,
Rewa – 486001, Madhya Pradesh, India,
E-mail: akjagcrewa@gmail.com (A. K. Jain)

[2]Department of Plant Breeding, JNKVV, College of Agriculture,
Rewa – 486001, Madhya Pradesh, India

ABSTRACT

Small millets comprising finger millet, little millet, Kodo millet, foxtail millet, barnyard millet, proso millet and brown top millet is important region-specific cereal crops of the semi-arid, tribal, and hill areas of the world. These are versatile, low water demanding, climate-resilient crops and adaptable for different cropping patterns. In India, small millets are cultivated in 1.64 m ha with a a production of 1.83 m tons and an average yield of 1038.5 kg ha^{-1} (2016–2017). These crops provide household food and nutritional security in resource-poor farmers. Small millets, now known as *nutricereals* are rich sources of dietary fiber, phytochemicals, micronutrients, and known to reduce diabetic Mellitus, celiac, and cardiovascular diseases. Banded blight caused by *Rhizoctonia solani* Kuhn [Basidial stage; *Thanatephorus cucumeris* (Fr.) Donk] is becoming a serious threat for a sustainable yield of small millets since last few years. The causal pathogen is a polyphagous, necrotrophic fungus, primarily soil-borne in nature, wide host range, and genetically diverse showing variability in cultural, morphological, and physiological characters. Small millets are attacked by the pathogen at all the stages of crop growth and symptoms appeared as characteristic banded blight pattern in the sheath and leaves. Significant yield loss was reported in small millets due to banded blight. A temperature of around 28–30C and relative humidity of 70% or above favors the rapid disease development. Most of the isolates of *R. solani* collected from small millets are reported cross infective with each other. The dearth of information is available on various aspects of banded blight, though few studies have been made to identify resistant sources, morphological characteristics of the pathogen, disease progression, host range, and management of the disease through botanicals, biocontrol agents, and chemicals in different small millets. In this chapter, a critical review on symptomatology, etiology, host plant resistance, and management of banded blight in small millets has been carried out.

7.1   INTRODUCTION

Small millets are small-seeded nutritious cereals generally grown by poor farmers in low fertile lands with fewer inputs under rainfed conditions for their own consumption as food and feed. Nowadays, these crops are suffering due to biotic stresses in changing climatic conditions. Among them, banded blight incited by *Rhizoctonia solani* (basidial stage; *Thanatephorus cucumeris*) is one of the emerging maladies in the successful cultivation of small millets. The widespread adoption of new susceptible high yielding varieties with a large number of tillers and changes in cultivation practices in small millets favor the development of banded blight. The pathogen overwinters as soil-borne sclerotia and mycelium in plant debris. Das and Girija (1989), for the first time reported sheath blight of finger millet from Vellayani in Kerala in a severe form and thought to be the first report of *Rhizoctonia solani* on finger millet from India. Later on the disease was observed in severe form at Vizianagaram, Ranchi, Uttarakhand, Karnataka, and Madhya Pradesh (Nagaraja and Anjaneya Reddy, 2010; Nagaraja et al., 2016; Kumar, 2019). The disease in proso millet was first time reported from Vellayani, Kerala (Rajgopalan et al., 1992) and observed that the disease initiated in the tillering stage and continued in the succeeding stages of crop growth. Further widespread occurrence of sheath blight in proso millet was reported from mid-hills of Uttarakhand by Kumar and Dinesh (2010) and Madhya Pradesh (Jain and Tiwari, 2013). Banded blight was also reported in Kodo millet, little millet, barnyard millet and foxtail millet from different small millet growing states of the country (Dubey et al., 1989; Kumar and Dinesh, 2009; Jain and Gupta, 2010; Kumar, 2016; Nagaraja et al., 2016; Kahar, 2017; Jain et al., 2018). In a roving field survey conducted at farmers field during 2013–2014 to 2016–2017 in five districts of Madhya Pradesh, the mean disease incidence of banded leaf and sheath blight and frequency of occurrence ranging from 6.9 to 18.8% and 26.1 to 75.4%, respectively were recorded by Jain et al. (2018). The study indicates that the disease is becoming a major problem for little millet growers.

7.2 LOSSES

Scanty information is available in the literature about yield losses due to *Rhizoctonia solani* in small millets. However, Chouhan (2014) reported average reduction in agro-morphological characters like plant height (14.7%), productive tillers per plant (15.7%), panicle length (14.2%), panicle weight (12.8%), grain yield per plant (17.8%), 1000 grain weight (11.3%) and fodder yield per plant (19.4%) on susceptible genotypes of little millet due to incidence of *R. solani*. Significant differences were also recorded in seed germination, shoot length, root length and seedling vigor index of seeds collected from healthy and *R. solani* infected plants of little millet. In Kodo millet, average reduction in plant height, tillers per plant, panicles per plant, length of panicle, panicle weight, grain yield per plant, 1000-grain weight, and fodder yield per plant was recorded 21.2., 32.1, 28.9, 17.5, 27.2, 30.5, 13.1, and 36.4%, respectively. Retarding effect on seed germination, shoot length, root length and seedling vigor index was 22.4%, 13.9%, 17.8% and 42.7%, respectively was recorded on seeds collected from banded leaf and sheath blight infected plants (Kahar, 2017).

7.3 SYMPTOMS

The symptoms produced by *Rhizoctonia solani* in small millets may vary on the different crops and even on the same host, depending on the stage at which the plant becomes infected and the prevailing environmental conditions. The most common symptoms caused by the *R. solani* are banded leaf and sheath blight, stem rot, foliage blights and spots. Dubey (1995) reported that the symptom of banded sheath blight on finger millet is characterized by oval to irregular, light gray to dark brown lesions on the lower leaf sheath. Its central portion subsequently turns white to straw with narrow reddish-brown border and gives a characteristic banded irregular appearance, due to which the disease has been named as banded blight (Figure 7.1). Nagaraja and Anjaneya Reddy (2010) observed the regularly distributed irregular spots on leaf blades. Under favorable conditions, the lesions enlarged rapidly and coalesced to cover large portions of the sheath and leaf lamina. On peduncles, fingers, and glumes irregular to oval dark brown to purplish necrotic lesions are formed. The pathogen isolated from different diseased parts of the plant yielded *R. solani*.

Jain and Gupta (2010) reported that the symptom of banded sheath blight on foxtail and barnyard millet begins with the development of irregular round, water-soaked, straw-colored lesions on leaf base and leaf sheaths. Within 12–15 days almost all sheaths below the ear were found infected and girdle, culms giving a characteristic banded appearance. Patro and Madhuri (2013) observed a series of copper or brown color bands across the leaves giving a very characteristic banded appearance in foxtail millet. Mycelial growth along with white to brown sclerotia can be observed on and around the lesions. Later on, the leaves dry up and plants appear blighted. Dwarf and early maturing landraces of foxtail millet were found susceptible to banded leaf and sheath blight as compared to tall and late-maturing landraces (Jain et al., 2014b). Kumar (2016) recorded the classic blight symptoms on leaves of barnyard millet. The new symptom was novel, manifested itself as oval to irregular and light gray to dull brown lesion on the leaves. Earlier, the symptoms of the disease were reported only on the leaf sheaths, but later the symptoms are recorded on leaves. According to Patro et al. (2018a) the symptom of banded sheath blight on proso millet is characterized by a series of copper or brown colored bands across the leaves, giving a characteristic banded appearance. The mycelia growth along with white to brown sclerotia can be observed on and around the lesions. Later on, the leaves dry up and plants become blighted.

FIGURE 7.1   Banded blight: (left to right) Kodo millet, little millet, barnyard millet, foxtail millet, finger millet and proso millet.

7.4   PATHOGEN

In nature, *R. solani* exist as different strains that exhibit variation in their morphological and cultural characters, pathogenicity, and virulence pattern.

7.4.1   *MORPHOLOGICAL CHARACTERS*

Morphological characters namely hyphal width, distance between two septa and mycelia weight were studied by Krishnaveni (2019) in 18 foxtail millet, 13 finger millet, 12 proso millet and 10 little millet isolates of *R. solani*. Hyphal width ranged 3.47 to 11.04 μm in foxtail millet, 5.17 to 10.16 μm in finger millet, 2.74 to 10.76 μm in proso millet and 2.51 to 10.82 μm in little millet isolates. The isolates were categorized into narrow, moderate to wide and wide on the basis of hyphal width. Majority of the isolates belongs to moderate to wide group (5–10 μm). Four isolates of foxtail millet, 6 isolates of proso millet and 3 isolates of little millet have narrow hyphal width, whereas 4 isolates of foxtail millet, one of finger millet, one of proso millet and two of little millet were grouped under wide(>10 μm) category. The distance between two septa varied from 19.31 μm to 541.85 μm and the maximum distance between the septation was observed in the isolate foxtail millet (FOX-13) and minimum distance was recorded in proso millet isolate (PM-11). On the basis of septal distance, the isolates were categorized into four groups viz., short, short to medium, medium to long and long. Out of 53 isolates, the distance between two septa was short (<50 μm) in 6 isolates and short to medium (50–100 μm) in 12 isolates. Eighteen isolates were categorized in medium to long group (100–200 μm) while the remaining 17 isolates were found to have very long septal distance (>200 μm). On the basis of the dry weight of mycelium, the isolates were categorized into three groups. Sixteen isolates had mycelial weight in the range of 0–0.10 g (Group-1), 24 isolates in the range of 0.10–0.20 g (Group-2) and 13 isolates had more than 0.20 g (Group-3). Majority of the isolates were in Group-2. Among all the isolates, foxtail millet isolate FM-12 (0.34 g) showed maximum dry weight of mycelium followed by FM-05 (0.31 g) whereas minimum was obtained in proso millet isolate PM-01 (0.02 g).

7.4.2   *CULTURAL CHARACTERS*

The cultural characteristics studies showed significant variation among the isolates of *R. solani* collected from different small millets. Chouhan (2014) recorded maximum mycelial growth in little millet isolate of *R. solani* in potato dextrose agar (PDA) followed by Czapek's dox agar (CDA) and Richards medium. Cottony regular and fast-growing whitish colony without pigmentation was recorded in PDA. Jain et al. (2017a) studied the cultural characteristics of *R. solani* from little millet in four culture media viz. PDA, CDA, Richards agar and corn meal agar (CMA) media. Cottony regular and fast-growing whitish colony without pigmentation was recorded in PDA. Whereas, cottony regular creamy colony with light pink pigmentation was recorded in CDA. Pinkish white cottony regular colony with dark pink pigmentation and moderate growth was recorded in Richards medium. Fungal colony was transparent, submerged, slow, and without pigmentation in CMA. Kahar (2017) observed regular slow-growing whitish regular colony of *R. solani* without pigmentation isolated from Kodo millet on PDA. The characteristic non-sporulating hyphae with 90° branches and distinct constriction at branch points were observed. Krishnaveni (2019) studied the cultural characters of foxtail millet, finger millet, proso millet and little millet isolates of *R. solani* and significant variation observed among the isolates. The colony diameter varied from 41.33–90 mm at 72 hrs. of incubation, and most of the isolates of foxtail millet and finger millet were slow-growing, whereas proso millet and little millet isolates were fast and medium growing. Majority of the isolates of foxtail millet and finger millet were light brown white proso millet and little millet isolates showed both light brown and brown color. Concentric circle is present in most of the isolates of foxtail millet and finger millet and it was completely absent in proso millet and little millet isolates. Texture of the isolates varied from flat plain to fluffy. Dispersion of mycelium varied from subdued to aerial growth in which most of the isolates exhibited subdued growth with flat plain texture while only finger millet isolates showed strandy and subdued growth with slightly fluffy texture. Prajapati (2019) recorded maximum radial growth in finger millet isolates of *R. solani* (91.3 mm) on PDA which was significantly at par in CDA (85.8 mm) medium. Least growth (71.4 mm) was observed in finger millet seed extract agar and Richards agar. The excellent fast abundant and white mycelial growth of *R. solani* was recorded on PDA. Whereas fast abundant and pale brown growth was observed on CDA.

7.4.3 SCLEROTIAL CHARACTERS

Krishnaveni (2019) studied the sclerotial characteristics in foxtail millet, finger millet, proso millet and little millet isolates of *R. solani*. A number of sclerotia per plate varied 11–138, in which foxtail millet isolate produced more number of sclerotia with minimum weight and size. The finger millet isolate produced a lesser number of sclerotia with maximum weight and size, whereas proso millet and little millet isolate showed the lesser number of sclerotia with minimum size and weight. Majority of the isolates have brown colored sclerotia with round shape and smooth surface, while most of the isolates of proso millet and little millet showed dark brown color sclerotia. Time taken for initiation of sclerotia varies from 4 to 8 days. Foxtail millet and finger millet isolates showed greater variation with different patterns of arrangement viz., peripheral, central, central + peripheral, concentric circle or uniformly or scattered distribution. Prajapati (2019) studied the sclerotial characters of finger millet isolate of *R. solani* and found variation in initiation of sclerotia formation from 2 to 3 days in different culture media. Maximum number of sclerotia was produced in oatmeal agar (57.3) closely followed by PDA (55.0) and czapeck dox agar (48.7). Maximum weight of 10 sclerotia was recorded in PDA (1040.7 mg) followed by czapeks dox agar (925.7 mg) and oatmeal agar (899.7 mg). Maximum size of sclerotia was formed in PDA followed by finger millet seed extract agar (1.0 mm) and CMA (0.9 mm).

7.4.4 PATHOGENIC VARIATION

Kumar (2018) recorded significant linear increase in relative lesion height (RLH %) of banded leaf and sheath blight (BLSB) at crop age of 20, 30, 40, and 50 days in Kodo millet, but progression and development of BLSB was maximum in 20 days and 30 days old crops. Lower values of area under disease progress curve (AUDPC) were recorded in resistant and moderately resistant (MR) cultivars, where as higher values were recorded in susceptible cultivars. Significant positive correlation of RLH with total rainy days ($r = 0.973*$) and minimum temperature ($r = 0.975*$) was found. Total rainfall with rainy days ($r = 0.967*$) and minimum temperature ($r = 0.960*$) were found positively correlated. A significant positive correlation between rainy days and minimum temperature ($r = 0.997*$) was also observed. Patro and Madhuri (2014) reported that a temperature around 28–30°C and a RH of >70% favors rapid development of banded blight in finger millet. Nagaraja et al. (2016) reported that temperature range from 23–30°C and a relative humidity of 80% or above favors rapid disease development in finger millet, where these enlarge rapidly and coalesce to cover large proportions of the sheath and leaf lamina. Krishnaveni (2019) studies on pathogenic variations among isolates of *R. solani* and the result revealed that variation observed in accordance with the latent period, lesion size and lesion shape. Latent period of all *R. solani* isolates varied from 3 to 8 days. Size of the lesion varied from 0.8 to 5.77 cm. In respect to lesion shape, most of the isolates produced either elliptical or elongated lesions while 4 isolates produced spherical/circular lesions and 8 isolates produced irregular lesions. Cross inoculation studies revealed that most of the isolates of *R. solani* were cross-infective with each other and all the proso millet isolates were more virulent on foxtail millet host. Prajapati (2019) reported that the development and spread of banded leaf and sheath blight was more in 40 and 50 days old crop of finger millet as compared to 60 and 70 days old crops. Number of lesions and vertical spread of disease were decreased as the crop approaches maturity. Significant variation in number of lesions, length of lesion, RLH, incubation period, apparent infection rate, and values of AUDPC was recorded in different genotypes of finger millet.

7.5   HOST RANGE

The pathogen has a wide host range and the intensity of symptom expression varied in the host plant. Sheath blight of rice causing pathogen was reported to infect the finger millet, little millet, Kodo millet, foxtail millet, barnyard millet, and proso millet (Kannaiyan and Prasad, 1978, 1981; Meena and Muthuswamy, 1998; Goswami et al., 2010; Lenka et al., 2014). Maize isolate of *R. solani* f. sp. *sasakii* was reported to infect finger millet, foxtail millet (Rathore et al., 1998) and browntop millet (Trivedi and Rathore, 2006). Isolates of foxtail millet, finger millet, proso millet, and little millet were reported to successfully infect the rice, sorghum, bajra, and soybean, but their virulence on different crops were quite different. Slow growing isolates of foxtail millet, finger millet, and little millet were more aggressive than medium and fast-growing isolates. Most of the isolates of *R. solani* were cross-infective with each other, but all proso millet isolates were more virulent on foxtail millet host (Krishnaveni, 2019).

7.6   MANAGEMENT

Management of disease by host resistance, biological, and chemical procedures is extremely important to minimize the crop losses.

7.6.1 HOST RESISTANCE

Management of *R. solani* is difficult due to its ecological behavior, broad host range, and high survival rate of sclerotia under different environmental conditions. The use of host resistance is the cheapest and feasible option for the management of any plant disease. A number of studies were carried out to screen different small millets cultivars against *R. solani* under sick plots and artificial inoculation conditions to identify resistant sources. Patro et al. (2017a, 2018f) evaluated 10 and 11 Kodo millet varieties against *R. solani*. The incidence of banded blight was recorded ranging from 78.0 to 98.7% during 2017 and 53.3 to 92.0% during 2018. None of the evaluated Kodo millet varieties were found resistant or MR to *R. solani*. Ten and 23 barnyard millet genotypes were evaluated by Patro et al. (2017b, 2018i) and none of the genotypes were found resistant of MR to banded blight. Divya et al. (2017a) evaluated early and medium maturing varieties of finger millet and none was found resistant to sheath blight. None of the evaluated entries of little millet were observed resistant to banded blight (Divya et al., 2017b; Patro et al., 2016b). Similarly, Patro et al. (2017c, 2018e) evaluated 8 and 11 proso millet genotypes against *R. solani* and none were found resistant or MR to banded blight. Few resistant sources of small millets were identified against *R. solani* and are listed in Table 7.1.

TABLE 7.1 Resistant Sources of Small Millets Against *Rhizoctonia solani* Causing Banded Blight Disease

Crop	Resistant Sources	References
Finger millet	VL 352, VL 384, GPU 91, GPU 92, VR 708, TNEC 1234, GPU 45	Jain and Joshi (2018)
	VR 1101, KMR 650, GPU 99	Prajapati (2019)
Little millet	BL 4, KRI 10-03, RLM 186, OLM 203 and DLM 104	Chouhan (2014)
	RLM 103, 130, 160, 171, 181, 182, 218, 219	Jain et al. (2014a)
	IIMRLM 8437-17, RLM 208	Patro et al. (2019)
Kodo millet	TNAU 176, TNAU 178	Kumar (2018)
	KOPN 21, KOPN 8, BK 5 and JK 137	Kahar (2017)
	RK 390-25, DPS 118	Patro et al. (2016a)
Foxtail millet	TNAU 219, TNAU 248, SiA 2757, SiA 326, SiA 2723, SiA 3036, SiA 3085, GPUS 30	Jain and Gupta (2010)
	RFM 82, 83, 84, 85, 87, 88, 90, 93, 94, 95, 96, 97	Jain et al. (2014b)
	SiA 2863, ISC 1199, ISC 74A, ISC 789, GS 889	Patro et al. (2018g)

TABLE 7.1 *(Continued)*

Crop	Resistant Sources	References
Barnyard millet	RBM 9-4, TNAU 128, TNAU 130, RAU 8, VL 29, VL 220, RBM 12	Jain and Gupta (2010)
	ACM 10-082, VL 172, DHB 23-3	Divya et al. (2016)
	VL 172	Patro et al. (2018h)
Proso millet	TNAU 137, GPUP 22, RAUM 8	Jain and Tiwari (2013)
	DHprMV 2164, 2769	Patro et al. (2015)

7.6.2 BIOLOGICAL CONTROL

The use of antagonistic microorganisms is a potential non-chemical means of controlling plant diseases by reducing the primary inoculum of the pathogen. Besides, the biological agents are able to induce systemic resistance in host plant against plant pathogens. A number of microbes were evaluated alone and in combination against *R. solani* in different small millet *in Vitro* as well as *in Vivo* and were found effective against *R. solani*.

7.6.2.1 IN VITRO

Chouhan (2014) evaluated 12 local isolates of *Trichoderma* against little millet isolate of *R. solani*. Four isolates from Rewa (2), Indore (1) and Khargone (1) districts of Madhya Pradesh were identified to inhibit the radial growth (30.9 to 33.6%) of *R. solani* in a confrontation assay. Krishnaveni (2019) found maximum inhibition (88.9%) of mycelia growth of different small millets *R. solani* isolates in *Trichoderma harzianum*-H followed by *T. Viride*-1 (86.0%), whereas minimum inhibition (54.1%) was recorded in *T. Viride*-2.

7.6.2.2 IN VIVO

Patro and Madhuri (2014) tested *Pseudomonas fluorescens (P.f.)* and *Trichoderma harzianum (T.h.)* as seed treatment for the management of *R. solani* in finger millet. Lowest disease severity was recorded in seed treatment with *Pseudomonas fluorescens* @ 10 g kg^{-1}+ *Trichoderma harzianum* @ 4 g kg^{-1} followed by seed treatment with *P. fluorescents* (10 g kg^{-1}) and *T. harzianum*

(4 g kg⁻1) alone. Three biocontrol agents namely *Pseudomonas fluorescence* (*P.f.*), *Trichoderma viride* (*T.v.*) and *Bacillus subtilis* (*B.s.*) were evaluated as seed treatment and soil application alone and in combination to manage the banded leaf blight in Kodo millet, little millet, barnyard millet, foxtail millet and proso millet. Soil application of value-added *P. f. + T. v.+ B. s.* @ 335 g each of talc-based formulation was mixed in 25 kg farmyard manure (FYM) or vermicompost (VC), incubated for 15 days and applied over an area of 1 acre at the time of sowing was most effective to reduce the banded blight development in Kodo millet (Kahar, 2017; Patro et al., 2018b), proso millet (Patro et al., 2018c, k), barnyard millet (Patro et al., 2018j), little millet (Kumar, 2018; Krishnaveni, 2019) and foxtail millet (Krishnaveni, 2019) and increasing the grain yield significantly. Patro et al. (2018d) reported that soil application of value-added *P.f + T. asperellum + B.s.* @ 1 kg talk-based formulation mixed in 25 kg FYM or VC, incubated for 15 days and applied over an acre at the time of sowing showed maximum reduction in disease intensity in little millet with higher grain and fodder yield over control.

7.6.2.3 BOTANICALS

Inhibitory effect of plant extracts might be attributed to the presence of antimicrobial substances in them and recognized as an important factor in the disease management strategies. Jain et al. (2017b) studied the bioefficacy of eight plant species belonging to seven families for their antifungal effect on mycelia growth of the fungus and reported that bud extract of garlic and rhizome extract of ginger completely inhibited the radial growth of little millet isolate of *R. solani*. Onion bulb extract, neem leaf extract, dhatura leaf extract, tulsi leaf extract, parthenium leaf extract, and bel leaf extract (9%) were found to inhibit the mycelia growth from 51.4 to 92.5%. Krishnaveni (2019) found garlic (*Allium sativum*) and ginger (*Zingiber officinale*) extract @ 10% most effective to inhibit the mycelia growth of little millet, foxtail millet and finger millet isolates. Bulb extract of onion (*Allium cepa*) and rhizome extract of ginger were found to inhibit the sclerotial production in foxtail millet isolates.

7.6.3 CHEMICAL

Chouhan (2014) reported that foliar application of non-conventional chemicals namely salicylic acid (SA) and sodium fluoride (SF) @ 200 ppm were effective to induce resistance in little millet against *R. solani* and reduce the incidence of banded leaf and sheath blight by 38.2 and 36.2%, respectively. Seed treatment with propiconazole @ 1 ml kg^{-1} seed and hexaconazole @ 2 ml kg^{-1} seed were most effective in reducing banded blight incidence in finger millet. Seed treatment with validamycin @ 4 g kg^{-1} seed and carbendazim were also effective (Patro and Madhuri, 2014). Jain et al. (2018) evaluated two non-conventional chemicals SA and SF @ 100, 150, and 200 ppm as foliar spray along with one fungicide validamycin (0.2%) against the banded leaf and sheath blight of little millet. Lowest RLH (19.0%) and highest grain yield (967.5 kg ha^{-1}) were recorded in the treatment of validamycin. Both the non-conventional chemicals were found to reduce the severity of the disease, infection rate and enhancing the grain yield, possibly due to induction of resistance in host plant against the pathogen. Maximum reduction in RLH (50.5%) and highest grain yield (937.5 kg ha01) was recorded in foliar spray of SA (200 ppm) followed by SF (200 ppm).

7.7   FUTURE PROSPECTS

Banded blight caused by *R. solani* is becoming a major threat in small millet cultivation. As these crops are cultivated by resource-poor farmers, the use of resistant varieties is the better option to combat the disease. Hence, vigorous screening of small millet cultivars against *R. solani*, host range study, and survival of the pathogen are important areas for future research. The influence of cultural practices, epidemiological studies, and integrated disease management (IDM) approach must be investigated to prevent the losses.

KEYWORDS

- *Bacillus subtilis*
- banded blight
- corn meal agar
- diagnosis
- farmyard manure
- salicylic acid
- small millets

REFERENCES

Chouhan, S. S., (2014). *Studies on Banded Leaf and Sheath Blight of Little Millet (Panicum sumatrense) Caused by Rhizoctonia Solani Kuhn.* MSc (Agri.) plant pathology thesis, JNKVV, College of Agriculture, Rewa, M.P., India.

Das, L., & Girija, V. L., (1989). Sheath blight of ragi. *Curr. Sci., 58,* 681, 682.

Divya, M., Patro, T. S. S. K., & Ashok, S., (2016). Evaluation of resistant sources of barnyard millet varieties against banded blight (BB) disease incited by *Rhizoctonia solani* Kuhn. *Frontiers of Crop Improvement Journal, 4,* 99–100.

Divya, M., Patro, T. S. S. K., Sandhya, R. Y., Triveni, U., & Anuradha, N., (2017a). Reaction of *Eleusine coracana* (L.) Gaertn. early and medium duration varieties against major maladies. *Progressive Research: An International Research Journal, 12,* 30–32.

Divya, M., Patro, T. S. S. K., Sandhya, R. Y., Triveni, U., & Anuradha, N., (2017b). Detection of resistant entries of little millet against banded blight disease incited by *Rhizoctonia solani* Kuhn. *Frontiers of Crop Improvement Journal, 5,* 74, 75.

Dubey, S. C., (1995). Banded blight of finger millet caused by *Thanatephorous cucumeris. Indian J. Mycol. Pl. Pathol., 25,* 315, 316.

Dubey, S. C., Dwivedi, R. P., & Narain, U., (1989). Banded blight of Italian millet caused by *Thanatephorous cucumeris. Farm Sci. J., 4,* 1–6.

Goswami, B. K., Bhuiyan, K. A., & Mial, I. H., (2010). Morphological and pathogenic variations in the isolates of *Rhizoctonia solani* in Bangladesh. *Bangladesh J. Agril. Res., 35,* 375–380.

Jain, A. K., & Gupta, A., (2010). Occurrence of banded leaf and sheath blight on foxtail millet and barnyard millets in Madhya Pradesh. *Ann. Pl. Protec. Sci., 18,* 268–270.

Jain, A. K., & Joshi, R. P., (2018). *Progress of Research of AICRP on Small Millets, Rewa for Quinquennial Review.* JNKVV, College of Agriculture, Rewa, M.P., India.

Jain, A. K., & Tiwari, A., (2013). Evaluation of promising genotypes of proso millet against banded leaf and sheath blight disease caused by *Rhizoctonia solani* Kuhn. *Mysore J. Agric Sci., 43,* 648–650.

Jain, A. K., Chouhan, S. S., Kumar, A., & Tripathi, S. K., (2017b). Evaluation of plant extracts against banded leaf and sheath blight of little millet caused by *Rhizoctonia solani* Kuhn. *Ann. Pl. Protec. Sci., 25,* 156–159.

Jain, A. K., Dhingra, M. R., & Joshi, R. P., (2014a). Strategic approaches for the management of biotic stresses in little millet under rainfed ecosystem. In: *National Symposium on Dryland Farming and Food Security in India.* Organized by RVSKVV, Gwalior (M.P.).

Jain, A. K., Joshi, R. P., & Singh, G., (2014b). Identification of host resistance against banded leaf and sheath blight of foxtail millet. *JNKVV Res. J., 48,* 171–175.

Jain, A. K., Kumar, A., Chouhan, S. S., & Joshi, R. P., (2018). Status and management of banded leaf and sheath blight of little millet caused by *Rhizoctonia solani* Kuhn with chemicals. *Ann. Pl. Protec. Sci., 26,* 122–126.

Jain, A. K., Kumar, A., Chouhan, S. S., & Tripathi, S. K., (2017a). Cultural characteristics and evaluation of *Trichoderma* isolates against *Rhizoctonia solani* Kuhn causing banded leaf and sheath blight of little millet. *Ann. Pl. Protec. Sci., 25,* 140–143.

Kahar, L., (2017). *Banded Leaf and Sheath Blight of Kodo Millet (Paspalum scrobiculatum L.) and its Management.* Thesis MSc (Ag) plant pathology. JNKVV, College of Agriculture, Rewa, M.P., India.

Kannaiyan, S., & Prasad, N. N., (1978). Reaction of certain cereal crop plant to sheath blight disease of rice. *Indian Phytopath., 31*, 541.

Kannaiyan, S., & Prasad, N. N., (1981). Host range of sheath blight pathogen of rice. *Madras Agric. J., 60*, 672–674.

Krishnaveni, V., (2019). *Variability of Rhizoctonia solani Kuhn Infecting Small Millets and Their Management.* Thesis MSc (Ag) plant pathology, BAU, Kanke, Ranchi, Jharkhand, India.

Kukreti, A., Kurmanchali, N., Rawat, L., & Bisht, T. S., (2017). Evaluation of fluorescent *Pseudomonas* spp., against *Pyricularia grisea, Rhizoctonia solani,* and *Sclerotium rolfsii* causing blast sheath blight and foot rot diseases of finger millet (*Eleusine coracana* L.) crop in mid-hills of Uttarakhand: *In vitro* study. *Bulletin of Environment, Pharmacology and Life Sciences, 6*, 29–35.

Kumar, B., & Dinesh, P., (2009). First record of banded sheath blight disease of barnyard millet caused by *Rhizoctonia solani. J. Mycol. Pl. Pathol., 39*, 352–354.

Kumar, B., & Dinesh, P., (2010). A new record on banded sheath blight disease of proso millet from mid-hills of Uttarakhand, India. *J. Mycol. Pl. Pathol., 40*, 331–333.

Kumar, B., (2016). A new addition to sheath blight of barnyard millet caused by *Rhizoctonia solani. Inter. J. Pl. Sci., 11*, 383–385.

Kumar, B., (2018). *Progression and Development of Banded Leaf and Sheath Blight in Little Millet Caused by Rhizoctonia Solani Kuhn and its Biological Management.* MSc (Ag) Thesis (plant pathology), JNKVV, College of Agriculture, Rewa, M.P., India.

Kumar, B., (2019). A new record on sheath blight disease of finger millet from Uttarakhand, India. *Inter. J. Pl. Sci., 14*, 77–80.

Lenka, S., Pun, K. B., Saha, S., & Rath, N. C., (2014). Studies on the host range of *Rhizoctonia solani* Kuhn causing sheath blight disease in rice. *Oryza, 51*, 100–102.

Meena, B., & Muthuswamy, M., (1998). Host range of *Rhizoctonia solani,* the incitant of sheath blight disease of rice. *Indian J. Pl. Protec., 26*, 62, 63.

Nagaraja, A., & Anjaneya, R. B., (2010). Banded blight: A new record of finger millet in Karnataka. *J. Mycopath. Res., 48*, 169, 170.

Nagaraja, A., Kumar, B., Jain, A. K., & Sabalpara, A. N., (2016). Emerging diseases: Need for focused research in small millets. *J. Mycopathol. Res., 54*, 1–9.

Patro, T. S. S. K., & Madhuri, J., (2013). Identification of resistant sources for sheath blight in foxtail millet incited by *Rhizoctonia solani* (Kuhn). *Indian J. Pl. Sci., 2*, 159–162.

Patro, T. S. S. K., & Madhuri, J., (2014). Management of banded blight of finger millet incited by *Rhizoctonia solani* (Kuhn). *Indian J. Pl. Sci., 3*, 163–166.

Patro, T. S. S. K., Divya, M., & Ashok, S., (2016b). Evaluation of donor screening nursery (DSN) of little millet against *Rhizoctonia solani,* the cause of sheath blight. *Frontiers of Crop Improvement Journal, 4*, 97, 98.

Patro, T. S. S. K., Divya, M., Sandhya, R. Y., & Ashok, S., (2016a). Screening of Kodo millet (*Paspalum scrobiculatum*) varieties against *Rhizoctonia solani,* the incitant of banded blight disease. *Progressive Research: An International Research Journal, 11*, 297, 298.

Patro, T. S. S. K., Divya, M., Sandhya, R. Y., Triveni, U., & Anuradha, N., (2017a). Resistance of Kodo millet (*Paspalum scobiculatum*) varieties against *Rhizoctonia solani,* the incitant of banded sheath blight disease. *Frontiers of Crop Improvement Journal, 5*, 78–79.

Patro, T. S. S. K., Divya, M., Sandhya, R. Y., Triveni, U., & Anuradha, N., (2017b). Identification of resistant sources against *Rhizoctonia solani* Kuhn, the incitant of sheath blight of *Echinochloa frumentacea*. *Progressive Research: An International Research Journal, 12*, 109, 110.

Patro, T. S. S. K., Divya, M., Sandhya, R. Y., Triveni, U., & Anuradha, N., (2017c). Evaluation of proso millet (*Panicum miliaceum* L.) genotypes against emerging malady of sheath blight caused by *Rhizoctonia solani* Kuhn. *Progressive Research: An International Research Journal, 12*, 103, 104.

Patro, T. S. S. K., Divya, M., Sandhya, R. Y., Triveni, U., & Anuradha, N., (2018). Evaluation of biological control agents against *Rhizoctonia solani* Kuhn in proso millet. *Adv. Biores., 9*, 38–40.

Patro, T. S. S. K., Meena, A., Divya, M., & Anuradha, N., (2018a). Identification of resistant sources of proso millet varieties against *Rhizoctonia solani* Kuhn inciting banded blight (BB) disease. *J. Pharmacogn Phytochem., 7*, 2604, 2605.

Patro, T. S. S. K., Meena, A., Divya, M., & Anuradha, N., (2018b). Ecofriendly Management of banded blight using biological control agents against *Rhizoctonia solani* Kuhn in Kodo millet. *J. Pharmacogn Phytochem., 7*, 2631–2634.

Patro, T. S. S. K., Meena, A., Divya, M., & Anuradha, N., (2018c). Ecofriendly Management of *Rhizoctonia solani* Kuhn inciting banded blight using biological control agents in proso millet. *J. Pharmacogn. Phytochem., 7*, 2660–2663.

Patro, T. S. S. K., Meena, A., Divya, M., & Anuradha, N., (2018d). Management of banded blight using biological control agents against *Rhizoctonia solani* Kuhn in little millet (*Panicum sumatrense*). *J. Pharmacogn. Phytochem., 7*, 2664–2667.

Patro, T. S. S. K., Meena, A., Divya, M., & Anuradha, N., (2018e). Disease reaction of donor screening nursery (DSN) of proso millet against *Rhizoctonia solani*, the cause of sheath blight. *J. Pharmacogn. Phytochem., 7*, 2684, 2685.

Patro, T. S. S. K., Meena, A., Divya, M., & Anuradha, N., (2018f). Evaluation of Kodo millet varieties resistant sources against banded blight (BB) disease incited by *Rhizoctonia solani* Kuhn. *J. Pharmacogn. Phytochem., 7*, 3172, 3173.

Patro, T. S. S. K., Meena, A., Divya, M., & Anuradha, N., (2018g). Evaluation of donor screening nursery (DSN) of foxtail millet against *Rhizoctonia solani*, the cause of sheath blight. *International Journal of Chemical Studies, 6*, 2189–2191.

Patro, T. S. S. K., Meena, A., Divya, M., & Anuradha, N., (2018h). Evaluation of resistant sources of barnyard millet varieties against banded blight disease. *International Journal of Chemical Studies, 6*, 2453–2455.

Patro, T. S. S. K., Meena, A., Divya, M., & Anuradha, N., (2018i). Identification of resistant sources in donor screening nursery (DSN) of barnyard millet against *Rhizoctonia solani*, the cause of sheath blight. *International Journal of Chemical Studies, 6*, 2514–2516.

Patro, T. S. S. K., Meena, A., Divya, M., & Anuradha, N., (2018j). Management of banded blight disease using biological control agents against *Rhizoctonia solani* Kuhn in barnyard millet. *International Journal of Chemical Studies, 6*, 3149–3152.

Patro, T. S. S. K., Neeraja, B., Sandhya, R. Y., & Jyothsna, S., (2015). Studies on proso millet (*Panicum miliaceum*) genotypes against emerging malady banded sheath blight caused by *Rhizoctonia solani* Kuhn under *in vivo*. *Progressive Research: An International Journal, 10*, 1916–1918.

Patro, T. S. S. K., Neeraja, B., Sandhya, R. Y., Keetthi, S., & Jyothsna, S., (2014). Identification of resistant sources for sheath blight in little millet (*Panicum sumatrense* Roth ex Roemer and Schultes). *Pl. Dis. Res., 29*, 239.

Patro, T. S. S. K., Rajkumar, S., Meena, A., Anuradha, N., Triveni, U., & Joga, R. P., (2019). Identification of resistant sources of little millet varieties against banded blight disease incited by *Rhizoctonia solani* Kuhn. *International Journal of Chemical Studies, 7*, 984–986.

Prajapati, S., (2019). *Studies on Banded Leaf and Sheath Blight of Finger Millet Caused by Rhizoctonia Solani Kuhn.* MSc (Ag) Thesis (plant pathology), JNKVV, College of Agriculture, Rewa, M.P., India.

Rajgopalan, B., Balakrishanan, B., & Das, L., (1992). Sheath rot of *Panicum miliaceum*. *Indian Phytopath., 45*, 279.

Rathore, R. S., Singh, P., & Jain, M. L., (1998). Varietal resistance and host range of *Rhizoctonia solani* fsp. *sasakii*, the incitant of banded leaf and sheath blight of maize. *J. Mycol. Pl. Pathol., 28*, 71.

Trivedi, A., & Rathore, R. S., (2006). New grass hosts for banded leaf and sheath blight of maize. *Indian Phytopath., 59*, 253, 254.

CHAPTER 8

Maize Diseases and Their Sustainable Management in India: Current Status and Future Perspectives

M. K. KHOKHAR,[1] K. S. HOODA,[2] P. N. MEENA,[1] R. GOGOI,[3]
S. S. SHARMA,[4] REKHA BALODI,[1] and M. S. GURJAR[3]

[1]*ICAR-National Research Center for Integrated Pest Management,
Pusa Campus, New Delhi – 110012, India,
E-mail: khokharmk3@gmail.com (M. K. Khokhar)*

[2]*ICAR-Indian Institute of Maize Research, PAU Campus,
Ludhiana – 141 004, Punjab, India*

[3]*Indian Agriculture Research Institute, New Delhi – 110012, India*

[4]*Maharana Pratap University of Agriculture and Technology,
Udaipur – 313001, Rajasthan, India*

ABSTRACT

Information embodied in this review pertains to the common diseases affecting maize production in India. The major diseases in the country are three downy mildews, four foliar, leaf, and sheath blight and two pre-flowering, three post-flowering stalk rots. The ubiquitous incidence of diseases at the pre-harvest stage has been an important bottleneck in increasing production. Losses from these diseases may vary according to disease occurrence, congenial environmental conditions, host genotype, and time (growth stage of the crop) of infection. Loss estimates in India have been put at around 13.2%, but actual losses can be as high as 80% in natural and artificial created epiphytotic conditions. The book chapter provides latest information on epidemiology, disease cycle, genetics of disease resistance and integrated crop management methods. All of this information should lead to more efficient management of maize pathogens.

8.1 INTRODUCTION

Among major cereal crops in production, corn (*Zea mays* L.) is the world's third-leading crop after wheat and rice grown in different agro-ecologies of the world. It has highest genetic yield potential amongst the cereal crops. Diseases are one of the major constraints in realizing the potential yield of this crop. Maize is attacked by as many as 112 pathogens in the world, posing major constraints in realizing the potential yield of maize. In India, about 35 diseases reported from different locations are predominantly of fungal and bacterial origin. The yield loss of 13.2% has been recorded due to these diseases (Payak and Sharma, 1985). Their geographical distribution in the agro-climatic zones is influenced by temperature (high/low), humidity, cultural practices and the type and diversity of maize cultivars used. Under favorable environmental conditions, these diseases play havoc and cause immense losses both in quantity and quality as well (Ali and Yan, 2012). Globally, about 9% yield losses have been estimated in maize due to diseases (Oerke, 2005). This varied significantly from 4% in northern Europe and 14% in West Africa and South Asia (http://www.cabicompendium.org/cpc/economic.asp). *Turcicum* leaf blight (TLB), maydis leaf blight (MLB), post-flowering stalk rots, ear, and cob rot; and banded leaf and sheath blight are prevalent throughout the country. Bacterial stalk rot, brown stripe downy mildew, and brown spot are reported from northern India whereas downy mildews are confined to peninsular India (Karnataka and Tamil Nadu) and Udaipur region of Rajasthan. Polysora rust is emerging as a potential threat in Karnataka and central Andhra Pradesh. Plant diseases represent important preventable hazards to grain production, which limits the productivity adversely. The information on different aspects of maize diseases was reviewed keeping in view objective to assess the present status of diseases and researchable area to alleviate losses which would further necessitate the development of bio-rational and climate-resilient integrated disease management (IDM) schedule (Table 8.1 and Figure 8.1).

TABLE 8.1 Yield Losses (%) Due to Major Diseases of Maize in India

Disease	Distribution	Losses (%)	References
Turcicum leaf blight	Jammu and Kashmir, Himachal Pradesh, Sikkim, West Bengal, Meghalaya, Tripura, Assam, Rajasthan, Uttar Pradesh, Bihar, Madhya Pradesh, Gujarat, Maharashtra, Andhra Pradesh, Karnataka, Tamil Nadu	13–50	AICRP (2013)

TABLE 8.1   *(Continued)*

Disease	Distribution	Losses (%)	References
Maydis leaf blight	Jammu and Kashmir, Himachal Pradesh, Sikkim, Meghalaya, Punjab, Haryana, Rajasthan, Delhi, Uttar Pradesh, Bihar, Madhya Pradesh, Gujarat, Maharashtra, Andhra Pradesh, Karnataka, Tamil Nadu	15–46	AICRP (2013)
Common Rust	Jammu and Kashmir, Himachal Pradesh, Sikkim, West Bengal, Tripura, Meghalaya, Punjab (Rabi), Uttar Pradesh, Haryana, Assam, Madhya Pradesh, Andhra Pradesh, Rajasthan, Bihar, Maharashtra, Karnataka, Tamil Nadu	18–49	Hagan (2010)
Brown stripe downy mildew	Himachal Pradesh, Sikkim, West Bengal, Meghalaya, Punjab, Haryana, Rajasthan, Delhi, Uttar Pradesh, Bihar, Madhya Pradesh, Gujarat	10–30	Spencer and Dick (2002)
Sorghum downy mildew	Gujarat, Maharashtra, Andhra Pradesh, Karnataka, Tamil Nadu	Up to 100	AICRP (2012)
Rajasthan downy mildew	Rajasthan and Surrounding areas.	10–60	Siradhana (1980)
Brown spot	Jammu and Kashmir, Himachal Pradesh, Sikkim, West Bengal, Rajasthan, Punjab, Madhya Pradesh, Karnataka	6–20	Payak and Sharma (1985)
Banded leaf and sheath blight	Jammu and Kashmir, Himachal Pradesh, Sikkim, Punjab, Haryana, Rajasthan, Delhi, Uttar Pradesh, Bihar, Madhya Pradesh	Up to 60	Tang et al. (2004)
Pythium stalk rot	Himachal Pradesh, Sikkim, West Bengal, Uttar Pradesh, Bihar, Punjab, Haryana, Rajasthan, Delhi.	–	–
Bacterial stalk rot	Himachal Pradesh, Sikkim, West Bengal, Rajasthan, Uttar Pradesh, Bihar, Madhya Pradesh, Andhra Pradesh, Delhi, Punjab, Haryana	Up to 85	Thind and Payak (1985)
Fusarium stalk rot	Jammu and Kashmir, Punjab, Haryana, Delhi, Rajasthan, Madhya Pradesh, Uttar Pradesh, Bihar, West Bengal, Andhra Pradesh, Tamil Nadu and Karnataka, where water stress occurs after flowering stage of the crop	10–42	Harlapur et al. (2002)

TABLE 8.1   *(Continued)*

Disease	Distribution	Losses (%)	References
Charcoal rot	Jammu and Kashmir, West Bengal, Rajasthan, Punjab, Haryana, Uttar Pradesh, Bihar, Madhya Pradesh, Andhra Pradesh, Karnataka, Tamil Nadu, Delhi.	25–32	Krishna et al. (2013)
Late wilt	Hyderabad, Uttar Pradesh, Rajasthan	Up to 51	Johal et al. (2004)
Seed and seedling blights	Where soil temperature is low (below 13°C) during planting time	5–20	Dodd and White (1999)

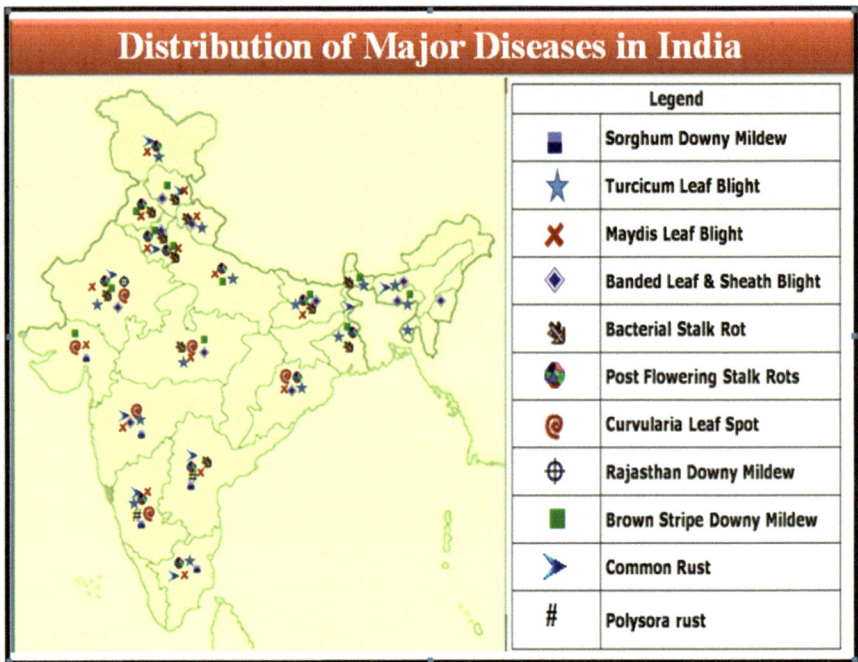

FIGURE 8.1   Distribution pattern of major diseases of maize in India.

8.2   TURCICUM LEAF BLIGHT (TLB)

Turcicum leaf blight of maize caused by *Setosphaeria turcica* is also known as "Northern Corn Leaf Blight" in the United States of America, "White blight" and "leaf stripe" in Karnataka (India), and is locally as "Urijinke" which meaning jumping deer with fire, because of its nature of occurrence and spread during the crop period. The disease was reported for the first time from New Jersey, USA in 1878. The symptom of the disease in maize is a severe defoliation, especially during the grain-filling period (Sibiya et al., 2013). The intensity of the disease is severe in mid-altitude tropical regions due to high humidity coupled with low temperature and cloudy weather. The first serious outbreak of TLB was occurred in Connecticut, USA during 1889. The disease also occurs whenever sorghum and maize are grown together.

8.2.1 DISTRIBUTION

It is serious problem in the Northeastern United States, in sub-Saharan Africa, and in areas of China, Latin America, and India (Hooda et al., 2017). In India, it was reported for the first time by Butler during 1907 from Bihar. Later it was reported from many parts of the country viz., Lalmardi, Srinagar (Kaul, 1957), Punjab (Mitra, 1981), Himachal Pradesh (Chenulu and Hora, 1962), Sikkim, Meghalaya, Tripura, and Assam (Reddy et al., 2013) and Kashmir valley (Payak and Renfro, 1968). The endemic areas (hot spots) of the disease in India have been identified, which are, Arabhavi, Nagenahalli in Karnataka, Kolhapur in Maharashtra, Karimanagar in Andhra Pradesh, Dholi in Bihar and Almora in Uttarakhand (Laxminarayan and Shankar-lingam, 1983). It is observed in most maize growing areas that have high humidity coupled with moderate temperatures.

8.2.2 ECONOMIC IMPACT

The disease affects maize from seedling stage to till harvest. Yield losses due to TLB can be high, but depending upon environmental conditions and geographical location. The losses were found to be directly proportional to disease intensity (Dharanendraswamy, 2003). However, yield losses under experimental conditions of artificially created disease epiphytotics were esti-mated to the extent of 66% in susceptible variety Basi and 56.0% in CM 202 (Payak and Sharma, 1985). The disease cause extensive defoliation during the grain filling period, resulting in grain yield losses up to 50% or more. Maize yield decreases up to 0.6 t ha^{-1} when its severity reaches more than 75% (Sibiya et al., 2013).

8.2.3 DISEASE SYMPTOMS

TLB starts as long, elliptical, gray-green lesions measuring 3–15 cm in length. As the lesions mature, they become tan with targeted, darkish zones of fungal sporulation as the disease develops further the leisons may coalesce, forming enormous blighted areas, and entire leaves may emerge as blighted (Vieira et al., 2014). The disease mainly harms the lamina, sometimes infects sheath and bract also, but does not harm fructose directly (Richards and Kucharek, 2006). The infection occurs from the lower lamina, gradually extends to the upsides (Zhao and Wang, 2009). The pathogen *Setosphaeria turcica* is a hemibiotrophic ascomycete fungus, living on live plant tissue before causing necrosis, drawing nutrition from dead tissue (Lim et al., 1974). The pathogen survives in temperatures ranging between 17 and 28°C with moderate to high humidity but can tolerate harsher conditions also. The pathogen allows penetration and colonization with the production of a range of secondary metabolites and toxins. The *S. turcica* genome includes two genes encoding xylanase enzymes, which degrade arabinoxylan in the plant cell wall causing loss of integrity and aiding penetration (Degefu et al., 2003; Figure 8.2).

FIGURE 8.2   Typical symptoms of *turcicum* leaf blight.

8.2.4 EPIDEMIOLOGY

The disease is favored by mild temperatures ranging from 20–28°C, relative humidity from 90 to 100%, and low luminosity (Natsvlishvilli, 1972; Sprague, 1977). The infection of *S. turcica* on maize occurs from seedling to harvesting, nevertheless, maximum severity used to be seen from tasseling and six to eight weeks after on silking, which resulted in heavy loss (Chenulu and Hora, 1962). Levy and Cohen (1983) reported that the disease is more aggressive in young susceptible plants with an optimum temperature for infection and lesion development at 28°C. Tsen (1981) studied the effect of sowing dates on the occurrence of disease and reported that optimum temperature of 20 to 26°C and over 80% RH were congenial to disease development. He concluded that RH plays a greater role than the temperature for the severe outbreak of the disease. Mallikarjuna et al. (2007) reported a linear relationship between disease severities and meteorological factors, namely temperatures and relative humidity.

8.2.5 DISEASE MANAGEMENT

The disease is difficult to manage by means of both fungicides and crop rotation alone. Thus, an integrated approach is essential for effective management, especially under favourable environmental conditions conditions (Kumar et al., 2005).

8.2.5.1 CULTURAL PRACTICES

De Leon and Pandey (1989) suggested crop rotation of one to two years or deep burying of infested maize residues before maize hybrids planted, reduce overwintering of the fungus and decrease disease pressure. Also, timely removal of overwintering infected crop residue will reduce the amount of available inoculums at the onset of the subsequent growing season. Patrick et al. (2005) showed that the severity of TLB is less seen in fields when adequate potassium in the form of potassium chloride has been applied.

8.2.5.2 BIOLOGICAL CONTROL

Biological control for TLB has not received much attention due to pathogen variation, thus preventing the development of suitable biological agents for control. However, potent control of *S. turcica* with BCAs has remained elusive. The use of plant products in disease management is a contemporary eco-friendly approach and gaining popularity considering that of its benefits over chemical compounds (Shivapuri et al., 1997). However, studies on control of TLB of maize with the aid of plant extracts are meager. The ultimate goal of reducing fungicide use in maize production will be accomplished by using different bio-rational fungicides in rotations with traditional fungicides.

8.2.5.3   CHEMICAL MANAGEMENT

Demethylation inhibitor (DMI) and quinine oxidation inhibitor (QoI) fungicides, and particularly propiconazole, have shown the efficacy in controlling TLB both *in vitro* and in field conditions (Kumar et al., 2009; Kemerait, 2012; Blandino et al., 2012; Wise, 2014; Veerabhadraswamy et al., 2014).

8.2.5.4   INTEGRATED DISEASE MANAGEMENT (IDM)

The disease is difficult to manage by means of both fungicides and crop rotation alone. Thus, an integrated approach is essential for effective management, especially under problematic conditions. Integration of resistant varieties, good agricultural practices, and use of recommended fungicides and biopesticides are necessary for the management of TLB. The seed treatment with carboxin + thiram or benomyl + thiram followed by foliar application of mancozeb (0.25%) thrice at 10 days interval improved the seed emergence and seedling vigor and reduced the initial inoculums. This was found very effective in integrated management approach against northern southern leaf blight of maize. Harlapur (2005) studied the IDM of TLB by using carboxin @2 g/kg for seed treatment followed by two sprays of mancozeb 0.25% as compared to control and observed significantly minimum percent disease incidence and maximum grain yield. The seed treatment with carboxin (2 g/kg) *Trichoderma harzianum* (6 g/kg) followed by three sprays of mancozeb (0.25%)/propiconazole (0.1%) at 30, 40, and 50 day after sowing reduced disease severity.

8.3 MAYDIS LEAF BLIGHT (MLB)

Maydis leaf blight (MLB) is an important disease caused by *Bipolaris maydis* (Nisikado and Miyake) Shoemaker. In India, the disease was first reported by Munjal and Kapoor (1960) from the specimens collected from Malda (West Bengal) by Butler in 1905 and now it is widespread in almost all maize growing regions where warm (20–30°C) and humid climate prevails.

8.3.1 DISTRIBUTION

MLB is found in all tropical and temperate maize growing regions where the growing season is characterized by warm and wet conditions. In India, MLB occurs in Punjab, Haryana, Delhi, Uttar Pradesh, Bihar, Madhya Pradesh, Gujarat, Jammu, and Kashmir, Sikkim, Meghalaya, Rajasthan, Andhra Pradesh and Maharashtra. Munjal and Kapoor (1960) reported its first presence in India. In India, the most widely prevalent race is 'O.' Although race 'T' has been recorded on seeds of other hosts such as *Phaseolus mungo, Vigna sinensis, Paspalum scrobiculatum. Drechslera maydis* on maize was reported for the first time from Punjab by Mitra in 1931. Khehra et al. (1976) reported the presence of race 'T' on maize experimental hybrids 2310 and 2420 from Ludhiana. The earlier record of race 'T' from India is from non-maize hosts from Delhi. However, the most prevalent race in the country still continues to be race 'O.'

8.3.2 ECONOMIC IMPACT

It has the potential to cause grain yield losses of more than 40% (Fisher et al., 1976). However, use of resistant germplasms, especially in the United States and Western Europe, has largely controlled yield losses due to MLB. It accounts for 20% or sometimes even more yield losses to maize crop in Pakistan (Hafiz, 1986). Regions with a warm (20 to 32°C) and damp growing season are at most risk from MLB. Long, dry, sunny periods during the growing season are unfavorable for the disease. Race 'O' pathotypes are widely distributed, whereas race 'T' pathotypes are only prevalent in varieties with Texas source of male sterility.

8.3.3 DISEASE SYMPTOMS

MLB of maize forms numerous lesions with a size up to 2.5 cm long, which are present mostly on the leaves. Initially, they are elliptical, later elongate, further becoming rectangular-shaped by restriction of the veins. The color of the spot is cinnamon-buff (sometimes with a purplish tint) with a reddish-brown margin and occasionally zonate coalescing and becoming grayish with conidia. The disease was first reported by Drechsler (1925) on maize in Florida, and later it was reported on teosinte in the Philippines. Initially, this pathogen was named *Helminthosporium maydis*. On the basis of the perfect stage, this fungus was renamed as *Ophiobolus heterstrophus* which was subsequently transferred to *C. heterostrophus* (Drechsl.) (Drechsler, 1934). Nisikado and Miyake (1926) found that *H. maydis* causing maize leaf spot in Japan was identical with the conidial stage of Drechslera is *C. hetrerostrophus*. Ito (1930) erected the genus *Drechslera* to which the maize pathogen was transferred. It was again transferred to genus *Bipolaris* by Shoemaker (1959). Currently, the fungus is designated as *Drechslera maydis* perfect stage *C. heterostrophus*. It however continues to be commonly referred to as *Bipolaris maydis* (Figure 8.3).

FIGURE 8.3   Typical symptoms of maydis leaf blight.

8.3.4   EPIDEMIOLOGY

The extent and severity of this disease varies from season to season. In warm (20–32°C) and moderately humid environment of the world, maydis blight is potentially damaging and may cause significant yield losses. Advance generation hybrids and unmaintained composites grown on farmer's field of Haryana were most susceptible to MLB (Dass and Dhanju, 2000). More than 80% prevalence and 43% severity was recorded in un-recommended varieties grown on the farmer's field of Haryana. Late sowing resulted in reduction of disease and yield. Severity of disease affected the phenology and plant growth. Dass and Dhanju (2000) also observed variation among the isolates (Hm-1 to Hm-16) w.r.t. pathogenicity, sporulation and size of lesions on maize plant, radial growth, colony color and spore production on media and size, shape, color, and septation of the conidia. While studying the physical factor that influence the growth and spread of *D. maydis*, observed that linear growth of the pathogen in Petri-plate was maximum at 33°C and minimum at 21 and 38°C. Below and above 33°C, mycelial growth gradually tended to decrease. Mycelial growth was good at 70–100% RH but was optimum at 90–100%. Below 90 RH, colony growth was slow. Conidial germination was also highest at 33°C, germination percentage was high between 22 and 32°C, but gradually decrease below 27°C and above 33°C. *In vitro* studies indicate that linear growth and conidial germination were affected by both temperature and RH.

8.3.5   DISEASE MANAGEMENT

8.3.5.1   CULTURAL PRACTICES

Field experiments conducted under artificial epiphytotic conditions at Kalyani, West Bengal during Kharif (1995–1998) revealed that disease incidence was favored by planting in July, while early planting in May or June, or late planting in August reduces the disease incidence (Pal and Kaiser, 2001). Disease incidence gradually increased with the increase in plant density and was maximum at a population of 7000 ha^{-1}, while it was minimum at 40,000 ha^{-1}. Nitrogen alone or in combination with phosphorus and potassium, or with both phosphorus and potassium reduces the disease incidence (Pal and Kaiser, 2001). There was a gradual increase in the disease severity with the increase in the dose of nitrogen at 160 kg ha^{-1}. *In vitro* studies, however, showed that nitrogen significantly increased the linear growth of the pathogen while both phosphorus and potassium individually or in combination with nitrogen reduces the disease incidence. Nitrogen also significantly increases the percentage of conidial germination, while both phosphorus and potassium individually or in combination with nitrogen reduces it.

8.3.5.2   BIOLOGICAL CONTROL

Protection of maize plants against *D. maydis* on sterilized soil by inoculation of non-pathogens (*Drechslera oryzae* and *Helminothosporium sativum*) followed by inoculation of pathogen as well as by mixed inoculation with both. Cultural filtrates of non-pathogen inhibited the conidial germination of the pathogen. Similar results were also recommended using extracts of maize leaves inoculated with non-pathogens.

8.3.5.3   PLANT EXTRACT

Garlic clove extract was found highly effective in inhibiting (66.5–83.9%) the growth of *H. maydis* followed by Neem leaf (37–65%) and Tulsi leaf extract (39–48%) (Kumar et al., 2009). However, onion bulb and mentha leaf extract produces only 38.6% and 25.4% inhibition respectively at 10% concentration. Disease can be adequately managed by spraying garlic and neem extract at 5% in the field.

8.3.5.4   CHEMICAL CONTROL

Kavach (60 microgram/ml) and mancozeb + thiophanate-methyl (100 microgram/ml) completely inhibited the growth and conidial germination of *D. maydis*. In field experiment, the application of captan resulted in the lowest disease index. Sitara and Akhter (2007) found ridomyl gold (MZ 68% WP) effective against seed-borne mycoflora of maize (including *Drechslera* spp.) followed by Aliette (80% w/w), Neem seed powder (0.1%, 0.2%, 0.3%), Antracol (70% w/w) and sodium hypochlorite (10%). Spraying of mancozeb or zineb at first appearance of disease @ 0.2% followed by 2 applications at 10 days interval was found effective in controlling the disease.

8.4   BANDED LEAF AND SHEATH BLIGHT

BLSB incited by the fungus *Rhizoctonia solani* f. sp. *sasakii* Exner [= *Thanetophoprus sasakii* (Shirai) Tu and Kimbro], has attained the status of the most economically important disease of maize. The disease was first reported from Srilanka (Bertus, 1927). Subsequently, it was reported from Malaysia as 'Banded sheath rot,' in the Philippines as 'Banded sclerotial disease' and as 'summer sheath blight' in Japan. In India, in the early sixties, the disease was of minor importance in the western and central Himalayan foothill region. However, it became increasingly severe during the successive two decades. At present, BLSB is considered as a major disease in India and several countries of Tropical Asia, wherever maize is grown.

8.4.1   DISTRIBUTION

BLSB was considered of minor importance by Bertus (1927), who reported it for the first time from Sri Lanka and described it as a sclerotial disease of maize caused by *Rhizoctonia solani* f. sp. *sasakii* (Kuhn) Exner. In recent years, the disease outbreaks have occurred in more countries and have assumed epidemic dimensions. In most of these countries, the disease has been a priority item in maize research programs due to its increasing incidence and economic damage. In India, the disease was first reported as banded leaf and sheath blight of maize caused by *Hypochonus sasakii* by Ullstrup in 1960 from *Tarai* region of Uttar Pradesh (Payak and Renfro, 1966) and occurred in epidemic form in 1972 in Mandi district of Himachal Pradesh (Thakur et al., 1973). The disease was described as ear rot caused by *Corticium sasakii* on maize in 1977 from Meghalaya State (Maiti, 1978). Since then the disease has been reported from states of Himachal Pradesh, Uttar Pradesh, Haryana, Punjab, Madhya Pradesh, Rajasthan, West Bengal, Meghalaya, Assam, and Orissa (Rani et al., 2013).

8.4.2   ECONOMIC IMPACT

The disease is more severe in rice-maize cropping pattern because the pathogen also causes sheath blight disease in rice. Although, presently, it is causing much more losses in yield comparing to earlier reports where 31.9% reduction in grain yield in popular maize cultivars with disease severity level up to 87.3% has been recorded (Lal et al., 1980). Lal et al. (1985) had suggested that grain yield loss can go up to an extent of 90%. Data regarding annual losses in grain yields are not available, but as per one estimate, it is more than 10% in India. However, the magnitude of grain loss may reach as high as 100% if the ear rot phase of the disease predominates. In India, out of 13.2% estimated annual losses in grain yield from all diseases taken together, sheath blight contributes 1% (Payak and Sharma, 1985). Recently, when the ear rot phase predominated, yield losses approached 100% in southern China (Tang et al., 2004), and the disease is spreading to new areas under intensive farming practices (Akhtar et al., 2009) (Table 8.2).

TABLE 8.2   Inoculation Techniques and Rating Scales of Major Diseases of Maize

SL. No.	Disease	Inoculation Technique	Rating Scales*	References
1.	*Turcicum* leaf blight (TLB)	Leaf whorl inoculation	1–5 (≤2.0 = R; 2.1–3.0 = MR; 3.1–4.0 = MS; >4.0 = S)	Payak and Sharma (1983)
2.	Maydis leaf blight (MLB)	Leaf whorl inoculation	–do–	–do–
3.	Banded leaf and sheath blight (BLSB)	Sheath inoculation	–do–	–do–
4.	Curvularia leaf spot (CLS)	Spray inoculation	–do–	–do–
5.	Brown stripe downy mildew (BSDM)	Spreader row and whorl inoculation	–do–	–do–
6.	Polysora rust (PR)	Spray inoculation of urediniospore solution	1–5 (≤1.0 = R; 1.1–2.0 = MR; 2.1–3.0 = MS, 3.1–4.0 = S, >4.0 = HS)	Lal and Singh (1984)
7.	Post flowering stalk rots (PFSR)	Toothpick method	1–9 (≤3.0 = R; 3.1–5.0 = MR; 5.1–7.0 = MS;>7.0 = S)	–do–
8.	Rajasthan downy mildew (RDM)	Spreader row and whorl inoculation	Per cent incidence (≤10.0% = R; 11–25% = MR; 25.1–50.0% = MS; >50.0% = S)	–do–
9.	Sorghum downy mildew (SDM)	Spreader row and whorl inoculation	–do–	–do–
10.	Bacterial stalk rot (BSR)	Hypodermic syringe inoculation	–do–	–do–

*R= Resistant, MR= Moderately resistant, MS= Moderately susceptible, S= Susceptible, HS= Highly susceptible.

8.4.3   DISEASE SYMPTOMS

The disease generally appears at a pre-flowering stage in 40–50 days old plants. The symptoms manifest on leaves, sheaths, stalk, and ear. As the disease is soil-borne, it starts from the leaf sheath or on leaves which are in context with soil and travel up to the ear. The disease lesion are characterized by the presence of alternate bleached area or zones, which are initially water-soaked and narrow purple-brown bends oriented perpendicular to long axis of leaves or leaf sheath. Banded leaf and sheath blight disease is caused by *R. solani = hypochonu ssasakii* (*Thanatephorus cucumeris* (Frank) Donk) is one of the most widespread, destructive, and versatile pathogen found in most parts of the world and is capable of attacking a wide range of host plants, including maize causing seed decay, damping-off, stem canker, root rot, aerial blight, and seed or cob decay (Singh and Sahi, 2012). The genus concept in *Rhizoctonia* was first established by De Candolle (1815). The identity of maize banded leaf and sheath blight pathogen has been controversial. Earlier, it was designated as anamorph of *Corticium sasakii*. However, based on perfect state (Warcup and Talbot, 1962), the isolate infecting maize has been designated as *Rhizoctonia zeae* with perfect state *Waitea circinata* (Figure 8.4).

FIGURE 8.4   Typical symptom of BLSB disease on the sheath and infected cobs: (A) highly susceptible reaction (disease score 5); (B) resistant reaction; and (C) infected cob bearing sclerotia.

8.4.4   EPIDEMIOLOGY

Environmental factors have a vital role in the development of the disease on aerial plant parts caused by *Rhizoctonia soiani*. A range of 25°C to 30°C coupled with an average relative humidity of 90–100% is most suitable for development of banded leaf and sheath blight disease in maize. Similarly, rainfall over 100 mm in the first two weeks of infection favored early infection and disease development (Ahuja and Payak, 1981). The highest level of the disease is induced at 90–100% relative humidity and optimum temperature of 28°C in the first week of infection. If the relative humidity goes below 70%, disease development and spread become very slow (Sharma, 2005). Additionally, high crop densities impact disease severity. Under favorable conditions, the disease generally appears at a pre-flowering stage in 40–50 days old plants (Saxena, 2002).

8.4.5   DISEASE CYCLE

The primary source of inoculum are the sclerotia in soil or in infected host debris, and the active mycelium on the other grass hosts that grow in the vicinity of maize plant in the field. In undistributed soil, sclerotia can survive at different moisture level of various depths up to 20 cm. Population of sclerotia in ploughed field is higher at 12 cm depth in soil have better buoyancy and germinability. Sclerotia which survive on plant debris often come up on the soil surface during field preparation and other operations. They come in contact with newly planted seedlings/plants and cause infection. Secondary spread is due to contact of healthy plants with infected leaves/sheaths. The infection continues to mid-dough stage covering the entire plant including the ear but not the tassel. Seeds are not considered to be a source of inoculums and may not play a major role in severe disease outbreaks.

8.4.6   DISEASE MANAGEMENT

Management of disease through chemical, biological, and cultural practices are still the mainstay for minimizing the devastation of BLSB in maize growing areas of South and Southeast Asia. For the cultural control of *R. solani*, selection of a well-drained field and planting on raised beds are important aspects to avoid contact of water with seeds and faster growth of seedlings. Composting of hardwood on *Rhizoctonia*-infested soil has been found to reduce disease severity, apparently by promoting the growth of *Trichoderma* and other antagonistic microorganisms (Hoitink, 1980). Cultural method like stripping of the second and third leaf-sheath from the ground level at the age of 35–40 days old crop is found effective in checking further BLSB development (Sharma and Hembram, 1990; Kato and Inoue, 1995). Till date there is no information on the availability or development of true resistant varieties in maize for BLSB disease. However, there have been limited reports on genetics of resistance to BLSB in maize (Sharma et al., 2002). Although development of resistant hybrids through classical plant breeding have been slow because of unavailability of BLSB resistance sources (Pan and Rush, 1997), there exists a hope as identified 11 QTLs for resistance to BLSB on the chromosome 1, 2, 3, 4, 5, 6, and 10 by composite interval mapping (CIM). Biocontrol agents like *Trichoderma* spp., *Gliocladium virens,* and *Pseudomonas fluorescens* have also been found promising in *R. solani* suppression *in vitro* and *in vivo* (Sivakumar and Sharma, 2000). Several fungicides like carbendazim, bengard, thiophanate-methyl, iprobenfos, captan, quintozene, mancozeb + thiophenate-methyl, copper oxychloride and thiram have been found effective in inhibiting the growth of pathogen under *in vitro* conditions. Under field conditions also, all the fungicides except thiram were effective in reducing disease severity. In China, the antibiotic Jinggangmycin (Validamycin) gave satisfactory control in field conditions. In India, the formulation of Validamycin had shown good control against banded leaf and sheath blight pathogen (Ahuja and Payak, 1988; Puzari et al., 1998; Rakesh et al., 2011). Other fungicides like carbendazim (Puzari et al., 1998; Sharma et al., 2002), hexaconazole, and Thiophenate M were very effective against *R. solani* and are potent enough to give high level of disease control. In a recent study, bavistin is reported as a highly effective seed dressing fungicide with 48.7% disease control and highest yield of 64.7 q/ha over control.

8.4.7 INTEGRATED DISEASE MANAGEMENT (IDM)

Integrated disease management (IDM) is a new concept of disease management which encompasses the strength of biological, cultural, physical-chemical and genetical disease management tactics. Isolates of *T. viride* were found highly antagonistic to *R. solani* and the effectiveness can be increased by 20% by combining bio-control with fungicides treatments. Some strains of *T. viride* mutants tolerant to fungicides like thiram and thiophanates methyl was found to be most effective against *R. solani*. chitosan can also be scheduled with *T. harzianum* in IPM strategies. Foliar application of carbendazim followed by *Trichoderma* sprays in sawdust and FYM amended field significantly reduces the diseases. In an IDM strategy, Singh and Singh (2011) found best performance of validamycin (0.25%) and *T. viride* as foliar spray than the fungicides like tilt (0.15) and bavistin (0.1%) and bioagent *Pseudomonas fluorescence* which contributed higher grain yield over check.

8.4.7.1   CURVULARIA LEAF SPOT

CLS of maize caused by *Curvularia lunata* var. *aeria* (Batista, Lima, and Vasconcelos) Ellis, is one of the most common disease of maize in tropical climate. In India, the disease was found in all states on maize during the rainy and winter season (Choudhary et al., 2011). In India, the disease was earlier considered to be of minor importance, but its severity is continuously increasing. Curvularia leaf spot on maize are seed borne and thus direct losses to the standing crop from early infection due to seed borne inoculums further secondary spread through air borne inoculums.

8.4.7.2   ECONOMIC IMPACT

This disease causes the reduction in yield both directly and indirectly since the pathogen can attack both leaves and seeds. The disease has been reported to cause yield loss in maize to tune of 21–23% in India (Rathore et al., 2005; AICRIP, 2013). CLS is found almost on every leaf of every plant during the rainy season. In China, reported yield loss in maize to the tune of 20–60%. Indirect losses are caused by the dissemination of the disease into newer areas through infected seeds.

8.4.7.3   DISEASE SYMPTOMS

On maize, the pathogen causes leaf spot, which usually appear as small light brownish circular or ovoid spots not more than 4 mm. in diameter, surrounded by brown ring. It may attack any part of the plant at any stage of growth. Curvularia leaf spot of maize is caused by *Curvularia lunata* var. *aeria* (Wakker) Boedijn, *C. pallescens*, *C. maculans*, *C. tuberculata*, and *C. clavata*. In Rajasthan, no information about involvement of *Curvularia* spp. other than *Curvularia pallescens*. It forms septate mycelium, measuring 2–5 micro m. in diameter. Conidiophore are dark brown, geniculated, Unbranched, measuring 70–270 × 2–4 μm. Conidia are usually attenuated towards the base with three partitions; the second cell from the top much larger and darken in color makes the conidium retroflexed and conidia measure 19–30 × 8–16 μm (Figure 8.5).

FIGURE 8.5   Typical symptoms of curvularia leaf spot.

8.4.7.4   EPIDEMIOLOGY

In the fields, the disease appears on 35 to 40 days old plants, but its severity usually increases on the older plants, when the crop reaches the tasseling stage. The disease is more prevalent in those areas where the climate is warm and humid. Dai-Fa-Chao et al. (1998) reported optimum temperature for growth of *C. lunata* at 28–32°C.

8.4.7.5   DISEASE MANAGEMENT

Many studies were conducted by the scientists with the objective to understand the inheritance of CLS resistance. Using generation mean analysis, Zhao et al. (2002) reported that resistance to CLS was inherited quantitatively and associated with additive and dominant genetic effects, which account for 70% of the total phenotypic variation across generations. HPR have been tried to manage the disease. CM 300, CM 111, Guidan 22, 23, and Jingzao are reported resistance at different levels (Wang et al., 2000).

8.4.7.5.1   Chemical Control

Fajemisin and Okuyemi (1976) found that copper oxychloride, copper oxychloride + zineb are effective in inhibiting of pathogen. Grewal and Payak found that control might be achieved by two application of 0.5% difoltan, The I[st] as preventive and IInd just after appearance of disease. Aqueous and acetone extract of *Azadirachta indica*, *Chromolaena odorata* and *Ocimum gratissimum* inhibited growth and sporulation of pathogen.

8.4.7.5.2   Biological Control

Antagonistic effect of *Streptomycin* spp. against the pathogen show strong inhibition.

8.5   COMMON RUST

Common rust caused by *Puccinia sorghi* Schw. was observed in several states of India. Losses in grain yields due to common rust disease ranging from 6.0 to 36% have been reported from different part of the country. Now the disease has attained damaging status and is economically very important compared to other foliar diseases. Like other rust diseases, the obligate biotrophic pathogen produces infectious uredospores, teliospores, and basidiospores. When the plants are infected, pustules are developed on maize leaf surfaces. Rust pustules are minute, round to elongated uredia, which occurs on both leaf surfaces and sometimes on the husk and other floral parts. These pustules are yellowish in the early stage but later stage become brown and surrounded by chlorotic haloes. Due to the noticeable form and shape of the pustules, their presence is the first indicator of the disease in maize fields. The disease is common in subtropical, temperate, and highland environment with moderate temperature (16–25°C) and is more common at a relative humidity of at least 98% (Figure 8.6).

8.6   DOWNY MILDEW

Downy mildews are an important factor limiting maize production in South and southeast Asia. Maize crop is uniquely attacked by ten different downy mildew pathogens world over. This belongs to three genera *Peronosclerospora*

(7 spp.) *Sclerophthora* (2 spp.) and *Sclerospora* (1 spp.) The major downy mildew prevalent in India are Rajasthan downy mildew *Peronosclerospora heteropogoni* (Siradhna, Dange, Rathore, and Singh), sorghum downy mildew *Peronosclerospora sorghi* (Weston and Uppal) Shaw. *P. phillippinensis* (Weston) Shaw and brown stripe downy mildew *Sclerophthora rayssiae* var. *zeae* (Payak and Renfro) (Pingali, 2001; Rathore et al., 2004) (Figure 8.7).

FIGURE 8.6   Typical symptoms of common rust of maize.

FIGURE 8.7   Typical symptoms of Rajasthan downy mildew.

The pathogen causing downy mildew in Rajasthan, *P. hetropogoni* was first reported in 1968 from regional research station, Vallabhnagar Udaipur as *S. sorghi*, and later on it was renamed as *P. heteropogoni*. Plant infected with *Peronosclerospora heteropogoni* showed chlorotic stunted and yellow to whitish striped leaves giving a "half disease leaf" appearance and were more erect and slender under congenial condition, profuse sporulation occurs nocturnally on both surface of the infected leaves. The pathogen did not form oospore on maize (Table 8.3).

The sorghum downy mildew is caused by *Perenosclerospora sorghi* and found in Gujarat, Maharashtra, Andhra Pradesh, Karnataka, and Tamil Nadu. Plant less than one-month-old is highly susceptible to this disease. The favorable condition for the disease is mild temperature (20–25°C) in presence of free water. Infected plants are chlorotic and the chlorotic area includes the base of the blade with transverse margin and easily defined between diseased and healthy tissue. Leaves of infected plants tend to be narrower and more erect than these healthy plants. A white downy mildew growth may appear on lower surfaces of infected leaves. In severe cases the tassels of diseased plants may exhibit phyllody. There is no seed set in such plants. In tolerant varieties, the plant show symptoms of infection but have normal seed setting. *Peronosclerospora sorghi* has a polycyclic disease cycle. It is capable of causing secondary infections on susceptible hosts throughout the growing season. Its resting structures, the structures that allow the pathogen to overwinter, are the oospores. These oospores are produced in the infected plants from the previous growing season. They are often disseminated by wind. The oospores can overwinter in the soil and in the debris on the surface of the soil. The oospores have very thick walls, which makes them capable of surviving in the soil for years under many different weather conditions. The major downy mildew prevalent in India are *Peronosclerospora hetropogoni*, Siradhna, Dange, Rathore, and Singh, *Peronosclerospora sorghi* (Weston and Uppal) Shaw. *P. phillippinensis* (Weston) Shaw and *Sclerophthora rayssiae* var. *zeae* (Payak and Renfro). The pathogen causing downy mildew in Rajasthan, *P. hetropogoni* was first reported in 1968 from regional research station, Vallabhnagar Udaipur as *S. sorghi* and later on it was renamed as *P. heteropogoni* by Sridhana et al. (1986).

TABLE 8.3 Downy Mildews of Maize and Their Distribution

Common Names of Diseases	Pathogen	Distribution	First Report
Rajasthan DM	*Peronosclerospora hetropogoni*	Rajasthan state, India	1968, RRS, Vallabh Nagar, Udaipur (Raj)
Sorghum DM	Siradhana, Dange, Rathore, and Singh	Worldwide: Asia, Middle East, Africa, Australia, North America, South America	1907, Tamil Nadu and Maharashtra 1913, Bihar
Philippine DM	*Peronosclerospora sorghi (Weston and Uppal) C. G. Shaw*	India, Philippines, Indonesia, Thailand, Nepal	India
Sugarcane DM	*P. philippinensis* (Weston) Shaw		1909 Taiwan
Java DM	*P. sacchari* (Miyake in Ito) Shirai and Hara *P. maydis* (Racib.) Shaw	India, Philippines, Australia, Thailand, Nepal, Fiji Java, Australia, Taiwan, Philippines	1897, Java, Indonesia 1921, Philippines
Spontaneum DM	*P. spontanea* (Weston) Shaw	Philippines, Thailand	1967, India
Leaf splitting DM			1902, Italy
Brown Stripe DM	*P. miscanthi* (Miyake apud sacc.) C. G. Shaw	Philippines, Thailand	
Crazy top DM		India, Nepal, Pakistan, Thailand	1909, Argentina
Graminicola DM	*Sclerophthora rayssiae* var. *zeae* Payak and Renfro	Worldwide	
	S. macrospora (Sacc.) Thirum, Shaw, and Naras	Argentina, Israel, USA	
	Sclerospora graminicola (Sacc.) Schroet		

8.6.1 ECONOMIC IMPACT

P. sorghi is most widely prevalent and is reported to cause severe losses in many regions of the world to be in tune of $ 25 billion. Rajasthan downy mildew cause heavy losses under condition favoring infection and cause 81.5% reduction in grain of maize.

8.6.2 DISEASE SYMPTOMS

Plant infected with *Peronosclerospora heteropogoni* showed chlorotic stunted and yellow to whitish striped leaves giving a "half disease leaf" appearance and were more erect and slender under congenial conditions, profuse sporulation occurs nocturnally on both surface of the infected leaves. The pathogen did not form oospore on maize. *P. sorghi* conidiophores are massive, wedge-shaped; short, dichotomously branched emerge through stomata singly or in-group of 2 conidiophores is 180–300 μm in diameter, reddish-brown in color, conidia are oval to spherical, 15–26.9 μm × 15–28.9 μm in size. Conidia and oospore germinate by germ tube. *Peronosclerospora hetropogoni* conidia is globose, thin-walled, varying from 14.3–22.4×14.3–20.4 (17.7×16.9 μm) in size, conidia germinate by elongated germ tube, mycelium is coenocytic hyaline branched septate intercellular and produced palm like houstoria (Figure 8.7).

8.6.3 DISEASE CYCLE

Pathogen of sorghum downy mildew *P. sorghi* is soil-borne (Oospore in the soil) as well as seed-borne. Oospore can remain viable in plant debris for 15 month. According to Rajasah and Ramatingam (1985), haystack and shady places near fields serve as sources of inoculums. Secondary spreads of the disease is carried out by conidia and the disease more severe in wetter part of the field.

8.6.4 INTEGRATED DISEASE MANAGEMENT (IDM)

Methods for controlling downy mildews are largely aimed at manipulation of the environment to the advantage of the host and to the detriment of pathogen since the pathogen survive in the form of oospore in the host tissue, removal, destruction, and burning of the infected plant debris along with weeds serves to reduce the primary inoculums.

8.6.4.1 CULTURAL METHODS

8.6.4.1.1 Avoidance of Monoculture

Growing the same crop and same variety over and over in particular fields helps the pathogen in disease build uproots of the non-host crop stimulate germination of oospore of *P. sorghi* thus, reducing primary source of inoculums load in the soil. The incidence of infection can be reduced by rouging and showing "bait crop."

8.6.4.1.2   Timely Sowing

Late planting increased the downy mildew. In case of Rajasthan downy mildew, no disease occurred when maize was sown with the onset of monsoon or pre-monsoon rains (Siradhana et al., 1978).

8.6.4.1.3   Sanitation and Rouging

Since the pathogen survives in the form of oospore in the host tissue, removal, destruction, and burning of infected plant debris along with weed. Rouging of any infected plants from the field in the early stage of crop reduced the disease severity in the field.

8.6.4.2   BIOLOGICAL CONTROL

The incidence of systematic infection of *Peronosclerospora sorghi* reduced up to 58% when a chytrid fungus (*Gaeumannomyces* spp.), effective at parasitizing oospore of *Peronosclerospora sorghi*, was added to soil.

8.6.4.3   HOST PLANT RESISTANCE

HPR provides a practical and economic method to control downy mildews and is environmentally sound. More than 5000 germplasms lines have been evaluated, of these several highly resistance source *viz.;* AH-742, AH-772, AH 776, ICI-701, PMZ-707 were found highly resistant (Rathore et al., 2004).

8.6.4.4   CHEMICAL CONTROL

Downy mildew of maize can be controlled by the application of metalaxyl as a seed treatment at 0.32–2.0 g.a.i. per kilogram seed to control downy mildew. Apart from metalaxyl, alietie, and akomin formulation of phosphoric acid can be used to manage downy mildews. Complete protection against

Rajasthan downy mildew with treatment of Apron 35 WS @2 g per kb of seed were observed.

8.7   *FUSARIUM* STALK ROT

Fusarium stalk rot of maize is caused by *Fusarium verticillioides* (Saccardo) Nirenberg (= *Fusarium moniliforme* (Sheldon), was first reported from the United States of America by Pammel in 1914 as a serious root and stalk diseases. Later Valleau (1920) indicated that *Fusarium moniliforme* was a primary cause of root rot and stalk rot of maize. In India, *Fusarium* stalk rot was first reported from Mount Abu, Rajasthan.

8.7.1   *DISTRIBUTION*

Stalk rot is one of the most devastating soil-borne diseases of maize, occurring in all continents of the world, including the USA (Koehler, 1960), Europe (Ledencan et al., 2003), Africa, Asia (Lal and Singh, 1984), and Australia (Francis and Burgess, 1975). In India, the disease is prevalent in most of the maize growing areas, particularly in rain-fed areas *viz.*, Jammu, and Kashmir, Punjab, Haryana, Delhi, Rajasthan, Madhya Pradesh, Uttar Pradesh, Bihar, West Bengal, Andhra Pradesh, Tamil Nadu, and Karnataka, where water stress occurs after the flowering stage of the crop (Singh et al., 2012).

8.7.2   *ECONOMIC IMPACT*

The stalk rot usually occurs after flowering stage and prior to physi-ological maturity, which reduces yields in two ways: (i) affected plants die prematurely, thereby, producing lightweight ears having poorly filled kernels and (ii) plants with stalk rot easily lodge, which makes harvesting difficult, and ears are left in the field during harvesting (Singh et al., 2012). Stalk rot reduces maize yield directly by affecting the physi-ological activity of the plants and finally results in lodging, which is the main cause of economic losses (Ledencan et al., 2003). Lal et al. (1998) reported that incidence of post-flowering stalk rot complex (Charcoal rot, *Fusarium* stalk rot, late wilt) varying from 5 to 40% at different parts of the country.

8.7.3 ASSOCIATED PATHOGENS

As different species of pathogens have been isolated from diseased maize stalks in different parts of the world, therefore, it appeared to be a complex disease (Chambers, 1987). Among the variety of pathogens, *Fusarium* is considered as a devastating fungal menace of the most prevalent fungus on maize. Reports of surveys conducted in African countries showed *Fusarium* as the most prevalent fungus on maize (Baba Moussa, 1998). Doko et al. (1996) reported *Fusarium verticillioides* as the most frequently isolated fungus from maize and maize-based commodities in France, Spain, and Italy. Likewise, Orsi et al. (2000) found *Fusarium verticillioides* as the predominant species on maize in Brazil. Dorn et al. (2009) surveyed the prevalence of *Fusarium* species and its impact between the north and the south regions of Switzerland and between kernel and stem piece samples. Several species of *Fusarium* have been reported to cause stalk rots like, *Fusarium subglutinans* (*Fusarium semitectum*), *Fusarium avenaceum*, *Fusarium sulfurcum*, *Fusarium acuminatum*, *Fusarium roseum*, *Fusarium merismoides*, *Fusarium nivale*, and *Fusarium solani* (Rintelen, 1965; Kommedahal et al., 1972; Nur Ain Izzati et al., 2011). In India, so far only *Fusarium moniliforme* and *Fusarium semitectum* are reported to be widespread in Western Uttar Pradesh, Punjab, and Rajasthan (Lal and Diwivedi, 1982). Macroconidia of the pathogen, *Fusarium moniliforme* are hyaline, curved near the tips, three to five septate and 2.5–5 × 15–60 μm. Micro conidia are abundant single-celled, 2–3 × 5–12 μm and borne in chains. Conidiophores are unbranched with branched manophialids (Leslie and Summerell, 2006).

8.7.4 DISEASE SYMPTOMS

The disease becomes apparent when the crop enters senescence phase and severity increases during grain filling stage. The stalk rot symptoms are observed during post-flowering and pre-harvest stage (Lal and Singh, 1984). The rotting extends from infected roots to the stalk and causes premature drying, stalk breakage and ear dropping, thus significantly reducing maize yields (Colbert et al., 1987). The disease causes internal decay and discoloration of stalk tissues, directly reducing yield by blocking translocation of water and nutrients, thus resulting in death and lodging of the plant (Dodd, 1980) (Figure 8.8).

FIGURE 8.8   Typical symptoms of *Fusarium* stalk rot of maize.

8.7.5   *EPIDEMIOLOGY*

Temperature may be one of the factors that determine the extent of invasion of the stalk rot fungi of maize (Williams and Munkvold, 2008). *Fusarium verticillioides* is more common in regions with hot and dry growing conditions (Doohan et al., 2003), especially before or during pollination (Pascal et al., 2002). Reid et al. (2002) observed that hot and dry conditions, especially at maize silking stage predisposes the plants to infection by *Fusarium moniliforme* and *Fusarium proliferatum.* Williams and Munkvold (2008) reported the role of high temperatures in promoting systemic infection of maize by *Fusarium verticillioides*, but plant-to-seed transmission may be limited by other environmental factors that interact with temperature during the reproductive stages. The water stress at flowering and high soil temperature help in increasing of the magnitude of the stalk rot symptoms at the post-flowering stage of maize crop (Smith and McLaren, 1997). The PFSR is more severe under moisture stress condition after flowering (Kumar and Shekhar, 2005). Schneider et al. (1983) observed that pre-tasseling moisture stage resulted in higher stalk rot incidence compared to moisture stress at post pollination and grain filling stages. Mews et al. (1988) opined that pre-tassel moisture stress reduced the stalk rot during later season by reduced photosynthetic sink because the plants are subjected to the highest moisture stress and did not produce any grains. Soil texture affected the incidence of *Fusarium monili- forme* on maize when it was grown alone or intercropped with cowpeas and soybeans. Disease incidence was greater in sandy soil than in loam or clay soils (Mohamed, 1991). In general, stalk rot incidence and severity increase with increased fertility. There is evidence that potassium fertilizers reduces the severity of stalk rot and that nitrogen fertilizers, especially if in excess compared with potash, increases the severity of stalk rot.

8.7.6   DISEASE CYCLE

The fungus, *Fusarium moniliforme* survives on crop residue in the soil or on the soil surface (Nyvall and Kommedahl, 1970). Under the favorable condi- tion, it may infect roots as well as stalk (Lipps and Deep, 1991). *Fusarium moniliforme* may be present throughout the life cycle of the plant, originating from infected seed (Headrick and Pataky, 1990).

8.7.7   GENETICS OF RESISTANCE

Owing to its soil-borne infection pathway, fungicidal control of *Fusarium* stalk rot is not effective. Alternatively, discovery and utilization of resis- tance genes to improve maize tolerance to stalk rot is a cost-effective and environment friendly approach to reduce the grain yield loss. A large body of efforts is being diverted toward the development of biotechnological tools for identification and tagging of genes conferring resistance to PFSR. The identification of quantitative trait loci (QTL) for resistance to PFSR is considered as an efficient tool in the development of disease-resistant maize hybrids. A major gene for *Fusarium* stalk rot resistance has been reported on chromosome 6 (Yang et al., 2004). Studies have also indicated that resistance to stalk rot is quantitatively inherited and controlled by multiple genes with additive effects. Pe et al. (1993) identified five resistance QTL to *Fusarium* stalk rot, located on chromosome 1, 3, 4, 5, and 10. Yang et al. (2010) detected two loci QTL *qrfg1* and *qrfg 2*, conferring resistance to *Fusarium* stalk rot. Report from Egypt indicated that resistance to *Fusarium* stalk rot was controlled by two genes and was dominant in expression. These two genes were located in the short arm of chromosome 7 and the long arm of 10. Resistance to *Fusarium* stalk rot in inbred 61 C was also attributed to two genes. Source of resistance against *Fusarium* stalk rot of maize identified were CM 103, CM 119, CM 125, CI 21 E, CML 31, 77, 79, 85, 90, and CML 381 (Kumar and Shekhar, 2005).

8.7.8   DISEASE MANAGEMENT

Since the stalk rot of maize is a complex disease involving more than one organism, it is very difficult to manage the disease with single control measure. Hence, efforts are needed to explore the feasibility of combination of various control measures for integrated management of stalk rots (Kulkarni and Anahosur, 2011). Trivedi et al. (2002) evaluated systemic and non systemic fungicides *viz.*, bavistin, dithane M-45, blitox 50, hex cap-75, TMTD, topsin-M, and apron-35SD against *Fusarium pallidoroseum* causing post-flowering stalk rot of maize *in vitro* at different concentration *viz.* 100, 250, 500, and 1000 ppm. All the fungicides completely inhibited the growth at 500 and 1000 ppm, though Topsin-M recorded the highest growth inhibition (70% and 98%) at low concentrations, i.e., 100 and 250 ppm, respectively. Chandra et al. (2008) evaluated two fungicides *viz.*, tebuconazole, *and* thiabendazole for their ability to inhibit the growth of toxigenic *Fusarium verticillioides* and found that tebuconazole 5% aqueous solution effectively reduced ear rot disease and fumonisins accumulation to a maximum extent compared to other fungicides. Bioagents are useful for the effective management of soil-borne pathogen propagules like chlamydospores of *Fusarium* species. The isolation and identification of effective biocontrol agents against PFSR is urgently required for use in IDM. For decades, various *Trichoderma* species had shown antagonistic activities against many pathogens, both *in vitro* and *in vivo* (Howell, 2003). Successful growth suppression of *Fusarium verticillioides (in vitro)* and its subsequent significant exclusion from internodes of maize (*Zea mays*) stem in the field (*in vivo*) by strains of *Trichoderma pseudokoningii* had been reported by Sobowale et al. (2005). Shekhar and Kumar (2010) reported the native isolate of *T. harzianum* resulted in good plant health and reduced post-flowering stalk rot of maize. Patil et al. (2003) reported the seed treatment with *T. harzianum* (4 g/kg seed) along with soil application of castor or neem cake (250 kg/ha), 15 days prior to sowing gave an effective control to stalk rot disease and gave better cost-benefit ratio.

8.7.8.1 INTEGRATED DISEASE MANAGEMENT (IDM)

Integration of biological and chemical control seems to be a promising way of controlling many pathogens with minimum interference in the biological equilibrium in soil (Papavizas, 1973). Since soil is highly complex and biologically active substrate through which the fungicide act against fungi, fungitoxicants often give viable success in controlling seedling disease of crops in diverse agro-climatic regions of the world (Khan et al., 2008). The use of fungicides and tolerant genotypes has been reported to be an effective method to manage stalk rot of maize which holds some promise. Kulkarni and Anahosur (2011) reported that application of farmyard manure and neem cake along with *Trichoderma harzianum* 15–20 days before sowing with two additional irrigation at tasseling and silking stage reduced the disease from 70.08 to 13.24%. Thori et al. (2011) reported that maximum germination (90%) with minimum mortality (0.0 and 2.5%) at 35 and 70 DAS and least percent disease index (PDI) of 23.2% was recorded by integration of *T. viride* (drenching), with bavistin seed treatment, followed by tebuconazole (ST) + *Trichoderma viride* (drenching). Among the individual treatments, seed treatment with bavistin and *Trichoderma viride* drenching showed good effects and resulted in 75% germination with 3.3% and 7.1% mortality after 35 and 70 days after sowing followed by 72.5% in tebuconazole seed treatment. Integration of plant resistance with these components was useful for reducing the losses caused by PFSR pathogen.

8.8 CONCLUSION

Plants and pathogens are continuously confronted with each other during evolution in a battle for growth and survival. In this rivalry, plants have evolved a stunning array of structural, chemical, and gene-based defenses, designed to combat pathogens of different nature and, so as the pathogens by developing new races. The overall destruction of maize diseases and the major diseases of maize crop has been documented. The interaction of pathogens with the resistance genes are just like a key to lock approach, while the virulence genes in the pathogens can cause disease in its host regardless of the genetic architecture of the host plant. Therefore, breeding for disease resistance will remain the most economical way of controlling maize diseases. Disease resistance in maize is reported to be conditioned by both major (qualitative) genes and minor (quantitative) genes or QTL. To date, 437 quantitative disease resistance loci or dQTL, 17 major resistance genes and 25 resistance gene analogs (RGAs) associated with resistance to 11 major maize diseases have been described in a few of these sources. However, management of these resistance genes to prolong their effectiveness or slow down their breakdown would require integration of other strategies, including the use of chemicals, biocontrol agents, and cultural practices. However, full exploitation of genomic approaches and information to develop and release maize cultivars more resilient to diseases will only be possible through (i) a deeper integration of genomic approaches with conventional breeding methodologies; (ii) a capacity to reliably and accurately phenotype diseases on a large scale; and (iii) a sound multidisciplinary knowledge of the biochemical and physiological processes determining crop yield and its stability under different disease stress regimes. Shared responsibility in the face of limited resources will allow us to generate more information that will bring about significant gains in our quest to use host resistance to durably manage maize diseases than single institutions would achieve alone.

8.8.1   FUTURE THRUST

1. The survey and surveillance of the diseases should be a regular feature covering wider areas under maize cultivation.
2. Studies on population biology of various foliar pathogens over time and space and to evaluate for stability of known resistance lines.
3. Develop comprehensive systems for race characterization and differential lines for different foliar pathogens.
4. The identified resistant inbred lines/genotypes may be used as the source of resistance in the development of single cross and other hybrids.
5. Determine the extent of seed transmission of different pathogen and standardize their acceptable limit for seed tech. and seed trade.
6. The proven resistant hybrids may be deployed in PFSR endemic areas after confirming their yield performance through multilocation yield trials.

7. Fingerprinting of inbreds and hybrids identified as donors of PFSR resistance and development of molecular markers for resistance to the disease can be undertaken.
8. Detection techniques to identify the pathogen and its variability using serological and molecular tools.
9. Use of isogenic lines in the race identification to be taken up.
10. Evaluate and develop resistance gene management strategies for durable resistance.
11. Integrated management approaches need to be refined and output-oriented research should be focused.
12. Develop location-specific integrated management modules by combining HPR, cultural practices and biological control for specific disease.
13. Develop technique and strategies to monitor virulence shifts in different pathogen through differential host lines as well as through molecular techniques.
14. Construction of refined molecular linkage maps using DNA analysis, understanding species relationship through the analysis of mitochondria and genetic transformation. The greatest gains from biotechnology in the near future can possibly come from work of defensive traits, especially for complex diseases like stalk rot.

CONFLICT OF INTEREST

All authors declare that they have no conflict of interest.

KEYWORDS

- **disease management**
- **disease resistance**
- **diseases yield losses**
- **epidemiology**
- **maize**
- **quantitative trait loci**

REFERENCES

Ahuja, S. C., & Payak, M. M., (1981). A laboratory method for evaluating maize germplasm to banded leaf and sheath blight. *Indian Phytopathol., 31,* 34–37.

Ahuja, S. C., & Payak, M. M., (1988). Banded leaf and sheath blight of maize. In: Agnihotri, V. P., Sarbhay, A. K., & Kumar, D., (eds.), *Perspectives in Mycology and Plant Pathology* (pp. 178–186). Malhotra Publishing House, New Delhi.

AICRP, (2013). *Annual Report of AICRP* (p. 148). Directorate of Maize Research, Maize Pathology New Delhi.

Akhtar, J., Jha, V. K., Kumar, A., & Lal, H. C., (2009). Occurrence of banded leaf and sheath blight of maize in Jharkhand with reference to diversity in *Rhizoctonia solani. Asian J Agri Sci., 1,* 32–35.

Ali, F., & Yan, J., (2012). Disease resistance in maize and the role of molecular breeding in defending against global threat. *Journal of Integrative Plant Biology, 54*(3), 134–151.

Asea, G., Vivek, B. S., Lipps, P. E., & Pratt, R. C., (2012). Genetic gain and cost efficiency of marker-assisted selection of maize for improved resistance to multiple foliar pathogens. *Mol. Breeding, 29,* 515–527.

Baba-Moussa, A. A., (1998). La microflora associee aux degats des lepidopteres foreurs de tigeset mi- neurs d'epis de mais (Zea mays) dans la region Sud du Benin avec reference speciale a *Fusarium moniliforme* Sheld. Memoired D'Ingenieur Agronomy. University du Benin, Lome, Togo 93.

Bertus, L., (1927). *A Sclerotial Disease of Maize Due to Rhizoctonia Solani* (pp. 46–48). Yearbook. Department of Agriculture, Ceylon.

Blandino, M., Galeazzi, M., Savoia, W., & Reyneri, A., (2012). Timing of azoxystrobin. Propiconazole application on maize to control northern corn leaf blight and maximize grain yield. *Field Crops Research, 139,* 20–29.

Chambers, K. R., (1987). Stalkrotofmaize: Host-pathogen Interaction. *J. Phytopath., 118,* 103–108.

Chandra, N. S., UdayaShankar, A. C., Niranjan, R. S., Ni-ranjana, S. R., & Prakash, H. S., (1987/2008). Tebuconazole and thiabendazole-novel fungicides to control toxigenic *Fusarium verticilloides* and fumonisin in maize. *J. Mycol. Plant Pathol., 38*(3), 430–436.

Chenulu, V. V., & Hora, T. S., (1962). Studies on losses due to *Helminthosporium* blight of maize. *Indian Phytopathology, 15,* 235–237.

Chodhary, O. P., Amit, T. R. N., Bunker, & Kusum, M., (2011). A new record in Involvement of *Curvularia andropogonis* in causing leaf spot on maize in India. *Indian Jr. Mycol. Pl. Pathol., 41*(1), 123–125.

Colbert, T. R., Kang, M. S., Myers, O., & Zuber, M. S., (1987). General and specific combining ability estimates for pith cell death in stalk internodes of maize. *Field Crop Res., 17,* 155–162.

Concibido, V. C., Denny, R. L., Boutin, S. R., Hautea, R., Orf, R. H., & Young, N. D., (1994). DNA marker analysis of loci underlying resistance to soybean cyst nematode (*Heterodera glycines Ichinohe*). *Crop Sci., 34,* 240–246.

De Leon, R., & Pandey, E. B., (1989). Durable resistance to two leaf blights in two maize inbred lines. *Theoretical and Applied Genetics, 80,* 542–544.

Degani, O., & Cernica, G., (2014). Diagnosis and control of *Harpophora maydis*, the cause of late wilt in maize. *Advances in Microbiology, 4,* 94–105.

Degefu, Y., & Hanif, M., (2003). Agrobacterium-tumefaciens-mediated transformation of *Helminthosporium turcicum*, the maize leaf-blight fungus. *Archives of Microbiology, 180,* 279–284.

Deise, I. D. C., & Walter, B., (2008). Aerial and ground applications of fungicide for the control of leaf diseases in maize crop (*Zea mayz* L.). *CIGR – International Conference of Agricultural Engineering.* XXXVII Congresso Brasileiro de Engenharia Agrícola, Brazil.

Dharanendraswamy, S., (2003). *Studies on Turcicum Leaf Blight of Maize Caused by Exserohilum Turcicum.* MSc (Agri.) Thesis. Dharwad: University of Agricultural Sciences.

Dodd, J. L., & White, D. G., (1999). Seed rot, seedling blight, and damping-off. In: White, D. G., (ed.), *Compendium of Corn Diseases* (pp. 10, 11). Saint Paul. The American Phytopathological Society.

Doko, M. B., Canet, C., Brown, N., Sydenham, E. W., Mpu-Chane, S., & Siame, B. A., (1996). Natural occurrence of fumonisin and Zearalenone in cereals and cereal-based foods from eastern and southern Africa. *J. Agric. Fd. Chem., 44,* 3240–3243.

Doohan, F. M., Brennan, J., & Cooke, B. M., (2003). Influence of climatic factors on *Fusarium* species pathogenic to cereals. *Euro J. Pl. Path., 109,* 755–768.

Dorn, B., Forrer, H. R., Schurch, S., & Vogelgsang, S., (2009). *Fusarium* species complex on maize in Switzer-land: Occurrence, prevalence, impact and mycotoxin in commercial hybrids under natural infection. *Eur. J. Plant Path., 125,* 51–61.

Drechsler, C., (1934). Phytopathological and taxonomic aspects of *Ophiobolus, Pyrenophora, Helminthosporium* and a new genus, Cochliobolus. *Phytopathology, 24,* 953–983.

Drori, R., Sharon, A., Goldberg, D., Rabinovitz, O., Levy, M., & Degani, O., (2012). Molecular diagnosis for *Harpophora maydis*, the cause of maize late wilt in Israel. *Phytopathology Mediterranea, 52,* 16–29.

Fisher, D. E., Hooker, A. L., Lim, S. M., & Smith, D. R., (1976). Leaf infection and yield loss caused by four *Helminthosporium* leaf diseases of corn. *Phytopathology, 66,* 942–944.

Francis, R. G., & Burgess, L. W., (1975). Surveys of *Fusarium* and other fungi associated with stalk rot of maize in Eastern Australia. *Aust. J. Agric. Res., 26,* 801–807.

Girma, T., Fekede, A., Temam, H., Tewabech, T., Eshetu, B., Melkamu, A., Girma, D., & Kiros, M., (2006). Review of maize, sorghum and millet pathology research. In: Tadesse, A., & Ali, K., (eds.), *Proceedings of the 14th Conference of the Plant Protection Society of Ethiopia* (pp. 39–47). Addis Ababa.

Gould, F., & Cohen, M., (2000). Sustainable use of genetically modified crops in developing countries. Agricultural biotechnology and the poor. In: Persley, G., & Lantin, M., (eds.), *Proceedings of an International Conference* (pp. 139–146). Washington, DC, USA.

Hafiz, A., (1986). *Plant Diseases* (p. 552). Directorate of Publication, Pakistan Agricultural Research Council, Islamabad, Pakistan.

Hagan, A., (2010). *Rust Diseases in Field Corn* (p. 14). The Alabama Cooperative Extension System.

Han, Y. P., Xing, Y. Z., Cheng, Z. X., Gu, S. L., Pan, X. B., Chen, X. L., & Zhang, Q. F., (2002). Mapping QTL for horizontal resistance to sheath blight in an elite rice restorer line, Minghui 63. *Acta Genet. Sin., 29*(7), 622–626.

Harlapur, S. I., Wali, M. C., Prashan, M., & Shakuntala, N. M., (2002). Assessment of yield losses in maize due to charcoal rots in Ghataprabha Left Bank Canal (GLBC) command area of Karnataka. *Karnataka J. agric. Sci., 15,* 590, 591.

Headrick, J. M., Pataky, J. K., & Juvik, J. A., (1990). Relationships among carbohydrate content of kernels, condition of silks after pollination, and the response of sweet corn inbred lines to infection of kernels by *Fusarium moniliforme. Phytophthora., 80,* 487–494.

Hoitink, H. A. J., (1980). Composted bark a light weight growth medium with fungicidal properties. *Plant Dis., 64,* 142–147.

Hooda, K. S., Khokhar, M. K., Shekhar, M., Chikkappa, G. K., Bhupinder, K., Mallikarjuna, N., Devlash, R. K., et al., (2017). *Turcicum* leaf blight-sustainable management of a re-emerging maize disease. *J. Plant Dis. Prot., 124,* 101–113.

Hooda, K. S., Sekhar, J. C., Chikkappa, G., Kumar, S., Pandurange, K. T., Sreeramsetty, T. A., Sharma, S. S., Kaur, H., et al., (2012). Identifying sources of multiple disease resistance in maize. *Maize Journal., 1*(1), 82–84.

Johal, L., Huber, D. M., & Martyn, R., (2004). Late wilt of corn (maize) pathway analysis: Intentional introduction of *Cephalosporium maydis.* In: *Pathways Analysis for the Introduction to the U.S. of Plant Pathogens of Economic Importance.* U.S. Department of Agriculture, Animal and Plant Health Inspection Service. Technical Report no. 503025.

Kato, A., & Incue, Y., (1995). Resistance to banded leaf and sheath blight (*Rhizoctonia solani* Kuhn) after fall of lower sheaths in maize (*Zea mays* L). *Bulletin of the National Grass/and Research Institute (Japan), 51,* 1–5.

Kaul, T. N., (1957). Food and agriculture organization. *Plant Prot. Bulletin., 5,* 93–96.

Kemerait, R. C., (2012). Corn disease and nematode update for 2013. In: *A Guide to Corn Production in Georgia.* Athens, GA.

Khan, M. S., Zaidi, A., & Wani, P. A., (2008). *Role of Phosphate Solubilizing Micro-Organisms in Sustainable Agriculture.* Nova Science Publishers, New York.

Khokhar, M. K., Sharma, S. S., & Gupta, R., (2014). Effect of plant age and water stress on the incidence of post-flowering stalk-rot of maize caused by *Fusarium verticillioides. Indian Phytopath., 67*(2), 143–146.

Kommedahal, T., Windels, C. E., & Stucker, R. E., (1972). Occurrence of *Fusarium* species in root and stalks symptoms in corn plants during the growing season. *Phytopathol., 69,* 61–966.

Krishna, K. M., Chikkappa, G. K., & Manjulatha, G., (2013). Components of genetic variation for *Macrophomina phaseolona* resistance in maize. *J. Res. ANGRAU, 41*(3) 12–15.

Kulkarni, S., & Anahosur, K. H., (2011). Integrated management of dry stalk rot disease of maize. *J. Pl. Dis. Sci., 6*(2), 99–106.

Kumar, S., & Shekhar, M., (2005). *Stress on Maize in Tropics* (pp. 172–194). Published by Directorate of Maize Research, Cummings Laboratory, Pusa Campus, New Delhi. Angkor Publisher (P) Ltd. Noida.

Kumar, S., Archana, R., & Jha, M. M., (2009). Efficacy of fungicide against *Helminthosporium* maydis of maize. *Annals Plant Protection Sciences., 17,* 255, 256.

Lal, S., & Diwivedi, B. R., (1982). Chephalosporium and *Fusarium* stalk rots of maize. *Recent Ads. in Pl. Path.*, 344–360.

Lal, S., & Singh, I. S., (1984). Breeding for resistance to downy mildews and stalk rots in maize. *Theor. Appl. Genet., 69,* 111–119.

Lal, S., Baruah, P., & Butchaiah, K., (1980). Assessment of yield losses in maize cultivars due to banded sclortial disease. *Indian Phytopathology, 33,* 440–443.

Lal, S., Leon, D. C., Saxena, V. K., Singh, S. B., Singh, N. N., & Vasal, S. K., (1998). Maize stalk rot complexes: Innovative breeding approaches. *Proc. Seventh Asian Regional Maize Workshop.* Los Banos, Philippines.

Laxminarayana, C., & Shankarlingam, S., (1983). *Turcicum* leaf blight of maize, techniques of scoring for resistance to important diseases of maize. In: *Proceedings of All India Coordinated Maize Improvement Project* (pp. 16–24). New Delhi: Indian Agricultural Research Institute.

Ledencan, T., Simic, D., Brkic, I., Jambrovic, A., & Zdunic, Z., (2003). Resistance of maize inbreds and their hybrids to *Fusarium* stalk rot. *Czech. J. Genet. Pl. Breed., 39,* 15–20.

Levy, Y., & Pataky, J. K., (1992). Epidemiology of Northern leaf blight on sweet corn. *Phytoparasitica, 20*(1), 53–66.

Li, Y., Dai, F. C., Jing, R. L., Wang, T. Y., Du, J. Y., & Jia, J. Z., (2002). QTL analysis of resistance to *Curvularia lunata* in maize. *Sci. Agr. Sinica., 35,* 1221–1227.

Lim, S. M., Kinsey, J. G., & Hooker, A. L., (1974). Inheritance of virulence in *Helminthosporium turcicum* to monogenic resistant corn. *Phytopathology, 64,* 1150, 1151.

Lin, H. J., Tan, D. F., Zhang, Z. M., Lan, H., Gao, S. B., Rong, T. Z., & Pan, G. T., (2008). Analysis of digenic epistatic and QTL x environment interactions for resistance to banded leaf and sheath blight in maize (*Zea mays*). *Int. J. for Agri. and Bio., 10,* 605–611.

Lipps, P. E., & Deep, I. W., (1991). Influence of tillage and crop rotation in yield, stalk rot and recovery of *Fusarium* and *Trichoderma* spp. from corn. *Pl. Dis., 75,* 828–833.

Maiti, A. P., (1978). Two new ear rots of maize from India. *Plant Dis. Report., 62,* 1074–1076.

Mallikarjuna, N., Pandurange, G. K. T., Manjunath, B., Kiran, K. K. C., & Sunil, K. N., (2007). Turcicum leaf blight disease severity of maize in relation to meteorological factors. *Environment and Ecology, 25*(4), 778–784.

Mehra, R., (2011). Disease management in maize. In: Mehla, J. C., Jaipal, S., Kamboj, M. C., Chand, M., & Mehra, R., (eds.), *Three Decades of Maize Research* (pp. 23–25). CCS HAU, RRS, Karnal.

Mews, M., & Rijkenberg, F. H., (1988). Moisture stress in the screening of maize cultivars for Stalk rot resistance and yield. *Pl. Dis., 72,* 1061–1064.

Misra, A. P., (1979). Variability, physiologic specialization and genetics of pathogenicity in graminicolous *Helminthosporia* affecting cereal crops. *Indian Phytopath., 32,* 1–22.

Mitra, M. A., (1981). comparative study of species and strains of *Helminthosporium* on certain Indian cultivated crops. *Transactions British Mycological Society, 15,* 254–293.

Mohamed, M. S., (1991). Effect of soil texture on incidence of maize stalk rot caused by *Fusarium moniliforme* in intercropping planting. *Asian Jagric. Sci., 22,* 3–12.

Natsvlishvilli, A. A., (1972). The basis of long term forecast of *Helminth sporiosis* of maize. *Mikol Fitopatol., 6,* 62–65.

Nur, A. I. M. Z., Azmi, A. R., Siti, N. M. S., & Norazlina, J., (2011). Contribution to knowledge of diversity of *Fusarium* associated with maize in Malaysia. *Plant Protect Sci., 47,* 20–24.

Nyvall, R. F., & Kommedahl, T., (1970). Saprophytes and survival of *Fusarium moniliforme* in corn stalks. *Phytopath., 60,* 1233–1235.

Oerke, E. C., (2005). Crop losses to pests. *The Journal of Agricultural Science, 144,* 31–43.

Orsi, R. B., Correa, B., Possi, C. R., Schammass, E. A., Nogueira, J. R., Dias, S. M. C., & Mallozzi, M. A. B. (2000). Microflora and occurrence of fumonisin in freshly harvested and stored hybrid maize. *J. Stored Prod. Res., 36,* 75–87.

Pal, D., & Kaiser, S. A. K. M., (2001). Effect of agronomic practices on maydis leaf blight disease of maize. *Journal of Mycopathological Research, 39,* 77–82.

Pan, Y. B., & Rush, M. C., (1997). Studies in the U.S. on genetics and breeding of resistance to rice sheath blight. *J. Jiangsu Agric. College, 18,* 57–63.

Pandurange, G. K. T., Shekara, S., Jayaramegowda, B., Prakash, H. S., & Sangamlal, (1993). Genetics of resistance to *Turcicum* leaf blight of maize. *Mysore J. Agric. Sci., 27*, 262–267.

Papavizas, G. C., (1973). Status of applied biological control of soil-borne plant pathogens. *Soil Biol. Biochem., 5*, 709.

Pascale, M., Visconti, A., & Chelkowsky, J., (2002). Ear rot susceptibility and mycotoxin contamination of maize hybrids inoculated with *Fusarium* species under field conditions. *Eur. J. Plant Pathol., 108*, 645–651.

Pataky, J. K., (1999). Rusts. In: Donald, G. W., (ed.), *Compendium of Corn Diseases* (pp. 35–38). St. Paul, Minnesota: *The American Phytopathology Society*.

Patil, R. K., Goyal, S. N., Patel, B. A., Patel, R. G., Patel, D. J., Singh, R. V., Panakaj, et al., (2003). Integrated management of stalk rot disease and phyto-nematodes in rabi maize. *Proceedings of National Symposium on Biodiversity and Management of Nematodes in Cropping Systems for Sustainable Agriculture* (pp. 250–254). Jaipur.

Patrick, J. K., Lafitte, H., & Redmeads, G. O., (2005). Association between traits in tropical maize inbred lines and their hybrids under high and low soil fertility. *Maydica., 47*, 259–267.

Payak, M. M., & Renfro, B. L., (1966). Diseases of maize new to India. *Indian Phytopathol Soc. Bull., 3*, 14–18.

Payak, M. M., & Sharma, R. C., (1985). Maize diseases and approaches to their management in India. *Trop. Pest Mgmt., 31*, 302–310.

Pe, M. E., Gianfranceschi, L., Taramino, G., Tarchini, R., Angelini, P., Dani, D., & Binelli, G., (1993). Mapping quantitative trait loci QTLs for resistance to *Gibberella zeae* infection in maize. *Mol. Gen. Genet., 241*, 11–16.

Pingali, P. L., (2001). *CIMMYT 1999–2000: World Maize Facts and Trends.* Meeting world maize needs: Technological opportunities and priorities for the public sector. CIMMYT, Mexico, D.F.

Rakesh, D., Guleria, S. K., & Thakur, D. R., (2011). Evaluation of seed dressing fungicides for the management of banded leaf &sheath blight of maize. *Plant Dis. Res., 26*, 169.

Rani, V. D., Reddy, P. N., & Devi, G. U., (2013). Banded leaf and sheath blight of maize incited by *Rhizoctonia solani* f. sp *Sasakii* and its management: A review. *Int. J. App. Biol. Pharm. Technol., 4*, 52–60.

Rathore, R. S., Bohra, B., Trivedi, A., & Mathur, K., (2005). *An Overview of Fusarium Stalk Rot of Maize in Rajasthan: Progress and Future Perspective in the 9th Asian Regional Maize Workshop* (pp. 175–178). Held on Sept 5–9, 2005 at Beijing, China, organized by CAAS and CIMMYT. Section-II biotic and abiotic stresses.

Reddy, T. R., Reddy, P. N., & Reddy, R. R., (2013). *Turcicum* leaf blight of maize incited by *Exserohilum turcicum*: A review. *International Journal of Applied Biology and Pharmaceutical Technology, 5*, 54–60.

Reid, L. M., Woldemariam, T., Zhu, X., Stewart, D. W., & Schaafsma, A. W., (2002). Effect of inoculation time and point of entry on disease severity in *Fusarium graminearum, Fusarium verticillioides,* or *Fusarium subglutinans* inoculated in maize ears. *Canadian J. Pl. Path., 24*, 162–167.

Richards, R., & Kucharek, T., (2006). *Florida Plant Disease Management Guide.* University of Florida.

Rintelen, J., (1965). *Fusarium culmorum* and *Fusarium* arten als erreger einer stengelfaule an reifen- den maispflanzen. *Z Pflanzenkr (Pflanzenpathol) Pflanzenschutz, 72*, 89–91.

Saxena, S. C., (2002). Bio-intensive integrated disease management of banded leaf and sheath blight of maize. In: *Proceed of 8th Asian Regional Maize Workshop: New Technologies for the New Millennium* (pp. 380–388). Bangkok, Thailand.

Schechert, A. W., Welz, H. G., & Geiger, H. H., (1999). QTL for resistance to *Setosphaeria turcica* in tropical African maize. *Crop Science, 39*, 514–523.

Schneider, R. W., & Pendery, W. E., (1983). Stalk rot of corn: Mechanism of predisposition by in early season water stress. *Phytopath., 73*, 863–871.

Sharma, R. C., & Hembram, D., (1990). Leaf stripping: A method to control banded leaf and sheath blight of maize. *Curr. Sci., 59*, 745, 746.

Sharma, R. C., Lilaramani, J., & Payak, M. M., (1978). Outbreak of a new pathotype of *Helminthosporium maydis* on maize in India. *Indian Phytopath., 31*, 112, 113.

Sharma, R. C., Rai, S. N., & Batsa, B. K., (2005). Identifying resistance to banded leaf and sheath blight of maize. *Indian Phytopathology, 58*, 121, 122.

Sharma, R. C., Srinivas, P., & Batsa, B. K., (2002a). Banded leaf and sheath blight of maize its epidemiology and management. In: Rajbhandari, N. P., Ransom, J. K., Adhikari, K., & Palmer, A. F. E., (eds.), *Proceedings of a Maize Symposium Held* (pp. 108–112). NARC and CIMMYT, Kathmandu.

Shehata, A. H., & Salem, A. M., (1971). Genetic analysis of resistance to late wilt of maize caused by *Cephalosporium maydis.* In: *7th Inter Asian Corn Improvement Workshop* (pp. 60–65). Los Banos. Philippines.

Shekhar, M., & Kumar, S., (2010). Potential biocontrol agents for the management of *Macrophomina phaseolina*, incitant of charcoal rot in maize. *Archives Phytopath. Pl. Prot., 43*, 379–383.

Shivapuri, A., Sharma, O. P., & Jhamaria, S. L., (1997). Fungi- toxic properties of plant extracts against pathogenic fungi. *Journal of Mycology and Plant pathology, 27*, 29–31.

Sibiya, J., Tongoona, P., Derera, J., & Makanda, I., (2013). Smallholder farmers' perceptions of maize diseases, pests, and other production constraints, their implications for maize breeding and evaluation of local maize cultivars in KwaZulu-Natal, South Africa. *African Journal of Agricultural Research, 17*, 1790–1798.

Singh, A., & Ashwani, B., (2012). Important diseases of maize diseases and their eco-friendly management. In: Vaibhav, K. S., Yogendra, S., & Akhilesh, S., (eds.), *Eco-Friendly Innovative Approaches in Plant Disease Management, Chapter: 16* (pp. 357–386). Publisher: International Book Distributors, Dehradun.

Singh, A., & Shahi, J. P., (2012). Banded leaf and sheath blight: An emerging disease of maize. *Maydic, 57*, 215–219.

Singh, A., & Singh, D., (2011). Integrated disease management strategy of banded leaf and sheath blight of maize. *Plant Dis. Res., 26*, 192.

Siradhana, B. S., Dange, S. R. S., Rathore, R. S., & Singh, S. D., (1978). Ontogenic predisposition of *Zea mays* to sorghum downy mildew. *Pl. Dis. Reptr., 62*(5), 467, 468.

Siradhana, B., Dange, S. R. S., Rathore, R. S., & Singh, S. D., (1980). A new downy mildew of maize in Rajasthan, India. *Curr. Sci., 49*, 316, 317.

Sivakumar, G., Sharma, R. C., & Rai, S. N., (2000). Biocontrol of banded leaf and sheath blight of peat-based *Pseudomonas fluorescens* formulation. *Indian Phytopath., 53*, 190–192.

Smith, E., & McLaren, M., (1997). Effect of water stress on colonization of maize roots by root infecting fungi. *African Pl. Prot., 3*, 47–51.

Sobowale, A. A., Cardwell, K. F., Odebode, A. C., Bandyo-Padhyay, R., & Jonathan, S. G., (2005). Growth inhibition of *Fusarium verticillioides* (Sacc.) Nirenberg by isolates of

Trichoderma pseudokoningii strains from maize plant parts and its rhizosphere. *J. Plant Prot. Res., 45*(4), 249–266.

Spencer, M. A., & Dick, M. W., (2002). Aspects of graminicolous downy mildew biology: Perspectives for tropical plant pathology and *Peronosporomycetes* phylogeny. In: Watling, R., Frankland, J. C., Ainsworth, A. M., Isaac, S., & Robinson, C. H., (eds.), *Tropical Mycology: Micromycetes* (Vol. 2, pp. 63–80). CAB International, Wallingford, UK.

Sprague, G., (1977). *Corn and Corn Improvement* (p. 774). American Society of Agronomy: Inc., Publisher Madison, Wisconsin, USA.

Tang, H. T., Rong, T. Z., & Yang, J. P., (2004). Research advance on sheath blight (*Zea mays* L.) in maize. *J. Maize Sci., 12*(1), 93–96, 99.

Tewabech, T., Dagne, W., Girma, D., Meseret, N., Solomon, A., & Habte, J., (2012). Maize pathology research in Ethiopia in the 2000s. In: Worku, M., Twumasi-Afriyie, S., Wolde, L., Tadesse, B., Demisie, G., Bogale, G., Wegary, D., & Prasanna, B. M., (eds.), *Meeting the Challenges of Global Climate Change and Food Security Through Innovative Maize Research* (pp. 193–201). Proceedings of the 3rd national maize workshop of Ethiopia. Addis Ababa, Ethiopia.

Thakur, S. M., Sharma, S. L., & Munjal, R. L., (1973). Correlation studies between incidence of banded sclerotial disease and ear yield in maize. *Indian J. Mycol. Plant Patho., 3,* 180, 181.

Thind, B. S., & Payak, M. M., (1978). Evaluation of maize germplasm and estimation of losses to Erwinia stalk rot. *Plant Dis. Rep., 62,* 319–323.

Thori, H., (2011). *Investigation on Biology and Management of Fusarium Moniliforme Sheldon causing Post Flowering Stalk Rot of Maize (Zea mays L.).* MSc thesis, Department of plant pathology RCA (MPUAT) Udaipur.

Trivedi, A., Jain, K. L., & Kothari, K. L., (2002). Efficacy of some fungicides against *Fusarium pallidoroseum* causing stalk rot in maize. *Plant Dis Res., 17*(2), 332, 333.

Tsen, C. M., (1981). The effect of sowing date on the occurrence of northern corn leaf blight and on the yield of disease. *Report of Corn. Res. Center Tainan DAIS, 15,* 31–36.

Veerabhadraswamy, A. L., Pandurangegowda, K. T., & Prasanna, K. M. K., (2014). Efficacy of strobilurin group fungicides against *Turcicum* leaf blight and polysora rust in maize hybrids. *Int. J. Agric. Crop Sci., 7*(3), 100–106.

Vieira, R. A., Mesquini, R. M., Cleiltan, N. S., Fernando, T. H., Dauri, J. T., & Carlos, A. S., (2014). A new diagrammatic scale for the assessment of northern corn leaf blight. *Crop Protection, 56,* 55–57.

Vivek, B. S., Odongo, O., Njuguna, J., Imanywoha, J., Bigirwa, G., Diallo, A., & Pixley, K., (2010). Diallel analysis of grain yield and resistance to seven diseases of 12 African maize (*Zea mays* L.) inbred lines. *Euphytica., 172,* 329–340. doi: 10.1007/s10681-009-9993-5.

Warcup, J. H., & Talbot, P. H. B., (1962). Ecology and identity of mycelia isolated from soil. *Trans Br. Mycol. Soc., 45,* 495–518.

Williams, M. A., & Munkvold, G. P., (2008). Systemic infection by *Fusarium verticilloides* in maize plants grown under three temperature regimes. *Pl. Dis., 92,* 1695–1700.

Wise, K., (2014). *Fungicide Efficacy for Control of Corn Diseases.* Purdue Extension Publication BP-160-W. West Lafayette: Purdue University.

Yang, D. E., Zhang, C. L., Zhang, D. S., Jin, D. M., Weng, M. L., Chen, S. J., Nguyen, H., & Wang, B., (2004). Genetic analysis and molecular mapping of maize (*Zea mays* L.) stalk rot resistant gene *Rfg1. Theor. Appl. Genet., 108,* 706–711.

Yang, Q., Yin, G., Guo, Y., Zhang, D., Chen, S., & Mingliang, X. M., (2010). A major QTL for resistance to gibberella stalk rot in maize. *Theor. Appl. Genet., 121,* 673–687.

Zhao, J., Wang, G. Y., Hu, J., Zhang, X. H., & Dai, J. R., (2002). Genetic analysis of maize resistance to *curvularia* leaf spot by ADAA model. *Acta Agron. Sinica., 28,* 127–130.

Zhao, M. J., Zhang, Z. M., Zhang, S. H., Li, W., Jeffers, D. P., Rong, T. Z., & Pan, G. T., (2006). Quantitative trait loci for resistance to banded leaf and sheath blight in maize. *Crop Sci., 46,* 1039–1045.

Zhao, Y., & Wang, Z., (2009). Research progress on northern leaf blight in corn. *Journal of North-East Agricultural University, 16*(2), 66–71.

CHAPTER 9

Recent Advances in Detection, Diagnosis, and Management of Finger Millet Diseases

PARDEEP KUMAR,[1] SHRVAN KUMAR,[2] JIWAN PAUDEL,[2] and D. P. SINGH[3]

[1]*KVK, Sohna (ANDUA&T, Ayodhya), Siddharthnagar, Uttar Pradesh, India, E-mail: drpardeepviro@gmail.com*

[2]*Rajiv Gandhi South Campus, BHU, Barkachha, Mirzapur, Uttar Pradesh – 231001, India*

[3]*KVK, Maharajganj (ANDUA&T, Ayodhya), Uttar Pradesh, India*

ABSTRACT

Ragi, *Eleusine coracana*, is a tall grass that is also well-known finger millet, African millet. It has tufted stalks, each with 4–6 spikes. Both upland and irrigated forms are cultivated from Northern Africa to Indonesia. Ragi gives a very high yield, frequently exceeding 1,500 pounds per acre. In India, it is a major food crop, especially during the rainy season. The grain is free from insects and can be stored for lengthy periods. Its flour is used for puddings and cakes, and a fermented drink is prepared from the grain. Many pathogen species cause diseases that cause major economic and production losses in the agricultural industry worldwide. Early monitoring of plant health and detecting the pathogen are important to decrease the disease intensity and spread and eco-friendly management practices. Molecular diagnostic techniques used in plant disease diagnostic clinics want to be inexpensive, robust, reliable, and easy to handle that they can compete with, and accompaniment traditional techniques. Recently, effective intensification platforms, probe improvement, several quantitative PCR (qPCR), DNA barcoding, and RNA-Seq-based next-generation sequencing have modernized the research in fungal detection field, and diversity area. Though the molecular diagnostics techniques have grown extensively over the last couple of decades still there is an extensive way to go in the development and tender of molecular diagnostics to support the plant disease diagnosticians.

9.1   INTRODUCTION

Finger millet-*Eleusine coracana* (L) Gaertn [4x, 2n = 36; Family: Poaceae Subfamily: Chloridoideae] commonly known as, kapai mandua, madua, nagli, and ragi is a broadly cultivated cover from Orissa in the East to Gujarat in the West; Uttarakhand in the North to Tamil Nadu in the South. Millets are one of the oldest foods acknowledged to humans, but they were discarded in admiration of rice and wheat with urbanization and industrialization (http://www.millets.res.in). Millets are the imperious food and fodder crops in semi-arid regions that are mainly ahead more significance in the world (http://www.millets.res.in). They are mostly grown where major cereals would fail to give sustainable yields (Global Facilitation Unit for Underutilized Species, 2014). In the World, the millet production 30.73 million tons, out of which 11.42 million tons (37%) are produced in India (http://www.fao.org). Millets produce multiple safekeeping, i.e., food, fodder, health, nutrition, and ecological making them the crops of agricultural (Millet Network of India-Deccan Development Society-FIAN, 2009). Minor millets (finger millet, foxtail, Kodo millet, proso millet, little millet, and barnyard millet) have acknowledged far less research and development than other crops about crop improvement, cultivation practices and employment (Global Facilitation Unit for Underutilized Species, 2014). Finger millet is grown in India, Japan, Malaysia, Nepal, parts of Africa, Srilanka, Madagascar, and Uganda (http://agritech.tnau.ac.in). India is the largest producer of several types of millets. In which, finger millet is 85% production in India (Divya, 2011). In India, finger millet is cultivated over an area of 1.19 million hectares with a production of 1.98 million tons and average productivity 1661 kg per ha. Percentages area and production of finger millet are Karnataka (56.21 and 59.52), Tamil Nadu (9.94 and 18.27), Uttarakhand (9.40 and 7.76) and Maharashtra (10.56 and 7.16), respectively (http://www.indiastat.com). The crop is grown in different seasons and parts of the country, mostly as a rain-fed crop. This crop is less infected diseases but blast is a major constraint at times causing heavy yield losses.

Basic methods mostly be determined by cultural, microscopic, and morphological approaches that require classical taxonomy knowledge, and extensive time labor for detect the organism (Nilsson et al., 2011; Chalupová et al., 2014). Attributable to the boundaries of the conventional methods, molecular techniques came in use for the study of identification and classification difficulties. A high variety of molecular methods are immunological methods, nucleic acid-based probe technology and polymerase chain reaction (PCR) technology and becoming valuable tools in all aspects of fungal diagnostics. The previous methods depend on upon phenotypic characters, although the latter based on genotypic characters gives fast, effective, highly specific, and more accurate results. In disparity to the basic methods, isolation of organism do not require culturing (Spring and Thines, 2010; Badali and Nabili, 2012). Nucleic acid-based methods countenance the determination of closely related species and detect the microscopic quantity pathogen when no visible sign is present (Aslam et al., 2017). In molecular approaches, DNA/RNA probe technology includes fluorescence *in situ* hybridization (FISH), *in situ* hybridization, Southern hybridization, microarray, and macroarray. Isothermal amplification technology includes loop-mediated isothermal amplification (LAMP), nucleic acid sequence-based amplification (NASBA) and rolling circle amplification (RCA). PCR technology includes DNA barcoding, multiplex-PCR, nested PCR, real-time PCR, and reverse transcriptase (RT)-PCR have been used.

9.2 BLAST

The disease was first reported by McRae (1920) from Tanjore delta (TN) in India. It is most destructive widely distributed in all the finger millet growing regions of the world. In India, the disease is recurring yield losses wherever finger millet is grown in all the states. The disease appears causing an average yield loss of around 28% every year. Sundaram et al. (1972) considered blast of finger millet as the 'number one' loss causing disease in Andhra Pradesh, Haryana, Karnataka, Madhya Pradesh, and Maharashtra state.

9.2.1 SYMPTOMS

The disease occurs as leaf, neck, and finger blast at all stages. If infected seeds are sowing and the death of the germlings is common in the nursery or the direct sown crop. The young seedlings give a burnt appearance due to severe leaf blight and death that effect into burnt patches:

1. **Leaf Blast:** The typical spindle-shaped spots appear on leaf lamina. Such spots enlarge, coalesce, and leaf blades, especially from the tip to base, give a blasted appearance under congenial conditions. If the temperature and humidity are favorable for severe leaf blast and permit a fungicidal application. Well-developed lesions maybe 0.5 cm × 2 cm measure.

2. **Neck Blast:** The pathogen attacks the culms at the nodal region resulting in a blackening area. However, when the pathogen attacks at the neck region and it is most damaging stage of the disease. Two to four inches of the peduncle turns brown and later black due to fungal infection below the ear. This area may appear an olive-gray growth of the fungus. The seed setting stage infection may result in sterility and while infection delayed may produce underdeveloped seeds. The ears hang down at the point of infection and sometimes may break away from the stalk.

3. **Finger Blast:** The infection usually begins from the apical portions on fingers and runs towards the base. The extent of destruction depends on the stage of infection and the weather conditions. The entire length of the ear is affected at times. The pathogen attacks the seeds resulting in shriveled and blackened seeds and otherwise healthy ears.

9.2.2 PATHOGEN

Pyricularia grisea (Cke.) Sacc. (Per. Stage: *Magnaporthe grisea*) Hypha is septate, hyaline but older becomes brown and hyphal cell ranges from 1.5–6.0 µ in length. The large number of conidia is produced on conidiophores giving a dirty brown color lesion under high humidity. Generally, pathogen growth is relatively more on the upper surface and conidiophores may emerge either through a stomatal opening or directly from the epidermal layer. Conidiophores are simple, septate, a basal portion being relatively darker. Conidia produces three celled acrogenously, hyaline, obpyriform in shape and measure 19–31 µ × 10–15 µ. The end cells germinate, giving out germ tubes. The formation of terminal or intercalary chlamydospores is globose, thick-walled, olive-brown measuring 4–10 µ in diameter. The pathogen grows abundantly on oatmeal, potato dextrose, bean meal, and ragi meal agar and produces abundant dark colored chlamydospores. The pathogen produces fertile perithecia *in vitro* condition.

> ➢ **Epidemiology:** The primary inoculum of the fungus generally comes from infected seeds as like perithecia, and dormant mycelium. The severity of the disease depends on weather conditions. A temperature of 25–30°C, the humidity of 90% and above, cloudy days with intermittent rainfall, are favorable for the development and rapid spread of disease. If there are constant rains at the time of heading heavy losses to the crop occur.

9.2.3 MANAGEMENT

- In India, some resistant cultivars developed for the disease are GPU 28, GPU 26, GPU-45, GPU-48, CO RA (14), Paiyur (RA)-2, L-5 and VL 149.
- Uses of anyone fungicide, i.e., blasticidin (Bla-S) @ 0.1 g/L, tricyclazole (Beam 75WP)@0.75 g/L, edifenphos (Hinosan 35WP) @1 g/L, iprobenfos (Kitazin 48EC) 1.0 g/L, difenconazole (Score 25 EC) @ 1.0 g/l, hexaconazole (Contaf 25 EC) @ 1.0 g/L, propiconazole (Tilt 25 EC) @ 1.0 g/L, mancozeb (Dithane M-45) @ 2.5 g/L, thiophanate-methyl (Neotopsin 70WP)@ 0.5 g/l have the potential to be used as highly effective against rice blast disease (Arun et al., 2011; Singh et al., 2011; Hajano et al., 2012).
- Application of fungicides as seed treatment with carbendazim @ 2 g/kg + spraying of tricyclazole @ 0.06% + spraying of plant extract of *Ocimum sanctum* @ 15%, 7 days of first spray + spraying of *Pseudomonas fluorescens* @ 0.4 g/l after 7 days of first spray (Varaprasada et al., 2018).
- Application of carbendazim @ 2 g/kg seed treatment +first spraying of tricyclazole @ 0.06% + second spray of tricyclazole @ 0.06% after 7 days of first spray (Varaprasada et al., 2018).

9.3 BROWN LEAF SPOT (SEEDLING BLIGHT OR LEAF BLIGHT)

Brown leaf spot of finger millet is next only to blast in terms of severity and distribution. The disease was first noticed in ragi by Butler (1918) at different parts of India.

9.3.1 SYMPTOMS

When infected seeds used for sowing that showing pre-emergence rot and post-emergence rot of the seeds is common. Where healthy seedlings are subject to attack by the inoculum from outside. The characteristic symptom on the leaf lamina is the appearances of brown to dark brown spots. The pathogen infects leaves more rapidly from the upper surface and infection takes place through the stomata and direct penetration. The fungus affects leaf sheath and culm especially at nodal joints. The area at the junction of leaf sheath and leaf blade is usually affected resulting in dark brown discoloration. The spots are generally oval in shape and measure 8–10 mm in length and 1 to 1.5 mm in breadth. These spots coalesce, giving the appearance of blighting of leaf, especially towards tip, that are ultimately killed prematurely. Footrot symptom is appearing severely infected plant under favorable weather conditions. The pathogen attacks the ear head; fingers as well as grains that affected grains may not develop fully and shrivel. The disease in such a situation results in heavy losses in yield.

9.3.2 PATHOGEN

Drechslera nodulosum Berk and Curt. [Perfect Stage *Cochliobolus nodulosus* Subram (Berk and Curt). and Jain]. The conidiophores are upright and curved at periods, unbranched, produced in abundance in the older part of the spot and infrequently seen in the expanding area. Conidiophores normally emerge from stomata, but it is not rare to see them arising from epidermal cells directly. These will range between 80 and 250 μ in length and 5–7 μ in width with several septa at either place. Conidia are either sub-cylinder or obclavate, 3–10 septate straight or curved, and measure 40–114 μ × 11–21 μ. They are carried on either individually or one after the others at the tip of the conidiophore. On one conidiophore can form as many as 11 conidia. They are thick walled and light russet-in green color. The pathogen produces sexual fruiting bodies that contain long cylindrical beaks, black in color and spherical in shape. These ascocarps (perithecia) range in diameter from 276 to 414 μ. The asci are small and straight, with a curved apex of 1 to 8 ascospores. They measure 120–193 μ × 14–17 μ.

The pathogen sustains for over 18 months in unsterilized soil, and the spores on grains are found to be effective for a year. The best possible infection temperature is 30–32°C although the disease can occur from

10–37°C. During the process of ear development and before grain forming, high humidity and intermittent rains cause heavy ear infection and yield losses.

9.3.3 MANAGEMENT

The damage by the pathogen to the seedlings, especially in nursery under warm, humid conditions or to the developing seeds in the ear, is quite considerable. Need-based spraying of Mancozeb 0.2% control the disease.

9.4 CERCOSPORA LEAF SPOT (CLS)

Cercospora leaf spot (CLS) finger millet is one of the most critical foliar diseases in India and Nepal confined to the Himalayas. Munjal et al. (1961) examined J.N's collected samples during October 1959. Kapoor from Kathgodam from Nainital provided the fungus elucidations and nomenclature. The disease will diminish yield by up to 40% and 1000 seed weight by 21% if it occurs soon after heading (Pradhanang, 1994). Nonetheless, yield reduction does not occur when the occurrence of the disease is about 25% (Pradhanang and Abington, 1993).

9.4.1 SYMPTOMS

The disease appears most serious in the early sown crop in the month of June. Infection typically starts at the older leaves and progresses to the young leaves. While disease incidence drops from older leaves to younger leaves. Initial signs occur as yellow halo reddish-brown specks and are easily misinterpreted with those of Helminthosporiose or blast. A few such specks eventually clump together to form massive lesions. For certain cases, the lesions increase to become 15 × 3 mm eye-shaped spots and appear like blast spots. These leaves offer a burnt outlook. Extreme infected leaves turn fully necrotic, shrivel, and dry at crop maturity. The plants appear full blighted at this point. These signs are seen on the stem, sheath of the leaf and fingers, too. Scattered dark brown spots are visible on the base. Such spots later coincide on the stem or fingers to develop a necrotic lesion.

9.4.2 PATHOGEN

Cercospora eleusines Conidiophores olive-brown, diluted tip, straight to bent, geniculate, septate at long intervals, unbranched, noticeable spore scar; measuring 4–5 × 27–300 µ. Acicular conidia, hyaline to sub-hyaline, indistinguishably multiseptate, simple to angled, base truncate, tip sub-acute and measure 3–4 × 50–260 µ. The disease is typically limited to colder areas. In Nepal, the disease is confined to mid hills where the mean daily temperature does not exceed 20°C and relatively high rainfall.

9.4.3 MANAGEMENT

Late sowing, i.e., in July lowers the incidence and prevalence of the infection. Fields from contaminated crop refuge should be cleaned. Carbendazim spray @ 0.05% decreases infection.

9.5   DOWNY MILDEW

Venkatarayan (1946) first documented this infection on ragi from the former Mysore state in India. Thirumalachar and Narasimhan (1949) gave a detailed description of the symptomatology which described the pathogen as *Sclerospora macrospora*. However, the disease could be devastating for its intermittent nature, which could escalate to complete crop failure due to malformation of the infected ears.

9.5.1 SYMPTOMS

The development of white cotton, hallmark of many downy mildews, is not necessarily seen in the downy mildew of finger millet. As a consequence the asexual process goes mostly unnoticed. The plants affected by downy mildew are usually stunted by shortened internodes and fulsome tillering. The leaves are crowded, which give infected plants a bushy appearance. The infected seedlings can be damaged early in the season. Pale yellow translucent spots can often be seen on impacted leaf surface. The disease's most distinctive characteristic, however, is partial or full spike proliferation let into leafy structures which often result in a brush-like appearance.

9.5.2 PATHOGEN

Sclerophthora macrospora (Sacc.) Thirum., Shaw, and Naras (Syn. *Sclerospora macrospora* Sacc.) The fungal mycelium is coenocytic and hyaline. By nature, sporangiophores mimic vegetative hyphae, sympodiously, and successively branched, and haploid. Sporangia are lime-shaped, measure 60–100 × 43–64 μ and borne individually at sporangiophore apices. They germinate indirectly by forming 24–48, unequally biflagellate zoospores. Oospores have a spherical diameter of 35–70 μ, with a granular material. The germination of oospore is indirect through the development of a large lemon-shaped sporangium, which releases 24 to 48 zoospores. A temperature of 25–30C is suitable for the development of diseases. Most sporangia are formed during the night when the temperature is around 22–25°C which releases zoospores. The pathogen is internally seed borne.

9.5.3 MANAGEMENT

Since systemic disease starts at the seedling level, as in most other downy mildews, seed treatment with chemical substances such as Apron 35 SD @ 2.5–3.0 g/kg will manage systemic infection. Sprays may not be cost effective.

9.6 FOOT ROT

Coleman (1920) was the first to record the occurrence of Footrot disease (*Sclerotium rolfsii*) from the princely state of Mysore in India. It was later recorded from the former presidency of Madras, Coimbatore, and Orissa. Rampur, Nepal, reported a loss of up to 50% (Batsa and Tamang, 1983).

9.6.1 SYMPTOMS

In the field the disease occurs at random. Disease starts throughout the region of the neck, the affected area being limited to two or three inches above ground level. Normally, the plants are targeted at a point when plants are flowering or setting seeds because of the stem's vulnerability as the movement of photosynthetic material is towards sink. The basal part of the infected plant slightly above ground initially tends to be saturated with water due to pathogen infection. It later turns brown and then dark brown with a consequent shrinking of the stalk in the affected area. Profuse mycelial growth of white cotton happens in this region. Then small roundish white velvety grain-like structures starts emerging in the fungal matrix. They grow and assemble mustard seed-like structure, turn brown, and these are the sclerotial bodies. The leaves, meanwhile, lose their luster, droop down, and dry. The crop, in the end, dries up prematurely.

9.6.2   PATHOGEN

Sclerotium rolfsii (Sacc.) Curzi. (Per. Stage *Pellicularia rolfsii*). The pathogen has a very wide variety of host and is therefore prevalent in almost every soil. Extremely large sclerotial bodies are developed at the end of the growing seasons from the growth that had developed on the host plant. The sclerotia find their way to soil, from field to field, more by rainwater. On the whole, *S. rolfsii* is a weak pathogen or an opportunist, capable of striking a host plant when it is vulnerable.

9.6.3   EPIDEMIOLOGY

Sandy loam soils promote disease occurrence, as the pathogen must thrive at low levels of soil moisture. The occurrence of the disease is more during warm and dry months. Control: Resistant hosts may be the right way to handle disease. Chemical management cannot be economical, because the soil is occupied by the pathogen. Vitavax may, however, be effective in controlling finger millet foot rot.

9.6.4   MANAGEMENT

- PPR 1735 showed stable resistance to footrot over three consecutive years while GPU 16 and PPR 2350 were moderately resistant (MR) (Jain et al., 1994).
- Soil application of 1 kg *T. viride* talc formulation or *P. fluorescens* + *T. viride* 500 g each mixed with 25 kg compost incubated for 15 days and spread over an acre at the time of weeding resulted in least footrot and higher ragi yields (AICSMIP, 2013).

- Drenching of *Bacillus pumilus* MSTA8 and *Bacillus amyloliquefa-ciens* MSTD26 @(1×10^7 CFU/mL) @5 ml/L because they were effective against the disease (Dheeman et al., 2020).

9.7 SMUT

In 1918, Kulkarni (1922) documented this disease from Malkapur from the then princely Kolhapur district, naming the pathogen and recognizing it as *Ustilago eleusine*. The disease was subsequently discovered from the then Bombay Presidency's Surat, Nasik, and Ratnagiri districts, and ultimately from the states of Mysore and Madhya Pradesh. Although there is no calculation of crop failure, heavy damage can happen if more grains are damaged, in addition to blackening the total product.

9.7.1 SYMPTOMS

The smut causes it appears, typically just a few days after it flowers. One can see the smutted grains distributed randomly in the ear. Normally, the magnitude shall be below-approximately 1% of the grains. The ovaries affected are converted into greenish gall-like bodies which are many times bigger than usual healthy grains. In the early stages, greenish flattened grains are evident from 2–3 mm in diameter which project beyond the glumes. As the disease progresses, the infected grains are enlarged and attain a diameter of 16 mm. The sorus' greenish outer tunica slowly transforms pink green and eventually on drying to dirty black. The affected grains are sometimes single or grouped into spots of variable sizes and are often restricted to one side or to the bottom or apex of the head and exhibit signs of rupture across several places.

9.7.2 PATHOGEN

Melanopsichium eleusinis Mundk. (Kulk.) and of course Thirum. (Syn, *Ustilago eleusine* Kulk.) Spores are globose or subglobose, auburn-colored, and measuring an average of 9.5 µ at 711 µ. The epispore (Mundkur, 1939) is coarse and meticulously pitted. They germinate through the formation of a septate promycelium which produces both lateral and marginal sporidia.

Promycelium first appears as a tiny papilla, slowly expanding into a solid germ pipe. The optimal spore germination temperature is at 25°C.

9.7.3   MANAGEMENT

The disease is more common in delay sown summer crops. Sowing in January prevents infection.

9.8   DAMPING OFF

The disease was documented from Annamalai Tamil Nadu, India (Raghunathan, 1968). The disease occurs especially during rainy months in poorly drained nursery/fields.

9.8.1   SYMPTOMS

The disease's symptoms are distinguished at ground level by yellowish-brown discoloration of the hypocotyl region. Such discoloration eventually extends to stems and also to roots, and the seedlings crash. These seedlings get destroyed at the collar region when lightly pulled.

9.8.2   THE PATHOGEN

The mycelium of *Pythium aphanidermatum*, is hyaline, coenocytic, extremely granular, copiously branched, and is 1.1 to 8.6 μ wide. The oospores are spherical, soft, and thin-walled, ranging in diameter between 14.1 and 19.2 μ. The oospore walls are mostly of thickness 1.5 to 3 μ. Sporangia is terminal or intercalary, plain or digitately branched, loculate structures having different shape and size.

9.8.3   MANAGEMENT

In the case of transplanted ragi, seedlings can be guarded in nursery by growing with adequate drainage on raised beds. Practicing mild irritation or watering and drenching soil with appropriate fungicides including Copper

Oxy Chloride, Captain, Hiram, or Metalaxyl compounds proves efficient but costly in the control of diseases.

9.9   BANDED BLIGHT

Lulu Das and Girija (1989) first recorded sheath blight of ragi from Vellayani in Kerala, India, in which it appeared at extreme form. The disease was subsequently reported in extreme form in the Birsa Agricultural University, Ranchi, Bihar, experimental plots in 1993.

9.9.1   SYMPTOMS

The disease on the lower sheath of the leaf is marked by oval to irregular light gray to dark brown lesions. Afterwards, the central portions of the lesions turn white to straw with a thin, reddish, brown boundary. These later-stage spots are unpredictably scattered across the leaf lamina. A temperature of about 28–300°C and humidity level of 70% or higher encourages the accelerated growth of infection as these lesions spread quickly and coalesce to cover significant sections of sheath and leaf lamina. At this stage, the symptom of the disease is characterized by a series of bands of copper or brown color across the leaves that give a very distinctive banded impression. The mycelial development can be seen along with white to brown sclerotia on and around lesions. Later, the leaves run dry and blighted plants appear.

Necrotic lesions develop on peduncles, fingers, and glumes irregular to oval, dark brown to purplish-brown. Initial infection on peduncle or at the base of the finger is very close to neck rot lead to poorer filling of grains. When the sheath becomes infected until the appearance of the peduncle, the fingers become disorganized and diminished in size. Diseased glumes contain shriveled, smaller seeds. Therefore, the symptoms developed on each section of the plant give a distinctive banded look, as a consequence of which the disease is called banded blight (Dubey, 1995).

9.9.2   PATHOGEN

Rhizoctonia solani Kuhn. [Basidial stage: Thanatephorus cucumeris (Fr.) Donk.] Dubey (1995) witnessed perfect state development in nature on infected plants as dirty white growth of hymeneal layer during

September-October, when high humidity (above 80%) and mild temperature (26 ± 20°C) prevailed.

9.9.3 MANAGEMENT

The disease can be avoided by healthy planting, pumping out surface water and clearing field weeds on bunds. Spraying of propiconazole @ 1 ml^{-1} water was effective for the management of banded blight pathogen in finger millet (Patro, 2008).

9.10   SHEATH BLIGHT

In Coimbatore, Tamil Nadu, India, the very first occurrence of a pathogenic mushroom on finger millet causing sheath ragi blight was recorded in 1974 (Parambaramani et al., 1975).

9.10.1 SYMPTOMS

The initial symptom of the infection is the presence of typical circular or elliptic, necrotic spots on the sheaths approximately 5 to 15 cm above ground surface The sheaths become trapped or tied together as the disease spreads, with the mycelium from the infection to the stem ultimately contributing to the plants wilting and death. The outer leaves discolored or dried out may clearly differentiate the diseased plants in the field. Tiny mushroom sporophores are detected on bottom sheaths of dead plants.

9.10.2 PATHOGEN

Marasmius candidus Bolt. The mushroom is white, fragile, leathery, and seasoned, it does not rot quickly, but it dries up in hot seasons and rejuvenates in wet weather or in water.

9.10.3 MANAGEMENT

- Treat the seeds with Captan or Thiram @4 g/Kg;

- Spray Mancozeb @ 1.25 Kg/ha;
- Spray 1% Bordeaux mixture or Copper oxychloride or Dithane Z-78 (2 g/lit. water).

9.11 LEAF SPOT

Shaw (1921) reported *Curvularia lunata* (*Acrothecium lunatum*) occurring on several small millets, such as finger millet from Pusa. In Chennai (Tamil Nadu), Kerala, and Assam, it was recorded later.

9.11.1 SYMPTOMS

The fungus creates minute spots on finger millet, *Setaria italica* and *Panicum frumentaceum*.

9.11.2 PATHOGEN

Curvularia lunata (Walker) Boedijin (Syn. *Acrothecium lunatum* Walker); the conidia are curved, 2–4 septate, knee-shaped, with large mid-curved cell. It measures 18.22 μ × 5.7 μ and it grows on a wide range of media. There is higher production of aerial hyphae in non-synthetic media, but the spores are typically smaller, while there is less production of aerial hyphae in synthetic media as the spores are larger.

9.11.3 MANAGEMENT

Treat the seeds with Captan or Thiram @4 g/Kg and Spray of Mancozeb @ 1.25 Kg/ha.

9.12 RUST

The incidence of rust on finger millet from Meerut in Uttar Pradesh was recorded only recently in 1976 (Dublish and Singh, 1976). There was almost no evidence from any part of the globe on this disease until 1996, when it was documented in Bangalore, Karnataka (Channamma et al., 1996).

9.12.1　SYMPTOMS

The signs occur as tiny, brown, fragmented pustules grouped sequentially on absolute top of the leaves. The Uredeniospores are pedicellate, globoid, or broadly ellipsoid with thick, smooth walls and dark cinnamon in color. The spores measure 24 μ × 26.25 μ, with approximately 3–4 germpores.

9.12.2　PATHOGEN

Uromyces eragrostidis Tracy. Additional facts of the disease have not been comprehensively worked out.

9.12.3　MANAGEMENT

Uganda developed four varieties *viz.*, SEC 915, 314, 712, and ICMV-221 which are resistant to finger millet rust.

9.13　BACTERIAL DISEASES

The first recorded incidence of bacterial disease in finger millet is possibly the one identified from Uttar Pradesh by Mehta and Chakravarty (1937). They did not recognize the pathogen though. Rangaswami et al. (1961) reported that Tamil Nadu had a bacterial disease that they identified as *Xanthomonas eleusinae*. A further species of *Xanthomonas* from Gujarat reported later on by Patel and Thirumalachar (1965). Billimoria and Hegde (1971), from Karnataka, confirmed a species of *Pseudomonas*.

9.14　LEAF SPOT

Across the rainy season of 1960, a leaf spot was observed at Chidambaram taluk, Tamil Nadu's South Arcot district, which was suspected to be affected by bacterium. Rangaswami et al. (1961) who researched the disease and causative agent confirmed that it was linked to previously identified Xanthomonas species which they called X. eleusine. Symptoms: Linear spots are found on the upper and lower surfaces of the blade of the leaf, which extend across the veins. The patches are 2 to 4 mm in length but frequently reach up to 1 inch or more. Spots are soft yellowish-brown at the beginning but then turn dark brown. The leaf breaks up along the streak in developed stage, giving a shredded look. All the leaves in a plant are damaged even the delicate shoots. The bacterium specifically targets the leaves but distinctive streaks can also be observed on the peduncle of the ear head. Such streaks are thin, 5 to 10 mm long, and are subcuticular in appearance. Pathogen: *Xanthomonas eleusinae* Rangaswami, Prasad, Eswaran The bacterium is a short string, $1.8–2.7 \times 0.8–1.0$ μ with a single polar flagellum, non-spore-forming, gram-negative, aerobic, non-capsulated, and non-acid forming. On nutrient agar, it produces dull yellow slimy and glossy colonies, and the development in nutrient broth is turbid with pellicle production. Gelatine liquefies easily but starch is not used. Litmus milk becomes neutral, and then coagulates. Reduced nitrate produced H_2S but does not produced ammonia and indole. It provides positive lipolytic activity and negative tests on MR and VP. This utilizes lactose as a source of carbon, generating acids and releasing next to no gas.

9.15 BACTERIAL BLIGHT

Desai et al. (1965) maintained that the bacterial pathogen identified on ragi by Rangaswami et al. (1961) was not a species of *Xanthomonas*, and identified ragi bacterial blight disease which was extremely common in Gujarat.

9.15.1 SYMPTOMS

The plants are prone to infection at all phases of development. When infection takes places during the early stages of development, the plants turn yellow and exhibit early wilting. Infection first occurs as water soaked, glossy, longitudinal, pale yellow to dark greenish-brown stripes, 5 to 10 mm long and extending parallel to the midrib of the lamina. The hyaline streak later grows into a large yellowish lesion measuring 3 to 4 cm and turns brown. Once the infection is strong, especially in the early stages, the whole leaf turns brown and fades away.

Pathogen: *Xanthomonas coracanae* Desai, Thirumalachar, and Patel. Bacterium occur as small rods with rounded edges, usually single, sometimes in pair, measuring $1.1–1.8 \times 0.5–0.7$ μ, motile by a polar flagellum, gram-negative, encapsulated, no endospore and non-acid fast. Territories on potato dextrose agar (PDA) plates are circular with an entire border, smooth, pulvinate, butyrous, and glistening yellow. Development on nutrient agar and PDA slants is mild to plentiful, filiform, convex, glistening, smooth opaque, butyrous, and lemon yellow; medium untouched.

9.16   LEAF STRIPE

Billimoria and Hegde (1971) reported an infection on ragi during 1969–1970, which was believed to be attributed to bacterium. The disease emerged in a huge portion in and around Bangalore district, Karnataka, India. Symptoms: The primary characteristic of the infection is brown coloration of the leaf sheath, particularly from bottom upwards. The affected part of the lamina generally includes the midrib and is straw colored. The sign extends to about three-quarters of the lamina and then suddenly ends or, in some cases hits the leaf tip. Infrequently the stripes of contaminated areas are seen to venture down along the edge of the lamina, leaving the central section, including the midrib safe. The bacteria are found in abundance in the phloem vessels. Infected plants can be identified from a distance by typical droopy leaves. Infected culms reveal a light brown discoloration on one side. In some cases, this discoloration starts from the bottom, but in other situations, it starts two or three inches above the bottom and spreads to the leaf sheath. There is, nevertheless, no significant decrease in girth or turgidity of the infected culms as contrasted to the healthy ones. Plants below a month old are generally free from the infection. The bacterium is systemic and soil borne.

9.16.1   CAUSAL ORGANISM

The causal organism of this disease is *Pseudomonasel eusineae*. The bacterium is a small rod, either singly or in pairs, measuring 0.83–2 µ by o.31–0.42 µ, capsulated by single or two monotrichous flagella. Gram-negative, non-sporing, and non-acid fast. It intensely hydrolyzes starch, generates acid, however, no gas from glucose transforms pure milk alkaline and demonstrates slight lipolytic activity. The bacterium doesn't really liquefy gelatine decreases nitrate, V.P. and Methyl red test positive. Somehow doesn't yield indole and therefore does not hydrolyze casein.

9.16.2 MANAGEMENT

1. Field sanitation, diseased leaves, twigs, stems, and crop stable should be destroy from infected field.
2. Selection of recommended resistant cultivars.
3. *Bacillus coagulans* formulation (1×10^7 CFU/mL) @5 mL/L sprayed has been found very effective against *Xanthomonas* spp. (Kishun, 1994b).
4. *B. subtilis* or *B. amyloliquifaciens* (1×10^7 CFU/mL) @5 ml/l sprayed have been very effective (Pruvost and Luisett, 1991).
5. Two sprays of streptocycline (200–300 ppm) at 10 days intervals (Bose and Singh, 1980).
6. Streptomycin sulfate (250 ppm) followed by Aureofungin (Prakash and Raoof, 1985).
7. Streptocycline (300 ppm) and Copper oxychloride (0.3%) were found more effective in controlling bacterial diseases (Prakash et al., 1994).
8. Mixture of 2 chemicals, such as Copper oxychloride (3000 ppm) + Agrimycin-100 (100 ppm) or bavistin (1000 ppm) + Agrimycin-100 (100 ppm) (Kishun and Sohi, 1984).

9.17 VIRUS DISEASES

Two catastrophes of viral disease problems prevailed on finger millet in Karnataka, first in 1940's and then in 1960's causing nearly complete loss of the grain production. In 1982, Nagaraju, and co-workers documented yet another but very unique virus that is considered to be infective to finger millet. Thereafter, three viruses' viz., *Sugarcane mosaic, Maize streak,* and *Mottle streak* have been documented to appear on ragi in India. *Sugarcane mosaic virus*, a *Potyvirus* transmitted by sap and aphids; *Maize streak virus*, a *Geminivirus* transmitted by hoppers; and *Ragi mottle streak virus*, a *Rhabdovirus* vectored by hoppers, have been reported to infect finger millet in India (Saveetha et al., 2007).

9.18 RAGI MOTTLE STREAK

Throughout Southern Karnataka, there was a severe disease concern in alarming rates during the mid-1960s. Govindu et al. (1966) studied this condition and thought it was due to a combination of a Virus and *Helmintho-sporium* sp. Ater, Mariappan et al. (1973) in Tamil Nadu reported ragi streak disease, which was transmitted by Sogatella sp. and found this virus to be a strain of the Karnataka. Maramorosch et al. (1977) reported reductions of 50–100% due to mottle streaks in certain regions.

9.18.1   SYMPTOMS

Whenever the plants are 4–6 weeks old, the infected plants display usual dark-green areas all along the leaf veins. Other Leaf symptoms include chlorosis and streaking. Irregular yellowing to near albino symptoms is also noted in certain situations. However, the indications are of the mottle type in the form of white specks in the lower leaves, and the infected plants are usually stunted with tiny ears.

9.18.2   PATHOGEN

Ragi mottle streak virus, small rod-like, bacilliform fragments seen in perinuclear position in cells from across all sections analyzed, comprising the mesophyll epidermis, and the conducting element. The particles were quite widespread and hence easy to spot. The particles measure 80 nm in cross-sections and 285 nm in longitudinal direction. The particles were engulfed, bacilli formed, and spiked according to the rhabdovirus structure. Two species of *jassids* transmit the virus viz., *Cicadulina bipunctella*, *C. Chinese*, and *C. bipunctella* may transfer as much as 82%. Just a portion of the vector *C. bipunctella* population transferred the virus in large proportion, and the virus is persistently carried in the leafhopper.

9.18.3   EPIDEMIOLOGY

Large cumulative extreme weather circumstances-minimum temperatures, comparatively low rainfall and lower average relative humidity in all affected districts from August to November between 1945 and 1965 is believed to have contributed to an unprecedented rise in the community of vectors.

9.19 RAGI SEVERE MOSAIC

An extreme mosaic in Kharif 1966 significantly impacted the ragi crop in Southern Karnataka and the Andhra Pradesh boundary districts. In certain pockets, such as Hiriyur Taluk in the district of Chitradurga and Devanahally taluk in the district of Bangalore, the disease was so extreme that the farmers deserted their crop because of heavily infected plants did not set seed (Joshi et al., 1966). The outbreak that arose in the 1960s was primarily due to persistent ragi cropping, combined with irregular weather conditions that had a favorable effect on the population of the vectors (Keshavamurthy and Yaraguntaiah, 1977).

9.19.1 SYMPTOMS

The virus causes signs of mosaics which are sharper and more noticeable on young leaves. Infected plants remain constricted, and the ears are malformed in seriously infected plants. These plants generate few smaller-sized seeds which drastically reduces production. Moreover, due to serious chlorosis, the severely impacted plants appear pale yellow, and in severe cases, become brownish-white. Thus the whole field appears yellow and can be readily distinguishable from distant non-infected stands. Stunted plants don't really retrieve, develop roots at nodes, typically produce no ears, and persist mostly sterile when produced.

9.19.2 PATHOGEN

Sugarcane mosaic virus the virus has a 50–55°C thermal inhibition point, an endpoint of dilution between 1:500 and 1:750, and an *in vitro* lifespan of 10–12 h at room temperature (30–36C), 30 h at 24–26°C and 5 days at 7–10°C. This can tolerate continuous freezing in contaminated leaves for 8 days, and desiccation for 34 h. Particles were flexuous rods with an average 667 ± 8 mμ length and a roughly 12–14 mμ diameter. Hence, the virus was recognized as a *sugarcane mosaic virus* strain (Subbayya and Raychaudhuri, 1970).

The virus is neither from seed, nor from soil. Paul Khurana et al. (1973) studied the relationship between virus and vector, using *Longiunguis sacchari* as vector. It was observed that the maximum acquisition-feeding cycle was 5 minutes, and that the optimum distribution feeding time was one hour.

The aphid will obtain the virus within one minute. Fasting pre-acquisition enhances the vector's efficiency, and the higher transmission is achieved within 11/2 to 2 h fasting. The virus is even transmitted by a single aphid, and the optimum is 10 aphids per plant. Post-acquisition fasting reduces vector effectiveness, and it is discovered that the virus is non-persistent in *R. maidis* since they were only maintained for one hour after the acquisition. The virus' incubating duration is considered to be affected by temperature but not through the host's age. Ragi plants of most generations are vulnerable but infection magnitude decreases with increasing in the host's age (Raychaudhuri and Subbayya, 1970).

9.20   RAGI STREAK

A virus disease-causing streaking and yellowing of leaves and adversely affecting of ragi plants in the fields surrounding Bangalore was detected during the time 1974–1975, and the virus was reported not to be spread either by mechanical sap inoculation or by aphid species being examined (Anon, 1975). In a broad coverage area viz. in the districts of Chitradurga, Mandya, Bangalore, Tumkur, and Hassan in successive surveys during 1977–1978 and 1978–1979, Ragi plants impacted with such symptomatology were observed. The prevalence varied between 5 and 45%. The reduction in yield of grain depends on the age the virus infects the crop. The loss in weight of 1000 grains was 84, 63, 27, and 24% whenever the disease took place at seedlings 30, 40, 50, and 60 days old. There's been no substantial improvement in the number of tillers and where infection occurred on 10-day-old seedlings when the number of seedlings increased dramatically (Nagaraju et al., 1982).

9.20.1   SYMPTOMS

Symptoms occur as pale specks or stripes of varying size when young leaves unfold. The specks coincide that included larger areas resulting in chlorotic bands running parallel to the midrib through almost the entire length of the leaf. However, sometimes, dark green areas interrupt those bands. Both the main shoot and the tillers' new emerging leaves portray a number of well-defined chlorotic streaks with relatively uniform width running parallel to the midrib across the length of the leaf lamina. The infected field plants produce relatively more tillers and carry yellowish sick ear heads, often carrying few shriveled seeds. The diseased plants during the very early stage die prior to their bloom.

9.20.2 PATHOGEN

Maize streak virus strain *Eleusine*, it can be spread by *Cicadulina chinai*. The viruliferous leafhopper stays infective across the course of its existence. The total feeding time for the inoculation is 30 minutes. The virus is able to transmit by single viruliferous leafhopper.

9.20.3 MANAGEMENT

Spray Methyldemoton 25 EC 500 ml/ha on noticing symptoms and repeat twice if necessary at 20 days interval for control of insect vectors.

9.21 NEMATODE PARASITES

Several nematodes *viz.*, *Pratylenchus, Rotylenchulus, Helicotylenchus, Trichodorus, Heterodera, Criconemoides, Macroposthonia, Pratylenchus, Rotylenchulus,* and *Meloidogyne* have been reported to parasitize finger millet.

9.22 HETERODERA SPP.

During 1972, Setty (1975) reported the existence of a cyst nematode for the first time on ragi at Hebbal, Bangalore, India. *Heterodera Marioni* had been recorded on this crop earlier by Ayyar (1933, 1934).

9.22.1 SYMPTOMS

The major signs are plant stunting, and partial yellowing of the leaves. The infected plants exhibit unthrift development even under the best possible moisture and nutritional conditions, and it can be easily removed. Naked eye may show the cysts incorporated or attached to the roots of the infected plants.

9.22.2   PATHOGEN

Heterodera gambiensis, the cysts are lemon-shaped, measure 800–960 μ × 450–600 μ whereas the second stage larvae measure 400–540 μ long.

9.23   ROTYLENCHULUS RENIFORMIS LINFORM AND OLIVEIRA

In his study Chandrasekaran (1964) found that finger millet was vulnerable to *R. reniformis.* As per Krishnappa et al. (2002), 4.8% of the ragi crop region is infected in Karnataka by *Rotylenchulus reniformis,* and green manuring has been highly effective at reducing nematode populations. Narayana Swamy and Govindu (1966) confirmed the natural occurrence of *Helicotylenchus, Trichodorus,* and *Pratylenchus* species from many other locations in Karnataka, in addition to *Heterodera* and *Rotylenchulus.* Mohanty and Das (1976) examined *Criconemoides ornatus,* the ring nematode in ragi.

9.23.1   SYMPTOMS

Reniform nematode damage involves dwarfing, browning, and cortical cell necrosis at nematode entry point accompanied by early decay, secondary root degradation. Since the nematodes, yield reduction is a loss in crop production. Lower populations of this nematode had favorable interactions with plant height decline, top weight, root weight, and yield. Rajagopal (1965) noticed correlation of high *Reniformis* population with stunted grassy finger millet patches.

9.23.2   MANAGEMENT

The nematodes are best controlled by soil amendments like poultry manure or neem cake or by applying granular insecticides viz., Phorate 10G or Carbofuron 3G.

9.24   RECENT MOLECULAR METHODS FOR THE DETECTION OF PLANT PATHOGENS

We discuss the advancements in molecular techniques for the identification of the pathogens which caused disease in ragi, as shown in Table 9.1.

TABLE 9.1 Several Detection Methods of Ragi Pathogens

Pathogens	Hosts	Detection Methods	References
Finger blast: *Pyricularia grisea* (Cke.) Sacc. (Per. Stage: *Magnaporthe grisea*)	Cereal	DNA barcodes-LSU GenBank: MH877665.1	Vu et al. (2019)
	Eleusine coracana	DNA barcodes-nrDNA GenBank: AB031341.1	Kusaba et al. (1999)
	Eleusine coracana, Wheat	RT-PCR-cDNA GenBank: AB025252.1	Urashima et al. (1999)
Brown spot (seedling blight or leaf blight: *Drechslera nodulosum* (*Cochliobolus nodulosus* Berk and Curt.)	Cereal	RT-PCR-cDNA GenBank: AB011648.1	Shimizu et al. (1998)
Downy mildew: *Sclerophthora macrospora* (Sacc.) Thirum., Shaw, and Naras.)	*Zea mays*	RT-PCR-cDNA-(cox2) gene GenBank: EU116059.1	Thines et al. (2008)
	Poaceae	RT-PCR-cDNA-rRNA gene GenBank: KT248939.1	Thines et al. (2015)
Footrot: *Athelia rolfsii* (Anamorph: *Sclerotium rolfsii* (Sacc.) Curzi.)	Soybean	DNA barcodes-ITS1 and ITS4 GenBank: JN081867.1	Zheng et al. (2020)
	Groundnut	DNA barcodes-ITS, LSU GenBank: KT878487	Paul et al. (2017)
Damping-off: *Pythium aphanidermatum.*	Cereals	DNA barcodes-ITS1 and ITS4, rDNA GenBank: KR095341.1 KF561235.1	Liu et al. (2018)
Banded blight: *Rhizoctonia solani* Kuhn. (*Thanatephorus cucumeris* (Fr.) Donk.)	Cereals	Multiplex PCR	Okubara et al. (2008)
Curvularia leaf spot: *Curvularia lunata* (Walker) Boedijin	*Brassica rapa* subsp. *Pekinensis*	DNA barcodes-ITS1 and ITS4, rDNA GenBank: KU844328.1, JX256430.1 and JX256429.1	Wonglom et al. (2018)

TABLE 9.1   *(Continued)*

Pathogens	Hosts	Detection Methods	References
Rust:	*Echeandia*	RT-PCR-cDNA-rRNA gene	Liu et al. (2015)
Uromyces eragrostidis Tracy	*flavescens*	GenBank: HQ317561.1	
Bacteria blight:	*Eleusine*	RT-PCR-cDNA-rpoD gene	Constantin et al.
Xanthomonas coracana Desai, Thirumalachar, and Patel	*coracana*	GenBank: KJ491600.1 and KJ491690.1	(2016)
Ragi Severe Mosaic:	Cereals	RT-PCR-SCMV-VER1 and SCMV-CAM6 gene	Chaves-Bedoya et al. (2011)
Sugarcane mosaic virus		GenBank: EU091075.1	
Ragi Streak disease:	Wheat/	RT-PCR-Amplified cDNA	Rybicki and
Maize streak virus strain *Eleusine*	Eleusine	GenBank: U20871.1	Hughes (1990)
Rotylenchulus reniformis Linform and Oliveira	Cotton	cPCR-18S rRNA gene GenBank: KR153037.1	Nyaku et al. (2016)

9.24.1   PCR-BASED METHODS

1. **Conventional PCR:** This approach enables the synthesis of the specific component of DNA in millions of copies by means of alternating denaturation cycles, annealing, elongation using some unique primers. PCR was initially extremely specific to the identification of diseases caused by bacteria and viruses. Today, this is also commonly used for identification of plant pathogen. PCR relies on the effectiveness of DNA extraction and the deoxynucleoside triphosphate intensity. Inhibitors present in sample assays often disrupt efficiency (Fang and Ramasamy, 2015; Mancini et al., 2016). In addition, PCR technology involves the use of a primer to initiate the DNA replication for the identification of pathogens, which may restrict the functional usefulness of the method for field sampling of disease. Often the precise and accurate results are not provided by a single pair of primers to address this constraint DNA probes and nested primers are often used nowadays (Compton, 1991).

2. **Nested PCR:** For fungal detection, nested PCR efficiency is 1000 times higher than single PCR (Yeo and Wong, 2002). Two successive rounds exist in this process, in which a single pair of primers are used to amplify a large area of DNA, then this replicated sequence of

DNA serves as a guide for the second round employing two internal primers. A significant risk of infection is assessed in this form of PCR as two amplification processes are to be carried out in different tubes (Rahman et al., 2013).

3. **Multiplex PCR:** It can be utilized to preserve time and money using a couple of pairs of primers in the same reaction to allow synchronized identification of different targeted DNA sequences. Various fragments unique to pathogenic fungi were amplified and identified on the agarose gels depending on their molecular sizes. The accuracy of DNA-synthesizing is seriously impaired by the amplicon scale. To prevent this pitfall, the primers should be specifically crafted along with the relative concentration and temperature for annealing (Dasmahapatra and Mallet, 2006). Padlock probes (PLPs) are used today for the detection of pathogenic fungi in the multiplex technique. The multiplex PCR technique has been used to investigate fungal pathogens concurrently, such as *F. oxysporum* (Choet et al., 2016).

4. **RT-PCR:** mRNA is destroyed in dead cells; hence mRNA can be identified by RT-PCR to test the viability of the cells (Capote et al., 2012). Firstly, in this step, the RNA is reversely transcribed by random primers and RT enzyme into cDNA and then amplified by any technique based on PCR. So, RT-PCR is often used for the identification and treatment of RNA-containing viruses (such as retroviruses). Treatment of RNA-containing viruses may be beneficial in designing or testing antimicrobial vaccines or therapy efficacy. The *Fusarium graminearum* fungi were quantified using RT-PCR (Brown et al., 2011).

5. **Real-Time PCR (RT-PCR):** The DNA including the amplification could be analyzed in this method (Mackay, 2004). Surveillance of reactions while amplification phases was made easier through the use of fluorescent colors including SYBR Green I or fluorescent-labeled sequence-specific probes such as the Taq Manprobe (Badali and Nabili, 2012). Fluorescent signal is produced when the fluorescent dye intercalates to DNA. This message is increasing as the amount of targeted DNA raises after each amplification cycle (McCartney et al., 2003). Use the fluorescent dye as a testing agent is less costly but has drawbacks because of its non-specific aspect. Apparently the binding of the intercalating dye to any of the present DNA will give the incorrect outcome in terms of priming dimer. Therefore,

fluorogenic samples came into effect because of their high sensitivity (Atkins and Clark, 2004; Buet et al., 2005). These samples are associated with two forms of fluorescent dyes, one is the reporter dye which is attached to the 5′end, and the other is quencher dye on the 3′ end. Close reporter proximity and the quenching dye inhibit fluorescence emission. Because of Taq polymerase's exonuclease function, reporter dye is separate from quenching dye and started to fluorescence (Dasmahapatra and Mallet, 2006).

9.24.2   *SERIAL ANALYSIS OF GENE EXPRESSION*

The approach focused on sequencing and quantification of the 15 bp or longer oligonucleotides and sequence similarity against the usable genome sequences to identify the associated expressed genes (Velculescu et al., 1995). This process exploits two samples, ligated, and labeled with the respective primers, and then amplified. Then, it eliminates the primers, creating sticky ends that form the concatemers. They are copied into vector, and a sequenced study followed by the large computation. SAGE does have certain disadvantages. Firstly, it requires a significant volume of mRNA. Second, 15 bp tag often are not adequate to precisely mark the gene of origin for the more complicated genomes. SAGE was first used to detect *B. graminis* on the barley leaves (Dawei and Peng, 2014).

9.24.3   *DNA BARCODING*

DNA barcoding is a molecular identification technology that involves a small portion of DNA to recognize the species from all eukaryotic life spheres. Standardized series of 500–800 base pairs can be used in DNA barcoding to recognize species with markers relevant for the broad range of taxonomic classes (Krishnamurthy and Francis, 2012). For microbial organisms, including fungi, bacteria, and algae, this PCR-based approach is useful not only in detecting the cultivated species but also in detecting species of uncultured taxa from natural habitat. The usefulness of barcoding is based on the presumption that genetic differences within a population are considerably fewer than differences between populations.

Active markers in this technique are very important in describing the poorly recognized fungal species variation in the natural environment (Roe et al., 2010). There are many scientific benefits of DNA barcoding, including: (i) enabling species recognition at any point in life; (ii) allowing organisms to be selected depending on the phylogenetic study in nucleic acid sequences; (iii) offering insights into life differentiation; (iv) supporting the creation of DNA sequencing methods that are significant in the context of biodiversity (Savolainen et al.). Molecular fungal systematics is mainly based on the study of the small (18S) and huge (28S) substituents of nuclear ribosomal RNA (rRNA) cistron (Hebert and Gregory, 2005). The ITS area can be expanded using a small collection of primers from several fungal taxa (Schoch et al., 2012).

This would be the most practical method of recognizing phytopathogenic fungi by DNA sequence. Clarifying the ITS region either to the ITS-1 spacer or the ITS-2 spacer has also received significant interest in the field of study based on bar code. Again, ITS2 is not a coding area, but it has a retained center of the secondary structure that supports release the data-handling systems. ITS-2 spacer proves to be a rather useful secondary plant barcode (Xu, 2016). Use ITS-2 for metabarcoding may potentially stretch the comparative research between plants and fungi. Therefore, ITS, SSU, LSU, EF-1α, and RPB are barcoding markers being used in fungal barcoding of DNA (Monteiro et al., 2015).

9.24.4 DNA/RNA PROBE-BASED METHODS (SEQUENCING INDEPENDENT METHODS)

DNA/RNA test methods use the probe to diagnose the plant diseases caused by microbes such as fungi with good effectiveness and pace. For most current knowledge, this innovation is known as the foundation. For such methods, the probe is used without its amplification to test nucleic acid. Probes are the single-stranded smaller DNA fragments identified with the molecule of the chemiluminescent reporter, or with radio-labeled isotopes such as 32P, 33P, and 35S. These can be used to recognize the homologous series to the targeted DNA-DNA probes are often used in conventional approaches as an alternative to PCR to detect fungi. Yet these are still used in modern approaches in combination with PCR (McCartney et al., 2003).

9.24.4.1 NORTHERN BLOTTING

Northern blot, also known as the RNA blot and is used to pass the RNA into a transporter for pathogens detection. The Northern blot is the same as to the Southern blot, but instead of the DNA, the RNA component is used. Initially, the RNA will be purified in each tissue to analyze the expression of the gene of interest. The content containing RNA is then placed into agarose gel. Later on the gel is transferred through an electrical current that drifts sRNA to the bottom of a bottle. Smaller RNAs are travel quicker than greater RNA (Kim et al., 2010). Distinct RNA fragments are then blotted onto a unique filter paper; thus each RNA molecule retains its location relative to all other molecules (Berg, 2007). Afterwards, the filter is subjected to radioactive probes to hybridize it to complementary sequences. The filter is then positioned for autoradiography so the film can form. Ultimately, if the probe has hybridized to a segment of RNA on the filter, a band will be observed on the autoradiograph. The Northern blot is effective for analyzing gene expression. Compared to the RT-PCR, northern blotting technique's sensitivity is relative low. *Magnaporthe grisea* has been detected in rice plants by using Northern Blotting and the real-time PCR (Qi and Yang, 2002).

9.24.4.2   IN SITU HYBRIDIZATION

Histochemistry of hybridization is also called *in situ*. This offers useful information for pathogens getting identified and enumerated. Single-stranded RNA probe is adopted in this approach which is also termed as rib probes. Such probes are 35S marked. Hybridization *in situ* is very similar to Northern blots. All rely on the hybridization of labeled DNA/RNA samples to the homologous mRNA sequences (Jensen, 2014). The key benefit of this technique is the optimum use of low supply tissue such as cultured cells and clinical biopsies. One can perform hundreds of distinct hybridizations on a single tissue. Tissue libraries can be produced and stored in the freezer for further use. There are other ways to do hybridization *in situ*, like cDNAs, cRNAs, and screening with synthetic oligonucleotides. However, rib probe testing provides the most precise and reliable tests. Samples can also be marked with radioactive or non-radioactive nucleotides. 35S rib probes are the most important tool for the detection of mRNA from such radioactive probes (Hayden et al., 2002). ISH is a method for visualizing plant tissue infection from the rust fungi. This approach has been used to generalize the ISH technique for locating rust fungi in the portion of plant tissue embedded in paraffin (Ellison et al., 2016).

9.24.4.3   FISH

Because of the disadvantage of radio-labeled probe-based hybridization, *in situ* fluorescence hybridization (FISH) was formulated, that is employed for the quick classification of microorganisms such as fungi. FISH offers better pace, precision, and protection, and clears the way for the simultaneous and quantitative phylogenetic study of multiple targets development (Tsui et al., 2011). For that, assessing the spatial structure of fungal communities is an essential part. This cytogenetic approach was used to identify different chromosomal DNA sequences. A fluorescent probe is often used in FISH that only connects certain parts of the chromosome in which it displays a higher level of homology or complementation. Fluorescent probes are designed over the entire length of the probe by enzymatic incorporation of fluorophore-modified base (Baschien et al., 2001). Traditional strategies for microorganism detection include active division of the cells. FISH can, however, be utilized for the non-dividing cells, making it a highly flexible approach. In FISH, the method of probe preparation is very complex, as it is important to tailor the probes to identify the different DNA sequences (Volpi and Bridger, 2008).

9.24.5   POST AMPLIFICATION TECHNIQUE

9.24.5.1   MICROARRAY

Through two technological developments originated the idea of microarray technology. Firstly, the DNA sequencing activities, and secondly, the emphasis on expressed genome portion. Microarray DNA chip technology enables simultaneous study of thousands of mRNAs and is used for detecting gene expression patterns. Through this context, this approach is quite different from the above approaches since it also gives the measurement of expression on specified gene sets (Eshaque and Dixon, 2006). Within a limited surface area, thousands of DNA probes are neatly arranged in this technique onto a support matrix which contains nylon filters or glass chip. The location of the probe is referred to as spot on the sensor (Robinson et al., 2000). Such cDNAs implemented for the hybridization as a solution to the chip.

The bounded cDNA quantitative analysis is done by using the radio-labeled probes or fluorescent tags. Signal is generated by hybridizing the probe to the targeted mRNA, which the dedicated software can identify and integrate. For every biological sample, the specialized software creates the Gene Expression Profile (Russo et al., 2003). By using this approach, one can obtain a detailed understanding of the fungus cell in a single array. Microarray technology is quite simple to do because the wide scale DNA sequencing is not needed. This technique is termed destructive experiment, since it involves physical cell interruption to obtain entry to its patterns of gene expression; inaccurate microarray data can be generated through mRNA deterioration. The microarray technique is being used to classify the species *Aspergillus candida* (Singh and Kumar, 2013).

9.24.5.2   MACROARRAY

Macroarray is also referred to as hybridization of the DNA array, or Reverse dot plot. This approach uses the DNA amplification sensitivity, which does not involve radioisotopes (Singh and Kumar, 2013). It functions on the basis of concurrent PCR amplification of the associated cells. With this, it also analyzes multiple amplified sequences in one hybridization reaction. This is a method which is more robust than just PCR. PCR amplification is paired with hybridization in this test, which improves sensitivity up to 1000-fold or better over PCR alone (Taoufik et al., 2004). In this method, the inner probes are developed to distinguish the species. These probes are placed to the nylon membrane support. The oligos is strongly linked by UV cross-linking to the membrane. The PCR-amplified items would then be hybridized with the spotted sequence of species-specific probes on the plates (Tsui et al., 2011; Leinberger et al., 2005).

9.24.6   ISOTHERMAL AMPLIFICATION-BASED METHODS

9.24.6.1   ROLLING CIRCLE AMPLIFICATION (RCA)

Single nucleotide polymorphism (SNP) identification is becoming common for detecting pathogens. The traditional real-time PCR has many complexities and pitfalls in detecting SNPs amongst various genotypes. Therefore, species-specific PLPs (Circularizing Oligonucleotide Samples) are used to address these disadvantages. Circularizing oligonucleotide samples are single-stranded DNA molecules with 20-nucleotide target recognition sequences present at both 3' and 5' ends bound by the long connector sequence of 40-nucleotide (Tsui et al., 2011). The concept of PLPs was first documented in 1994 by Nilsson et al. When hybridized to a selected area, one side ends up coming close to the other side and, through the use of ligase, become circularized, producing no gaps. The circulated probe crosses the whole target area in a way close to that of the padlocks, guided by the double-stranded DNA's helical structure (Wang and Yang, 2010).

Hence, the Padlock probe is ideally suited to mimic a RCA reaction. RCA is focused on the rolling duplication of small ring-shaped single-stranded DNA molecules. The RCA method is isothermal, and is also known as rolling circle replication (RCR). This technique involves DNA polymerase, a primer to begin duplication, dNTPs, and proteins to bind and unwind DNA. In the amplification of signals, most groups use the RCA reaction, where the tiny ring-shaped probes behave as the model. Primer has double roles in RCA reaction by being complementary to the DNA-targeted sequence, as both RCA signal amplifier and unique identifier (Kuhn et al., 2002). A second primer complementary to the RCA component can be utilized in RCA; this will produce a hyper-branched RCA (HRCA) reaction. HRCA is often utilized as an alternative DNA amplification tool to the PCR (Gusev et al., 2001; Ahmed et al., 2014).

9.24.6.2   *LOOP-MEDIATED ISOTHERMAL AMPLIFICATION (LAMP)*

It is a vigorous and innovative nucleic acid amplification method and considered an alternative to the PCR. It amplifies the targeted nucleic acid under the isothermal conditions with the high specificity. LAMP does not require the temperature changes (Isothermal temperature) and DNA amplify at a single temperature without thermal cycler (Tsui et al., 2011). Consequently, it is constructed on auto-cycling strand displacement amplification of DNA. In this technology, the best DNA polymerase and asset of four primers that consist of two inner and two outer primers are used, which identify a total of six unique sequences on the targeted DNA. Two inner primers are denoted to as forwarding inner primer (FIP) and backward inner primer (BIP), although outer primers are F3 and B3 (Fakruddin, 2011). LAMP technology does not require costly equipment to get a high level of precision, and less number of preparation steps requires than the conventional, and qPCR. Due to high-amplification efficacy, up to 1039 copies of a target part can be attained in < 1 h of incubation. In many cases inhibition reaction occurred less in LAMP than that in PCR, and no time is lost for thermal changes in LAMP as it happened in PCR. An expensive thermal cycler is not required in the LAMP reaction while need just a single tube (Fakruddin, 2011). Moreover, binding to LAMP product sequences changes the MB spatial configuration, which separate fluorophore and quencher at both the ends of a single strand of nucleic acids and desorbing fluorescence (Liu et al., 2017).

9.24.6.3   NUCLEIC ACID SEQUENCE-BASED AMPLIFICATION (NASBA)

NASBA is an isothermal, very sensitive and transcription based amplification system that is specially designed for RNA detection. Some NASBA systems can also amplify the DNA and also well known as self-sustained sequence replication (3SR). This tool is useful in basic research as well as in the application-oriented fields, i.e., clinical medicine development, and infectious diseases diagnosis (Sergentet-Thevenot et al., 2008). Numerous amplification methods have already been established, such as PCR (Saiki et al., 1992), LAMP (Lee et al., 2009), and RCA (Lizardi et al., 1998). These methods cannot amplify the RNA directly with high sensitivity. NASBA delivers several advantages over the other techniques of mRNA amplification. It can amplify $> 10^9$ copies of the nucleic acid sequence in just one and half an hour with the help of three enzymes. NASBA is performed at 41°C (isothermal reaction), and omits the need for a thermal cycler which can facilitate the production of point-of-test devices (Fakruddin et al., 2012).

Several studies have informed that the amplification power of NASBA is better than that of the RT-PCR (Chang et al., 2012). RNA being the genomic material of many RNA viruses, an RNA-based amplification technique in contrast to the PCR keeps away from an additional reverse transcription step, thus minimizing the contamination risk and lowering hands-on time. It assists in better RT-PCR reaction as it provides the faster amplification kinetics that is especially suitable for the detection of Retroviruses. It can measure the replication of DNA viruses by detecting late mRNA expression. It supports the detection of human mRNA sequences lacking DNA contamination risk (Lauri and Mariani, 2009; Fakruddin et al., 2012). Gene expression studies can be supported without the intron-flanking primers or DNases. There are also some disadvantages to NASBA.

9.24.7   RNA INTERFERENCE (RNAI)

RNA interference (RNAi) is used as the modern technology to recognize and control the plant pathogenic fungi. This technique can be used in a vastly tissue-specific way to fight toxigenic fungi and causes the infection in crop plants (Panwar et al., 2012). Positive transgenic RNAi execution depends on many factors, which includes: (i) designing the vectors in such a way, to produce the double-stranded RNAs (dsRNAs) that will make the small interfering RNA (siRNA) species for the ideal gene silencing; (ii) accessibility of plentiful target siRNAs at infection place; (iii) efficient uptake of siRNAs by a fungus; (iv) siRNA half-life; (v) amplification of the silencing effects. RNAi eliminates the negative consequences of current disease control and fights the alarming rise of the fungicide resistant plant pathogens (Ishii and Holloman, 2015). RNAi knocks down the genes by using the event of intrinsic cellular defense. Detection of the dsRNA or hairpin RNA (hpRNA) by the fungal cells leads to the targeted transcripts by using the sequence homology important for the degradation or silencing (Nakayashiki et al., 2005). To identify the unique fungal targets, cell-specific, and dual RNA sequencing data should be provided. Then hpRNA or dsRNA can be modified for a definite transcript that can directly limit the fungal pathogenesis. RNAi technology has a benefit of the cell's natural machinery, which is assisted by the short-interfering RNA molecules, to successfully knock down the expression of a gene of interest. The major disadvantage of RNAi is the possibility of off-target effect, which leads to genes silencing, which tolerate the partial complimentarily to the sense or antisense strand of the targeted gene (Chen et al., 2006).

9.24.8   RNA-SEQ-BASED NEXT-GENERATION SEQUENCING

RNA-Seq is a newly developed deep-sequencing technology. Usually, a large population of RNA is altered to a cDNA library with converters that linked to one or both ends. Every fragment with or without amplification is sequenced in a high-input manner to get small sequences from one end as in single-end sequencing or both ends as in pair-end sequencing. Reads are usually 30–400 bp, depending on the DNA sequencing procedure used. The library is prepared to understand how closely the RNA sequencing results make known the original RNA transcripts are mostly determined in the library preparation step. To develop an RNA-seq library, the fragmentation of either the RNA or the cDNA is necessary to allow the processing through next-generation sequencing. Developed mRNA should be primed for RT response by the use of either random primers or oligo primers. The benefit of using oligo (dT) is that the majority of cDNA produced should be poly-adenylated mRNA; henceforth more of the sequences obtained should be informative (Mortazavi et al., 2008). Through this technology, the fungus

Magnaporthe oryzae is identified, which causes rice blast disease in rice (Soanes et al., 2012).

RNA-seq could be used to find the inclusive changes in fungi within a plant or could be used to identify the new pathogens. The potential relevance of mRNA-seq data for the recognition of nucleotide differences can make known the plant pathogenic fungal pathogenicity genes that are mutant in their protein-coding transcriptase. Technologies are making us closer to the capability to use RNA measurements for the plant disease diagnostics (Metzker, 2009).

9.25   CONCLUSION

In recent years, progressively investigative laboratories are exhausting molecular techniques to perceive and identify the diseases caused by plant pathogens (Ahmed et al., 2014; Kim et al., 2010). Better understanding of accurate detection of pathogens, rapid and pathogenicity factors to the species or strain level are the critical precondition for disease surveillance and development of novel disease control strategies. Current technologies such as DNA barcoding and RNA-Seq-based next-generation sequencing helped to face and overcome these challenges. We expectation that new progresses will enhancement the implementation of these new technologies for the diagnosis and study of plant disease (Chang et al., 2012; Tsai et al., 2006). Molecular techniques can also be established based on the different fungicide targeted mechanisms to rapidly detect resistant isolates. Furthermore, a timely detection of resistance levels in populations of the plant pathogens in a field and resulted help the growers formulate proper decisions on resistance management programs to control plant diseases (Chalupová et al., 2014).

KEYWORDS

- **backward inner primer**
- ***Eleusine coracana***
- **finger millet**
- ***Magnaporthe grisea***
- **management**
- ***Rhizoctonia solani***

REFERENCES

Ahmed, S. A., Van, D. E. B. H. G., Fahal, A. H., Van, D. S. W. W., & De Hoog, G., (2014). Rapid identification of black grain Eumycetoma causative agents using rolling circle amplification. *PLoS Negl. Trop Dis., 8*(12), e3368.

AICSMIP, (2013). *Annual Report of the All India Coordinated Small Millets Improvement Project, Project Coordinating Unit.* ICAR, GKVK, Bangalore.

Aslam, S., Tahir, A., Aslam, M. F., Alam, M. W., Shedayi, A. A., & Sadia, S., (2017). Recent advances in molecular techniques for the identification of phytopathogenic fungi: A mini-review. *Journal of Plant Interactions, 12*(1), 493–504.

Atkins, S. D., & Clark, I. M., (2004). Fungal molecular diagnostics: A mini-review. *J. Appl. Genet., 45*(1), 3–15.

Ayyar, P. N. K., (1933). Some experiments on the root gall nematode *Heterodera Marioni* in South India. *Madras Agric. J., 21,* 97–107.

Ayyar, P. N. K., (1934). Further experiments on the root gall nematode, *Heterodera Marioni* in South India. *Indian J. agric. Sci., 3,* 1064–1071.

Badali, H., & Nabili, M., (2012). Molecular tools in medical mycology; where we are! Jundishapur. *J. Microbiol., 6*(1), 1–3.

Baschien, C., Manz, W., Neu, T. R., & Szewzyk, U., (2001). Fluorescence in situ hybridization of freshwater fungi. *Int. Rev. Hydrobiol., 86*(4/5), 371–381.

Batsa, B. K., & Tamang, D. B., (1983). Preliminary report on the study of millet diseases in Nepal. In: *Maize and Finger Millet.* 10th Summer Workshop 2328 Jan, 1983, Rampur, Chitwan, Mysore.

Brown, N. A., Bass, C., Baldwin, T. K., Chen, H., Massot, F., Carion, P. W., Urban, M., et al., (2011). Characterization of the *Fusarium* gramine arum-wheat floral interaction. *J. Pathogens,* 10.4061/2011/626345.

Bu, R., Sathiapalan, R. K., Ibrahim, M. M., Al-Mohsen, I., Almodavar, E., Gutierrez, M. I., & Bhatia, K., (2005). Monochrome light cycler PCR assay for detection and quantification of five common species of Candida and Aspergillus. *J. Med. Microbiol., 54*(3), 243–248.

Butler, E. J., & Bisby, G. R., (1918). *Fungi of India* (p. 547). Sci. Monogram., XVIII, Calcutta. Butler, E. J., 1918, Fungi and Diseases in Plants, Thacker Spink and Co. Calcutta.

Capote, N., Aguado, A., Pastrana, A. M., & Sánchez-Torres, P., (2012). *Molecular Tools for Detection of Plant Pathogenic Fungi and Fungicide Resistance.* Valencia: INTECH Open Access Publisher.

Chandrasekaran, N. M., (1964). *Studies on the Reniform Nematode Rotylenchulus Reniformis with Special Reference to its Pathogenicity on Castor (Ricinus communis) and Ragi (Eleusine coracana).* MSc (Agri.) thesis, Univ. of Madras.

Chang, C. C., Chen, C. C., Wei, S. C., Lu, H. H., Liang, Y. H., & Lin, C. W., (2012). Diagnostic devices for isothermal nucleic acid amplification. *Sensors, 12*(6), 8319–8337.

Channamma, K. A. L., Viswanath, S., & Mantur, S. G., (1996). New record of *Uromyces* sp. on ragi from Karnataka. *Curr. Res., 25,* 97.

Chaves-Bedoya, G., Espejel, F., Alcala-Briseno, R. I., Hernandez-Vela, J., & Silva-Rosales, L., (2011). Short distance movement of genomic negative strands in a host and nonhost for Sugarcane mosaic virus (SCMV). *Virol. J., 8,* 15.

Chen, X., Steed, A., Harden, C., & Nicholson, P., (2006). Characterization of *Arabidopsis thaliana-Fusarium graminearum* interactions and identification of variation in resistance among ecotypes. *Mol. Plant Pathol., 7,* 391–403.

Cho, H. J., Hong, S. W., Kim, H. J., & Kwak, Y. S., (2016). Development of a multiplex PCR method to detect fungal pathogens for quarantine on exported cacti. *Plant Pathol. J., 32*(1), 53, 54.

Coleman, L. C., (1920). The cultivation of ragi in Mysore. *Bull. Dep. Agric. Mysore. Gen. Ser., 11.*

Compton, J., (1991). Nucleic acid sequence-based amplification. *Nature, 350*(6313), 91, 92.

Constantin, E. C., Cleenwerck, I., Maes, M., Baeyen, S., Van, M. C., De Vos, P., & Cottyn, B., (2016). Genetic characterization of strains named as *Xanthomonas axonopodis* pv. *dieffenbachiae* leads to a taxonomic revision of the *X. axonopodis* species complex. *Plant Pathology, 65*(5), 792–806.

Dasmahapatra, K., & Mallet, J., (2006). DNA barcodes: Recent successes and future prospects. *Heredity., 97*(4), 254, 255.

Dawei, W., & Peng, Y., (2014). Assessing the impact of dominant sequencing based gene expression profiling techniques (SGEPTs) on phytopathogenic fungi. *Chiang Mai. J. Sci., 41,* 922–9444.

Desai, S. G., Thirumalachar, M. J., & Patel, M. K., (1965). Bacterial blight disease of *Eleusine coracana* Gaertn. *Indian Phytopath., 28,* 384–386.

Dheeman, S., Baliyan, N., Dubey, R. C., Maheshwari, D. K., Kumar, S., & Chen, L., (2020). Combined effects of rhizo-competitive rhizosphere and non-rhizosphere bacillus in plant growth promotion and yield improvement of *Eleusine coracana* (Ragi). *Canadian Journal of Microbiology, 66*(2), 111–124.

Divya, G. M., (2011). *Growth and Instability Analysis of Finger Millet Crop in Karnataka.* Unpublished master's thesis, University of Agricultural Sciences, Bengaluru, India.

Dubey, S. C., (1995). Banded blight of finger millet caused by *Thanatephorus cucumeris*. *Indian J. Mycol. Pl. Pathol., 25,* 315, 316.

Dublish, P. K., & Singh, P. N., (1976). Phytopathogenic fungi of Meerut: Some new records for India. *Curr. Sci., 45,* 168.

Ellison, M. A., McMahon, M. B., Bonde, M. R., Palmer, C. L., & Luster, D. G., (2016). *In situ* hybridization for the detection of rust fungi in paraffin-embedded plant tissue sections. *Plant Methods, 12*(1), 37.

Eshaque, B., & Dixon, B., (2006). Technology platforms for molecular diagnosis of cystic fibrosis. *Biotechnol. Adv., 24*(1), 86–93.

Fakruddin, M., (2011). Loop-mediated isothermal amplification (LAMP): An alternative to polymerase chain reaction (PCR). *Bangladesh Res. Pub. J., 5*(4), 425–439.

Fakruddin, M., Mazumdar, R. M., Chowdhury, A., & Mannan, K., (2012). Nucleic acid sequence-based amplification (NASBA)-prospects and applications. *Int. J. Life Sci. Pharma. Res., 2,* 106–107.

Fang, Y., & Ramasamy, R. P., (2015). Current and prospective methods for plant disease detection. *Biosensors, 5*(3), 537–561.

Govindu, H. C., Shivanandappa, N., & Renfro, B. L., (1966). Observations on diseases of *Eleusine coracana* with reference to *Helminthosporium* disease. *Abstr. Internat. Symp. Pl. Path.,* (pp. 48, 49). New Delhi.

Gusev, Y., Sparkowski, J., Raghunathan, A., Ferguson, H., Montano, J., Bogdan, N., Schweitzer, B., et al., (2001). Rolling circle amplification: A new approach to increase sensitivity for immunohistochemistry and flow cytometry. *Am. J. Pathol., 159*(1), 63–69.

Hayden, R., Qian, X., Procop, G., Roberts, G., & Lloyd, R., (2002). In situ hybridization for the identification of filamentous fungi in tissue section. *Diagn. Mol. Pathol., 11*(2), 119–126.

Hebert, P. D., & Gregory, T. R., (2005). The promise of DNA barcoding for taxonomy. *Syst. Biol., 54*(5), 852–859.

Ishii, H., & Holloman, D. W., (2015). *Fungicide Resistance in Plant Pathogens*. Tokyo: Springer.

Jain, A. K., Gupta, J. C., & Yadava, H. S., (1994). Stability of resistance to footrot in finger millet. *Bhartiya Krishi Anusandhan Patrika., 9,* 109–112.

Jensen, E., (2014). Technical review: *In situ* hybridization. *Anat. Rec., 297*(8), 1349–1353.

Joshi, L. M., Raychaudhuri, S. P., Batra, S. K., Renfro, B. L., & Ghosh, A., (1966). Preliminary investigations on a serious disease of Eleusine coracana in the states of Mysore and Andhra Pradesh. *Indian Phytopath., 19,* 324–325.

Keshavamurthy, K. V., & Yaraguntaiah, R. C., (1969). Further studies on the transmission of the virus component of the ragi disease complex in Mysore. *Mysore J. Agric. Sci., 3,* 480.

Kim, S. W., Li, Z., Moore, P. S., Monaghan, A. P., Chang, Y., Nichols, M., & John, B., (2010). A sensitive non-radioactive northern blot method to detect small RNAs. *Nucleic Acids Res. 38,* 98.

Kishun, R., & Sohi, H. S., (1984). Control of bacterial canker of mango by chemicals. *Pesticides, 18,* 32, 33.

Kishun, R., (1994b). Evaluation of phylloplane micro-organisms from mango against *Xanthomonas campestris* pv. *Mangifera indica. Indian Phytopathology, 47,* 313.

Krishnamurthy, P. K., & Francis, R. A., (2012). A critical review on the utility of DNA barcoding in biodiversity conservation. *Biodivers Conserv., 21*(8), 1901–1919.

Kuhn, H., Demidov, V. V., & Frank-Kamenetskii, M. D., (2002). Rolling-circle amplification under topological constraints. *Nucleic Acids Res., 30*(2), 574–580.

Kulkarni, G. S., (1922). The smut of nachani or ragi (*Eleusine coracana*). *Ann. Appl. Biol., 9,* 184–186.

Kusaba, M., Eto, Y., Don, L. D., Nishimoto, N., Tosa, Y., Nakayashiki, H., & Mayama, S., (1999). Genetic diversity in *pyricularia* isolates from various hosts revealed by polymorphisms of nuclear ribosomal DNA and the distribution of the MAGGY retrotransposon. *Jpn. J. Phytopathol., 65,* 588–596.

Lauri, A., & Mariani, P. O., (2009). Potentials and limitations of molecular diagnostic methods in food safety. *Genes. Nutr., 4*(1), 1–12.

Lee, D., La, M. M., Allnutt, T. R., & Powell, W., (2009). Detection of genetically modified organisms (GMOs) using isothermal amplification of target DNA sequences. *BMC Biotechnol., 9*(1), 7.

Lee-Du, H., (1997). Morphological characters and seed transmission of *Bipolaris panici-miliacei* causing leaf spot of common millet. *Korean J. Pl. Pathol., 13*(1), 18–21.

Leinberger, D. M., Schumacher, U., Autenrieth, I. B., & Bachmann, T. T., (2005). Development of a DNA microarray for detection and identification of fungal pathogens involved in invasive mycoses. *J. Clin. Microbiol., 43*(10), 4943–4953.

Liu, B., Feng, C., Matheron, M. E., & Correll, J. C., (2018). Characterization of foliar web blight of spinach, caused by *Pythium aphanidermatum*, in the Desert Southwest of the United States. *Plant Disease, 102*(3), 608–612.

Liu, M., McCabe, E., Chapados, J. T., Carey, J., Wilson, S. K., Tropiano, R., Redhead, S. A., et al., (2015). Detection and identification of selected cereal rust pathogens by TaqMan real-time PCR. *Can. J. Plant Pathol., 37*(1), 92–105.

Liu, W., Huang, S., Liu, N., Dong, D., Yang, Z., Tang, Y., Ma, W., et al., (2017). Establishment of an accurate and fast detection method using molecular beacons in loop-mediated isothermal amplification assay. *Sci Rep., 7,* 40125.

Lizardi, P. M., Huang, X., Zhu, Z., Bray-Ward, P., Thomas, D. C., & Ward, D. C., (1998). Mutation detection and single-molecule counting using isothermal rolling-circle amplification. *Nat. Genet., 19*(3), 225–232.

Lulu, D., & Girija, V. K., (1989). Sheath blight of ragi, *Curr. Sci., 58,* 681–682.

Mackay, I. M., (2004). Real-time PCR in the microbiology laboratory. *Clin. Microbiol. Infect., 10*(3), 190–212.

Mancini, V., Murolo, S., & Romanazzi, G., (2016). Diagnostic methods for detecting fungal pathogens on vegetable seeds. *Plant Pathol., 65,* 691–703.

Maramorosch, K., Govindu, H. C., & Kondo, F., (1977). Rhabdo virus particles associated with a mosaic disease of naturally infected Eleusine coracana (finger millet) in Karnataka state (Mysore) South India. *Plant Dis. Reptr., 61,* 1029–1031.

McCartney, H. A., Foster, S. J., Fraaije, B. A., & Ward, E., (2003). Molecular diagnostics for fungal plant pathogens. *Pest Manag. Sci., 59*(2), 129–142.

McRae, W., (1920). *Detailed Administration Report of the Government Mycologist for the Year 1919–20* (p. 285). McAlpine, D. 1910. The smuts of Australia, Melbourne. Government Printer.

Mehta, P. R., & Chakravarty, S. C., (1937). A new disease of *Eleusine coracana. Indian J. Agric. Sci., 7,* 793–796.

Metzker, M. L., (2009). Sequencing technologies-the next generation. *Nat. Rev. Genet., 11,* 31–46.

Mishra, A. P., & Mishra, B., (1969). *Helminthosporium holmii* on graminaceous hosts. *Indian Phytopath., 22,* 412–414.

Mohanty, K. C., & Das, S. N., (1976). Free amino acids in the roots of finger millet plants infected with ring nematodes. *Indian Phytopath., 29,* 434–436.

Monteiro, F., Romeiras, M. M., Figueiredo, A., Sebastiana, M., Baldé, A., Catarino, L., & Batista, D., (2015). Tracking cashew economically important diseases in the West African region using metagenomics. *Front Plant Sci., 6,* 482–483.

Mortazavi, A., Williams, B. A., McCue, K., Schaeffer, L., & Wold, B., (2008). Mapping and quantifying mammalian transcriptomes by RNA-Seq. *Nat. Methods., 5,* 621–628.

Mundkur, B. B., (1939). A contribution towards knowledge of Indian *Ustilaginales. Trans. Brit. Mycol. Soc., 23,* 105.

Munjal, R. L., Lall, G., & Chona, B. L., (1961). Some *Cercospora* species from India-VI. *Indian Phytopath., 14,* 179–190.

Nagaraju, Viswanath, S., Reddy, H. R., & Lucy, C. K. A., (1982). Ragi streak, a leafhopper transmitted virus disease in Karnataka. *Mysore J. Agric. Sci., 16,* 301–305.

Nakayashiki, T., Kurtzman, C. P., Edskes, H. K., & Wickner, R. B., (2005). Yeast prions [URE3] and [PSI+] are diseases. *Proc. Natl. Acad. Sci., 102*(30), 10575–10580.

Nyaku, S. T., Kantety, R. V., Cebert, E., Lawrence, K. S., Honger, J. O., & Sharma, G. C., (2016). Principal component analysis and molecular characterization of reniform nematode populations in Alabama. *The Plant Pathology Journal, 32*(2), 123.

Okubara, P. A., Schroeder, K. L., & Paulitz, T. C., (2008). Identification and quantification of *Rhizoctonia solani* and *R. oryzae* using real-time polymerase chain reaction. *Phytopathology, 98*(7), 837–847.

Panwar, V., McCallum, B., & Bakkeren, G., (2012). Endogenous silencing of *Puccinia triticina* pathogenicity genes through in planta-expressed sequences leads to the suppression of rust diseases on wheat. *Plant J., 73*, 521–532. doi: 10.1111/tpj.12047.

Parambaramani, C., Subramanian, C. L., & Kandaswamy, T. K., (1975). Sheath blight and wilt or ragi (*Eleusine coracana* Gaertn) caused by *Marasmius candidus* Bolt. *Curr. Sci., 44*, 358.

Patel, M. K., & Thirumalachar, M. J., (1965). Notes on some *Xanthomonas* species described from South India. *Curr. Sci., 34*, 436, 437.

Patro, T. S. S. K., Rani, C., & Kumar, G. V., (2008). *Pseudomonas fluorescens*, a potential bioagent for the management of blast in *Eleusine coracana. J. Mycol. Pl. Pathol., 38*(2), 298–300.

Paul, K. S. M., Raychaudhuri, S. P., & Sundaram, N. V., (1973). Further studies on ragi mosaic in Delhi. *Indian Phytopath., 26*, 554–559.

Paul, N. C., Hwang, E. J., Nam, S. S., Lee, H. U., Lee, J. S., Yu, G. D., & Yang, J. W., (2017). Phylogenetic placement and morphological characterization of *Sclerotium rolfsii* (Teleomorph: *Athelia rolfsii*) associated with blight disease of ipomoea batatas in Korea. *Mycobiology, 45*(3), 129–138.

Pradhanang, P. M., & Abington, J. B., (1993). Cercospora leaf spot disease (*Cercospora eleusine*) of finger millet in Nepal. In: *Advances in Small Millets*. Riley.

Prakash, O., & Raoof, M. A., (1985). *Bacterial Canker in Mango* (p. 59). (Abs.) IInd International Symposium on mango, Bangalore.

Prakash, O., Misra, A. K., & Raoof, M. A., (1994). Studies on mango bacterial canker disease. *Bio. Memoirs., 20*, 95–107.

Pruvost, O., & Luisetti, J., (1991). Effect of time of inoculation with *Xanthomonas campestris* pv. *Mangiferae indicae* on mango fruits susceptibility, epiphytic survival of *X. c.* pv. *Mangiferae indicae* on mango fruits in relation to disease development. *Journal of Phytopathology, 133*, 139–151.

Qi, M., & Yang, Y., (2002). Quantification of *Magnaporthe grisea* during infection of rice plants using real-time polymerase chain reaction and northern blot/phosphoimaging analyses. *Phytopathology, 92*(8), 870–876.

Raghunathan, V., (1968). Damping-off of green gram, cauliflower, daincha, ragi and cluster beans. *Indian Phytopath., 21*, 456, 457.

Rahman, M. T., Uddin, M. S., Sultana, R., Moue, A., & Setu, M., (2013). Polymerase chain reaction (PCR): A short review. *Anwer Khan Mod. Med. Coll. J., 4*(1), 30–36.

Rajagopal, B. K., (1965). *Further Studies on the Damage Caused by the Reniform Nematode Rotylenchulus reniformis Linford and Oliveira 1940 to Ragi and Castor* (p. 62). MSc (Agri.) thesis, Univ. of Madras.

Rangaswami, G., Prasad, N. N., & Eswaran, K. S. S., (1961). Two new bacterial diseases of sorghum. *Andhra Agric. J., 8*(6), 269–272.

Robinson, B., Erle, D., Jones, D., Shapiro, S., Metzger, W., Albelda, S., Parks, W., & Boylan, A., (2000). Recent advances in molecular biological techniques and their relevance to pulmonary research. *Thorax., 55*(4), 329–339.

Russo, G., Zegar, C., & Giordano, A., (2003). Advantages and limitations of microarray technology in human cancer. *Oncogene, 22*(42), 6497–6507.

Rybicki, E. P., & Hughes, F. L., (1990). Detection and typing of maize streak virus and other distantly related Gemini viruses of grasses by polymerase chain reaction amplification of a conserved viral sequence. *J. Gen. Virol., 71*(11), 2519–2526.

Saiki, R. K., Scharf, S., Faloona, F., Mullis, K., Horn, G., Erlich, H., & Arnheim, N., (1992). Enzymatic amplification of beta-globin genomic sequences and restriction site analysis for diagnosis of sickle cell anemia. *Biotechnology (Reading, Mass), 24*, 476.

Saveetha, K., Sankaralingam, A., Pant, R., & Ramanathan, A., (2007). Etiology and transmission of mottle streak disease of finger millet (*Eleusine coracana* Gaertn.). *Archives of Phytopathology and Plant Protection, 40*(1), 53–60.

Savolainen, V., Cowan, R. S., Vogler, A. P., Roderick, G. K., & Lane, R., (2005). Towards writing the encyclopedia of life: An introduction to DNA barcoding. *Philos. Trans. R Soc. B., 360*(1462), 1805–1811.

Schoch, C. L., Seifert, K. A., Huhndorf, S., Robert, V., Spouge, J. L., Levesque, C. A., Chen, W., et al., (2012). Nuclear ribosomal internal transcribed spacer (ITS) region as a universal DNA barcode marker for fungi. *Proc. Natl. Acad. Sci. U.S.A., 109*(16), 6241–6246.

Sergentet-Thevenot, D., Montet, M. P., & Vernozy-Rozand, C., (2008). Challenges to developing nucleic acid sequence-based amplification technology for the detection of microbial pathogens in food. *Rev. Med. Vet., 159*, 514–527.

Setty, K. G. H., (1975). *Studies on Ragi Cyst Nematode Heterodera Sp.* Report of UNDP/ ICAR Research Project, University of Agril. Sciences, Hebbal, Bangalore.

Shaw, F. J. F., (1921). *Report of the Imperial Mycologist Scient* (pp. 34–40). Reports Agric. Res. Inst. Pusa, 1920–21.

Shimizu, K., Tanaka, C., Peng, Y. L., & Tsuda, M., (1998). Phylogeny of bipolar is inferred from nucleotide sequences of Brn1, a reductase gene involved in melanin biosynthesis. *J. Gen. Appl. Microbiol., 44*(4), 251–258.

Singh, A., & Kumar, N., (2013). A review on DNA microarray technology. *Int J. Curr. Res. Rev., 5*(22), 01–05.

Soanes, D. M., Chakrabarti, A., Paszkiewicz, K. H., Dawe, A. L., & Talbot, N. J., (2012). Genome-wide transcriptional profiling of appressorium development by the rice blast fungus *Magnaporthe oryzae*. *PLoS Pathog., 8*, 1002514.

Subbayya, J., & Raychaudhuri, S. P., (1970). A note on a mosaic disease of ragi (*Eleusine coracana*) in Mysore, India. *Indian Phytopath., 23*, 144–148.

Sundaram, N. V., Palmer, C. T., Nagarajan, K., & Presscot, J. M., (1972). Disease survey of sorghum and millet in India. *Pl. Dis. Reptr., 56*, 740–743.

Taoufik, A., Nijhof, A., Hamidjaja, R., Jongejan, F., Pillay, V., Sonnevelt, M., & De Boer, M., (2004). *Reverse Line Blot Hybridization in the Detection of Tick-Borne Diseases.* BTi.

Thines, M., Goker, M., Telle, S., Ryley, M., Mathur, K., Narayana, Y. D., Spring, O., & Thakur, R. P., (2008). Phylogenetic relationships of graminicolous downy mildews based on cox2 sequence data. *Mycol. Res., 112*(PT 3), 345–351.

Thines, M., Telle, S., Choi, Y. J., Tan, Y. P., & Shivas, R. G., (2015). Baobabopsis, a new genus of graminicolous downy mildews from tropical Australia, with an updated key to the genera of downy mildews. *IMA Fungus, 6*(2), 483–491.

Thirumalachar, M. J., & Narasimhan, M. J., (1949). Downy mildew of *Eleusine coracana* and *Iscilema laxum* in Mysore. *Indian Phytopath., 2*, 46–51.

Tsui, C. K., Woodhall, J., Chen, W., Lévesque, C. A., Lau, A., Schoen, C. D., Baschien, C., et al., (2011). Molecular techniques for pathogen identification and fungus detection in the environment. *IMA Fungus., 2*(2), 177.

Urashima, A. S., Hashimoto, Y., Don, L. D., Kusaba, M., Tosa, Y., Nakayashiki, H., & Mayama, S., (1999). Molecular analysis of the wheat blast population in Brazil with a homolog of retrotransposon MGR583. *Jpn. J. Phytopathol., 65*, 429–436.

Velculescu, V. E., (1995). Serial analysis of gene expression. *Science, 270,* 484–487.

Venkatarayan, S. V., (1946). Diseases of ragi (*Eleusine coracana*). *Mysore Agri. J., 24,* 50–57.

Volpi, E. V., & Bridger, J. M., (2008). Fish glossary: An overview of the fluorescence in situ hybridization technique. *Biotechniques., 45*(4), 385–386.

Vu, D., Groenewald, M., De Vries, M., Gehrmann, T., Stielow, B., Eberhardt, U., Al-Hatmi, A., et al., (2019). Large-scale generation and analysis of filamentous fungal DNA barcodes boosts coverage for kingdom fungi and reveals thresholds for fungal species and higher taxon delimitation. *Stud. Mycol., 92,* 135–154.

Wang, Z., & Yang, B., (2010). Padlock-probes and rolling-circle amplification. In: *MicroRNA Expression Detection Methods* (pp. 241–247). Berlin: Springer. doi: 10.1007/978-3-642-04928-6_16.

Wonglom, P., Ito, S., & Sunpapao, A., (2018). First report of *Curvularia lunata* causing leaf spot of *Brassica rapa* subsp. Pekinensis in Thailand. *New Disease Reports, 38,* 15.

Xu, J., (2016). Fungal DNA barcoding 1. *Genome, 59*(11), 913–932.

Yeo, S. F., & Wong, B., (2002). Current status of nonculture methods for diagnosis of invasive fungal infections. *Clin. Microbiol. Rev., 35*(3), 465–484.

Zheng, B., He, D., Liu, P., Wang, R., Li, B., & Chen, Q., (2020). Occurrence of collar rot caused by *Athelia rolfsii* on soybean in China. *Canadian Journal of Plant Pathology,* 1–5.

CHAPTER 10

Recent Approaches for Diagnosis and Management of Economically Important Diseases of Field Pea (*Pisum sativum* L.) in India

SONIKA PANDEY, R. K. MISHRA, MONIKA MISHRA, A. K. PARIHAR, and G. P. DIXIT

ICAR-Indian Institute of Pulses Research, Kanpur – 208024, Uttar Pradesh, India, E-mail: rajpathologist@yahoo.com (R. K. Mishra)

ABSTRACT

Field pea (*Pisum sativum* L.) is one of the most important cool season food legume crops in the world and occupies a very prominent place because of its production and multiple uses. In India, the crop is cultivated in the northern and central parts of the country. The reason behind the low productivity of pea is its susceptibility to several biotic and abiotic stresses at different stages of its growth. Among the biotic stresses, diseases cause a heavy reduction to both quality and quantity of pea seed and grain production. Some environmental conditions like high humidity and moderate temperature enhance the possibility of development of diseases. Annual losses differ from year to year and variety to variety, depending upon the local weather conditions. In order to minimize the disease incidence and to maximize the productivity, accurate, sensitive, and specific detection and diagnostic techniques are required. There are various direct and indirect disease diagnosis methods which are currently employed in agriculture such as PCR, RT-PCR, ELISA, IF, FCM, GC-MS, thermography, fluorescence imaging (FI), hyperspectral techniques, and biosensors. Modern nanofabrication techniques have increased the sensitivity of biosensors. With the help of bioinformatics, we can identify the specific motifs, DNA/RNA sequences, protein sequences which increase the specificity and accuracy of disease diagnosis tools. Daily monitoring of plant health is essential to detect and reduce plant disease severity. DNA, RNA, and serological-based techniques provide accurate and early diagnosis compared to visual observation.

10.1   INTRODUCTION

Field pea (*Pisum sativum* L.) is one of the most important cool season food legume crops in the world and occupies a very prominent place because of its production and multiple uses. In India, the crop is cultivated in northern and central parts of the country. It is an important *Rabi* pulse crop grown in about 0.97 m ha with annual production of 0.89 m tones. The major field pea growing states are Uttar Pradesh, Madhya Pradesh, Jharkhand, Bihar, Assam, and Maharashtra. The average productivity of this crop has increased considerably over the years which is now to the tune of 1 t/ha that is very much lower as compared to other countries likely France (>4300 kg/ha), Germany (>3000 kg/ha), USA (>2100 ka/ha) and Canada (>2100 kg/ha). The reason behind low productivity of pea is its susceptibility to several diseases and insect pests. Diseases cause heavy reduction to both quality and quantity of pea seed and grain production. Some environmental conditions like high humidity and moderate temperature enhance the possibility of development of diseases. The major diseases and insects of pea are wilt, root rot, powdery mildew, rust, aphids, stem fly, leaf miner and pod borer.

Interaction of disease and crops is a never-ending process. Early farmers were in practice to use superstitious and supernatural ideas for the crop damage. Modern plant pathological detection systems have made the detection of plant disease based on scientific approaches. Filippo Re and Carlo Berti Pichat first started to classify plant disease according to the symptoms. For the rapid management of disease, we should know the severity of the disease. Therefore, without estimating the disease severity, we cannot control it. Thus, analysis of disease and symptoms severity is very important for understanding plant stress biology. Visual recognition of symptoms is essential for the diagnosis of Plant disease. However, this method is not very specific. Newer technologies offer disease assessment with great accuracy and precession. Visible light photography and digital image analysis have been used over the last 30 years (West et al., 2003). Hyperspectral imagery is a recent technique, and it is not using widely in plant pathology. Plant diseases are the main cause of crop destruction in the world. For example, rice blast disease every year causes a great loss in rice production (Dean et al., 2005). Plant disease causes a great loss in all food crops, including fruit crops also. In South Western Europe Flavescence Doree, grapefruit disease cause a great loss in grape production (Martinellli et al., 2014). Early detection and diagnosis of disease is of key importance to prevent disease spread (Yang et al., 2014). Traditional identification of disease is not enough for the early disease diagnosis and its management. Introduction of PCR techniques has a profound impact on the disease diagnosis. Through nucleic acid-based techniques we can detect the pathogens that have not been cultured. According to the European and Mediterranean Plant Protection Organization (EPPO), the protocol for detecting plant pathogen are-integrating phenotypic, serological, and molecular techniques. For newly discovered pathogens, molecular assays can be developed easily. Nucleic acid-based techniques are easy to develop and are very accurate and sensitive. Present methods for plant disease detection include proximate detection, immunological, and DNA-based approaches, analysis of volatile compounds and genes, remote sensing, spectroscopic methods sensors, biophotonics, etc. There are several reports which emphasis the importance of biochemical and serological methods (Hampton et al., 1990; Schaad and Frederick, 2002) or applied RS techniques (West et al., 2003; Bock et al., 2010; Sankaran et al., 2010; Mahlein et al., 2012a). However, there are many other methods which can be used for the disease diagnosis. Here we will discuss about those techniques.

10.2 ECONOMICALLY IMPORTANT DISEASES OF FIELD PEA

10.2.1 RUST

Rust is one of the most economically important diseases of field pea. The disease is caused by *Uromyces fabae* (Pers. de Bary). It is an autoecious fungus completing its entire life cycle on the same host. The pathotypes develop particularly in regions with warm and humid climate and the disease generally appears at flowering or podding stage. In the tropical and subtropical regions such as India and China, the principal causal agent of pea rust has been reported as fungus *U. viciae-fabae*, where warm humid weather is suitable for the appearance of both the uredial and the aecidial stage. This disease is serious in North India (Mishra et al., 2009). The stem of the plant become malformed, and the affected plant dies out. The earliest symptoms are the yellow spot having aecia in round or elongated clusters. Then the uredo pustules develop which are powdery and light brown in appearance. High humidity and cloudy weather with temperatures of 20–22°C favor disease development. The plants give dark brown or blackish appearance visible as patches in the field (Mishra et al., 2010).

10.2.1.1   MANAGEMENT STRATEGIES

1. The affected plant debris and trash should be burnt after harvesting.
2. Integrated management of rust includes control of volunteer plants over the summer and removal of infected plant debris. It is advisable to use clean seeds without rust contaminations, and to treat the seed with a suitable fungicide such as diclobutrazole. Preventive fungicide sprays of mancozeb (0.25%), Propiconazole (0.1%) and Hexaconazole (0.1%) at early disease development stage have been recommended (Mishra and Pandey, 2009). The use of host plant resistance is the best means of rust management (Mishra et al., 2009).

10.2.2   RHIZOCTONIA SEEDLING BLIGHT

It is caused by *Rhizoctonia solani*, soilborne fungus. It is also known as Rhizoctonia tip blight or stem rot of pea. Characteristic symptoms of the disease are the death of the growing tip of very young pea seedlings.

10.2.2.1   MANAGEMENT

Rotate peas with crops know to be relatively poor hosts of the fungus, such as grain crops and corn.

10.2.3   BACTERIAL BLIGHTS

Causative agent of disease is *Pseudomonas syringae*. Characteristic symptoms of the disease are the appearance of leaf and pod lesions (approx.: 3 mm in diameter and angular shape). Lesions are brown shiny and translucent.

10.2.3.1 MANAGEMENT

1. Use of healthy and disease-free seeds for planting helps to minimize the disease spread.
2. At least 3-year rotation with non-host crops.

10.2.4 ASCOCHYTA DISEASE

Phorma medicaginis var. *pinodella* and *Ascochyta pisi* are the main causal agents of aschyomycota disease. There are three main Ascochyta diseases-Mycospharella blight, Ascochyta foot rot, and ascochyta leaf and pod spot. Out of all three, Mycosphaerella blight is most important.

10.2.4.1 MYCOSPHAERELLA BLIGHT

Occurrence of small purple spots on the stem, pods, and foliage are the typical symptoms of the disease. Leaf spot may enlarge up to 5–6 nm. Small lesions have no definite shape, but larger lesions are circular with distinct margins.

10.2.4.2 ASCOCHYTA FOOT ROT

Similar to Mycosphaerella blight; however, this is most often apparent on lower stem and upper root.

10.2.4.2.1 Management

1. Use of clean seeds and destroys old pea crop residues.
2. Rotate crops so there is at least 3 years between pea crops, and often 4 or 5 years may be required. Follow the recommended sowing dates for your district.
3. Avoid early sowing at high seeding rates. Don't mix fungicide with the inoculants as this will reduce the number of rhizobia.
4. Application of foliar fungicides before rain. *Pantoea agglomerans* and *Bacillus* species have been found effective in the control of Ascochyta disease of field pea.

10.2.5   LEAF AND POD SPOT

Lesions of this disease are different from the above-mentioned disease they are about 5–8 mm in diameter and sunken and tan with a dark and distinct border. Lesions are oval on foliage and pods and are elongated on stems.

10.2.5.1   MANAGEMENT

1. Use western-grown disease-free seed.
2. Use a 3- or 4-year rotation. Mold board-plow pea profuse deeply and completely.
3. Isolate new pea fields ¼ mile (0.4 km) from old ones.

10.2.6   ANTHRACNOSE

The causative agent of this disease is *Colletotrichum capsici* Characteristic symptoms of the disease are oval lesions on the leaf and stipule, which are around 2–8 mm in diameter with brown margins and gray tan centers.

10.2.6.1   MANAGEMENT

Practice Crop rotation; use Chlorothalonil for the management of disease.

10.2.7   DOWNY MILDEW

Downy mildew is caused by *Peronospora viciae*. On the upper side of the foliage, lesions appear as irregularly shaped yellow to brown areas with indistinct margins. On the underside of foliage, these discolored areas have patches of mouse color. Sometimes the symptoms of disease may appear on pods and not on foliage. Young pods are more susceptible to disease as compared to old ones. Yellow to brown sunken blotches are the characteristic symptoms of the disease. Oospores of the fungus sometimes present on the pod wall.

10.2.7.1 *MANAGEMENT*

1. Since the causal organism survives from season to season in the form of oospores in the plant debris, the destruction of previous year's plant debris and following crop rotation of two or three years are very effective control measures.
2. Spraying and dusting pea plants with fungicides are effective in limiting the spread of the disease. Deep tillage to bury crop residues.
3. Use tolerant cultivars.
4. Use metalaxyl for seed treatment and application of Chlorothalonil just appearance of the symptoms for the management of disease.

10.2.8 *POWDERY MILDEW*

Powdery mildew is the most important disease of field pea. This air borne disease is caused by *Erysiphe pisi,* which is a biotrophic ascomycete fungus. Usually, it develops at the pod formation stage or just before harvest of the crop. As the name implies, a white powdery coating covers the aerial plant parts including leaves and pods. Eventually, the speckles turn into spots and the plant turns purple, bluish-green, and lastly brown. Warm-dry days and cool nights are critical for the development of diseases. Late-sown crop is typically at higher risk. Frequent heavy dews in the absence of rain accelerate the disease's progression. The number and weight of the pods are reduced. Yield reduction due to these diseases is very high within a short period of time. It appears in severe form almost every year when the plants are in the pod stage towards the end of January and in February. The losses in yield in a 100% infected crop were estimated by Munjal et al. (1963) to be 21–31% in pod number and 26–47% in pod weight. However, several powdery mildew resistant varieties have been developed by the NARS.

10.2.8.1 *MANAGEMENT STRATEGIES*

1. Late planting should be avoided the disease severity.
2. Collect and burn the plant debris/stubbles that left on the field after crop harvest.
3. Grow early maturing short duration varieties in high disease prone areas.

4. The diseases can be managed by alternate spray of any of the wettable sulfur (0.3%), karathane (0.05%) or carbendazim (0.05%). Give the first spray after the appearance of the diseases in the crop. The second spray should be given 15 days after the first spray and the third spray only if there is a need for it.

5. Use of diseases resistant varieties and to avoid excess irrigation in the field during the flowering stage of the crop.

6. Biocontrol agents like *Acremonium alternatum, Irpex lacteus, Paecilomyces fumosoroseus, Verticillium lecanii, sporothrix regulosa, Trichoderma harzianum, Stephanoascus* spp., and *Tilletiopsis* spp. have been used for management of powdery mildew. Bacterization by *P. fluorescens* and *P. aeruginosa* is also effective for management of powdery mildew disease (Mishra and Pandey, 2010). Other fungi such as *Acrodontium crateriform, Dissoconium acicularae* and *Ramichloridium apiculatum* have also been reported to be the antagonist of powdery mildew pathogen but their efficacy as biocontrol agent have not been studied (Sara Fondevilla et al., 2012).

10.2.9   STEM ROT/WHITE ROT

Stem rot caused by *Sclerotinia sclerotiorum* (Lib.) de Bary, is a cosmopolitan necrotrophic fungal pathogen with a very broad host range (field pea, chickpea, pigeon pea, lentil, rajmash, bean, mustard, potato, sunflower, etc.), and it infects a variety of broadleaf crops and weeds. It often enters fields by means of contaminated seed lots or equipment's. It survives as a mass of fungal hyphae covered by an environmentally resistant black rind called sclerotia. When conditions are favorable, such as at canopy closure when humidity levels in the canopy are high and the crop begins flowering, an inverted mushroom structure called an apothecium produces ascospores which then infect the flowers. Hyphae can also grow directly into plant tissues, especially when plant-soil contact is established. The fungus kills tissue and fills the stem with white hyphae and sclerotia. In this way it survives in the soil for many years.

This fungus is a soil-borne fungus that attacks plant in all stages of growth, including young seedlings, mature plants and their harvested products distributed worldwide (Agrios, 2005; Purdy, 1979). It is a major disease of several leguminous crops such as field pea, chickpea, pigeon pea, lentil, soybean, faba bean, and alfalfa (Nene et al., 1996; Mishra et al., 2015). It is a major disease of field pea under favorable conditions; it may cause serious losses to the crop. Symptoms appear as wet, soft, and white rotting of the tissues on all arial part of the plants, including pods. In India, regular occurrence of the disease has been reported in some districts of Himachal Pradesh, where it causes considerable yield losses (Grewal and Pal, 1986). The yield loss ranges from 21 to 47% at pod bearing stage of the crop (Sharma, 1990). Limited variety resistance may be available in dry pea and lentil cultivars.

10.2.9.1 MANAGEMENT STRATEGIES

1. Seed cleaning is essential to remove the sclerotial mixer during the threshing and processing period.
2. Deep plowing, low land with continuous stagnation of water reduces the sclerotial population.
3. Alternate spray of recommended fungicides at 10–15 days intervals is essential.
4. All infected plant parts must be collected and burnt before drying of the plants.
5. Removal of the weeds from the fields.

10.2.10 FUSARIUM ROOT ROT

This is the most important disease of pea. The causative agent of the disease is *Fusarium solani* f. sp. *pisi.* Primary symptoms of the disease appear along the taproot and side roots of the plant as brown slender lesions. Red discoloration of vascular tissue is also visible up to the three nodes above the soil line.

10.2.10.1 MANAGEMENT

1. Soil amended with decomposed *Sesbenia aculata*, mustard cake, and Farmyard manure.
2. Deep summer plowing during the month of May-June followed by cultural practices like long crop rotation is recommended to reduce the severity of the disease.
3. Seed treatment with fungicides (Carbendazim @2.5 g/kg seeds) is feasible and protects the seedlings from infection and ensures better plant stand.

4. Treatment of field pea seeds with bioagents (*Trichoderma harzianum* @ 6 g/kg) and fungicide (vitavax 1.0 g/kg).
5. Early sowing should be avoided in badly infested areas.

10.2.11 ENATION MOSAIC

This disease is caused by enation mosaic virus. The character symptom of the disease is blister like outgrowth on the underside of foliage and pods. Infected pods are malformed and stunted. Infected plants die prematurely.

10.2.11.1 MANAGEMENT

Use of resistant (Oregon pioneer) or tolerant cultivars. It using susceptible varieties, planting before March 31 helps avoid aphid infestation. Spray water to knock aphids off plants.

10.2.12 PEA MOSAIC

There are two causative agents responsible for pea mosaic virus, pea common mosaic virus and bean yellow mosaic virus (bean virus 2). Molting of the foliage is the characteristic symptoms of this disease. Veins of the tissue become yellow, leaving the green patches scattered irregularly over the surface of leaves and stipules. In young plants infected with the virus, they become stunted pods may be fewer in number as compared to the healthy plants. The severity of the symptoms depends upon the virus and pea varieties involved.

10.2.12.1 MANAGEMENT

Use resistant cultivars and monitor aphids closely. Apply insecticide when aphid populations reach two per plants.

10.2.13 PEA STREAK

Pea streak is caused by Wisconsin pea streak virus, western peak streak virus, alfalfa mosaic virus and bean yellow mosaic virus. The characteristic symptoms of the disease are stem streaks. The streak begins as light brown to purple oblong lesion along the length of the stem and petiole. On terminal foliage vein necrosis and leaf yellowing are common. Pods develop brown to purple dead areas which are sunken and are of different size.

10.2.13.1 MANAGEMENT

Monitor aphids closely. Control when aphid populations reach two per plant.

10.2.14 SEED BORNE MOSAIC

Vein clearing and mosaic, stunting, and downward leaf curling, internode failure to elongate, pods become very short and stubby.

10.2.14.1 CONTROL

Plant pea seed lots that have been tested and found free of the virus. Control aphids. Use resistant or preferably immune pea cultivars.

10.2.15 DETECTION AND DIAGNOSIS OF DISEASES

10.2.15.1 SEROLOGICAL ASSAYS

Serological assays were developed for the detection of viruses. Using polyclonal and monoclonal antisera techniques such as ELISA, western blots, immunostrip assay, and dot blot immune binding assays and serologically specific electron microscopy (SSEM) (VanVuurde et al., 1987; Hampton et al., 1990). In 1970, ELISA was first employed, it is the most widely used immunodiagnostic techniques. The sensitivity of ELISA depends upon the organism, sample, and titer value. For viruses and fungi, polyclonal antisera have been developed and they are been using for numerous protocols. Monoclonal antibodies (MAbs) recognize one epitome only and are expensive, while polyclonal antibodies recognize multiple epitomes and are less expensive. In present time for phytopathogenic bacteria, ELISA procedures using polyclonal and MAbs are been employed.

10.2.15.2   NUCLEIC ACID-BASED METHODS

DNA-based detection methods are-fluorescence *in situ* hybridization (FISH) and PCR variants (PCR, nested PCR, CO-PCR multiplex PCR, real-time PCR DNA fingerprinting) RNA Based techniques are reverse transcriptase PCR, nucleic acid sequence-based amplification (NASBA) and Aplidet RNA. Molecular detection methods are based on the accurate design of oligonucleotides and probes. PCR offers several advantages over immunoassays. In PCR-based diagnostic techniques, primers are designed to pair with the target DNA. Specific amplification of the target DNA sequence is widely used to detect and identify the plant pathogens (Mumford et al., 2006). Amplification confirms the presence of organism in the tested sample. PCR reliability for pathogen specificity has been improved by the use of dye-quenched probes (Morris et al., 1996; Thelwell et al., 2000). For plant virus detection RT-PCR is most reliable. By using n-PCR we can increase the specificity and sensitivity. Co-PCR techniques were especially developed for plant virus and bacteria detection. Co-PCR is carried out in a single reaction, it reduces the rate of contamination, and sensitivity is also similar to the n-PCR and RT-PCR. Colorimetric detection increases the sensitivity of virus detection up to 1000 times. In M-PCR, two or more than two target sequences are amplified simultaneously.

M-PCR is useful because several pathogens infect the same plant at a time (Davino et al., 2012), so with M-PCR we can easily detect more than one pathogen. Unlike conventional PCR, RT-PCR allows pathogen quantification. In RT-PCR Mainly two chemistry works SYBR green and Taq Man. Taq man used most commonly. In DNA fingerprinting identification is done through the identification of unique patterns present in DNA. There are many fingerprinting techniques such as RFLP, SNP, STR, rep-PCR, amplified 16S ribosomal DNA (rDNA) restriction analysis (ARDRA) (Scortichini et al., 2001), and amplified fragment length polymorphism (AFLP) (Clerc et al., 1998; Manceau and Brin., 2003). Technique choice depends upon the aim.

NASBA is a technique used to amplify RNA sequences. This technique was developed in 1990s, it does not require PCR machine it works on water bath only. This technique commonly used to detect bacteria and viruses (Klerks et al., 2001; Rodriguez-Làzaro et al., 2006). Specificity and speed of this technique is very high. Loop-mediated isothermal amplification (LAMP) this technique is based on photometry (Mori et al., 2001). This technique is a simple screening assay which eliminates the need of thermocycler. This technique is used for the plant virus detection (Varga and James, 2006). DNA microarray technique came into the light in 1990, and it creates a revolution in nucleic acid detection. Microarray techniques have been used extensively for Plant pathogen detection. There are many factors on which choice of molecular techniques is dependent such as budget, time, and no. of species (Lopez et al., 2009). Molecular methods are highly efficient and specific. However, there are many limitations in molecular techniques also such as sometimes PCR give false results due to impairing and primer dimerization, dead pathogens can also give false-positive results. Costing of these techniques is also a very important issue.

10.2.16   NEW APPROACHES FOR MANAGEMENT OF FIELD PEA DISEASES

Lateral flow microarrays-This technique allows colorimetric visualization of hybridized sequences. This technology is based on the interaction of host and pathogen biomarkers (Carter and Cary, 2007).

10.2.16.1   VOLATILE MARKER BASED TECHNIQUE

10.2.16.1.1   Techniques Based on Volatile Compounds as Biomarkers

Plants secrete many volatile organic compounds in their surroundings; these volatile metabolites have low molecular weight, high pressure, and low boiling point. These compounds help the plant in growth, communication, defense, and survival. These compounds exist in the gaseous phase under standard temperature and pressure and present in very low concentration. Voc Profiling is a new field with numerous applications. In 1990, VOC profiling was used for peanut plants. In the VOC based studies profiling of emitted VOCs is done, and it compared with the healthy treatment. Through VOC profiling CMV has also been identified. E-nose is a platform through which profiling of VOCs has done. In this system, an array of metal oxide sensors is used which is specific for a particular class of VOCs. In-plant health physiology monitoring e-nose has been used extensively.

10.2.16.1.2   Remote Sensing of Plant Disease

Remote sensing is a tool through which we obtain information without physical contact through the reflection of electromagnetic radiation or energy. Radar and Lidar are the instruments which are being used for the remote sensing process.

Visual diagnostic of plant-pathogen is the primary requisition of disease diagnosis. However, the accuracy of the same is not less. Serological detection of pathogens is based on the identification of disease based on the linking of antigen and antibody linking. ELISA is the most commonly used serological method used in diagnostics. The main principle of ELISA is the binding of antibody with antigen. For Phytophthora *infestans*, *Ralstonia solanacearum*, Tomato Mosaic Virus, the Potyvirus group, and many other pathogens lateral flow-ELISA is widely used for site-specific detection tissue print ELISA and other lateral flow devices are used (Danks and Baker, 2000). There are many drawbacks associated with ELISA such as it cannot access the viability of pathogen. RNA interference (RNAi) is a newer technique that is used to control the phytopathogenic fungi. It is an efficient technique to identify tissue way.

10.2.16.1.3   Flow Cytometry (FCM)

This is an optical technique in which LASER beam is used. This technique is used primarily for cell sorting, counting, protein engineering and biomarker detection. The main advantage of FCM is that it can process multiple samples simultaneously. In-plant pathology research it is not commonly used as detection tool but these techniques has the potential to be used in pathogen detection, assessment of genome size and in checking viability of pathogens (D'Hondt et al., 2011). Serological diagnostic techniques like ELISA can be used for detection, but it cannot be used for the viability assessment. Therefore, FCM has the potential to be used in pathogen detection and viability assessment.

10.2.16.1.4   Role of Bioinformatics in Disease Detection

Bioinformatics is an interdisciplinary branch of biology which uses concepts of statics, mathematics, chemistry, physics, biochemistry, and other branches of biology and have application in mapping, alignment of DNA and protein sequences, formation of 3D structures.

10.3 CONCLUSIONS

The basic method of disease detection involves visual detection of signs and symptoms. However, recent advancements in researches have provided many accurate and sensitive methods for disease detection and diagnosis. These advancements in plant pathology have been made by coupling biotechnology, bioinformatics, and molecular biology together. Plating assay provide information about the presence and viability of pathogens, but these methods are time-consuming and have low specificity. PCR, qPCR, ELISA, flow cytometry are the modern methods which are less time-consuming and are very accurate. However, these methods require proper standardization for specificity and sensitivity. Global acceptance for standardized protocols is necessary for the management of the diseases of field pea.

DISCLOSURE STATEMENT

Authors declare that they have no conflict of interest.

KEYWORDS

- **amplified fragment length polymorphism**
- **diagnosis**
- **field pea**
- **loop-mediated isothermal amplification**
- **nucleic acid sequence-based amplification**
- ***Pisum sativum* L.**

REFERENCES

Abhishek, S., & Sobita, S., (2017). Eco-friendly management of powdery mildew and rust of garden pea (*Pisum sativum* L.). *Journal of Pharmacognosy and Phytochemistry, 6*(5), 90–93.

Agrios, G. N., (2005). *Plant Pathology* (5th edn.). Academic Press, London.

Bock, C. H., Poole, G. H., Parker, P. E., & Gottwald, T. R., (2010). Plant disease severity estimated visually, by digital photography and image analysis, and by hyperspectral imaging. *Crit. Rev. Plant Sci., 29,* 59–107, doi: 10.1080/07352681003617285.

Carter, D. J., & Carym, R. B., (2007). Lateral flow microarrays: A novel platform for rapid nucleic acid detection based on miniaturized lateral flow chromatography. *Nucleic Acids Res., 35,* 74, doi: 10.1093/nar/ gkm269.

Clerc, A., Manceau, C., & Nesme, X., (1998). Comparison of randomly amplified polymorphic DNA with amplified fragment length polymorphism to assess genetic diversity and genetic relatedness within genospecies III of *Pseudomonas syringae. Appl. Environ. Microbiol., 64,* 1180–1187.

D'hondt, L., Hofte, M., Van, B. E., & Leus, L., (2011). Applications of flow cytometry in plant pathology for genome size determination, detection and physiological status. *Mol. Plant Pathol., 12,* 815–828.

Danks, C., & Baker, I., (2000). On-site detection of plant pathogens using lateral-flow devices. *EPPO Bull., 30,* 421–426.

Davino, S., Miozzi, L., Panno, S., Rubio, L., Davino, M., & Accotto, G. P., (2012). Recombination profiles between tomato yellow leaf curl virus and tomato yellow leaf curl Sardinia virus in laboratory and field condition: Evolutionary and taxonomic implications. *J. Gen. Virol., 93,* 2712–2717. doi: 10.1099/vir.0.045773-0.

Dean, R. A., Talbot, N. J., Ebbole, D. J., Farman, M. L., Mitchell, T. K., & Orbach, M. J., (2005). The genome sequence of the rice blast fungus *Magnaporthe grisea. Nature, 434,* 980–986. doi: 10.1038/nature03449.

Hampton, R., Ball, E., & De Boer, S., (1990). *Serological Methods for Detection and Identification of Viral and Bacterial Plant Pathogens.* A laboratory manual. APS Press St. Paul.

Klerks, M. M., Leone, G., Lindner, J. L., Schoen, C. D., & Van, D. H. J. F. J. M., (2001). Rapid and sensitive detection of Apple stem pitting virus in apple trees through RNA amplification and probing with fluorescent molecular beacons. *Phytopathology, 91,* 1085–1091. doi: 10.1094/21 PHYTO.2001.91.11.1085.

López, M. M., Llop, P., Olmos, A., Marco-Noales, E., Cambra, M., & Bertolini, E., (2009). Are molecular tools solving the challenges posed by the detection of plant pathogenic bacteria and viruses? *Mol. Biol., 11,* 13–46. swfrec.ifas.ufl.edu/hlb/database/pdf/00002423. pdf.

Mahlein, A. K., Oerke, E. C., Steiner, U., & Dehne, H. W., (2012a). Recent advances in sensing plant diseases for precision crop protection. *Eur. J. Plant Pathol., 133,* 197–209. doi: 10.1007/s10658-011-9878-z.

Manceau, C., & Brin, C., (2003). Pathovars of *Pseudomonas syringae* are structured in genetic populations allowing the selection of specific markers for their detection in plant samples. In: Iacobellis, N. S., et al., (eds.), *Pseudomonas Syringae and Related Pathogens* (pp. 503–512). Kluwer.

Martinellim, F., Uratsum, S. L., Albrechtm, U., Reaganm, R. L., & Phum, M. L., (2012b). Transcriptome profiling of citrus fruit response to huanglongbing disease. *PLoS One, 7,* 38039. doi: 10.1371/journal.pone.0038039.

Mishra, R. K., & Pandey, K. K., (2010). Effect of application of PGPR and neemazal on management of pea rust (*Uromyces fabae*). *J. Basic and Appl. Mycobiol., 1, 2,* 102–106.

Mishra, R. K., & Pandey, P. K., (2009). Effect of rust on nodulation and nitrogenase activity of pea (*Pisum sativum*). *Pest Management in Hort. Ecosystem, 15*(1), 60–62.

Mishra, R. K., & Pandey, P. K., (2009). Efficacy of fungicides against pea rust caused by *Uromyces fabae*. *Annals of Plant Protection Sciences, 2,* 54, 55.

Mishra, R. K., Mishra, K. K., Pandey, K. K., & Singh, U. P., (2010). Biochemical changes in Pea infected by *Uromyces fabae*. *Indian Phytopathology., 65*(2), 98–101.

Mishra, R. K., Naimuddin, A., & Saabale, P. R., (2015). First report of stem rot in pigeon pea from India. *Pulses News Letter,* 7.

Mishra, R. K., Pandey, K. K., & Pandey, P. K., (2009). Screening of different pea genotypes against rust of pea caused by *Uromyces fabae*. *Indian Journal of Agricultural Sciences, 79*(5), 403–405.

Mori, Y., Nagamine, K., Tomita, N., & Notomi, T., (2001). Detection of loop-mediated isothermal amplification reaction by turbidity derived from magnesium pyrophosphate formation. *Biochem. Biophys. Res. Commun., 289,* 150–154.

Morris, T., Robertson, B., & Gallagher, M., (1996). Rapid reverse transcription-PCR detection of hepatitis C virus RNA in serum by using the TaqMan fluorogenic detection system. *J. Clin. Microbiol., 34,* 2933–2936.

Mumford, R., Boonham, N., Tomlinson, J., & Barker, I., (2006). Advances in molecular phytodiagnostics: New solutions for old problems. *Eur. J. Plant Pathol., 116,* 1–19. doi: 10.1007/s10658-006-9037-0.

Munjal, R. L., Chenulu, V. V., & Hora, I. S., (1963). Assessment of losses due to powdery mildew *Erysiphe polygoni* on pea. *Indian Phytopath., 16,* 260–267.

Na, L., Shengchun, X., Xiefeng, Y., Guwen, Z., Weihua, M., Qizan, H., Zhijuan, F., & Yaming G., (2016). Studies on the control of *Ascochyta* blight in field peas (*Pisum sativum* L.) caused by *Ascochyta pinodes* in Zhejiang Province, China. *Front Microbiol., 7.*

Nene, Y. L., Sheila, Y. K., & Sharma, S. B., (1996). *A World List of Chickpeas and Pigeon Pea Pathogens* (5ᵗʰ edn.). Patancheru 502 324, Andhra Pradesh, India: International Crops Research Institute for the Semi-Arid Tropics.

Purdy, L. H., (1979). *Sclerotinia sclerotiorum*: History, diseases and symptomatology, host range, geographic distribution, and impact. *Phytopathology, 69*(8), 875–880.

Rodriguez-Làzaro, D., Hernàndez, M., D'Agostino, M., & Cook, N., (2006). Application of nucleic acid sequence-based amplification for the detection of viable foodborne pathogens: Progress and challenges. *J. Rapid Meth. Aut. Mic., 14,* 218–236. doi: 10.1111/j.1745-4581.2006. 00048.x.

Sankaran, S., Mishra, A., Ehsani, R., & Davis, C., (2010). A review of advanced techniques for detecting plant diseases. *Comput. Electron. Agric., 72,* 1–13. doi: 10.1016/j.compag.2010.02.007.

Sara, F., & Diego, R. A., (2012). Powdery mildew control in pea. A review. *Sustain. Dev., 32,* 401–409. doi: 10.1007/s13593-011-0033-1.

Schaad, N. W., & Frederick, R. D., (2002). Real-time PCR and its application for rapid plant disease diagnostics. *Can. J. Plant Pathol., 24,* 250–258. doi: 10.1080/07060660209507006.

Scortichini, M., Marchesi, U., Rossi, M. P., & Di, P. P., (2001). Bacteria associated with hazelnut (*Corylus avellana* L.) decline are of two groups: *Pseudomonas avellanae* and strains resembling *P. syringae pv. syringae*. *Applied Appl. Environ. Microbiol., 68,* 476–484. doi: 10. 1128/AEM.68.2.476-484.2002.

Thelwell, N., Millington, S., Solinas, A., Booth, J., & Brown, T., (2000). Mode of action and application of Scorpion primers to mutation detection. *Nucleic Acids Res., 28,* 3752–3761, doi: 10.1093/nar/28.19.3752.

Van, V. J. W., Ruissen, M. A., & Vruggink, H., (1987). Principles and prospects of new serological techniques including immunosorbent immunofluorescence, immunoaffinity isolation and immunosorbent enrichment for sensitive detection of phytopathogenic bacteria. In: Civerolo, E. L., Collmer, A., Davis, R. E., & Gillaspie, A. G., (eds.), *Plant Pathogenic Bacteria; Curr. Plant Sci. Biotech. Agric.* (Vol. 4, pp. 835–842).

Varga, A., & James, D., (2006). Use of reverse transcription loop-mediated isothermal amplification for the detection of plum pox virus. *J. Virol. Method, 138,* 184–190. doi: 10.1016/j.jviromet.2006.08.014.

West, J. S., Bravo, C., Oberti, R., Lemaire, D., Moshou, D., & McCartney, H. A., (2003). The potential of optical canopy measurement for targeted control of field crop diseases. *Ann. Rev. Phytopathol., 41,* 593–614.

Yang, W., Chen, J., Chen, G., Wang, S., & Fu, F., (2013). The early diagnosis and fast detection of blast fungus, *Magnaporthe grisea,* in rice plant by using its chitinase as biochemical marker and a rice cDNA encoding mannose-binding lectin as recognition probe. *Biosens. Bioelectron., 41,* 820–826. doi: 10.1016/j.bios.2012.10.032.

CHAPTER 11

Recent Advances in Bio-Intensive Management (BM) of Major Diseases of Pigeon Pea in India

MONIKA MISHRA,[1] R. K. MISHRA,[1] SONIKA PANDEY,[1] U. S. RATHORE,[1] RAJESH K. PANDEY,[2] and MANJUL PANDEY[3]

[1]ICAR-Indian Institute of Pulses Research, Kanpur – 208024, Uttar Pradesh, India, E-mail: rajpathologist@yahoo.com (R. K. Mishra)

[2]Bundelkhand University, Jhansi, Uttar Pradesh, India

[3]KVK, Banda University of Agriculture and Technology, Banda, Uttar Pradesh, India

ABSTRACT

Among the pulse crops grown in India, pigeon pea is the second most important *Kharif* legume crop after chickpea. India is the largest producer as well as consumer of pigeon pea, followed by Myanmar, Malawi, and Kenya. It is protein-rich, contains about 22% protein, which supplies a major share of the protein requirement of the vegetarian population of the country. For the management of these diseases, chemical pesticides are generally used. It is well established that agricultural chemicals cause not only environmental pollution but also pose serious health hazards as their injudicious use often results in toxicity to man, plants, domestic animals and wildlife and therefore are regarded as ecologically unacceptable. The management of wilt and PSB by chemicals has not yielded the anticipated results. Therefore, a renewed knowledge is needed to assess the current severity of these problems and to develop or refine bio-intensive management (BM) strategies for these biotic stresses to protect crops in an eco-friendly manner. It is a multidisciplinary approach that manages diseases effectively by integration of host plant resistance, cultural, physical, biological, safer chemical, biocontrol agents, and molecular approaches. These approaches can play a major role in reducing the losses due to the diseases under subsistence farming conditions. BM will definitely prove to be an effective strategy for enhancing pigeon pea production under the changing climate scenario.

11.1   INTRODUCTION

Pigeon pea [*Cajanus cajan* (L.) Millsp.], an important grain legume crop is predominantly cultivated in tropical and subtropical regions of the world. India is the largest producer as well as consumer of pigeon pea, followed by Myanmar, Malawi, and Kenya (FAO, 2016). In India, pigeon pea is the second most important legume after chickpea and the country alone contributes 63.4% of world production with 62.6% of world-cultivated area (FAO, 2016). Maharashtra, Karnataka, Gujarat, Andhra Pradesh, Madhya Pradesh, and Uttar Pradesh are the major pigeon pea growing States. The crop is affected by several biotic and abiotic stresses at various stages of the crop (Reddy et al., 2012). Heavy losses caused by this disease warrant for an urgent need to identify suitable and effective management strategies to mitigate the losses which would add to the pigeon pea production in the country. Several methods have been exploited and recommended by the various researchers for management of this important disease (Haware and Donald, 1996; Govil and Rana, 1994; Mishra et al., 2016–2019). However, for the management of these diseases, chemical pesticides are generally used across the world. It is well established that agricultural chemicals cause not only environmental pollution but also pose serious health hazards as their injudicious use often results in toxicity to plants, humans, animals, and wildlife. Further, the majority of the most important pathogens of pigeon pea are soil-borne in nature, and their management by chemicals has not yielded the anticipated results. Status and importance of various diseases have changed over the years in India. A renewed knowledge is needed to assess the current severity of these problems and to develop or refine bio-intensive management (BM) modules to protect crops in a cost-effective and eco-friendly manner. BM is a multidisciplinary approach that manages diseases effectively by integration of cultural, physical, biological, chemical, and molecular strategies. These are eco-friendly and sustainable tools besides directly fighting against the pathogens. The aim of this chapter is to review the distribution and symptomatology of major diseases of pigeon pea and their management through bio-intensive approaches, i.e., cultural, chemical, biological, and molecular.

Additionally, we outlined future research strategies on the issues related to disease management.

11.2   IMPORTANT DISEASES THAT AFFECT THE PIGEON PEA CROP

11.2.1   WILTS

11.2.1.1   CAUSAL ORGANISM

Wilt is a major problem in all over the pigeon pea growing areas of the country. The disease caused by *Fusarium udum* a major biotic stress leading to over 20–25% yield losses at various stage of the crop.

11.2.1.2   EPIDEMIOLOGY

Wilt disease has been reported from several other pigeon pea growing countries including India. Pathogen is soil-borne in nature, it penetrates the vascular bundles of roots portion of plants and stops or reduces water uptake to the foliage. The fungus can stay alive on infected plant debris in the soil for several years. Symptoms usually appear at the flowering and podding stage of the crop, but sometimes occur earlier when plants are 1–2 months old. Long and medium duration pigeon pea crop affected more than short duration types.

11.2.1.3   SYMPTOMATOLOGY

Initial symptoms of wilt-infected plants are dull green color of the leaves. At later mature plants showing typical wilt symptoms of drooping, yellowing, and gradual killing of plant, darkening of xylem vessels of infected plants. The roots of the wilted plants do not show any external rotting but when split open vertically, later stage of the disease dark brown discoloration of internal xylem is seen (Nene et al., 1991; Dubey, 2002). Pods from the wilted plants look normal, but seeds are generally smaller, wrinkled, and discolored. Infested soil is the main source of primary inoculum for the development of *Fusarium* wilt epidemics in pulses. Infected seeds are also a source of primary inoculum of the disease spread.

11.2.2   PHYTOPHTHORA STEM BLIGHT (PSB)

11.2.2.1   CAUSAL ORGANISMS

The causal organism of this disease is *Phytophthora drechsleri* f. sp. *Cajani.*

11.2.2.2   EPIDEMIOLOGY

The disease is soil-borne in nature. The fungus survives in soil and on infected plant debris. Continuous cloudy weather, drizzling rain with warm and humid weather favor the disease incidence. The first disease symptoms appear at the seedling stage of the crop and are more severe in low-lying areas of fields where water temporarily stagnates.

11.2.2.3   SYMPTOMATOLOGY

Initial symptoms of Phytophthora disease Infected plants have water-soaked lesions on the upper surface of leaves and later brown to black lesions/spots on their stems and petioles. Finally, the infected leaves and stems lose their turgidity. Ultimate the affected stems or branches are breakdown, causing the foliage above the lesion to dry up. Pigeon pea plants that are infected by blight, but not killed, often produce large galls on their stems, especially at the edges of the lesions. The pathogen infects the foliage and stems but not the root system.

11.2.3   STEM/WHITE ROT

11.2.3.1   CAUSAL ORGANISMS

The causal organism of this disease is *Sclerotinia sclerotiorum* (Lib.) de Bary.

11.2.3.2   EPIDEMIOLOGY

It is a cosmopolitan necrotrophic fungal pathogen with a very broad host range (bean, mustard, potato, sunflower, etc.), and it infects a variety of broadleaf crops and weeds. The fungus is a soil-borne in nature that attacks plant in all stages of growth (Agrios, 2005; Purdy, 1979).

11.2.3.3   SYMPTOMATOLOGY

It is a major soil-borne disease of several leguminous crops; causes almost 100% crop losses annually around the world (Purdy, 1979; Subbarao, 1998; Tahtamouni et al., 2006; Nene et al., 1996; Mishra et al., 2015). The pathogen can survive through sclerotia, and serve as the primary source of inoculum for infection. The yield loss ranges from 21 to 47% at pod bearing stage of the crop (Sharma, 1990). Limited variety resistance may be available in dry pea and lentil cultivars. No resistance has been observed among chickpea cultivars. White mold is more common in winter-sown lentils than spring lentils due to their exposure to prolonged wet and cool spring weather.

11.2.4   ALTERNARIA BLIGHT

11.2.4.1   CAUSAL ORGANISM

The causal organism of this disease are *Alternaria tenuissima* (Kunze ex Persoon) Wiltshire and *Alternaria alternata* (Fries) Keissler.

11.2.4.2   EPIDEMIOLOGY

It is an emerging disease of pigeon pea under adverse climate conditions. Generally it appears in late sown crop (August-September) under warm, humid conditions.

11.2.4.3   SYMPTOMATOLOGY

Disease symptoms develop as small, circular, necrotic spots on leaves that develop quickly forming typically concentric rings. The lesions appear on all aerial plant parts including flowers, flower buds, and pods. They cause blighting of leaves and severe defoliation and drying of infected branches.

11.2.5   DRY ROOT ROT

Root Rot: (Scelerotial stage: *Rhizoctonia bataticola*) (Pycnidial stage: *Macrophomina phaseolina*).

11.2.5.1   EPIDEMIOLOGY

It is a major problem in short-duration pigeon pea sown in early season. Hot (30°C and above) and dry weather conditions favor the disease development. Rain after a prolonged dry spell predisposes plants to the disease.

11.2.5.2   SYMPTOMATOLOGY

In pigeon pea, dry root rot is an emerging problem during early/seedling stage of the crops in the month of July and August. The pathogen is seed and soil-borne in nature and it has a wide host range. High day temperature (~30°C) with moisture stress conditions at podding and flowering stage rapidly aggravate disease incidence. Affected plant petiole and leaflets droops only at the top of the plant. Affected root system devoid of lateral roots and taproot become black. When we uproot the plant usually lower portions of the taproot remain in the soil. Dead roots become hard and brittle and show shredding of bark.

11.2.6   COLLAR ROT

11.2.6.1   CAUSAL ORGANISMS

The causal organism of this disease is *Sclerotium rolfsii Saccardo*.

11.2.6.2   EPIDEMIOLOGY

Infection of this disease may appear when the temperatures of about 30°C and soil moisture at sowing predispose seedlings of the crop. In India, the disease is more of a problem in early-sown (June) than in later-sown crops.

11.2.6.3 SYMPTOMATOLOGY

The initial symptom of the disease appears at seedling disease, when patches of dead seedlings at the primary leaf stage are seen. Symptoms appeared in scattered form over the field. The infected seedlings may turn slightly chlorotic before they die and later a rotting in the collar region appeared with white mycelial growth of the plants. Affected seedlings/plants can be uprooted easily, but the lower part of their root usually remains in soil. At a later stage of the disease infection, white or brown sclerotial bodies of the fungus can be found at the collar region of a dead seedling or in the soil around the plants.

11.2.7 STERILITY MOSAIC DISEASE (SMD)

11.2.7.1 CAUSAL ORGANISM

Sterility mosaic virus vector. *Eriophyid mite (Aceria cajani).*

11.2.7.2 EPIDEMIOLOGY

A single eriophyid mite vector is sufficient to transmit the sterility mosaic disease (SMD). Perennial crop of pigeon pea and the ratooned growth of harvested plants provide more severity of the vector. Diseases symptoms are suppressed during the hot summer months, but with monsoon rains, they reappear on the new growth. Shade and humidity encourage mite multiplication, especially in hot summer weather conditions.

11.2.7.3 SYMPTOMATOLOGY

Initial symptoms of the SMD infected pigeon pea plants are showing stunted and bushy appearance of the plants, reduces the leaves size with chlorotic rings or mosaic symptoms, and partial or complete cessation of flower production (i.e., sterility). Infected plants remain in the vegetative stage and it did not produce any flower. The causal agent of the disease is PPSMV, a virus with a segmented, negative-sense, single-stranded RNA genome.

11.2.8   YELLOW VEIN MOSAIC DISEASE (YMD)

11.2.8.1   CAUSAL ORGANISMS

This disease is caused by the mung bean yellow mosaic virus (MYMV) belonging to the Gemini group of viruses, which is transmitted by the whitefly (*Bemisia tabaci*).

11.2.8.2   EPIDEMIOLOGY

In pulses, YMV infect a wide host range including chickpeas, pigeon pea, faba beans, field peas, lentils, mung bean, and Urd bean. This viral disease is also found on several alternate and collateral hosts, which act as primary sources of inoculums.

11.2.8.3   SYMPTOMATOLOGY

Infected leaves showing yellowing mosaic-like spots on pigeon pea leaves. Infected plants having less flowering, fruiting, and pod development. At later stages of the disease symptoms, plants become dwarfed, and symptoms usually occur in patches, and along the edges of paddocks.

11.2.9   INTEGRATED APPROACHES FOR BIO-INTENSIVE MANAGEMENT (BM) OF PIGEON PEA DISEASES

11.2.9.1   CULTURAL PRACTICES

The use of healthy seeds, clean cultivation, and three-year crop rotation has been recommended for the management of this disease. Disease-free seed can be produced either by surface irrigation in semi-arid regions where conditions of high temperature and low humidity prevent infection of this fungus, or under a pedigreed seed program, in which seed plots were isolated and subjected to strict inspection for disease-free seed. Best sowing time of different pulses/legume crops in *Rabi, Kharif,* and Spring season for maximum yield and minimum diseases incidence.

11.2.9.2 CHEMICAL APPROACHES

Management of diseases through chemical is most effective but due to several environmental pollution and health hazards posing toxicity to man, plants, animals, and wildlife render these chemical-based interventions ecologically unacceptable for management of soil borne pathogens (Haware et al., 1996). Various systemic and non-systemic fungicides such as captan, thiram, carbendazim, etc., have been used by various workers to reduce seed infections (Sindhan and Bose, 1981; Trutman et al., 1992) of various diseases. Foliar sprays of carbendazim, benomyl, zineb, maneb, and mancozeb have been recommended for its management (Chakrabarty and Shyam, 1988). Benlate and Bavistin persisted in the plant up to 15–20 days after seed treatment (Sindhan and Bose, 1981). EBI fungicides like triadimefon and triforine were effective against the pathogen under *in vitro* conditions but were phytotoxic to *P. vulgaris* seedlings when used as seed treatment (Chakrabarty and Shyam, 1988). A combination of seed treatment with tebuconazole (0.1%) and foliar sprays of tricyclazole (0.03%) reduced the disease severity both on leaves and pods while maximum seed yield was recorded in case of seed treatment and foliar sprays of carbendazim (Gupta et al., 2000c).

Continuous use of systemic fungicides has been shown to be associated with the development of resistant biotypes (Tu and McNaughton, 1980). Maringoni et al. (2002) reported that all isolates of *C. lindemuthianum* collected from *P. vulgaris* fields located at Paranpanema Valley, Brazil were insensitive to benomyl, carbendazim, and thiophanate methyl showing the occurrence of cross-resistance to different benzimidazole fungicides while all isolates were sensitive to chlorothalonil. Recently, Oliveira and Oliveira (2003) reported that strobilurin fungicides like azoxystrobin and trifloxystrobin efficiently controlled this disease while chemicals like carbendazim and fluquinconazole applied without rotation were least efficient treatments. However, a combination of systemic fungicides with non-systemics and rotation of fungicides can be an effective approach for management of this disease instead of constant use of one systemic fungicide.

11.2.9.3 BIOLOGICAL APPROACHES

Considerable efforts were made to manage different diseases of pulses, including pigeon pea through various beneficial microbial antagonists.

Among the antagonistic fungi *Trichoderma* species are the most studied bioagents in pulse ecosystem (Singh, 2013; Mishra et al., 2016–2018; Chaudhary et al., 2004). They are successfully used to control wilts, root rots, collar rot and stem rot diseases incited by different *Fusarium* spp., *Rhizoctonia solani, R. bataticola, Sclerotium rolfsii, Sclerotinia sclerotiorum, Ascochyta, Alternaria* spp. *Phytophthora* spp. in different pulse crops. Chaudhary et al. (2004) and Mishra et al. (2018) compiled that many biocontrol agents have been effectively managed different phytopathogens of pulses, including pigeon pea (Table 11.1).

TABLE 11.1   Management of Major Diseases of Pigeon Pea by Bio-Control Agents*

Crops	Diseases	Pathogens	Effective Biocontrol Agents
Pigeon pea	Wilt	*Fusarium udum*	*Trichoderma harzianum, T. hamatum, T. asperellum, T. koningii, T. longibrachiatum, T. afroharzianum, B. subtilus*
	Phytophthora stem blight	*Phytophthora drechsleri* f. sp. *cajani*	*Trichoderma harzianum, T. hamatum, T. asperellum, T. koningii, T. longibrachiatum, T. afroharzianum, Glomusmosseae, Pseudomonas fluorescens, Bacillus subtilis*
	Alternaria blight	*Alternaria alternata*	*Trichoderma harzianum, T. koningii, T. asperellum, T. konningi, T. longibrachiatum, T. afroharzianum*
	Dry Root Rot	*Rhizoctonia bataticola*	*Trichoderma harzianum, T. hamatum, T. asperellum, T. konningi, T. longibrachiatum, T. afroharzianum,*
	Seed borne diseases	*Pseudomonas compestris* pv. *vinae*	*T. viride, T. harzianum*

Source: Chaudhary et al. (2004); Mishra et al. (2018).

11.3   CONCLUSIONS AND FUTURE STRATEGIES

Earlier research mainly focused on resistant sources and chemical control of few diseases. Now the major emphasis is placed on identifying, evaluating, and integrating location specific BM packages for diseases. These

BM packages of different pulse crops have been successfully refined and validated in partnership with stakeholders and end-users. Despite the development of various BM modules to tackle economically important diseases of pulse crops, a considerable knowledge gap persists between scientists and farmers, particularly in the developing countries. Therefore, urgent need is there for consistent refinement, validation, transfer, and adoption of these developed bio-intensive modules. In addition, refined BM strategies need to be in place for tackling emerging pathogens such as Phytophthora blight in pigeon pea. Experience over the last few decades has shown that plant viruses cannot be controlled by preventive measures only. The complicated ecology of many of the viruses which affect legume crops, and for this matter, crops in general, calls for a strategy that is long term, sustainable, economically acceptable to the resource-poor farmers, and friendly to the environment. The promising approach is to harness all available measures and deploy them in an integrated fashion. Future BM strategies will also account for the effects of climate variability-on pathogen dynamics that are contributing to the emergence and resurgence of new strains/pathogens. This in turn necessitates appropriate refinement for disease control. HPR, fungicides, natural plant products, biofungicides, botanicals, and agronomic practices will remain the potentially viable options for BM of different pulse crops. In addition, the future modules will incorporate the latest advancements witnessed towards the identification of native biocontrol agents and their products delivered through biotechnological interventions. In parallel, a renewed attention is needed on using information technology for disease modeling and utilization of remote sensing to up-scale and disseminates IDM technologies.

KEYWORDS

- **bio-intensive management**
- ***Cajanus cajan***
- **phytophthora stem blight**
- **pigeon pea**
- **sterility mosaic disease**
- **yellow vein mosaic disease**

REFERENCES

Agrios, G. N., (2005). *Plant Pathology* (5ᵗʰ edn.). Academic Press, London.

Chaudhary, R. G., Sharma, N., & Prajapati, R. K., (2004). Biological control of soil borne diseases: An update in pulse crops. In: Shahid, & Narai, (eds.), *Eco-Friendly Management of Plant Diseases* (pp. 178–200).

Haware, M. P., & McDonald, D., (1996). Integrated management of Botrytis gray mold of chickpea. In: Haware, M. P., Faris, D. G., & Gowda, C. L. L., (eds.), *Botrytis Gray Mold of Chickpea* (pp. 3–6). (ICRISAT: Patancheru, AP, India).

Haware, M. P., (1998). Diseases of chickpea. In: Allen, D. J., & Lenne, J. M., (eds.), *The Pathology of Food and Pasture Legumes* (pp. 473–516). ICARDA, CAB International: Wallingford, UK.

Kannaiyan, J., Ribeiro, O. K., Erwin, D. C., & Nene, Y. L., (1980). Phytophthora blight of pigeon pea in India. *Mycologia, 72*, 169–181.

Mishra, R. K., Bohra, A., Naimuddin, Krishna, K., Kiran, G., Sujayanand, G. K., Saabale, P. R., et al., (2018). Utilization of biopesticides as sustainable solutions for management of pests in legume crops: Achievements and prospects. *Egyptian Journal of Biological Pest Control.* doi: 10.1186/s41938-017-0004-1.

Mishra, R. K., Mishra, M., Naimuddin, & Kumar, K., (2018). *Trichoderma asperellum:* A potential biocontrol agents against wilt of pigeon pea caused by *Fusarium udum* Butler. *Journal of Food Legumes, 31*(1), 50–53.

Mishra, R. K., Naimuddin, Akram, & Saabale, P. R. (2016). First report of stem rot in pigeon pea from India. *Pulses News Letter*, 7.

Mishra, R. K., Naimuddin, Krishna, K., Sujaynand, G. K., Jagdeeswaran, R., Saabale, P. R., Akram, M., & Singh, N. P., (2016). *Production and Popularization of Biological Control Agents to Enhance Pulse Production: An Eco-Friendly Approach.* ICAR-Indian Institute of Pulses Research, Kanpur.

Mishra, R. K., Naimuddin, Monika, M., & Sandeep, S., (2016). *Trichoderma:* Potential biocontrol agents for pulses. *Dalhan Alok., 14*, 80–86.

Mishra, R. K., Naimuddin, Saabale, P. R., Satheesh, N. S. J., Abhishek, B., Singh, F., & Singh, I. P., (2016). Evaluation of promising lines of pigeon pea for resistance to wilt caused by *Fusarium udum* butler. *J. Food Legumes, 29*(1), 64–66.

Mishra, R. K., Naumuddin, Akram, M., Sachan, D. K., Mishra, M., & Kumar, K., (2016). Characterization of indigenous *Trichoderma* spp. through ITS based sequences. *Pulses News Letter, 27*(2), 5.

Mishra, R. K., Sonika, P., Monika, M., & Naimuddin, (2019). *Trichoderma afroharzianum:* New biocontrol agent identified from pulses rhizosphere of U.P. *Pulses News Letter, 30*(3), 5.

Nene, Y. L., Reddy, M. V., Haware, M. P., Ghanekar, A. M., & Amin, K. S., (1991). Field diagnosis of chickpea diseases and their control. In: *Information Bulletin No. 28. Ed. by Int. Crops Res. Inst. for the Semi-Arid Tropics.* Patancheru, India.

Nene, Y. L., Sheila, V. K., & Sharma, S. B., (1984). *A World List of Chickpeas (Cicer arietinum) and Pigeon Pea (Cajanus cajan (L.) Millsp.) Pathogens.* ICRISAT Pulse Pathology Progress Report 32, Patancheru, Andhra Pradesh, India.

Nene, Y. L., Sheila, Y. K., & Sharma, S. B., (1996). *A World List of Chickpeas and Pigeon Pea Pathogens* (5ᵗʰedn.). Patancheru 502 324, Andhra Pradesh, India: International Crops Research Institute for the Semi-Arid Tropics.

Purdy, L. H., (1979). *Sclerotinia sclerotiorum*: History, diseases and symptomatology, host range, geographic distribution, and impact. *Phytopathology, 69*(8), 875–880.

Reddy, M. V., Raju, T. N., Sharma, S. B., Nene, Y. L., Mcdonald, D., Pande, S., & Sharma, M., (2012). *Handbook of Pigeon Pea Diseases (Revised)* (p. 64). Inf. Bull. No. 42. Patancheru, A. P. India: ICRISAT.

Reddy, M. V., Reddy, D. V. R., & Sacks, W. R., (1994). Epidemiology and management of sterility mosaic disease of pigeon pea, In: *Proceedings of the International Symposium on Rose Rosette and Other Errophyrd Mite-Transmitted Plant Disease Agents' Uncertain Etiology* (pp. 29–32). Lowa State University, Ames, Iowa, USA.

CHAPTER 12

Innovative Approaches in Diagnosis and Management of Diseases in Ginger (*Zingiber officinale* Roscoe) and Turmeric (*Curcuma longa* L.)

AJIT KUMAR SINGH,[1] DEVENDRA KUMAR CHOUDHARY,[1] SHRIKANT SAWARGAONKAR,[1] and RAKESH KUMAR SINGH[2]

[1]College of Agriculture and Research Station, IGKV, Raigarh, Chhattisgarh, India, E-mail: ajitspices8@gmail.com (A. K. Singh)

[2]RVSKVV, College of Agriculture, Indore, Madhya Pradesh, India

ABSTRACT

Ginger and turmeric are major spices of India as well as the world. These crops are perennial in nature but grown as annual for the uses of spices and medicinal purposes. Due to the clonal nature of the crops, genetic variability is very narrow, and to create the variability in traits is limited only by mutation and somaclonal variations. The narrow variability in genetics makes fewer opportunities in crop improvement for yield augmentation, disease resistance, agronomic suitability, and biochemical enhancement. The crop yielded poor evolving many factors, i.e., weather, agronomic practices, soil, and disease, etc., but diseases are major limiting factors that affect the quality and quantity of the products under field and storage conditions. The major diseases are evolved in ginger crops are Rhizome rot caused by *Pythium* spp., Yellow disease caused by *Fusarium spp.*, Bacterial wilt caused by *Ralstonia solanacearum,* Leaf spot caused *by Phyllosticta zingiberi*. Major diseases of turmeric are Rhizome rot caused by *Pythium spp.*, Colletotrichum leaf spot caused by *Colletotrichum capsici,* and Taphrina leaf blotch caused by *Taphrina maculens*. These diseases affect crops at different stages of crop growth and cause huge economic loss to grower. Due to the importance of the crop and its nature, a strong innovative approach in diagnosis and management is required for the welfare of the farming and trade communities.

12.1   BACTERIAL WILT (*R. SOLANACEARUM*) OF GINGER

Wilt disease of ginger caused by *Ralstonia solanacearum* race 4 is a major cause of production constraint worldwide and exclusively in the subtropical and tropical regions (Kumar and Sarma, 2004). The disease inflicts somber economic losses to small and marginal farmers who depend on this crop for their subsistence, especially when farmers cut rhizomes before planting. The causal organism, *R. solanacearum* race 4 (*Pseudomonas solanacearum* Smith, 1896) is a soil and plant inhabiting bacterium.

12.1.1   SYMPTOMS OF BACTERIAL WILT

Bacteria primarily infect rhizome and spread in vascular tissue. First visible symptoms can be observed above ground and characterized as small water-soaked lesions at the neck region of the pseudostem. As pathogen hinder translocation of water and nutrient, which leads to mild drooping and downward curling of leaf margins because of loss of turgidity. Within three to four days of infection, yellowing and curling starts from the lowermost leaves and gradually progresses to the upper leaves and complete leaves dry up. The complex stage of the disease plants manifests rigorous yellowing and wilting symptoms. Dissection of the infected pseudo stem reveals darks streaks symptoms in vascular tissues. When the affected pseudostem and rhizome are pressed lightly it extrudes cloudy ooze from the xylem tissues which are indicative symptoms of the disease. The infected rhizomes start to rot and decompose due to the attack of other soil microorganisms. The underground-decomposed rhizomes emit a stinking smell and the infected plants wither and expire within a few weeks.

12.1.2   DETECTION AND DIAGNOSIS

12.1.2.1   ENRICHMENT TECHNIQUES

Modified SMSA (Selective medium South Africa) has been developed for the specific growth of *Ralstonia solanacearum* (Elphinstone et al., 1996).

The limit of detection reported was 10^2–10^3 CFU per mL, and colonies were countable in 72 hours. As like that on CPG-TTC (casamino acid-peptone-glucose-triphenyl tetrazolium chloride) medium (Kelman, 1954) can also be used for *Ralstonia* detection as it produces pink color with mucoid and diffused colonies. For isolation of pathogen *R. solanacearum* from fresh, symptomatic plants or soil, these media can also be used.

12.1.2.2   OOZE TEST

For quick field diagnosis and identification of *Ralstonia* wilt from other fungal wilt diseases, bacterial streaming from vascular tissue of affected plant material can be used. A pseudostem or rhizome showing discoloration is cut from a plant with a razor-sharp knife (Martin and French, 1985). The cut part is placed in a water-filled clear glass container so that the end of the section just touches the water surface. Milky pallid strands containing bacteria (Ooze) and extracellular polysaccharide (EPS) will stream from the cut ends of the vascular tissue (xylem) (Martin and French, 1985).

12.1.2.3   ELISA

Rajeshwari et al. (1998) developed ELISA for the identification of *Ralstonia* in tomato. Priou et al. (1999) developed a highly sensitive detection method of *Ralstonia* wilt in symptomless infected potato by post enhancement ELISA on membrane of nitrocellulose. The same technique was used and validated in ginger by Kumar et al. (2002). Indirect ELISA can be implemented to detect Ralstonia in soil or rhizome prior to symptom development. The detection limit of ELISA was reported 10^5 CFU/gm (Pradhanang et al., 2000). The sensitivity of detection reported to increase after incubation of the suspension in SMSA broth earlier to testing, where after populations of 10^4 CFU/gm were detected, and this technique, termed as bio-ELISA (Pradhanang et al., 2000).

12.1.2.4   PCR

Horita et al. (2004) developed two primers pair for precise detection of *Ralstonia* in ginger. One primer pair namely AKIF-AKIR produced a single band of 165 bp and another pair, namely 21F-21R produced one band of 125 bp. They reported that PCR detection limit for the pathogen was 2×10^2 cfu. Kumar and Anandaraj (2006) developed PCR primer to detect bacterial wilt pathogens in rhizomes of ginger and infected soil collected from the wilt affected area.

12.1.2.5   *REAL-TIME PCR*

Using primer RSAF1 and RSAR1 a real-time TaqMan PCR technique was developed to detect Ralstonia in ginger rhizome, which gives 329 bp DNA fragments (Thammakijjawat et al., 2006). By using this technique, *Ralstonia* can be detected in asymptomatic rhizome and prevent the spread of pathogen to the field.

12.1.3   MANAGEMENT OF GINGER WILT

Complete eradication or control of pathogen in specific and *Ralstonia solanacearum* in particular ruins the vision of most researchers. Due to lack of knowledge, economic stamina, and technology available to farmers, it's very difficult to control the disease. Different control methods have been evaluated to manage the disease with limited success. The strategies for wilt disease management are as follows:

12.1.3.1   *CULTURAL METHOD*

Pathogen can survive in water, soil, and crop debris for a longer time, so the residue of the previous crop should be destroyed or buried deep in the soil after harvest, down-slope, and far from irrigation canals, alternatively, they can be burned. Sanitation also limits build up of pathogen survival and dissemination.

Crop rotation of three to 5 years with gramineous crops can be implemented to eradicate soil inoculum; time for rotation may vary according to environmental conditions and soil character. *Ralstonia solanacearum* survives in the rhizosphere of weeds, so weed should be uprooted before planting. If the occurrence of wilt is low in a standing crop, diseased plants must be removed the instant they are observed, to avoid contagion of healthy bordering plants. To prevent the dissemination of pathogen through soil, restrict movement of human, animal, and machinery vehicles to the infested side. All tools should be sterilized using available bactericides or calcium hypochlorite or by flame. Also, avoid movement of irrigation water from infested field to neighboring field to prevent further spread.

12.1.3.2   PLANTING MATERIAL

Selection of healthy rhizome material from disease-free areas, with no history of bacterial wilt in the past, is necessary to avoid introduction of disease to new area and reduce economic loss.

12.1.3.3   BIOLOGICAL CONTROL

Biological control measures are self-sustaining once the biocontrol agent established in the environment, and it's becoming general in integrated disease management (IDM). *Pseudomonas fluorescence, B. amyloliquifaciens,* and *B. subtilis* have been reported as efficient biocontrol agents for *R. solanacearum,* and they can be used for rhizome treatment as well as sprayed in standing crop.

12.1.3.4   QUARANTINE

Once *Ralstonia* is introduced to an area; quarantine regulations must be applied to avoid further spreading of pathogen to non-infested areas. Avoid transport of seed rhizome from an infected area to wilt disease-free area to avoid dispersal the pathogen.

12.1.3.5   RHIZOME TREATMENT

Heat treatment of rhizome aided by solar radiation is known as "rhizome solarization." This method is very useful for disinfection of bulk rhizome. For preplant rhizome treatment 45–50°C temperature is beneficial and can be achieved after 4–5 hours of sunlight, either directly or after covering them with polythene sheets of 100 to 200 μm.

12.1.3.6   SOIL AMENDMENT

Composts and mulched promote microbial action in soil. The activity of *Ralstonia* can be reduced by other saprophytic soil microorganism competition and/or antibiosis.

12.1.3.7   BIO FUMIGATION

Green manures crop such as mustard, mint, palmarosa, eucalyptus, and lemongrass can be applied in the field, and they decompose and release the essential oils which are toxic to the wilt pathogen *Ralstonia*.

12.1.3.8   CHEMICAL CONTROL

The chemical control method should be the last option in IDM practices and only be exploited when other measures are not effective as expected level. Even though very few synthetic chemicals are effective to control *Ralstonia* wilt. Use of calcium or sodium chloro-oxide, ClO_2 (chlorine dioxide), antibiotics, lactic acids, and acetic acids had revealed some effectiveness in controlling bacterial pathogens.

12.2   SOFT ROT (*PYTHIUM SPP.*)

Soft rot is also known as rhizome rot or Pythium rot. The disease is common in almost all the ginger-growing regions across the world, for instance, Australia, Taiwan, Korea, China, Nigeria, Fiji, Hawaii, India, Japan, and Sri Lanka (Lin et al., 1971). Numerous species of *Pythium* such as *Pythium aphanidermatum*, *P. vexans*, *P. myriotylum*, *P. ultimum*, *P. deliense*, and *P. splendens* have been reported as causal organism of ginger rhizome rot worldwide. Rhizome rot is both seed and soil-borne, and most destructive amongst the diseases that distress ginger crop in the globe (Mundkar, 1949). The disease occurs mostly during the months of July and September. Elevated soil moisture and low ambient temperature of 25–28°C in rainy season are greatly favorable to the dissemination of the disease.

12.2.1   SYMPTOMS OF PYTHIUM ROT

The disease is prone to waterlogged situation. The plant is vulnerable to illness throughout all the stages of growth. All the underground parts stem, and emerging sprouts are prone to soft rot disease. Mostly pathogen enters through the buds, roots, budding underground stem, the rhizome, and collar regions. When the infected rhizome is used as planting material, the primary noticeable symptom appears at the neck region of the pseudostem as water-soaked and brown discoloration which progresses downwards and upwards that causes rotting of pseudostem as well as rhizome. The rhizome becomes soft and collapses easily on pressing. In the advanced stage, internal tissue rotted completely and only outer skin is left. Rotted rhizome also gives a putrefying smell. Rotting attracts saprophytic bacteria, fungi, insects, and, scavengers. Foliar symptoms emerge as slight fading of green color from tips of lower leaves which slowly extends to leaf blades. As disease progress, the yellowing/chlorosis extend to all leaves, followed by withering, drooping, and drying of the leaves. The drying of the top leaf extending to petiole then sheath portion of the stem then to entire plant. In case the infection has occurred early may cause a cent percent yield loss under favorable condition? Rotting may continue to storage conditions.

12.2.2   DETECTION AND DIAGNOSIS

12.2.2.1   SELECTIVE MEDIUM

For specific seclusion of *Pythium,* the P_5ARP medium can be used (Jeffers and Martin, 1986). Then culture medium incubated at approximately 20°C with ordinary daylight for 6 days then should be examined under the compound microscope at 100x magnification to check for characteristic features (Jeffers and Martin, 1986).

12.2.2.2   COLONY CHARACTERS

Pythium produces white, cottony mycelium on potato dextrose agar (PDA). Mycelium lack cross-walls and measure up to 10 μm wide (Plaats-Niterink, 1981). Pythium produces terminal, inflated, lobate sporangia with hyphal swellings up to 20 μm wide. Sporangia give rise to biflagellate zoospores, with tinsel and whiplash flagella. It also produces Oospores which is aplerotic, 18 to 22 μm (av. 20.2 μm) in diameter, with a 1 to 2 μm thick wall (Plaats-Niterink, 1981).

12.2.2.3   PCR

The use of molecular technique is a more reliable tool to detect pathogens present in planting material, weed host, or in soil. It can be utilized to detect simultaneous detection of two or more pathogen species or more disease. Wang et al. (2003) developed primer Pmy5 for PCR using rDNA ITS1 region together with universal primer ITS2, and able to detect *Pythium myriotylum*. Gel electrophoresis gave a 150 bp DNA product by booster PCR with primer Pmy5. The primer was able to detect pathogen from buds, and scales of infected rhizome. The specificity of primer proved with no amplification from *Xanthomonas zingiberi* and *Foz* DNA tested from rhizome of the ginger. Same primer can be exercised to detect pathogen present in soil prior to planting to assess inoculum level. Wang et al. (2003) also developed specific primer for 18 species of *Pythium*, primer sequence given in Table 12.1.

TABLE 12.1   Primer Sequence for 18 Species of *Pythium*

Species	Primer*	Primer Sequence	Annealing (°C)
P. acanthicum	Pac1	5′ GTGCCTCGTCTTGTTGAAAG	68
P. aphanidermatum	Pa1	5′ TCCACGTGAACCGTTGAAATC	67
	Pa3	5′ ATTTTTCAAACCCATTTACC	57
P. arrhenomanes	Par1	5′ AAGTGTAGTTAATTCTGTACGCTGC	58
P. dimorphum	Pdi1	5′ TTTATTATATTACATCAACTCT	56
P. graminicola	Pgr1	5′ ACGAAGGTGGGCTGCATGTA	68
P. heterothallicum	Phe1	5′ TTTGTATGAGATCAGCTGAT	58
P. hydnosporum	Phy1	5′ GTCTGCGTCTATTTTGGATGC	70
P. inflatum	Pinf1	5′ AAGGTGGGCGCATGTATGTG	68
P. intermedium	Pint1	5′ ATACTGCTGGCGGGTGCGAG	70
P. irregulare	Pir1	5′ AGCGGCGGGTGCTGTTGCAG	72
P. oligandrum	Po1	5′ TGCGTCTATTTTGGATGCGG	70
P. rostratum	Pro1	5′ TAGTGTAGCTTTTGTTGCGC	68
P. spinosum	Pspi1	5′ TGTGTGTTGTGATCGTGCCT	70
P. splendens	Pspl1	5′ GAAGGTCGGAGTAAAATCTGGC	62
P. sulcatum	Psu1	5′ CTAAACGAAGGTGGGCCGCT	64

TABLE 12.1   *(Continued)*

Species	Primer*	Primer Sequence	Annealing (°C)
	Psu3	5′ CACGTGAACCGTAATAATCA	62
P. torulosum	Pto1	5′ AGGTAGAGCTGCATGTAAAAGT	59
P. u. var. *sporangiiferum*	Pus1	5′ GTGGGTGTTGCGGGTGCTATT	60
P. vanterpoolii	Pva1	5′ GTGCTACAGTCTGCCGATGC	58

*Pdi1 paired with ITS4 in PCR, rest paired with ITS2.

Ishiguro et al. (2013) developed multiplex PCR primer for simultaneous detection of three species of *Pythium* species *P. helicoides*, *P. aphanidermatum*, and *P. myriotylum*.

12.2.2.4   ELISA

Kageyama et al. (2002) used monoclonal antibodies (MAbs) to differentiate *P. myriotylum*, *P. zingiberis* and *P. sulcatum* and from infected tissues or soil. ELISA can also be used with a combination of PCR using ITS1 primer to increase sensitivity, reliability, and to avoid false-positive reactions. Indirect ELISA was performed by Ray et al. (2016) to detect *P. aphanidermatum* using polyclonal antibody and got detection level 10 µg/ml.

12.2.2.5   MOLECULAR MARKER

Sagar et al. (2009) developed three different RAPD primers OPA-03 (50-AGTCAGCCAC), OPA-15(TTCCGACCC), and OPB-08 (GTCCA-CACGG) for the detection of *P. aphanidermatum* in soil and diseased plants. The same primers can also be used to check variability in *P. aphanidermatum*. Lee and Moorman (2008) developed SSR markers for specific detection of 22 different species of *Pythium*. Schroeder et al. (2013) employed RFLP to distinguish different species of *Pythium* by comparing DNA patterns achieved by processing of DNA with restriction enzymes. Wang and White (1997) reported that the PCR-RFLP assays were reliable in *Pythium* recognition by using four enzymes Taq I, Hinf I, Cfo I and Mbo I.

12.2.2.6   *LAMP (LOOP-MEDIATED ISOTHERMAL AMPLIFICATION)*

Fukuta et al. (2013, 2014) successfully developed a set of six primers, including two loop primers, to specifically detect *P. aphanidermatum* and *P. myriotylum* using LAMP (loop-mediated isothermal amplification). The advantage of using LAMP over PCR is no need of thermal cycler, efficiently, and fast (Notomi et al., 2000; Nagamine et al., 2002), much cheaper and easier to carry out (Fu et al., 2011). the benefit of LAMP over ordinary PCR is that pure DNA not required and it is more reliable and sensitive (Dai et al., 2012); Fukuta et al. (2013, 2014) amplified target DNA straight from rough tissue take out, and this technique used for the development of a kit for field survey (Fukuta et al., 2013, 2014).

12.2.3   *MANAGEMENT OF SOFT ROT*

12.2.3.1   *SELECTION OF SEED RHIZOME*

Infected rhizomes act as a key source of illness and spread *Pythium* rot in the new area. The best measure to manage seed-borne diseases like rhizome rot is to use healthy rhizome collected from disease-free field/area.

12.2.3.2   *SEED RHIZOME TREATMENT*

Treatment of rhizome in metalaxyl + Mencozeb (1.25 gm/liter) for 30 minutes earlier to storage and planting effectively checks the disease.

12.2.3.3   *HOT WATER TREATMENT*

To eradicate pathogen inoculum present in seed rhizome, rhizome be supposed to treated with 51°C hot water for 10 minutes, as this temperature kills pathogen but not affects the viability of seed rhizome.

12.2.3.4   *CULTURAL METHODS*

One of the favorable conditions for soft rot dissemination in ginger is poorly drained fields in wet weather. So keeping this in mind, well-drained sandy loam soil ensure better yield and should be adopted for cultivation. Hygienic farming, collection, and burning of dead leaves, stems, and other plant parts help in reducing the inoculum load of pathogens. Crop rotation leguminous crop such as ragi, maize, paddy, etc., keeps the illness under check.

12.2.3.5   SOIL SOLARIZATION

This method obtains the benefit of solar radiation by harnessing solar heat to raise soil temperature. After light irrigation, moist soil is covered with transparent polythene film during hot weather conditions in summer. It also improves soil property as well as helps to kill pathogen inoculum and diminish disease incidence. This practice must be adopted, wherever possible.

12.2.3.6   BIOLOGICAL CONTROL

Biocontrol agent, namely *T. hamatum, T. harzianum, T. virens, Bacillus*, and *Pseudomonas fluorescens* have been identified to suppress soil-borne pathogens of ginger. The biological agent rapidly grows on rhizome and suppresses the pathogen. The formulation of these biocontrol agents can be either used as seed treatment @ 5 gm/kg of rhizome or multiplied on substrate such as FYM, coir pith, decomposed coffee pulp, margosa cake or kitchen waste and the applied in field.

12.2.3.7   RESISTANCE

A wild ginger relative, *Zingiber zerumbet* (L. Smith), has been recognized as a probable *Pythium* rot resistance donor with disease resistance gene ZzR1 (Nair and Thomas, 2013).

12.2.3.8   CHEMICAL CONTROL

Spraying of the soil surface with Mancozeb (0.3%) or Bordeaux mixture (6:6:50 concentration) or Dithane Z-78 (0.75 kg in 500 lit of water) @ 7 liters per square feet before planting and then after germination at 2–3 weeks interval gives partial control of disease.

12.2.3.9   BIO FUMIGATION

Use bio fumigants as mustard crop 30 before planting help in reduce rhizome rot incidence.

12.3   DRY ROT OR *FUSARIUM* YELLOWS

Dry rot, also known as *Fusarium* yellows is caused by *F. oxysporum* f. sp. *zingiberi* (*Foz*). The disease is common and very serious fungal diseases and only specific to ginger.

12.3.1   SYMPTOMS OF THE DISEASE

Fusarium rhizome rot is mainly a seed or soil-borne disease. In three ways ginger plants infected by pathogen shows yellows symptoms:

- First, if the infected rhizome is used as seeds, decay of rhizome occurs, that prevents emergence of the shoots, or the shoots are dreadfully weeks and shortly die.
- Second, on lower leaves, symptoms start as leaf margins yellowing, which slowly extend over the whole leaf. Then older leaves dry up, followed by the younger leaves. Other identifiable symptoms in field are early drooping, wilting, and drying of plants in patches in the cultivated area or in the whole bed. As wilt development is slow plants usually do not lodge on the ground unlike soft rot or bacterial wilt. In rhizomes symptom develops as a brown discoloration and in the later stages of illness, only the fibrous tissues remain within the rhizomes.
- Third, a white cottony fungal tissue may build up on the outside of stored rhizomes. In these situations, the fungus continues to cause decay.

12.3.2   DETECTION AND DIAGNOSIS

12.3.2.1   SELECTIVE MEDIUM

For specific isolation of *Fusarium*, selective medium Fo-G2 and Fo-G1 were developed by Nishimura (2007), Komada's medium developed by Komada

(1975); malachite green agar (MGA) was developed in 1997 by Castella and coworkers. Any selective medium can be used for isolation/identification of *Fusarium* from ginger rhizome and soil.

12.3.2.2 COLONY CHARACTERISTICS

Foz develops mycelial color varied from white or slightly yellowish to pinkish tinge. It develops micro and macroconidia, microconidia are most abundant and are oval, elliptical or kidney-shaped, and produced on aerial mycelia, microconidia size varied from 5.20×4.00 μm to 12.30×5.70 μm. Whereas macroconidia are three to five cells and gradually pointed or curved edges, size varied from 16.20×4.70 μm to 32.0×5.7 μm. It also forms chlamydospore from macroconidia (Dohroo and Sharma, 1992; Siddiqui and Kaushal, 2000).

12.3.2.3 ELISA

Ray et al. (2016) developed polyclonal antibody for *Foz* using fungal antigen and developed an indirect ELISA technique to detect the pathogen. They got optimum results using antisera at a dilution 1:2000.

12.3.2.4 PCR

PCR primer using ITS1 and ITS4 sequence of *Foz* available in NCBI was developed by Li et al. (2014) for specific detection of pathogen. This primer set can selectively amplify *Foz* DNA from mixed infection.

12.3.3 MANAGEMENT OF THE DISEASE

12.3.3.1 HEALTHY RHIZOME SELECTION

As the disease is majorly seed-borne and spreads through contaminated rhizomes, collection of vigorous and disease-free rhizomes has been found to be an effectual protective method for illness control. The seed rhizomes must be chosen from disease-free areas.

12.3.3.2   HOT WATER TREATMENT OF RHIZOME

Rhizomes are exposed to 51°C temperatures for 10 minutes in hot water. Long exposure to hot water may affect the germination ability of rhizome. This is mainly suggested in places where the illness is endemic.

12.3.3.3   RHIZOME TREATMENT WITH CHEMICAL

Seed rhizomes can be treated with chemicals such as carbendazim (0.25%) or Mancozeb (0.3%) or in combination before planting, if biological control agent is not available at the right time.

12.3.3.4   SANITATION OF FIELD

The annihilation of unhealthy crop debris by burning is a vital practice to minimize the initial inoculum of the pathogen. It has been stated that disease appears less when ginger grows under the shade of trees, such as coconut trees.

12.3.3.5   SOIL SOLARIZATION

Solarization of bed for 20–30 days is beneficial in checking multiplication and survival of pathogen inoculum in soil.

12.3.3.6   BIOLOGICAL CONTROL

Biocontrol agents such as *T. harzianum*, *T. hamatum*, and *T. virens* as were found to control the disease when it is used during seed treatment and soil application. *T. harzianum* and *Gigaspora margarita* give the most excellent control of ginger yellows caused by *Foz*, ensuing higher yield of healthy ginger.

12.3.3.7   CULTURAL CONTROL

While cutting the seed rhizome, the pieces viewing discoloration or shriveling symptoms must be laid off, and the cutting knife must be frequently dipped in methyl alcohol or in any other disinfectant. Crop rotation with non-host helps in avoiding inoculum build-up. Crop rotation should be adopted in infested field for at least 2 to 5 years; will help reduce pathogen inoculum. Ginger cultivation should avoid in field with heavy nematode infestation. Waterlogging conditions augment humidity in crop canopy, so it must be avoided.

12.3.3.8   CHEMICAL CONTROL

Spray of Hexaconazole (@ 0.1%) or Propiconazole (@ 0.1%) or Carbendazim (@ 0.1%) when first disease appearance and then two times at 20 days interval can manage disease effectively.

12.4   PHYLLOSTICTA LEAF SPOT (*PHYLLOSTICTA ZINGIBERI*)

Phyllosticta zingiberi T.S. Ramakr is the cause of Phyllosticta leaf spot in turmeric. The disease is reported in most of ginger-growing countries, including India. In India, leaf spot disease was first time reported in the Kerala and Malabar and Godavari districts in Andhra Pradesh, India, by Ramakrishnan (1942). In the current past due to the increasing number of severity and incidence in many states of India, it is becoming increasingly important.

12.4.1   SYMPTOMS

The disease may be noticed on the leaves from July to October. Disease starts as a water-soaked lesion on younger leaves and later turns as a white, spindle to oval or elongated spots surrounded by dark brown margins and yellowish halo (Ramakrishnan, 1942). Size of lesions increase and coalesce to neighboring lesions to outline necrotic areas, which sooner or later decrease the effectual photosynthetic area on the leaf surface. In dry weather condition, such infected areas frequently dry up at the center and fallen off cause holes in leaves. In severe infection, the whole leaf dries up leads to complete failure of photosynthetic activity.

12.4.2   DETECTION AND DIAGNOSIS

12.4.2.1   CULTURING OF PATHOGEN

The pathogen can be grown in culture medium such as French bean agar, Quaker oats agar, PDA or Richards' agar. *Phyllosticta zingiberi* produces grayish-white mycelium, and pycnidia on culture media (Rai et al., 2017). Generally, pycnidia formation starts on the fourth or fifth day. The color of pycnidia is light brown in the beginning but later turn to deep brown (Ramakrishnan, 1942). The diameter of pycnidia ranging from 100–270 μm, and spore measuring from 3.7 to 7.4 × 1.2 to 2.5 μm (Ramakrishnan, 1942). The bigger pycnidia formed in cultures than those in field condition. The asexual spores are hyaline and oblong with rounded ends and often biguttulate (two oil droplets) (Ramakrishnan, 1942).

12.4.3   MANAGEMENT OF THE DISEASE

12.4.3.1   SELECTION OF SEED RHIZOME

The seed rhizomes must be chosen from pathogen-free areas, as the disease can disseminate through rhizomes and enter to area where pathogen not reported previously.

12.4.3.2   RHIZOME TREATMENT

To eradicate the pathogen inoculum present in seed rhizomes, it should be treated with Mancozeb + carbendazim combination or only carbendazim (@ 0.25%) before planting. Some other chemicals such as Chlorothalonil, Mancozeb, Prochloraz, Tebuconazole, and Captan, have been reported to give fair control and improved ginger yield (De Nazareno, 1995).

12.4.3.3   SANITATION AND SHADE

The annihilation of infected crop residue by burning is a vital practice to reduce the initial inoculum of the disease. It has been stated that in shade disease appear in less severity (Patiram Upadhaya et al., 1995). The disease severity and sunburn were statistically lower in heavy shade in comparison to open sun-grown ginger. In addition, the augmented quantity of tillers per plant and yield was observed. Therefore, cultivation of ginger in fractional shade may be optional to keep away from the fungicidal spray for managing *Phyllosticta* leaf spot and consequently avoiding fungicidal pollution (Singh et al., 2004).

12.4.3.4   SOIL SOLARIZATION

Solarization of bed for 20–30 days is beneficial in checking multiplication and survival of pathogen inoculum in soil.

12.4.3.5   BIOLOGICAL CONTROL

Biocontrol agent, namely *Trichoderma viride* (Bonord.), *T. harzianum* (Rifai.) and *T. koningii* (Oudem.) have been stated to suppress *Phyllosticta zingiberi* of ginger.

12.4.3.6   CULTURAL CONTROL

When ginger cultivation is taken in the same field for a longer period, it builds up pathogen inoculum with time. So to avoid inoculum build-up crop rotation with graminaceous crop (non-host crop) should be adopted. To maintain appropriate humidity in plant canopy, waterlogging condition should be avoided.

12.4.3.7   HOST RESISTANCE

Many researchers screened available genotypes against *Phyllosticta* leaf spot, but most of ginger genotypes were establish to be susceptible to the disease (Setty et al., 1995; Dohroo et al., 1986; Rao et al., 1995). The cultivars Narasapatom, Nadia, Tura, and Thingpani were identified as moderately resistant (MR) with illness index less than 5% in Karnataka (Setty et al., 1995). In Himachal Pradesh, none of the tested genotypes were rated resistant to the disease, however, 8 lines showed moderate resistance (Dohroo et al., 1986). Rao et al. (1995) screened 100 accessions of ginger for their reaction to leaf spot under field conditions and found

11 accessions as tolerant and 42 as moderately tolerant. Senapati et al. (2012) found that PGS-16, PGS-17 and Anamica as MR out of 135 ginger cultivars tested.

12.4.3.8   CHEMICAL CONTROL

Spray of Hexaconazole (@ 0.1%) or Propiconazole (@ 0.1%) or Carbendazim (@ 0.1%) when first disease appearance and then two times at 20 days interval have been reported effective in checking the disease.

12.5   RHIZOME ROT

It is a chief soil-borne disease in all major turmeric growing areas and the most destructive of all the diseases affecting turmeric. The infection was initially reported by Park (1934) in Sri Lanka (then Ceylon). In India, the disease was reported by Ramakrishnan and Sowmini (1954) from Krishnagiri, Tiruchirapally, and Coimbatore in Tamil Nadu, and from Kasaragod in Kerala (Anon, 1974; Joshi and Sharma, 1980), and also, from Assam, India (Rathaiah, 1982). *Pythium myriotylum* (Rathaiah, 1982), *Pythium graminicolum* (Ramakrishnan and Sowmini, 1954), and *Pythium aphanidermatum* (Park, 1934) are reported cause of the disease.

12.5.1   SYMPTOMS OF RHIZOME ROT

The early symptom of the disease is the manifestation of small water-soaked discoloration at the neck region of the pseudostem that enlarges rapidly, resulting in yellowing of the leaves. Symptom on roots appear as browning and rotting of the infected roots, as the disease advanced to the rhizomes, it changes rhizome color from bright orange to different shades of brown followed by softening and complete putrefaction. The illness afterward spreads to all pieces of the rhizome, including mother rhizome, and ultimately the plant succumbed to decease. When the infected rhizomes are cut open, brown to dark brown tissue discoloration can be seen. The internal tissue of the rhizome disintegrates, and outer skin gets depressed. Typical yellowing starts from lower leaves, which steadily spread to the upper region of the plant. The borders of the leaves turned necrotic and start drying from the margin to inward, resulting in partial or complete drying of the leaves (Rao, 1995).

12.5.2 DETECTION AND DIAGNOSIS

12.5.2.1 ISOLATION OF CAUSAL ORGANISM

The infected samples, including pseudostem, roots, and rhizomes collected from field should be used for isolation. The infected parts ought to wash carefully with running water to get rid of the adhered soil. Then cut small infected section along with some healthy section and surface sterilized with 10% NaClO or with 75% ethanol for 1–2 min and then washed in sterile distilled water (SDW) thrice and should transfer in 90 mm Petri plate of PDA medium (Rangaswami, 1958). Colony should observe regularly after 3 days of culture. *Pythium* produces white, cottony mycelium on PDA (Burr and Stanghellini, 1973; Blaschek et al., 1992).

12.5.2.2 ELISA

Detection of pathogen in seeds or planting material prevent spread of disease to new area/filed and prevent from further management and help farmer to minimize input cost. For early and accurate diagnosis of pathogen in turmeric using ELISA has been developed by Ray et al. (2016). Polyclonal antibody (pAb) has been raised for the protein take out from *Pythium aphanidermatum* for the identification of pathogen in soil, rhizome, or infected plant using serological assays. This has been further confirmed by using the Western blot technique (Ray et al., 2016). These serological approaches allowed developing a sensitive, rapid, accurate, and specific kit for early detection of *Pythium* in turmeric.

12.5.2.3 PCR

Rajalakshmi et al. (2016) developed a PCR marker for unique identification of *P. aphanidermatum* using specific primers ITS1 and ITS2, which yielded a single fragment of 210 bp. Forward primer (Pa1) sequence TCCACGT-GAACCGTTGAAATC from ITS2 and reverse primer (Pa3) sequence ATTTTTCAAACCCATTTACC was reported by White et al. (1990). The same work supports earlier work done by Klemsdal et al. (2008). Prabhukarthikeyan et al. (2015) developed PCR primer combination Paph54F/ITS2 for species-specific detection of *P. aphanidermatum*, which yielded 200 bp amplicon.

12.5.3   MANAGEMENT OF PYTHIUM ROT

12.5.3.1   RESISTANCE

Host resistance is the most economical and safest method for disease management and one of the components in IDM practice. Resistant cultivar *C. aromatic* (Ca 69) or tolerant cultivar such as Shillong, Tallkashi, Suvarna (PCT 8), Suguna (PCT 13), Sudharshana (PCT 14), Pragati, Narendra Hadi-1, Suvarna, Pratibha, Suranjana, etc., should be used for cultivation.

12.5.3.2   SELECTION OF RHIZOME

Infected rhizome can carry disease to new areas, so only disease-free seed rhizome should be collected from a disease-free area.

12.5.3.3   CULTURAL METHOD

Dense planting should be avoided. The rhizome should be dibbled in the side of the ridge, 45 cm apart at 15 cm spacing at the depth of 4 cm. The drainage of the field must be sound to shun water logging situation. Intercropping with maize also reported to minimize disease incidence.

12.5.3.4   BIOLOGICAL CONTROL

Biological control of *Pythium* species is very difficult because of rapid germinating of sporangia in response to seed or root exudates followed by immediate infection and the ability to cause long-term root rots (Whips and Lumsden, 1991). But use of consortia of *Pseudomonas fluorescent* (Pf1) and *Trichoderma viridae* as rhizome treatment at the rate of 4 g/kg and soil application at the rate of 2.5 kg/ha each as basal and top dressing at 150 days after planting gave positive result. Use of rhizobacterial cultures including

Pseudomonas fluorescens, and *Bacillus sp.* along with VAM fungi is also effective to manage disease incidence as they colonize on rhizosphere and prevent root infection by pathogen.

12.5.3.5   NUTRIENT MANAGEMENT

Spray micronutrient mixture at the rate of 500 ml/ha. Apply well-decomposed FYM (25 t/ha)/vermicompost (VC) (2 t/ha) or composted coir pit (10 t/ha) and groundnut cake (200 kg/ha), neem cake (500 kg/ha) as basal application. Topdressing of urea (55 kg/ha) and murate of potash (30 kg/ha) at 30, 60, 90, 120, and 150 days after planting.

12.5.3.6   CHEMICAL CONTROL

Rhizome rot diseases are both soil and seed-borne, and incidence of maggots (*M. coerulifrons*) in partly diseased rhizomes increase severity, so dipping of rhizomes in a combination of a solution containing both fungicides and insecticide prevent ailment occurrence. This operation should be carried out just before storage or sowing. The rhizomes so treated should be dried under shade before storage. The treatment of rhizome with a combination of Mancozeb (0.25%), and Quinalphos (0.75%) for half an hour has been proved to be the best practice to contain the incidence of the disease. Drenching of the soil with Metalaxyl + mancozeb twice at 15–20 days interval has been proved to be quite effective in containing the disease.

12.6   LEAF BLOTCH DISEASE OF TURMERIC

Leaf blotch is caused by *Taphrina maculans*, which is one of the common foliar diseases of turmeric in India. The disease affects various stages of susceptible plant during its growth and causes rigorous economic failure to the farmer.

12.6.1   SYMPTOMS

The earliest visible symptom of disease appears usually on the down surface of the leaves as tiny, scattered, oily-looking translucent spots. The individual spots are small and are mostly rectangular in shape. Spots arranged in rows along the leaf vein. Later the spots turn as pale yellow, followed by intensifies in color to golden yellow, and sometimes to bay shade. In severe condition, spots coalesce to each other covers most area of leaf lamina, and forming reddish-brown blotches leading to blighting of leaves. The chances of infection are more in lower leaves as compare to upper leaves, which are relatively younger in age. In severe infection, hundreds of spots appear on a single leaf on both surfaces; they coalesce on both sides of the leaf. Blighting of leaves severely reduces the photosynthetic activity, as almost all green parts turn color from green to yellow.

12.6.2   DETECTION AND DIAGNOSIS OF DISEASE

12.6.2.1   ISOLATION OF PATHOGEN

Taphrina maculans is dimorphic fungi in which mycelium stage is biotrophic on vascular plants and yeast-like stage is saprophytic in nature and only culturable stage (Bacigálová et al., 2003). So, the leaves bearing the final stage of the spot formation should select for isolation purposes. The leaves should wash with the SDW. Then the washings from the leaves should streak out on the plates poured with the acidified PDA (pH 4.5) (Ahmed and Kulkarni, 1968). The plates should be incubated at 20°C. After 5 days, pinkish colonies will develop. These pinkish colonies should immediately pick up and subculture on the slants. On microscopic examinations, typical yeast-like budded conidia can be observed (Ahmed and Kulkarni, 1968).

12.6.3 MANAGEMENT OF LEAF BLOTCH

12.6.3.1   RESISTANCE

Cultivation of resistant varieties such as PTS-62, ACC-360, ACC 361, BSR-1, Kasturi, Clone DKHT-6, Roma, Rashmi, Suguna, Suroma, Sudharshana, TCP 56, Pratibha, TCP 11 and tolerant varieties such as Suvarna (PCT 8), Suguna (PCT 13), Sudharshana (PCT 14), Suvarna should be cultivated.

12.6.3.2   CULTURAL METHOD

Select seed material from disease-free areas. When infection is seen in earlier stages of plant growth, removal of the infected plant and burning it away from the field is the most important practice in containing the disease. Removal and burning should not be restricted only to the infected leaves, but to the entire plant to assure complete disease spread control technique.

12.6.3.3 BIOLOGICAL CONTROL

Two sprays of *Pseudomonas fluorescens* along with rhizome treatment are very effective.

12.6.3.4 BOTANICAL

Azadirachta indica seed extract (5%) and Eucalyptus leaf extract (5%) are very effective in managing the disease.

12.6.3.5 CHEMICAL CONTROL

Rhizome should be treated with propiconazole (0.1%) to reduce disease incidence. Two foliar sprays of Propiconazole (0.1%) or mancozeb + carbendazim (0.1%) at 45 and 90 days after planting is very effective.

12.7 LEAF SPOT DISEASE OF TURMERIC

Fungus *Colletotrichum capsici*, *C. curcumae* (Kandaswamy, 1958), and *C. gloeosporioides* (Patel et al., 2005) are reported as the cause of leaf spot disease. In comparison of severity with other diseases, it is next to leaf blotch, and the disease is prevalent in almost all the turmeric-growing areas, and often it can occur in epidemic form.

12.7.1 DISEASE SYMPTOMS

The indicative symptom frequently appears on leaf, flowers, and rarely on leaf sheath. Initially, in leaves, elliptic or oblong spots of different size appear in both the surface, but spots are more in the upper surface as compare to the lower surface. The measures of spots are only a few centimeters in length and are 2–4 cm in width, which afterward may enlarge in size. Shortly, leaf spots coalesce to each other and expand into irregular patches covering the chief portion of the leaf blade and followed by drying to leaf. The center of the spot becomes grayish-white and thin, in which numerous black dots known as acervuli are arranged in concentric rings. The grayish-white area is bounded by a brown area with a yellow halo. The central area of the spot may become thin in dry condition and easily tear off.

12.7.2 DETECTION OF THE DISEASE

12.7.2.1 ISOLATION OF PATHOGEN

Colletotrichum capsici can be isolated using PDA. In PDA medium it produces zigzag cottony, ring or circular like with zigzag zonation colonies. The colors of colony vary from grayish-white to light brown on the ventral surface and black from the reverse. Conidia are hyaline, falcate with acute apex and narrow truncate base, aseptate, uninucleate, ranging from 18–23 × 3–5 µm (Than et al., 2008; Adhipathi et al., 2013).

12.7.2.2 PCR

Adhipathi et al. (2013) developed a PCR primer using 5.8S rDNA of ITS region (Forward sequence 5'-GTCCTAACAAGGTTTCCGTA-3 and reverse sequence 5'-TTCTCCGCTTATTGATATGC-3) and amplified the specific amplicon size of 590 bp. Thermal cycler was set as initial denaturation at 94°C for 5 min, 35 cycles of 45 sec at 94°C denaturation, 45 sec at 46°C annealing, and 1 min at 72°C extension and one cycle of final extension at 72°C for 10 min as recommended by Shenoy et al. (2007).

12.7.3 MANAGEMENT OF DISEASE

12.7.3.1 RESISTANCE

Resistant varieties such as Nallakatla, Duvvur, Gandikota, Clone DKHT-6, Roma, Rashmi, Suguna, Suroma, Sudharshana, TCP 56, Pratibha, TCP 11, and tolerant varieties such as Suvarna (PCT 8), Suguna (PCT 13), Sudharshana (PCT 14), Suvarna should be cultivated in disease-affected areas.

12.7.3.2   CULTURAL METHOD

Select seed material from disease-free areas. The infected and dried leaves of previous year crop act as primary source of inoculum, so dried leaves should be collected and burnt in order to reduce the disease inoculum. Crop rotations with non-host (cereals and legumes) should be followed whenever possible. Avoid sprinkler systems of irrigation and should adopt a drip system. Proper row-to-row and plant-to-plant spacing should be adopted.

12.7.3.3   BIOLOGICAL CONTROL

Trichoderma harzianum formulation can be used for the treatment of rhizome before planting, or it can be sprayed in standing crop at fifteen days intervals for 2–3 times.

12.7.3.4   BOTANICAL

Foliar pathogens like leaf spots can be effectively managed by spraying garlic extracts.

12.7.3.5   CHEMICAL CONTROL

Treat the seed material with Mancozeb (3 g/liter of water) or Carbendazim (1 g/liter of water) for 30 minutes and shade dry before sowing helps to minimize disease inoculum present in the rhizome. On the first appearance of disease, Propiconazole (0.1%) or Hexaconazole (0.1%) or Tricyclazole (0.1%) or Mancozeb + Carbendazim (0.1%) should be sprayed two or three times at fortnight intervals.

KEYWORDS

- dry rot
- extracellular polysaccharide
- ginger
- rhizome rot
- soft rot
- wilt

REFERENCES

Adhipathi, P. A., Nakkeeran, S. E., & Chandrasekaran, A. N., (2013). Morphological characterization and molecular phylogeny of *Colletotrichum capsici* causing leaf spot disease of turmeric. *The Bioscan., 8*(1), 331–337.

Ahmed, L., & Kulkarni, N. B., (1968). Studies on *Taphrina maculans* butler, inciting leaf spot of turmeric (*Curcuma longa* L.). I. Isolation of the pathogen. *Mycopathologia Et Mycologia Applicata., 34*(1), 40–46.

Bacigálová, K., Lopandic, K., Rodrigues, M. G., Fonseca, A., Herzberg, M., Pinsker, W., & Prillinger, H., (2003). Phenotypic and genotypic identification and phylogenetic characterization of *Taphrina* fungi on alder. *Mycological Progress, 2*(3), 179–196.

Bhai, R. S., Kishore, V. K., Kumar, A., Anandaraj, M., & Eapen, S. J., (2005). Screening of rhizobacterial isolates against soft rot disease of ginger (*Zingiber officinale* Rosc.). *Journal of Spices and Aromatic Crops, 14*(2), 130–136.

Blaschek, W., Käsbauer, J., Kraus, J., & Franz, G., (1992). *Pythium aphanidermatum*: Culture, cell-wall composition, and isolation and structure of antitumor storage and solubilized cell-wall (1→ 3),(1→ 6)-β-d-glucans. *Carbohydrate Research, 231*, 293–307.

Burr, T. J., & Stanghellini, M. E., (1973). *Pythium aphanidermatum* in field soil. *Phytopathology, 63*, 1499–1501.

Cabañes, (1997). Malachite green agar, a new selective medium for *Fusarium*. *Mycopathologia, 137*, 173–178.

Castellá, G., Bragulat, M. R., Rubiales, M. V., & Cabañes, F. J., (1997). Malachite green agar, a new selective medium for *Fusarium* spp. *Mycopathologia, 137*(3), 173–178.

Elphinstone, J. G., Hennessy, J., Wilson, J. K., & Stead, D. E., (1996). Sensitivity of different methods for the detection of *Pseudomonas solanacearum* in potato tuber extracts. *EPPO/OEPP Bulletin, 26*, 663–678.

Horita, M., Yano, K., & Tsuchiya, K., (2004). PCR-based specific detection of *Ralstonia solanacearum* race 4 strains. *Journal of General Plant Pathology, 70*(5), 278–283.

Ishiguro, Y., Asano, T., Otsubo, K., Suga, H., & Kageyama, K., (2013). Simultaneous detection by multiplex PCR of the high-temperature-growing *Pythium* species: *P. aphanidermatum, P. helicoids,* and *P. myriotylum*. *Journal of General Plant Pathology, 79*(5), 350–358.

Jeffers, S. N., & Martin, S. B., (1986). Comparison of two media selective for *Phytophthora* and *Pythium* species. *Plant Disease, 70*, 1038–1043.

Kageyama, K., Kobayashi, M., Tomita, M., Kubota, N., Suga, H., & Hyakumachi, M., (2002). Production and evaluation of monoclonal antibodies for the detection of *Pythium sulcatum* in soil. *Journal of Phytopathology, 150*(3), 97–104.

Komada, H., (1975). Development of a selective medium for quantitative isolation of *Fusarium oxysporum* from natural soil. *Review of Plant Protection Research, 8*, 114–124.

Kumar, A., & Anandaraj, M., (2006). Method for isolation of soil DNA and PCR based detection of ginger wilt pathogen, *Ralstonia solanacearum*. *Indian Phytopathology, 59*(2), 154–160.

Kumar, A., Sarma, Y. R., & Priou, S., (2002). Detection of *Ralstonia solanacearum* in ginger rhizomes using post-enrichment NCM-ELISA. *Journal of Spices and Aromatic Crops, 11*(1), 35–40.

Li, Y., Chi, L. D., Mao, L. G., Yan, D. D., Wu, Z. F., Ma, T. T., & Cao, A. C., (2014). First report of ginger rhizome rot caused by *Fusarium oxysporum* in China. *Plant Disease, 98*(2), 282–282.

Martin, C., & French, E. R., (1985). Bacterial wilt of potato *Ralstonia solanacearum*. *Technical Bulletin Information* (13).

Nair, R. A., & Thomas, G., (2013). Molecular characterization of ZzR1 resistance gene from *Zingiber zerumbet* with potential for imparting *Pythium aphanidermatum* resistance in ginger. *Gene., 516*(1), 58–65.

Nishimura, N., (2007). Selective media for *Fusarium oxysporum. Journal of General Plant Pathology, 73*(5), 342–348.

Plaats-Niterink, A. J., (1981). *Monograph of the Genus Pythium.* Studies in mycology, No. 21. Centraalbureau Voor Schimmelcultures, Baarn, Netherlands.

Prabhukarthikeyan, S. R., Karthikeyan, G., & Raguchander, T., (2015). Molecular characterization of *Pythium aphanidermatum* causing rhizome rot disease in turmeric. *Biochem. Cell. Arch, 15,* 265–269.

Pradhanang, P. M., Elphinstone, J. G., & Fox, R. T. V., (2000). Sensitive detection of *Ralstonia solanacearum* in soil: A comparison of different detection techniques. *Plant Pathology, 49*(4), 414–422.

Priou, S., Gutarra, L., & Aley, P., (1999). Highly sensitive detection of *Ralstonia solanacearum* in latently infected potato tubers by post enrichment enzyme-linked immune sorbent assay on nitrocellulose membrane. *EPPO Bulletin., 29*(12), 117–125.

Rai, B., Bandyopadhyay, S., Thapa, A., Rai, A., & Baral, D., (2017). Morphological and cultural characterization of *Phyllosticta zingiberi* (Ramkr.) causing leaf spot disease of ginger. *Journal of Applied and Natural Science, 9*(3), 1662–1665.

Rajalakshmi, J., Durgadevi, D., Harish, S., & Raguchander, T., (2016). Morphological and molecular characterization of *Pythium aphanidermatum* the incitant of rhizome rot in turmeric. *Int. J. Env. Ecol. Family and Urban Studies, 6*(4), 1–8.

Rajeshwari, N., Shylaja, M. D., Krishnappa, M., Shetty, H. S., Mortensen, C. N., & Mathur, S. B., (1998). Development of ELISA for the detection of *Ralstonia solanacearum* in tomato: Its application in seed health testing. *World Journal of Microbiology and Biotechnology, 14*(5), 697–704.

Ramakrishnan, T. S., (1942). A leaf spot disease of *Zingiber officinale* caused by *Phyllosticta zingiberi* n. sp. In: *Proceedings of the Indian Academy of Sciences-Section B.* (Vol. 15, pp. 167–171). Springer India.

Rangaswami, G., (1958). An agar blocks technique for isolating soil microorganisms with special reference to pythiaceous fungi. *Sci. Culture, 24,* 85.

Ray, M., Dash, S., Shahbazi, S., Achary, K. G., Nayak, S., & Singh, S., (2016). Development and validation of ELISA technique for early detection of rhizome rot in golden spice turmeric from different agroclimatic zones. *LWT-Food Science and Technology, 66,* 546–552.

Thammakijjawat, P., Thaveechai, N., Kositratana, W., Chunwongse, J., Frederick, R. D., & Schaad, N. W., (2006). Detection of *Ralstonia solanacearum* in ginger rhizomes by real-time PCR. *Canadian Journal of Plant Pathology, 28*(3), 391–400.

Than, P. P., Jeewon, R., Hyde, K. D., Pongsupasamit, S., Mongkolporn, O., & Taylor, P. W. J., (2008). Characterization and pathogenicity of *Colletotrichum* species associated with anthracnose on chili (*Capsicum* spp.) in Thailand. *Plant Pathology, 57*(3), 562–572.

Wang, P. H., Chung, C. Y., Lin, Y. S., & Yeh, Y., (2003). Use of polymerase chain reaction to detect the soft rot pathogen, *Pythium myriotylum*, in infected ginger rhizomes. *Letters in Applied Microbiology, 36*(2), 116–120.

Wang, P. H., Wang, Y. T., & White, J. G., (2003). Species specific PCR primers for *Pythium* developed from ribosomal ITS1 region. *Letters in Applied Microbiology, 37*(2), 127–132.

CHAPTER 13

Diagnosis and Diversity Analysis of *Alternaria brassicae* and *A. brassicicola* in Vegetable Cruciferous Crops

PRATIBHA SHARMA,[1] SHAILY JAVERIA,[2] SWATI DEEP,[3] MANIKA SHARMA,[4] and RAJA MANOKARAN[1]

[1]*Department of Plant Pathology, SKN Agriculture University, Jobner, Jaipur – 303329, Rajasthan, India, E-mail: psharma032003@yahoo.co.in*

[2]*Division of Seed Science and Technology, ICAR-Indian Agricultural Research Institute, New Delhi – 110012, India*

[3]*Technology and Innovation Center, International Panacea Limited, Gurugram, Haryana – 122003, India*

[4]*North American College of Pharmaceutical Technology, Mississauga, Ontario, Canada*

ABSTRACT

Alternaria black spot is an important disease of cruciferous vegetables, i.e., cauliflower and mustard caused by major species of *Alternaria*. Severe crop yield and quality loss are a major problems in the Brassicaceae family growing area of the world. *Alternaria brassicae* and *A. brassicicola* are major pathogens infecting cruciferous vegetable crops but in rapeseed *A. brassicae* causing major economic loss. The pathogens are mainly transmitted through seeds infected by spores. Both pathogenic fungi can seed-borne and over-winter on susceptible weeds or saprophytically surviving mycelium on crop debris and on seed plants, as well as on stockings. In this literature survey, knowledge on the epidemiology of dark leaf spot in seed production of Brassica is evaluated along with their culture, morphological, and pathogenic variability. The biochemical and molecular characterization of the pathogens are very important. Molecular markers like RAPD, RFLP, AFLP, SSR, and

SNP, etc., are used to assess and detect the pathogens from the infected samples. These markers are easy and rapid to assess the pathogens from the initial phase of infection and also provide detailed knowledge on diagnosis of the disease caused by this pathogen.

13.1   INTRODUCTION

Vegetables constitute is an important component of the balanced diet and act as supplemental food. India is the second-largest producer of vegetables in the world (ranks next to China) and accounts for about 15% of the world's production of vegetables. In India, the total area under vegetable crops was reported to be 9.2 million ha and the production was found to be 162.1 million MT with the productivity of 17.6 MT/ha (Indian Horticultural Database, 2013). Among vegetables the Crucifer family (Brassicaceae) is the economically important family of flowering plants in the vegetable plant kingdom, the family contains about 3500 species and 350 genera, is one of the 10 most economically important plant families (Warwick et al., 2000). A number of biotic and abiotic stresses pose major constraints in cauliflower production. The crop is subjected to attack by a variety of pathogens both in the nursery as well as in the field. Among the infectious disease black spot (*Alternaria brassicae, Alternaria brassicicola* and *Alternaria raphani*), blackleg and phoma root rot (*Leptosphaeria maculans* and *Phoma lingam*), anthracnose (*Colletotrichum higginsianum*), black mold rot (*Rhizopus stolonifer*), white rust and stag head (*Albugo candida*) and black rot (*Xanthomonas campestris* pv. *campestris*) are important. *A. brassicae* (Berk.) Sacc. *A. brassicicola* (Schwein.) wiltsh., *A. raphani* Groves and Skolko, and *A. alternata* (Fr.) Keissl are most common and destructive pathogens causing diseases to Brassicaceae crops. *A. brassicicola* (Schwein) Wiltshire (Maude and Humpherson-Jones, 1980; Humpherson-Jones, 1985; Maude et al., 1984; Kubota et al., 2006; Deep and Sharma, 2012; Sharma et al., 2013b, c) is most important pathogen which leads to considerable yield and quality loss both in terms of curds and seed yield in vegetable white cabbage and cauliflower. *Alternaria* has been reported from all continents of the world and in India also it was reported in most of the states. *Alternaria* blight appears every year in all the crucifer growing states of India, such as Rajasthan, Uttar Pradesh, Bihar, Odisha, West Bengal, Punjab, Haryana, Gujarat, and Madhya Pradesh (Saharan, 1992a). *Alternaria* causes yield loss of up to 47% in mustard (Kolte, 1985), on cabbage, cauliflower, broccoli, and other brassicas more than 50%. Decaying during transit and storage also causes considerable losses at market. The *Alternaria* causes seed losses up to 50% in cauliflower seed plants and from 70–90% of harvest on rape and seed cabbage plantings (Kear et al., 1977). Considerable damage occurs to cabbage and cauliflower in transit (Ellis, 2001).

In recent years, the application of molecular tools like PCR, DNA fingerprinting, etc., plays an important role in variability studies among the pathogens, which makes the quantification of genetic variation in pathogens (Brown, 1996; Michelmore and Hulbert, 1987). These tools are easy to identify the pathogens at the nucleotide level by using DNA fingerprint profiling. The DNA-based marker like Randomly Amplified Polymorphic DNA (RAPD), inter simple sequence repeats (ISSR), Simple Sequence Repeats (SSR), etc., are used widely to reveal the genetic variability among the pathogens. The biomolecules such as enzymes, proteins, amino acids, and carbohydrates also play a significant role in host-pathogen interactions.

To understand host-pathogen co-evolution, epidemiology, and resistance management, the study of genetic diversity of plant-pathogen populations is a basic need (Milgroom and Fry, 1997). PCR-based marker-assisted genotypic identification of phytopathogens have been successfully going on, at the species and subspecies level (Benali et al., 2011). Characterization of genetic variation in populations of *Alternaria* species pathogenic to crucifers by RAPD and nuclear ribosomal DNA (rDNA) sequences (Sharma and Tewari, 1995, 1998; Roberts et al., 2000; Tigano et al., 2003; Jasalavich et al., 1995) was reported with high level of similarity among the isolates of *A. brassicae* from different origin (Cooke et al., 1998; Sharma et al., 2013a). Occurrence of substantial genetic variability in *A. solani* and *A. alternata* infecting tomato and potato (Weir et al., 1998; Morris et al., 2000) suggested diversity can be maintained by mutation, uniform host selection, extensive dispersal or the existence of a cryptic sexual stage. While resolving this, Bock et al. (2002, 2005) found moderate levels of genetic diversity in the form of substantial polymorphism between isolates from five populations of *Alternaria brassicicola* attacking *Cakile maritima* along the New South Wales coast of Australia but the inter/intra species-level variation among the Indian isolates of *A. brassicicola* (Schwein) Wiltshire is still lacking.

13.2 HISTORY AND OCCURRENCE

Alternaria is a dictoyosporae genus of the family Dematiaceae, order Hyphomycetes, fungi Imperfecti. The genus was established in 1817 by

Nees, as *A. alternata* (originally *A. tenius*) as the type species (Ellis, 1971). *Macrosporium brassicae* Berk was causal fungus on plants belonging to the Brassicaceae identified in 1836 by Berkeley and later renamed as *A. brassicae* (Berk.) Sacc. by Saccardo (1886). Studies were conducted on the same fungi during different intervals by Milbraith (1922); Weimer (1924); and Rangel (1945). Wiltshire (1947) separated two species based on small and big spore's namely *A. brassicicola* (Sch.) Wiltshire and *A. brassicae* (Berk.) Sacc.

Two species are probably more confused than any other species in this genus of *Alternaria* (Wiltshire, 1947). The first is *Alternaria brassicae* (Berk.) Sacc., described as *Macrosporium brassicae* Berk., in 1836 and the second is the fungus commonly known as *A. circinans* (Berk. and Curt.) Bolle. or *A. oleracea* Milbraith or, incorrectly as *A. brassicae* (Berk.) Sacc. for which Wiltshire (1947) proposed the name *A. brassicicola* (Sch.).

A. brassicicola (Sch.) Wiltshire and *A. brassicae* (Berk.) Sacc. have been reported from almost every continent on Brassicaceae hosts. In vegetable Brassica seeds, especially white cabbage and cauliflower [*Brassica oleracea* (L.) var botrytis], *Alternaria brassicicola* is the dominant pathogen (Maude and Humpherson-Jones, 1980; Humpherson-Jones, 1985; Maude et al., 1984; Kubota et al., 2006; Deep and Sharma, 2012; Sharma et al., 2013a–c) whereas in oilseed rape, especially mustard [*Brassica juncea* (L.)], *A. brassicae* (Berk) Sacc. is dominant (Kolte, 1985; Humpherson-Jones and Phelps, 1989; Sharma et al., 2013a, b). Black spot disease caused by *Alternaria brassicicola* is of worldwide economic importance (Humpherson-Jones, 1985; Humpherson-Jones and Phelps, 1989; Rotem, 1994; Sigareva and Earle, 1999). They are widely occurring in Asian countries (Bangladesh, China, Bhutan, India, Japan, Nepal, Malaysia, Pakistan, Singapore, Sri Lanka, Thailand, etc.), and other cruciferous growing areas like Argentina, Australia, Brazil, Canada, Denmark, England, France, Germany, Ireland, Italy, New Zealand, Russia, Spain, South Africa, USA, etc. In India, it is prevalent in all crucifer-growing states.

13.3   SYMPTOMATOLOGY

Symptoms of the disease are characterized by the formation of spots on leaves, stem, curd, and silique. Leaf spots appear on the upper surface of the leaf as bright to pale yellow or tan flecks, surrounded by light green or yellow halos. Circular to irregularly lobed and light brown-black in color were observed in older spots and may or may not have characteristic concentric rings. Lesions are dark brown, often coalesce, girdle the stems, spots increase in size and number, coalesce, and cause leaf blights (Figure 13.1). Leaf spots of *A. brassicicola* appear as small, dark-colored areas. They spread and form circular lesions up to 1.0 cm in diameter. Dark stem lesions immediately after germination that can result in damping-off, or stunted seedlings at nursery conditions. In humid weather, dark conidiospores are seen on the surface of the lesion in concentric rings. Linear spots appear on petioles, stem, and seedpods. In general, lesions produced by *A. brassicicola* are black sooty velvety compared to gray lesions produced by *A. brassicae* (Tahvonen, 1979). The spots appear on the middle and upper leaves with smaller sized spots, when defoliation of lower leaves occurs. Then, round black conspicuous spots appear on silique and stem. Spots on cauliflower silique are brownish-black with a distinct gray center. The infection of *Alternaria* blight on leaves and silique reduces the photosynthetic area drastically. A brown discoloration of the cauliflower a curd occurs which darkens to an olivaceous color with age. *A. brassicicola* attacks cabbage heads mostly after harvest. In long storage, they overrun the outer leave and sporulate profusely, giving a black, moldy appearance. In this phase, they commonly associated with the incidents of rhizopus soft rot and gray mold rot (Chupp and Sherf, 1960).

FIGURE 13.1 Symptoms of *Alternaria* blight of cauliflower.

13.4   LOSSES CAUSED BY THE DISEASE

Alternaria leaf spot is one of the severe diseases of vegetable crucifers affecting crops as well as seed crops and also infecting the leaves, stems, pods, which influence both quantity and quality of yield of brassica crops (Butler, 1918). In Indian conditions after winter rains, yield losses in *Alternaria* infected plants increase considerably (Dey, 1948) and due to severe infections of the disease shriveling of seed and reduction in quantity of oil content are the major problem (Vasudeva, 1958; Chahal and Kang, 1979; Chohan, 1978; Kaushik et al., 1984). The seed production of brassicas has been seriously affected by this disease by damaging the tissues of the stem and leaves (Bandhopadhyay et al., 1974; Nielson, 1933).

In vegetable crops, losses occur from damping-off of seedlings and spotting of lower leaves and heads of cabbage. The disease can be destructive in seedbeds, especially in cabbage and cauliflower. Quality and market value of cauliflower and cabbage heads reduced by spotting and browning (Sherf and Macnab, 1986). Maximum 80% reduction in seed yield caused by the fungi which severely depress germination (Smith et al., 1988). Gorshkov (1976) reported 80–100% losses by damping-off of cabbage caused by *A. brassicae*. More than 50% yield losses were observed due to *Alternaria* leaf spot on cabbage, cauliflower, broccoli, and other brassicas in the USA during wet seasons (Ramsey and Smith, 1961). The *Alternaria* causes seed losses up to 50% in cauliflower seed plants and from 70–90% of harvest on rape and seed cabbage plantings (Kear et al., 1977). Considerable damage occurs to cabbage and cauliflower in transit (Ellis, 2001). In India, these disease losses were observed of 15 to 71% in all crucifers' crops (Kadian and Saharan, 1983; Singh and Bhowmik, 1985; Kumar, 1986; Ram and Chauhan, 1998). In canola or rape, the yield reduction was observed up to 20–50% (Rotem, 1994). *A. brassicicola* is not only infect the leaves and also infect all parts of plants, including seeds, stems pods, and particular importance as postharvest disease (Rimmer and Buchwaldt, 1995).

13.5   DISEASE CYCLE

The primary infection results from the wind-borne spores produced on debris of previous season crops, weeds, and another collateral host. Seed is also act as primary sources, especially in temperate regions. Conidia readily germinate in the presence of moisture by giving rise to a germ tube that emerges from any cell of the spores. Theoretically only one spore can cause infection in the brassicaceous seedling indicating a threshold infection value at a very low level (Czyzewska, 1971). Germ tubes from germinated spores of *A. brassicicola, A. brassicae, A. raphani,* and *A. alternata* generally penetrate the undamaged tissues of many brassicaceous hosts directly (Czyzewska, 1971; Changsri and Weber, 1963), although indirect penetration through stomata has been reported by both *A. brassicicola* and *A. brassicae* (Changsri, 1961; Tsuneda and Skoropad, 1978a, b).

Cuticular penetration is most common with *A. brassicicola* and stomatal penetration occurs in *A. brassicae*. Once within the host, the epidermal cells are fully invaded and mycelia ramify through and between the mesophyll and palisade cells, the entire leaf is soon parasitized. Early in the post-penetration phase, the invaded epidermal cells become necrotic, and parenchyma tissues ahead of the advancing hyphae often collapse (Dixon, 1981). The conidia germination in host tissues needs suitable environmental condition like moisture and produce toxin during penetration in the host. It will be occurring through stomata (Dehpour et al., 2007). Sharma et al. (2014) demonstrated interactions among cauliflower – *A. brassicicola* and mustard – *A. brassicae* by using light and electron microscope. The infection process involves conidia germination, penetration, and colonization and several germ-tubes from conidia were colonized extensively on the surface of leaves. Penetration of these spores in plant surface with occasional appressoria formation via epidermis or stomata.

The wind spread 'dry conidia' of these pathogens can survive at different temperatures and relative humidity in seeds for several months (Kumar and Gupta, 1994; Abul-Fazal et al., 1994) and disease spreads during the moist and cool weather by wind-blown or rain-splashed spores (Rotem, 1994; MacKinnon et al., 1999; Oliver et al., 2001).

The *Alternaria* spp. infects mainly seed, leaf, inflorescence, stems, and fruits of the cruciferous and releases the phytotoxins through chlorotic effect (Jung et al., 2002). After penetration the leaf spot causing pathogen becomes subcuticular followed by colonization of the epidermal and the mesophyll cells (Tewari, 1983, 1986), of the necrotic center. Leaf chlorosis of the leaf induced by the pathogen by producing diffusible metabolite nearby chlorotic area and these metabolites primarily targets the plasma membrane of the leaf followed by chloroplast is either directly or indirectly affected leading to leaf chlorosis. In Mitochondria, these effects were observed at later stage. The cells surrounded by necrotic area completely devoid of cellular organelles and reveal electron-dense lamellar deposits (Tewari, 1983). According to Suri and Mandahar (1982, 1985), cytokinin like substances appears to be actively involved in the infection and pathogenesis of *Alternaria* spp. The correlation in susceptibility differences due to amino acids contents were studied in the Brassicaceous host to *A. brassicae, A. brassicicola,* and *A. raphani* (Changsri, 1961). Secretion of lytic enzymes viz. cellulase, pectin lyase, pectin methylesterase and polygalacturonase (PG) or mycotoxins either host-specific toxin (HST) or non-host specific toxin were responsible for the chlorotic effect in host tissues (Nozaki et al., 1997; Berto et al., 1997; Jahangeer et al., 2005; Gautam et al., 2012). In some infected cell, deposition of lignin to the cell wall was found (Von, 1962).

The population of the pathogen varies with the age and variety of the host, maturity of the leaves and climatic conditions. The younger leaves have a lower incidence of disease of either pathogen compared to older leaves. In susceptible plants, spores remain dormant until they gets its favorable conditions like moisture or rain for its germination. The germinated spores can cause lesion on the surface of the host tissues by penetration and further spread through wind or rain to other parts of the crop (Sharma et al., 1986). The leaf spots are primarily occurring in the older leaves than spread to towards the tip of the new leaves (Chattopadhyay, 1999; Meena et al., 2010). Hostage is an important predisposing factor for both the *A. brassicicola* and *A. brassicae* for infecting crucifers is also been reported by Deep and Sharma (2012).

A. brassicicola and *A. brassicae* also survive in the form of microsclerotia and chlamydospores can be isolated from infected leaf samples (Tripathi and Kaushik, 1984) and these were formed at lower temperature (3°C) within conidial cells. *In vitro* studies show that these microsclerotia and chlamydospores are resistant to freezing and desiccation and develop in natural soil and room temperature (Tsuneda and Skoropad, 1977).

13.6   CULTURAL VARIABILITY

Variations in cultural characteristics and pathogenesis of different isolates of four *Alternaria* species infecting Brassicaceae have been observed, however information on the existence of distinct pathotypes using stand host differentials is rather limited (Saharan, 1992a, b). The variability among the pathogens was studied based on the colony, spore morphology, sporulation, mycelial growth in different media, and other cultural characteristics (Ansari et al., 1989a, b; Patni et al., 2005; Kaur et al., 2007). The cultural characteristics such as growth behavior and colony character were studied by Singh et al. (2007) and describe the isolates having lot of morphological and cultural variation. The alternate dark and light cycles of 12 h resulted in maximum growth of pathogen mycelium up to 85 mm. The colony morphology studies were conducted by Ramegowda and Naik (2008), among 14 isolates of *Alternaria* spp. the mycelia were initially light gray later turns in to blackish color. In potato dextrose agar (PDA), eight isolates were observed with smooth margin and remaining six isolates shows irregular margin.

Based on the culture shape, pigmentation, and colony growth on PDA Prasad et al. (2009) classified twenty-six isolates of *A. helianthi* from India. Kumar et al. (2008) growth of studied eleven isolates of *A. solani* on PDA at 25°C±1°C. The radial growth of the mycelium was ranged from 14.9 mm to 57.7 mm and significant growth of mycelia was observed in PDA (Kumar et al., 2008).

The variability in morphology and cultural characterizers were observed among the isolates of *A. brassicae* collected from rapeseed-mustard and cauliflower of different agro-climatic and geographical regions of India (Meena et al., 2005; Kaur et al., 2007). In a similar way, Goyal et al. (2011) observed variation in mycelial growth and sporulation among *A. brassicae* isolates. Sharma et al. (2013a, b) reported cultural and morphological variability within *A. brassicae* isolates infecting cauliflower and mustard of Indian origin. The effect of different parameters likes culture media; inorganic nitrogen sources, pH, and temperature were studied on the growth and sporulation of *A. brassicicola* collected from cabbage. The host leaf extract media and PDA show good results on the growth of the pathogen (Khatun et al., 1996). However, variability study in *A. brassicicola* infecting vegetable crops like cauliflower and cabbage are still under research.

13.7 MORPHOLOGICAL VARIABILITY

Morphological variability among the *Alternaria* spp. has been widely studied by many workers (Verma and Saharan, 1994; Varma et al., 2006). Conidia morphology can be varied in wide range in regards to septation, size, shape, color, and ornamentation of the conidia of *Alternaria* spp. (Simmons, 1992). Keeping in this view, the development of subgeneric grouping based on the diversity of conidia includes shape and size have been studied. Simmons (1992) reported fourteen group of *Alternaria* spp. based on the conidia chain formation. In continuation, Simmons and Roberts (1993) analyzed further species grouping concept with certain reference groups namely *alternata* species group, the *tenuissima* species group and the *infectoria* species group.

In addition to this work, the following species groups namely the *arborescens, brassicicola, porri*, and *radicina* groups were studied by different workers (Simmons and Roberts, 1993; Pryor and Gilbertson, 2000; Roberts et al., 2000).

Based on the morphology, most of the *Alternaria* spp. are exhibits considerable changes depends upon the substrate and pH of the media, temperature, light, and humidity (Misaghi et al., 1978; Nishimura and Kohmato, 1983). The conidia of *A. solani* causal agent of early blight on tomato were studied by Ellis (1971). The dark muriform, pale golden or olivaceous brown colored conidia length of 150–300 µm and width of 15–19 µm were observed with 9–11 transverse and 1–4 longitudinal or oblique septa.

Kumar et al. (2003) studied the variability in *A. brassicae*, founds that significant variation in conidia length and septation. Meena et al. (2005) studied the spore behavior in atmosphere and trends in variability of *A. brassicae* population in India. Kaur et al. (2007) documented the variation in *A. brassicae* isolates based on conidial morphology, fungicidal sensitivity, and molecular profile. Similarly, Singh et al. (2007) also studied cultural and morphological variability in *A. brassicae* isolates of Indian mustard. Goyal et al. (2011); and Sharma et al. (2013a) studied the morphological variations among the *A. brassicae* isolates by using conidial length, conidial width, beak length and number of septa in conidia (Figures 13.2 and 13.3).

Small beakless conidia of *A. brassicicola* was studied for variability in morphology by some earlier worker like Ellis (1971); Pattanamahakul and Strange (1999); and Jung et al. (2002). The variability among the Indian isolates of *A. brassicicola* infecting vegetable crops are still lacking behind.

Isolate Code	CaAbU2	CaAbU3	CaAbU4	CaAbU5	CaAbU6	CaAbU7	CaAbU8	CaAbU9
Culture Characters On PDA plates and Microscopic images								
Spore Structure								

FIGURE 13.2   Spore morphology and cultural characters on PDA plates and microscopic images of *A. brassicae* isolates.

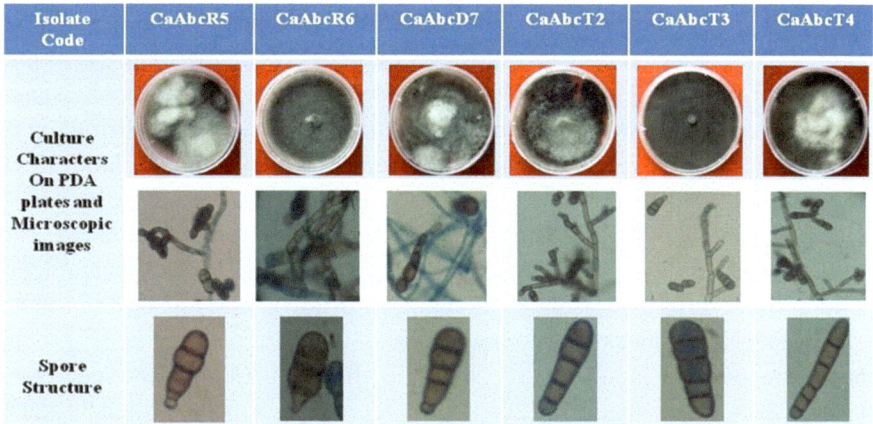

FIGURE 13.3 Spore morphology and cultural characters on PDA plates and microscopic images of *Alternaria brassicicola* isolates.

13.8 PATHOGENIC VARIABILITY

In plant disease management, virulence, and aggressiveness of the pathogen is most important aspect. Therefore, knowledge on pathogenic variability can be useful in integrated disease management (IDM) program. In general, aggressiveness of the pathogen defined as ability to colonize host and cause the damage to the host and degree of virulence that causes disease (Shurtleff and Averre, 1997). On the basis of pathogenesis of the three isolates of *A. brassicae* (A, C, and D) infecting rapeseed and mustard collected from Pantnagar was characterized by Awasthi and Kolte (1989). These isolates were further resembled with Bihar and Kanpur isolates (Kolte et al., 1991). The virulence of these isolates was tested by Viswanath and Kolte (1997) in fourteen host genotypes of five different *Brassica* species and suggested that isolate A was more virulent, C was moderate, and D was avirulent. Saharan and Kadian (1983) observed wide variations in virulence among the races of *A. brassicae* through studies were based on isolates collected from the same location. Hong et al. (1996) differentiated the isolates of *A. brassicae* on the basis of their virulence. Mehta et al. (2003) studied the morphological and pathological variability in *A. brassicae* in rapeseed and mustard. Pathogenic diversity of isolates of *A. brassicae* were studied by Kumar et al. (2003), all those isolates were tested on a set of seventeen host differentials of different species of brassicae in green house conditions, and results revealed that all those isolates behaved differently in host differentials. The studies indicated the existence of variability among isolates of *A. brassicae* in Haryana. Gupta et al. (2003) selected eleven genotypes of *Brassica juncea* (L.) Czern and Coss were selected as host differentials to identify pathotypes of *A. brassicae*. Meena et al. (2012) studied the aggressiveness, diversity, and distribution of *A. brassicae* isolates infecting oilseed *Brassica* in India and similarly Singh et al. (2012) found morphological and pathological variations in isolates of *A. brassicae* causing leaf blight of rapeseed and mustard. Sharma et al. (2013a) studied variation at morphological, cultural, pathogenic, and molecular level among the *A. brassicae* isolates infecting cauliflower and mustard in India.

Twenty-six isolates were categorized into three pathogenicity groups based on disease incidence recorded on different *Helianthus* species as low (< 20% PDI), medium (20–50% PDI) and high pathogenicity (>50% PDI) groups (Prasad et al., 2007). The pathogenic variability of *A. alternata* infecting senna was studied by Tetarwal et al. (2008), and noted the significant variation among the isolates and their virulence pattern. Similarly, Varma et al. (2006) found variability among *A. solani* isolates associated with early blight of tomato. Reis and Boiteux (2010) studied the geographical distribution, host range, and specificity of different *Alternaria* species infecting Brassicaceae in the Brazilian neotropics. Sofi et al. (2013) characterized *A. mali* associated with *Alternaria* leaf blotch of apple based on cultural, morphological, pathogenic, and molecular characterization (Figure 13.4).

FIGURE 13.4   Appearance of black spot symptoms on cauliflower.

13.9   BIOCHEMICAL CHARACTERIZATION

Fungal plant pathogens are a group of microorganisms that show a very high versatility during their infection cycles (Fernandez-Acero et al., 2007), which allows them to infect a wide variety of crops (Garrido et al., 2010*)*. There are complex of interaction between the host and fungal community during the infection (Idnurm and Howlett, 2001; Odds et al., 2001; Tivoli et al., 2006). Common strategies of phytopathogenic fungi include forming specialized infection structures (haustoria and appressoria, etc.), and to release cell wall degrading enzymes (CWDEs) necessary for penetrating and colonizing their plant host (Schulz et al., 2002). Enzymatic degradation host cell wall is one of the important processes during the pathogenicity; the pathogenic microbes should have the capability to produce the cell wall degradation enzymes. The wall is a nutritional source for microorganisms containing mainly carbohydrate, proteins, and sugars. Nutrition is an essential factor or fundamental basis of pathogenesis which influence in majorly during the pathogen and host interaction. Most microbial cell-wall-degrading enzymes are extracellular, i.e., secreted, which are subject to substrate induction and catabolite repression.

Pathogenesis depends on the secretion of CWDEs which are most important for the penetration of pathogens into host tissues. These enzymes are required by pathogens and not specific on the host, race or cultivar. These are important for cell wall depolymerization for obtaining nutrients from the host. These enzymes play a vital role during adverse circumstances of the pathogens (Walton, 1994). In some endophytes, the following enzymes include pectinases, xylanase, cellulases, and lipases, whilst proteases and phenol oxidase have also been documented. These enzymes are not related to the pathogenesis (Tan and Zou, 2001).

13.9.1 PECTINASES

Pectin is a high molecular weight polysaccharides compound present in higher plants. They are the main component for primary cell wall and middle lamella (Alkorta et al., 1998). The degradation of pectin was done by synergistic action of several enzymes (Gummadi and Panda, 2003). Polygalacturanase and pectinase are major enzymes for the cell maceration and death of plant tissue (Fernando et al., 2001). The pectic enzymes are majorly involved in the cell wall and middle lamella degradation reported in several types' diseases such as wilt, blight, soft rot, dry rot and leaf spot caused by pathogenic fungi, bacteria, and nematodes (Ramos et al., 2010). These enzymes are majorly affecting the pectic substances of the host and plant an important role in pathogenicity (Cole et al., 1998) and also determine the virulence of many phytopathogens (Rogers et al., 2000).

The widely studied pectinolytic enzymes are protopectinases, PGs, lyases, and pectin esterases. Protopectinases catalyze the solubilization of protopectin. PGs are the most abundant pectinolytic enzymes that hydrolyze the polygalacturonic acid chain by the addition of water. Galacturonic acid polymers were catalyzed by lyases. Pectin and methanol liberated by pectin-esterases by de-esterifying the methyl ester linkages of the pectin backbone. Application of biotechnological tools, these pectinolytic enzymes are playing an important role of industrial sectors like food industries, paper industries, textile industries, wastewater treatment plants, coffee, and tea fermentation and processing industries, breweries, and distilleries industries (Jayani et al., 2005). Among the family of pectinolytic enzymes, polygalacturonases (PGases) are the most extensively studied enzyme which are present in the form of endo-PGase (E.C. 3.2.1.15) and exo-PGase (E.C. 3.2.1.67). PGases are widely applied in the food processing industries and also play an important role in fungal-host interactions.

Exo-PGases have been reported in *A. mali* (Nozaki et al., 1997). Similarly, Endo PG was reported in *A. alternata* (Chaudhri and Suneetha, 2012). High pectinase activity was found in *A. citri, A. raphani,* and *A. tenuissima,* and moderate activity was exhibited by *Alternaria alternata* while screening dematiaceous hyphomycetes fungi infecting leaves of broad bean. The highest production of pectinase was observed in pH 6 at 30°C after 8 days in the liquid medium amended with carbon and nitrogen source, respectively, citrus pectin and ammonium sulfate. It was also suggested that pectinase activity was related to the pathogenicity of the fungal pathogen (Saleem et al., 2012). For example, endopolygalacturonase is playing an important role in citrus block rot caused by *A. citri* but not brown spot caused by *A. alternata* (Isshiki et al., 2001).

13.9.2   CELLULASES

Cellulose is the most abundant component of plant biomass, exclusively in plant cell walls (Lynd et al., 2002). This linear, unbranched homopolysaccharide is insoluble in water (Lederberg, 1992) consisting of glucose linked by β-1,4-glycosidic linkages. These polymer molecules vary individually by length and arranged by fibrils or bundles (Walsh, 2002) and occurs in the form of crystalline or paracrystalline (amorphous) structures (Walter, 1998).

Fungi and bacteria are majorly producing cellulolytic enzymes among the other microbes (Lederberg, 1992). Due its decomposition ability, fungi are most important for degradation of cellulosic organic substrates (Lynd et al., 2002). Enzymatic hydrolysis of the polymers is basic and initial step in cellulose degradation. Cellulase enzyme or its complex is involved in the hydrolysis of cellulose into simple, water-soluble products (Alexander, 1961). Depolymerization of cellulose requires three types of enzymes namely, endoglucanase (EG or CX), Exoglucanase, and Cellobiase or β-glucosidase (BGL). Endoglucanase (EG or CX) are randomly hydrolyzes the internal β-1,4-glucan chain of cellulose (Grassin and Fauquembergue, 1996; Walsh, 2002). Initially within the amorphous regions and display low hydrolytic activity toward crystalline cellulose. An enzyme Exoglucanase, i.e., exoacting cellobiohydrolases (CBH) removes cellobiose from the non-reducing end of cello-oligosaccharide and of crystalline, amorphous, and acid or alkali-treated cellulose. Depolymerization of cellulose completed by hydrolysis of cellobiose done by cellobiase or BGL and yields two glucose molecules (Himmel et al., 1994). Cellulases play an important role in various industries (Walsh, 2002) like food processing, water recycling, textile, feed, beverages preparation and biofuel (Philippidis, 1994).

CWDEs were reported from *Alternaria helianthi, Alternaria triticina, and Alternaria sesame* (Bhaskaran and Kandaswamy, 1978; Jha and Gupta, 1988; Marimuthu et al., 1974). Screening of cellulytic ability of fungi from native environmental source including soil, air, and infected plant were done, in which *Alternaria* spp. also found to possess cellulose-degrading ability. Highest yield of enzyme was recorded at 37°C while maximum activity was found after 7 days of incubation in the range of pH 4–4.8. In the availability of cellulose, the synthesis of cellulase has been increased by ~10 folds, whereas in the presence of glucose it was suppressed (Jahangeer et al., 2005). Cellulose production also observed from *A. alternata* causing fruit rot in chili (Anand et al., 2008). Hubballi et al. (2011) finds that the activity of cellulolytic enzymes increased with the increase in the age of the culture of *A. alternata* when infected to Noni plant. Similarly, the enzymes activity produced by the pathogens isolates which are avirulent did not decrease, and these activities of enzymes increased with the age of the culture. *Alternaria sps* and *A. alternata* were found among the cellulase producers by helping in the biodegradation of municipal solid waste (Gautam et al., 2012). Similarly, *Alternaria alternata* along with several other fungal species isolated from anise and cumin seeds in Upper Egypt were found to produce cellulase (exo- and endo-β-1,4-glucanase) on solid media (Saleem et al., 2013).

Macris (1984) also reported similar result in *Alternaria alternata*, extra-cellular BGL, and carboxymethylcellulose performed optimally at pH 5 to

4.5 and 5 to 6 and at temperatures of 70 to 75°C and 55 to 60, respectively. Both optimal temperature and P-glucosidase thermostability were among the highest ever recorded for the same enzyme excreted by BGL and cellulase hyper producing microorganisms. To understand the mechanism on the change in the CWDEs activities *viz.* polygalacturonase (PGU), cellulase, and β-1,3-glucanase, induced by *Alternaria brassicae* was studied by Jain and Dhawan (2008) and finds that role of both PG and cellulase in pathogenesis. Garg et al. (1999) studied the CWDEs in *Alternaria brassicae* and finds that polygalacturonase and cellulase decreased in leaf blight resistant cultivar RC-781 and increased in the susceptible cultivar Varuna up to 3 days. Cellulase and Polygalacturonase (PG) activities progressively increased after infection with *Alternaria helianthi* in susceptible Sunflower leaves up to 9 and 7 days, respectively. In contrast, the cellulose and PG activities were decreased in the resistant cultivar of sunflower leases up to 9 days after inoculation. The results showed that cellulase and PG activities may play an important role in pathogenesis and β-1,3-glucanase in the expression of resistance (Dawar and Jain, 2010).

13.9.3   LIPASES

Lipases (triacylglycerol acylhydrolase, E.C.3.1.1.3) catalyze the hydrolysis at ester bonds of triglycerides at the interface between oil and water. This reaction is reversible, and some lipases also catalyze the synthesis of esters and transesterification in microaquesous condition. Lipases are found in microorganisms, plants, and animals (Kamimura et al., 2001; Burkert et al., 2004). The secretion of lipases is influenced by different factors including temperature, pH nitrogen, and carbon (George et al., 1999). All microorganisms have optimal growth temperature characteristic at which they introduced their highest growth and a tolerable range of temperatures (Cho et al., 2007). Lipases have potential industry applications as the detergents (hydrolysis of fat), additive in foods (flavor modification), fine chemicals (synthesis of esters), leather (removal of lipids from animal skins), and pharmaceuticals (digestion of fat and oil in foods) (Fariha et al., 2005). Many genera as *Penicillium, Rhizopus, Aspergillus, Humicola, Mucor, Candida,* and *Fusarium* have been noted as producers of lipases with desirable properties.

Fungi are widely used to produce a higher quantity of lipases for biotechnological purposes (Iwai and Tsujisaka, 1984), but not the phytopathogenic fungi, which are scarcely chosen to be lipase sources even if they are able to destroy lipidic structures of host plants (Trail and Köller, 1993). Ghosh et al. (1996) reviewed about the microbial lipase production, their properties, thermal, and stability, substrate specificity, and activity in organic solvents, especially from fungi and bacteria. Lipases were produced by *Alternaria* sp., *Acremonium* sp., *Fusarium* sp., *Aspergillus* sp. and *Pestalotiopsis* sp., while protease and amylase were produced by some of them (Silva et al., 2006). Nutritional sources like nitrogen, carbohydrates, sulfur, and phosphorous on protease, amylase, and lipase action of six *Alternaria* species viz. *A. citri, A. alternata, A. crassa, A. macrospora, A. dianthicola* and *A. tenuissima* were studied (Rathod and Chavan, 2010). Screening and production of lipases were studied in *Acremonium* sp., *Alternaria chlamydospores, Alternaria* sp., *Aspergillus* sp., *Fusarium* sp. and *Pestalotiopsis* sp. (Panuthai et al., 2011). Extracellular lipases biosynthesis by several local fungal cultures isolated from various lipid-rich habitats of Faisalabad were done. The isolated strains of *Penicillium chrysogenuma, Aspergillus niger, Mucor mucedo, Rhizopus microsporus, Alternaria alternata, Trichophyton* sp., *Fusarium* sp., *Curvularia* sp., *Aspergillus flavus* were screened for the production of extracellular lipases. Several environmental parameters such as temperature, pH, inoculum size, incubation time, and amount of substrate were optimized for the selected hyper producer (Iftikhar et al., 2011).

Lipase activity of fungal stains *(Aspergillus niveus, Alternaria humicola, Bipolaris spicifera* and *Fusarium moniliforme)* isolated from otitis media (OM) of agricultural workers and reported that isolates *Fusarium moniliforme* and *Alternaria humicola* recorded highest lipase activity, i.e., 44 and 30 unit mg-1 protein respectively in 10 days at 30°C. Fungal virulence in relation to lipase activity was discussed in the pathogenicity of otomycoses (Srivastava and Gautam, 2013).

The work done by Yao and Koller (1995); Berto et al. (1999); and Cho et al. (2006) reveal the functional redundancy of lipases in regards to pathogenicity. An extracellular lipase was found to be produced by *A. brassicicola in vitro* (Berto et al., 1997, 1999). By *Alternaria brassicicola* higher quantities (3.2 U/ml) of an inducible extracellular lipase (EC 3.1.1.3) supplemented with 20 mM methyloleate in shaken synthetic medium was produced. SDS-PAGE analysis determined the molecular weight of lipase to be 80 kDa; optimum pH and temperature for activity of the enzyme were 9.0 and 25°C, respectively. Anti-lipase antibodies were found to significantly decline the ability to cause disease on cauliflower leaves by *A. brassicicola*. However, disruption of four predicted *A. brassicicola* lipase genes expressed during plant infection did not result in reduced virulence on cabbage (Cho et al., 2006).

13.10   MOLECULAR CHARACTERIZATION RANDOM AMPLIFIED POLYMORPHIC DNA (RAPD) AND INTER SIMPLE SEQUENCE REPEAT (ISSR) MARKER CHARACTERIZATION

Generally, population of pathogen characterization has been based on growth characteristics, morphology, and disease reaction on the host. However, these methods lack precision, demand more tissue, and are not so trustable due to component of interaction with the environment.

In recent years, molecular tools have been used in the study of variability which quantifies a variation in genes as a relatively straightforward endeavor (Brown, 1996; Michelmore and Hulbert, 1987). Recent improvement has been done in the field of DNA technology which provides exciting avenues for characterization of pathogens diversity and host genes, phylogenetic relationship within and outside the population of the pathogen (Bielikova et al., 2002; Xu et al., 2003; Zamani et al., 2004; Gouveia et al., 2005). The use of DNA profiling systems reveals variation in the nucleotide sequence of DNA. With improved molecular techniques, several studies have examined variation in isolates of *Alternaria* species (Pryor and Michailides, 2002; Gherbawy, 2005; Mercado et al., 2006). Quick assessment of genetic variability in various taxa was allow by using RAPD analysis and has been used to study inter and intraspecific variability among fungal isolates which showed an effective method for identifying genetic variability of *A. alternata* and *A. cassiae* isolates occurring in *Senna obtusifolia* and also for differentiate *Alternaria* sp. (Tigano et al., 2003).

Fifty-five isolates of *Alternaria* spp. has been profiled by using RADP which belongs to 13 small spored sps. and 3 large spored sps. were carried out using 12 arbitrary primers. The large spored species viz. *A. lecunthemi, A. solani,* and *A. porri* were differentiated from small-spored species by 0.44 genetic distances and from each other by 0.25 genetic distance which indicates that analysis by RAPD can be used to discover the phylogenetic relationship of *Alternaria* sp. (Wang and Zhang, 2003).

The similarity matrix of RAPD analysis of 26 isolates *A. helianthi* indicated that most of the isolates exhibited 83% similarity coefficient. Isolates were grouped into six groups in the dendrogram generated with the RAPD markers at genetic similarity of 0.34 (Prasad et al., 2007). Roberts et al. (2000) reveals the RAPD analyzes of *A. alternata, A. tenuissima, A. infectoria,* and small spored host-specific species revealed distinctive RAPD fragment patterns for all species, and cluster analysis did resolve these species into distinct clades, which suggests that these taxa constitute well defined species.

Genetic variation of twenty isolates of *Alternaria brassicicola, A. raphanin* and *A. brassicicola* with RAPD markers were studied by Sharma and Tiwari (1998). Analysis of RAPD showed by UPGMA reported that all 20 isolates which were isolated from different geographical regions could be classified into four distinct groups. Intra-regional variation within isolates was less apparent. Variation was higher in *A. brassicicola*, as based on RAPD analysis. Cooke et al. (1998) revealed that the 13 phytopathogenic *Alternaria* species which were profiles by RAPD-PCR and closely related two out-groups were explored using six distinct primers. Each species generated a different pattern of DNA fragments which were utilized as a compute of the degree of relatedness within species. *A. brassicae* isolates of various origin found high levels of similarity, but little similarity was noted within other species. The nearest interspecific genetic distances were recorded between *A. alternate, A. longipes,* and *A. citri.* Weir et al. (1998) were evaluated the genetic diversity of several species of *Alternaria* by RAPD analyzes but did not assess the conformity of morphological and genetic characters.

The RAPD questioned to its lack of total reproducibility (Lamboy, 1994), but it can been utilized to generate unique PCR products or amplicons in any living organism especially filamentous fungal species or strains of interest to be converted into species- or strain-specific sequence-characterized amplified region (SCAR) markers (Abbasi et al., 1999; Lecomte et al., 2000; Li et al., 1999; Schilling et al., 1996). SCAR primers are designed on the basis of known DNA sequences of the particular organisms of the study. Therefore, both the marker, i.e., SCAR and RAPD are differ from each other, which develop the specificity and sensitivity to diagnosed the specific fungal species in the laboratory cultures as well as the mixed DNA sample from field because annealing of SCAR primers are very specific to particular fungal sequences.

No clear grouping of isolates from different geographical regions because of the formation of mixed or poorly resolved clusters. RAPD and ISSR analysis proved to be an effective method for diagnose the genetic variability amongst different *Alternaria* species viz., *A. brassicicola, A. raphani, A. brassicae* and (Sharma and Tewari, 1998), *A. alternata* and *A. cassiae* (Tigano et al., 2003; Iram and Ahmad, 2005; Pusz, 2009), *A. tenuissima, A. alternata* and *A. arborescens* (Hong et al., 2006) *A. carotiincultae* and *A. radicina* (Park et al., 2008), *A. solani* (Varma et al., 2006; Kumar et al., 2008; Leiminger et al., 2010).

Significant genetic variability among *Alternaria* isolates, even between the same species group were reported by Gherbawy (2005). These results were also well endorsed by detecting the other species of *Alternaria* infecting crucifer (Sharma et al., 2013a; Meena et al., 2012; Sangwan and Mehta, 2007) and other crops (Kale et al., 2012).

Characterization of genetic variation in populations of *Alternaria species* pathogenic to crucifers by nuclear rDNA sequences and RAPD (Sharma and Tewari, 1995, 1998; Roberts et al., 2000; Tigano et al., 2003; Jasalavich et al., 1995) was recorded with high level of comparability among the isolates of *A. brassicae* from different origin (Cooke et al., 1998; Sharma et al., 2013b). Occurrence of substantial genetic variability in *A. alternata* and *A. solani* infecting potato and tomato (Weir et al., 1998; Morris et al., 2000) suggested diversity can be retained by mutation, uniform host selection, extensive dispersal or the existence of a cryptic sexual stage. Varma et al. (2006) reported similar results while studying genetic diversity among the *A. solani* isolates infecting tomato crop. While resolving this, Bock et al. (2002) got genetic differences among a small sample of isolates of *A. brassicicola* using AFLP markers. They studied 18 isolates of *A. brassicicola* from distinct regions also indicated at least modest levels of genetic diversity between *A. brassicicola* along the New South Wales coast. 16.7–27.9% were polymorphic out of 43–66 markers scored per primer combination. Using data from four primer combinations, most isolates were identified as separate genotypes on the basis of an unweighted paired group method of arithmetic averages analysis. Similarly, Bock et al. (2005) found modest levels of genetic diversity in the form of considerable polymorphism within isolates from 5 populations of *Alternaria brassicicola* attacking *Cakile maritima* along the New South Wales coast of Australia, with a maximum of two genotypes being shared between population pairs, but the inter/intra species-level variation among the Indian isolates of *A. brassicicola* (Schwein) Wiltshire is still lacking.

13.10.1   SIMPLE SEQUENCE REPEATS (SSR)

Genetic diversity of all genotypes was estimated using simple sequence repeats (SSR) markers. The genotypes were estimated for polymorphisms after amplification with 41 pairs of SSR markers. Nineteen pair of SSR markers in total was found to be very informative with a polymorphic information content value 0.50. The evaluation of genetic similarity and cluster analysis together with disease resistance provides some useful guides for assisting plant breeders in selecting suitable genetically diverse parents for the crossing program as well as for the development of mapping population (Ghosh et al., 2019).

13.10.2 INTERNAL TRANSCRIBED SPACER (ITS) REGION AND RESTRICTION FRAGMENT LENGTH POLYMORPHISM (RFLP) ANALYSIS

In modern systematics analysis of rDNA sequences has become a common tool and also has been used to construct and established molecular phylogenetic relationships between many groups of fungi (White et al., 1990). The inter transcribed spacer (ITS) region is extensively used in classifying fungi (Chillali et al., 1998) due to its nature of variability among species (Jung et al., 2002). This region is appropriate for PCR amplification, restriction analysis and sequencing procedure (Pryor and Gilbertson, 2000). Molecular systematic studies among *Alternaria* spp., on the basis of nuclear rDNA which have been previously analyzed by Kusaba and Tsuge (1994, 1995); Pryor et al. (2002). *Alternaria* species pathogenic to crucifer varied from each other by rDNA sequence analysis (Jasalavich et al., 1995).

For example, analysis of the comparatively conserved nuclear 18S rDNA sequence has been utilized to determine the genus and family relationships between the *Pleosporales* isolates, and disclosed a close phylogenetic relationship within *Alternaria* and *Pleospora* (anamorph *Stemphylium*) (Morales et al., 1995; Berbee, 1996). For analysis of fungal taxa at or under the level of species, the more variable ITS region is mostly used. This ITS region has been utilized to observed phylogenetic relationships among *Alternaria* species that generate host-specific toxins (Kusaba and Tsuge, 1995) or *Alternaria* sp. that are pathogenic on crucifers (Jasalavich et al., 1995), as well as among species of the *Pleosporales* (Morales et al., 1995; Khashnobish and Shearer, 1996).

The sequences coding for the nuclear 5.8S rRNA, 18SrRNA, ITS1 and ITS2 were amplified by the PCR and sequenced for one isolate each of *Alternaria brassicicola, A. brassicae, A. alternata, A. raphani* and *Pleospora herbarum.* The 5.8S rDNA sequences from all four *Alternaria* species were the same and only differed for one base pair from that of *P. herbarum.* The ITS sequences, especially ITS1, were highly variable in both base length and composition. The 18SrDNA sequences were very conserved, but sufficient variability was available to differentiate genera clearly. Phylogenetic analysis of the sequence data sets by both parsimony and maximum likelihood methods clearly segregate genera and species. All of the *Alternaria* species were firmly associated. *Pleospora* also appeared to be very firmly associated to *Alternaria* than to *Leptosphaeria* (Jasalavich et al., 1995).

Nucleotide sequences of the internal transcribed spacers (ITS) 1 and 2, rDNA and a 1068 bp section of the beta-tubulin gene sorted seven selected species of *Alternaria* into 5 taxa. *Stemphylium botryosum* formed a 6th closely related taxon. Isolates of *A. linicola* possessed the same ITS sequence to one group of *A. solani* isolates and two clusters of *A. linicola* isolates, to show minor variation which revealed from beta-tubulin gene data, which were as genetically identical to *A. solani* isolates they were to each other. The investigation was suggested that *A. linicola* falls between *A. solani* isolates. Similar studies suggested that *A. linicola* falls between *A. alternate* species. RAPD analysis of the complete genomic DNA from the *Alternaria* spp. accorded with the analysis of nucleotide sequence. An oligonucleotide primer (ALP) was choosing from the rDNA ITS1 region of *A. solani. A. linicola*. PCR with primers ALP and ITS4 (from a conserved region of the rDNA) amplified 536 bp fragments from *A. linicola* and *A. solani* isolates but not from other isolates of *Alternaria* sp. and also not from other fungi which may be connected with linseed. These primers amplified similar fragment which was confirmed by Southern hybridization, from DNA released from infected linseed leaf tissues and seed. These primers have the potential to be used also for the diagnosis of *A. solani* in host tissues material (McKay et al., 1999).

To elucidate the connection between *Alternaria, Stemphylium,* and *Ulocladium* species, mitochondrial small subunit (SSU) nuclear ITS and rDNA sequences from four *Ulocladium,* four *Stemphylium* spp. and 18 *Alternaria* were determined and compared. Phylogenetic analysis of the SSU rDNA and ITS sequences were carried out by the maximum parsimony and neighbor-joining methods which revealed that the *Stemphylium* spp. were phylogenetically distinct from the *Ulocladium* and *Alternaria* spp. Most *Ulocladium* spp. and the *Alternaria* spp. were placed together in a large *Ulocladium-Alternaria* clade. Within this large clade, the *Alternaria* spp. clustered into several distinct species-clades, most of which correlated with species-groups previously established based upon morphological characteristics. The *Ulocladium* spp. were placed into two species-clades, each of which also included *Alternaria* spp. *A. longissima* was distantly associated to the other *Alternaria* spp., as well as the *Stemphylium* and *Ulocladium* spp. Based upon nuclear 18S rDNA and ITS sequence identities, *A. longissima* was most firmly associated with *Leptosphaeria*. (Pryor and Gilbertson, 2000).

Molecular characteristics of five *Alternaria* isolates such as *A. alternata*, *A. tenuissima*, *A. arborescens*, and *A. infectoria* were determined using, polymerase chain reaction-restriction fragment length polymorphism (PCR-RFLP), sequence analysis of nuclear ITS rDNA, analysis of nuclear intergenic spacer (IGS) rDNA and random amplified polymorphism DNA (RAPD) analysis. Based on analysis of ITS sequence data, the infectoria species-group was phylogenetically distinct from the other species-groups. Based on cluster analysis of PCR-RFLP and RAPD data, three distinct clusters were evident; the infectoria cluster, the arborescens cluster, and a combined alternata/tenuissima cluster (Pryor and Michailides, 2002).

The fungal strain SW-3 having antimicrobial activity was isolated from soil of crucified plants in Pocheon, Kyungki-Do, Korea. Isolate SW-3 was diagnosed as *Alternaria brassicicola* by its morphological characteristics, and confirmed by the ITS regions of rDNA and 18S gene analysis. The fungus showed 99% similarities with *Alternaria brassicicola* in the 18S rDNA sequence analysis (Jung et al., 2002).

The ITS sequences within the rDNA region were targeted to described genetic variability between 8 *Alternaria* species which cause economically important diseases in crops. The rDNA regions of *Alternaria* species comprising of ITS regions and the rRNA genes were cloned and sequenced. Phylogenetic relationship based on the rDNA sequences and PCR-RFLP of amplified rDNA sequences clustered eight species of *Alternaria* into three major groups.

13.11 CHARACTERIZATION OF *NIT* MUTANT PHENOTYPES

Nit mutants (*nitM, nit1,* and *nit3)* were recovered from *A. brassicicola* isolates. The nit mutant recovery rates were unstable among isolates. The largest group of mutants proved to *nit1* (36.8%), followed by *nit3* (32.6%), *nitM* (15.3%), wild type (9.3%) and finally unknown (6.0%). All three *nit* mutant classes were recovered from 13 out of 32 isolates, and in rest of all, at least two different nit phenotypes were isolated. There were 122 nit mutants out of 144 sectors (Javeria et al., 2018). Similar results were reported by Hawthorne and Rees-George (1996) where they reported *nit1* and *nit3* mutants were readily recovered mutants out of which *nit1* mutants were of higher frequency than *nit3* and *nitM* from all 89 strains in *F. solani* on MMC containing 3.0% $KClO_3$.

13.12   VEGETATIVE COMPATIBILITY ASSAYS

Complementary heterokaryons evidently paired on the Petri-plates containing minimal media within the same VCG. Pairing response was treated as positive if aerial mycelium was produced on the border within two complementary mutants. When *nit* mutants from distinct parent isolates were discovered to be complementary, then they were placed in the same VCG. Based on complementary reactions the thirty-two isolates were grouped into five VCGs, which were capriciously designated as VCG 1-A, VCG 1-B, VCG 1-C, VCG 1-D and VCG 1-E (Javeria et al., 2018). In previous investigation of *F. oxysporum* indicated that the strains which were pathogenic were *usually* grouped in fewer VCGs as compared to strains which were non-pathogenic (Najafinia and Sharma, 2011). Sexual reproduction of *Fusarium* spp. was suggested as contributing to the degree of genotypic diversity. Segregation of genes during meiosis would generate genetically diverse genotypes in nature (Britz et al., 2005).

13.13   MICROSCOPIC OBSERVATION OF HETEROKARYONS

Stable heterokaryosis were distinctly visible under the compound microscope within compatible *nit* mutants after anastomosis. Same isolates were mostly produced self-hyphal fusions than exist in compatible or incompatible isolates (Figure 13.5). The fusions of hyphae were easily visible under compound microscopy where they concluded that there are 5 capriciously designated groups of VCG which may be further used for the diversity analysis of *Alternaria brassicicola* isolates.

13.14   DETECTION AND DIAGNOSIS

In India, oilseed and vegetable crucifers viz., mustard, cauliflower, cabbage, and broccoli are commonly grown which are affected by different fungal diseases amongst which leaf spot caused by *Alternaria brassicae* (Berk) Sacc and *Alternaria brassicicola* (Schwein) Wiltshire is a prevalent foliar disease widespread in all cruciferous crops (Meena et al., 2010) thus reducing the yield and quality. In vegetable brassica seeds, *A. brassicicola* (Maude and Humpherson-Jones, 1980; Humpherson-Jones, 1985; Maude et al., 1984; Kubota et al., 2006) is the dominant pathogen, whereas in oilseed rape *A. brassicae* (Kolte, 1985; Humpherson-Jones and Phelps, 1989) is more common. Both the pathogens are seed-borne in nature and dispersed by wind or rain during the growing season. For the production of healthy seed, prevention of dark leaf spot in seed crops is a pre-requisite. The diagnostic leaf spot symptoms are quite similar for both *A. brassicae* and *A. brassicicola* therefore; it becomes difficult to distinguish them in a mixed infection, where confusion occurs for detecting the species of the pathogen.

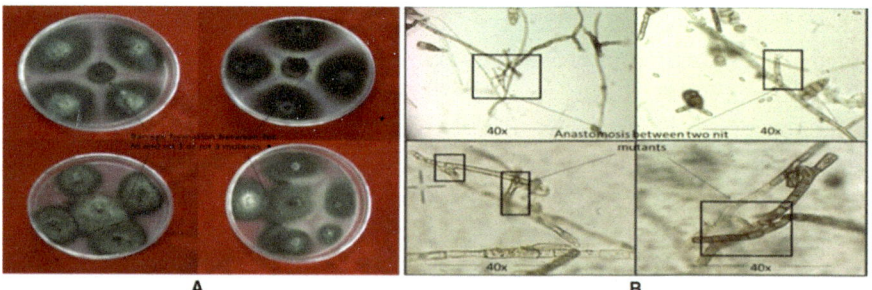

FIGURE 13.5 (A) Compatibility test of *nit* mutant; (B) showing anastomosis between *nit1* and *nit3* mutants with *NitM* mutants of known phenotype.

The conventional detection of pathogenic *A. brassicae* and *A. brassicicola* infection in crucifers, including cultural, morphological, and pathogenic characterization, are time consuming and also need artificially controlled conditions that are temperature and humidity dependent. So true symptoms or exact detection is not possible between these two leaf spot pathogens in crucifers. In case of seed quarantine, crucifer seeds face serious difficulty as diagnosis is diffident because so many saprophytic *Alternaria* species are found on them, thus further the pathogen identification gets complicated.

PCR based on amplification of ITS region of the rDNA is one of the most sensitive and rapid method of fungal pathogen detection. Specific primer from non-ribosomal peptide synthase (NRPS) gene of *A. brassicae* were developed for detecting *A. brassicae* infection in crucifer seeds by conventional and real-time PCR-based assay (Guillemette et al., 2004). PCR-based assay for *A. brassicicola* (Iacomi-Vasilescu et al., 2002) was demonstrated, and the epidemiological and disease management aspects of *A. brassicicola* in India have been worked out but pathogen detection techniques need more extensive research as Indian isolates of *A. brassicicola* was not detectable through the pre-existing detection protocols.

The objective of the present study was to develop specific primer from the conserved ITS rDNA regions for the PCR-based assay to detect *A. brassicicola* in infected leaf and seed samples of cruciferous crops including both vegetables and oilseed crop.

The PCR-based assay is specific, sensitive, and fast in detecting the organism at the species level. These methods have already been reported by various workers for detection of soil-borne pathogens (Taylor, 1993; Smith et al., 1996) including *Alternaria* sp. in different crops *viz*, carrot (Konstantinova et al., 2002; Pryor and Gilbertson, 2001), linseed (McKay et al., 1999) and cruciferous seed (Guillemette et al., 2004; Iacomi-Vasilescu et al., 2002; McKay et al., 1999) and food product (Zur et al., 1999). The conserved ITS regions of rDNA within a species have been used to design specific primers for identification of plant pathogens (Iacomi-Vasilescu et al., 2002). *Alternaria* species of which had been distinguished morphologically was distinctly separated from each other by variation of sequences of ITS1 and ITS2 (Kusaba and Tsuge, 1995). Similarly, ITS regions and 58S genes were used to distinguish 4 *Alternaria species*: *A. brassicicola, A. brassicae, A. alternata* and *A. raphani* isolated from crucifer crops (Jasalavich et al., 1995). Therefore, in this study, the ITS region was used to design specific primers for *A. brassicicola*. In earlier studies PCR- based detection of *A. brassicicola* was described which detected the pathogen in seeds with infection level of 10% or higher but the minimum template DNA level was not discussed (Iacomi-Vasilescu et al., 2002).

In the present study, a primer set of highly specific for *A. brassicicola* was designed from the less conserved ITS region viz., 3′end position of 100–110 bp and at 5′end position of 520 bp, which showed non-specificity with the other pathogenic *Alternaria* sp. including *A. brassicae* and *A. alternata* and similar results were observed for other seed contaminants like *Fusarium oxysporum* and *Aspergillus niger*. The primer set prove to be highly specific for detecting *A. brassicicola* in the inoculated seed samples with the infection level of 85–100%. The fungal DNA was isolated from 24–48 hrs incubated seed samples of four cruciferous crops. This approach of obtaining DNA from the seed macerates was different from the method developed in which fungal biomass was enhanced on nutrient media (Guillemette et al., 2004). DNA of 50 ng concentration was used in the PCR-reaction where amplicons of ~600 bp was obtained on an agarose gel in infected seed samples. For further clarification, the primer set was tested with a pure culture of *A.*

brassicicola and *A. brassicae* as positive and negative control in which specificity was confirmed with *A. brassicicola*. The sensitivity of the designed primer set was tested for detecting the lowest concentration of pathogen-targeted DNA detection, which showed sensitivity of up to 0.1 ng or 100 pg.

13.15   COMPETITIVE LATERAL-FLOW ASSAY

A competitive lateral flow assay of 6 min for in-field diagnosis of A. brassicae from the brassica crop canopy has been introduced by Wakeham et al 2016. This method is a combination weather-driven infection, in-field risk assessment by using an automated seven days multi vial air cyclone chamber. This model has the potential for infield detection for disease management system.

13.16   CONCLUSION

Vegetable are the most important sources of many vitamins, fibers, phytochemicals, and minerals. Brassicaceae family crops are economically important flowering plants in the vegetable plant kingdom. In cabbage and cauliflower, fungal diseases are responsible for economic losses in terms of production. Black spot or *Alternaria* leaf spot is one of the important fungal diseases of cruciferous family crops caused by *A. brassica* and *A. brassicicola*. It is widely occurring in the Brassicaceae growing like US, Europe, and Asian countries. Detection and diagnosis is an important component in plant pathology which helps to understand the actual causal agent of the disease. The precise detection of pathogens can achieve by the integration of different tools likes morpho, molecular, and biochemical characterization based on the growth characteristics, morphology, and disease reaction on the host. Recent advances in the application of DNA-based markers like ITS region, Randomly Amplified Polymorphic DNA (RAPD), ISSR, SSR, etc., are used widely to study the genetic variability among the pathogens. Biochemical analysis of CWDEs can help in understating the pathogenicity and virulence of the phytopathogens. Vegetative compatibility and lateral-flow assay are keys in pathogen detection and diagnosis. The combination of these tools can be useful for disease diagnosis and also an important component in plant disease management.

KEYWORDS

- **Brassicaceae**
- **cell wall degrading enzymes**
- **cellobiohydrolases**
- **chromatographic lateral flow device**
- **molecular markers**
- **vegetative compatibility grouping**

REFERENCES

Abbasi, P. A., Miller, S. A., Meulia, T., Hoitink, H. A. J., & Kim, J. M., (1999). Precise detection and tracing of *Trichoderma hamatum* in compost amended potting mixes by using molecular markers. *Appl. Environ. Microbiol., 65*(12), 5421–5426.

Abul-Fazal, M., Khan, M. I., & Saxena, S. K., (1994). The incidence of *Alternaria species* in different cultivars of cabbage and cauliflower seeds. *Indian Phytopathol., 47,* 419–421.

Alexander, M., (1961). Microbiology of cellulose. *In: Intro. Soil Microbio.* (2nd edn.). John Wiley and Son, Inc. New York and London.

Alkorta, I., Garbisu, C., Liama, M. J., & Serra, J. L., (1998). Industrial applications of pectic enzymes. *Process Biochem., 33,* 21–28.

Anand, T., Bhaskaran, R., Raguchander, T., Karthikeyan, G., Rajesh, M., & Senthilraja, G., (2008). Production of cell wall degrading enzymes and toxins by *Colletotrichum capsici* and *Alternaria alternata* causing fruit rot of chilies. *Plant Prot. Res., 48*(4), 437–451.

Ansari, N. A., Khan, M. W., & Muheet, A., (1989a). Effect of some factors on growth and sporulation of *Alternaria brassicae* causing *Alternaria* blight of rapeseed and mustard. *Acta Bot. In., 17,* 49–53.

Ansari, N. A., Khan, M. W., & Muheet, A., (1989b). Survival and perpetuation of *Alternaria brassicae* causing Alternaria blight of oilseed crucifers. *Mycopathol., 105,* 67–70.

Awasthi, R. P., & Kolte, S. J., (1989). Variability in *Alternaria brassicae* affecting rapeseed and mustard. *Indian Phytopathol., 42,* 275.

Bandyopadhya, D. C., Saha, G. N., & Mukherjee, D., (1974). Note on variations in quantitative composition of seeds of "B-9" variety of yellow sarson caused by *Alternaria* blight. *Indian J. Agr. Sci., 44,* 406–407.

Benali, S., Mohamed, B., Eddine, H. J., & Neema, C., (2011). Advances of molecular markers application in plant pathology research. *Eur. J. Sci. Res., 50,* 110–123.

Berbee, M. L., (1996). Loculoascomycete origins and evolution of filamentous ascomycete morphology based on *18S rRNA* gene sequence data. *Mol. Biol. Evol., 13,* 462–470.

Berto, P., Belingheri, L., & Dehorter, B., (1997). Production and purification of a novel extracellular lipase from *Alternaria brassicicola*. *Biotechnol. Lett., 19(6),* 533–536.

Berto, P., Commenil, P., Belingheri, L., & Dehorter, B., (1999). Occurrence of a lipase in spores of *Alternaria brassicicola* with a crucial role in the infection of cauliflower leaves. *FEMS Microbiol. Lett., 180,* 183–189.

Bhaskaran, R., & Kandaswamy, T. K., (1978). Production of a toxic metabolite by *Alternaria helianthi in vitro* and *in vivo*. *Madras Agr. J., 65*, 801–804.

Bielikova, L., Lamda, Z., Osborne, L. S., & Curn, V., (2002). Characterization and identification of entomopathogenic and mycoparasitic fungi using RAPD-PCR technique. *Plant Protect. Sci., 38*, 1–12.

Bock, C. H., Thrall, P. H., & Burdon, J. J., (2005). Genetic structure of populations of *Alternaria brassicicola* suggests the occurrence of sexual recombination. *Mycol. Res., 109*, 227–236.

Bock, C. H., Thrall, P. H., Brubaker, C. L., & Burdon, J. J., (2002). Detection of genetic variation in *Alternaria brassicicola* using AFLP fingerprinting. *Mycol. Res., 106*, 428–434.

Britz, H., Coutinho, T. A., Wingfield, B. D., Marasas, W. F. O., Wingfield, M. J. (2005). Diversity and differentiation in two populations of Gibberella circinata in South Africa. Plant Pathol., 54, 46–52.

Brown, J. K. M., (1996). The choice of molecular marker methods for population genetic studies of plant pathogens. *New Phytol., 133*, 183–195.

Burkert, T. F., Maugeri, F., & Rodrigues, M. I., (2004). Optimization of extracellular lipases production by *Geotrichum sp.* using factorial design. *Bioresour. Technol., 910*, 77–84.

Butler, E. J., (1918). *Fungi and Diseases in Plants*. Thacker Spink & *Co.*, Calcutta, India.

Chahal, A. S., & Kang, M. S., (1979). Different levels of *Alternaria* blight in relation to grain yield of brown sarson. *Indian J. Mycol.Pl. Path., 9*, 260, 261.

Changsri, W., & Weber, G. F., (1963). Three Alternaria species pathogenic on certain cultivated crucifers. *Phytopathol., 53*, 643–648.

Changsri, W., (1961). Studies of *Alternaria* spp. pathogenic on Cruciferae. *Diss. Abstr., 21*, 1698.

Chattopadhyay, C., (1999). Yield loss attributable to Alternaria blight of sunflower (*Helianthus annuus* L.) in India and some potentially effective control measures. *Int. J. Pest Manage., 45*, 15–21.

Chaudhri, A., & Suneetha, V., (2012). Microbially Derived pectinases: A review. *IOSR J. Pharm. and Biol. Sci., 2*(2), 1–5.

Chillali, M., Ladder-Ighili, H., Guillaumin, J. J., Mohammed, C., Lung, E. B., & Botton, B., (1998). Variation in the ITS and IGS regions of ribosomal DNA among the biological species of *European armillaria*. *Mycol. Res., 102*, 533–540.

Cho, H., Bancerz, R., Ginalska, G., Leonowicz, A., Cho, N. S., & Ohga, S., (2007). Culture conditions of psychrotrophic fungus, *Penicillium chrysogenum* and its lipase characteristics. *J. Fac. Agr. Kyushu U., 52*, 281–286.

Cho, Y., Davis, J. W., Kim, K. H., Wang, J., Sun, Q. H., & Cramer, R. A., (2006). A high throughput targeted gene disruption method for *Alternaria brassicicola* functional genomics using linear minimal element (LME) constructs. *Mol. Plant-Microbe Interact., 19*, 7–15.

Chohan, J. S., (1978). Diseases of oilseeds crops, future plans and strategy for control under smallholdings. *Indian Phytopathol., 31*, 12–22.

Chupp, C., & Sherf, A. F., (1960). Crucifer diseases. In: *Vegetable Diseases and Their Control* (Vol. 8, pp. 237–288). Ronald Press Co., New York.

Cole, L., Dewey, F. M., & Hawes, C. R., (1998). Immuno-cytochemical studies of the infection mechanisms of *Botrytis fabae* II. Host cell wall breakdown. *New Phytol., 139*, 611–622.

Cooke, D. E. L., Forster, J. W., Jenkins, P. D., Gareth, J. D., & Lewis, D. M., (1998). Analysis of intraspecific and interspecific variation in the genus *Alternaria* by the use of RAPD-PCR. *Ann. Appl. Bio., 132*, 197–209.

Czyzewska, S., (1971). The pathogenicity of *Alternaria* spp. isolated from *Crambe abyssinica*. *Acta Mycologia., 7*, 171–240.

Dawar, V., & Jain, V., (2010). Cell wall degrading enzymes and permeability changes in sunflower *(Helianthus annuus)* infected with *Alternaria helianthi*. *Int. J. Agric. Environ. Biotechnol., 3*(3), 321–325.

Deep, S., & Sharma, P., (2012). Hostage as predisposing factor in black leaf spot of cauliflower caused by *Alternaria brassicae* and *Alternaria brassicicola*. *Indian Phytopathol., 65*(1), 71–75.

Dehpour, A. A., Alavi, S. V., & Majad, A., (2007). Light and scanning electron microscopy studies on the penetration and infection processes of *Alternaria alternata*, causing brown spot-on *Minneola tangelo* in the West Mazandaran-Iran. *World Appl. Sci. J., 2*(1), 68–72.

Dey, P. K., (1948). *Plant Pathology* (Vol. 1946, No. 47, pp. 39–42). Adm. Rep. Agric. Dept. U. P.

Dixon, G. R., (1981). *Vegetable Diseases* (p. 409). Avi Publishing Co., Inc., Westport, CT.

Ellis, M. B., (1971). *Dematiaceous Hyphomycetes* (pp. 464–497). Commonwealth Mycological Institute, Kew, Surrey, England. No-608.

Ellis, M. B., (2001). *Dematiaceous Hyphomycetes* (p. 608). CABI Publishing, Wallingford, Oxford Shire, U.K.

Fariha, H., Aamer, S., & Adul, H., (2005). Industrial applications of microbial lipases. *Enzyme Microb. Technol., 39*, 235–251.

Fernandez-Acero, F. J., Carbu, M., Garrido, C., Vallejo, I., & Cantoral, J. M., (2007). Proteomic advances in phytopathogenic fungi. *Curr. Proteomics, 4*, 79–88.

Fernando, T. H. P. S., Jayasinghe, C. K., & Wijesundera, R. L. C., (2001). Cell wall degrading enzyme secretion by *Colletotrichum acutatum*, the causative fungus of secondary leaf fall of *Hevea brasiliensis* (rubber). *Mycol. Res., 105*, 195–201.

Garg, S., Dhawan, K., Chawla, H. K. L., & Nainawatee, H. S., (1999). *Alternaria brassicae* induced changes in tile activity of cell wall degrading enzymes in leaves of *Brassica juncea*. *Biol. Plantarum, 30(5)*, 387–392.

Garrido, C., Cantoral, J. M., Carbu, M., Gonzalez-Rodriguez, E. V., & Fernandez-Acero, F. J., (2010). New proteomic approaches to plant pathogenic fungi. *Curr. Proteomics, 7*, 306–315.

Gautam, S. P., Bundela, P. S., Pandey, A. K., Jamaluddin, Awasthi, M. K., & Sarsaiya, S., (2012). Diversity of cellulolytic microbes and the biodegradation of municipal solid waste by a potential strain. *Int. J. Microbiol.*, 1–12.

George, E., Tamerler, C., Martinez, C., Martinez, M. J., & Keshavarz, T., (1999). Influence of growth composition on the lipolytic enzyme activity of *Ophiostoma piliferm. J. Chem. Technol. Biot., 74*, 137–140.

Gherbawy, Y. A. M. H., (2005). Genetic variation among isolates of *Alternaria* spp. from selected Egyptian crops. *Arch. Phytopathol., 38*, 77–89.

Ghosh, P. K., Saxena, R. K., Gupta, R., Yadav, R. P., & Davidson, W. S., (1996). Microbial lipases: Production and application. *Sci. Progress, 79*, 119–157.

Ghosh, S., Mazumder, M., Mondal, B., Mukherjee, A., De, A., Bose, R., Das, S., Bhattacharyya, S., & Basu, D., (2019). Morphological and SSR marker-based genetic diversity analysis of Indian mustard (Brassica juncea L.) differing in Alternaria brassicicola tolerance. Euphytica 215, 206.

Gorshkov, A. K., (1976). *Infectious Damping-off of Cabbage Seedlings Under Glass* (Vol. 117, pp. 25–29). Trudy Vses. S.-kh. Inst. Zaoch. Obrazo.

Gouveia, M. M. C., Ribeira, A., Varzea, V. M. P., & Rodrigue, C. J., (2005). Genetic diversity of *Hemileia vastatrix* based on RAPD markers. *Mycologia., 97*, 396–404.

Goyal, P., Chahar, M., Mathur, A. P., Kumar, A., & Chattopadhyay, C., (2011). Morphological and cultural variation in different oilseed *Brassica* isolates of *Alternaria brassicae* from different geographical regions of India. *Indian J. Agr. Sci., 81*, 1052–1058.

Grassin, C., & Fauquembergue, P., (1996). Wine and fruit juices. In: Godfrey, T., & West, S., (eds.), *Industrial Enzymology* (2nd edn.). Macmillan Press Ltd.

Guillemette, T., Iacomi-Vasilescu, B., & Simoneau, P. (2004). Conventional and real-time PCR-based assay for detecting pathogenic *Alternaria brassicae* in cruciferous seed. *Plant Dis., 88* (5), 490–496.

Gummadi, S. N., & Panda, T., (2003). Purification and biochemical properties of microbial pectinases: A review. *Process Biochem., 38*, 987–996.

Gupta, R., Awasthi, R. P., & Kolte, S. J., (2003). Influence of sowing dates and weather factors on development of *Alternaria* blight on rapeseed-mustard. *Indian Phytopathol., 56*(4), 398–402.

Hawthorne, B. T., & Rees-George, J., (1996). Use of nitrate non-utilizing mutants to study vegetative incompatibility in *Fusarium solani (Nectria haematococca)*, especially members of mating populations I, V and VI. *Mycol. Res., 100*(9), 1075–1081.

Himmel, M. E., Baker, J. O., & Overend, R. P., (1994). Approaches to cellulose purification. In: *Enzymatic Conversion of Biomass for Fuel Production*. ACS symposium series 566.

Hong, C. X., Fitt, B. D. L., & Welham, S. J., (1996). Effects of wetness period and temperature on development of dark pod spot (*Alternaria brassicae*) on oilseed rape (*Brassica napus*). *Plant Pathol., 45*, 1077–1089.

Hong, S. G., Maccaroni, M., Figuli, P. J., Pryor, B. M., & Belisario, A., (2006). Polyphasic classification of *Alternaria* isolated from hazelnut and walnut fruit in Europe. *Mycol. Res., 110*, 1290–1300.

Hubballi, M., Sornakili, A., Nakkeeran, S., Anand, T., & Raguchander, T., (2011). Virulence of *Alternaria alternata* infecting noni associated with production of cell wall degrading enzymes. *J. Plant Prot. Res., 51*(1), 87–92.

Humpherson-Jones, F. M., & Phelps, K., (1989). Climatic factors influencing spore production in *Alternaria brassicae* and *Alternaria brassicicola*. *Ann. Appl. Biol., 114*, 449–459.

Humpherson-Jones, F. M., (1985). The incidence of *Alternaria spp.* and *Leptosphaeria maculans* in commercial brassica seed in the United Kingdom. *Plant Pathol., 34*, 385–390.

Iacomi-Vasilescu, B., Blanchard, D., Guenard, M., Molinero-Demilly, V., Laurent, E., & Simoneau, P., (2002). Development of a PCR based diagnostic assay for detecting pathogenic Alternaria species in cruciferous seeds. *Seed Sci. Technol., 30*(1), 87–96.

Idnurm, A., & Howlett, B. J., (2001). Pathogenicity genes of phytopathogenic fungi. *Mol. Plant Path., 2*, 241–255.

Iftikhar, T., Niaz, M., Anwer, M., Abbas, S. Q., Saleem, M., & Jabeen, R., (2011). Ecological screening of lipolytic cultures and process optimization for extracellular lipase production from fungal hyperproducer. *Pak. J. Bot., 43*(2), 1343–1349.

Indian Horticultural Database (2013). National Horticulture Board, Ministry of Agriculture, Government of India.

Iram, S., & Ahmad, I., (2005). Analysis of variation in *Alternaria alternata* by pathogenicity and RAPD study. *Polish J. Microbiol., 54*, 13–19.

Isshiki, A., Akimitsu, K., Yamamoto, M., & Yamamoto, H., (2001). Endo-polygalacturonase is essential for citrus black rot caused by *Alternaria citri* but not brown spot caused by *Alternaria alternata*. *Mol Plant-Microbe Interact., 14*(6), 749–757.

Iwai, M., & Tsujisaka, Y., (1984). In: Borgström, B., & Brockman, H. L., (eds.), *Lipases* (pp. 443–466). Amsterdam: Elsevier Sci.

Jahangeer, S., Khan, N., Jahangeer, S., Sohail, M., Shahzad, S., Ahmad, A., & Khan, S. A., (2005). Screening and characterization of fungal cellulases isolated from the native environmental source. *Pak. J. Bot., 37*(3), 739–748.

Jain, V., & Dhawan, K., (2008). Major cell wall degrading enzymes in two contrasting cultivars of *Brassica juncea* infected with *Alternaria brassicae*. *Cruciferae Newsletter, 27*, 20–21.

Jasalavich, C. A., Morales, V. M., Pelcher, L. E., & SeGuin-Swartz, G., (1995). Comparison of nuclear ribosomal DNA sequences from *Alternaria species* pathogenic to crucifers. *Mycol. Res., 99*, 604–614.

Javeria, S., Deep, S., Prasad, L., & Sharma, P., (2018). Vegetative compatibility grouping of *Alternaria brassicicola* causing black leaf spot in cauliflower. *Indian Phytopathol., 71*, 43–47.

Jayani, R. S., Saxena, S., & Gupta, R., (2005). Microbial pectinolytic enzymes: A review. *Process Biochem., 40*, 2931–2944.

Jha, D. K., & Gupta, D. P., (1988). Production of pectinolytic enzymes by *Alternaria triticina*. *Indian Phytopathol., 41*, 652.

Jung, D. S., Na, Y. J., & Ryu, K. H., (2002). Phylogenic analysis of *Alternaria brassicicola* producing bioactive metabolites. *J. Microbiol., 40*, 289–294.

Kadian, A. K., & Saharan, G. S., (1983). Symptomatology, host range, and assessment of yield losses due to *Alternaria brassicae* infection in rapeseed and mustard. *Indian J. of Mycol. Pl. Path., 13*, 319–323.

Kale, S. M., Pardeshi, V. C., Gurjar, G. S., Gupta, V. S., Gohokar, R. T., Ghorpade, P. B. & Kadoo, N., (2012). Inter simple sequence repeat markers reveal high genetic diversity among *Alternaria alternata* isolates of Indian origin. *J. Mycol. Pl. Pathol., 42*, 194–200.

Kamimura, E. S., Medieta, O., Rodrigues, M. I., & Maugeri, F., (2001). Studies on lipase affinity adsorption using response-surface analysis. *Biotechnol. Appl. Biochem., 33*, 153–159.

Kaur, S., Singh, G., & Banga, S. S., (2007). Documenting variation in *Alternaria brassicae* isolates based on conidial morphology, fungicidal sensitivity and molecular profile. In: *Proceeding of the 12th International Rapeseed Congress* (Vol. 4, pp. 87–89). Wuhan, China.

Kaushik, C. D., Saharan, G. S., & Kaushik, J. C., (1984). Magnitude of losses in yield and management of *Alternaria* blight in rapeseed-mustard. *Indian Phytopathol., 37*, 398.

Kear, R. W., Williams, D. J., & Stevens, C. C., (1977). The effect of iprodione on the fungal deterioration of stored white cabbage. *Proc. 1977. British Crop Prot. Confr., Pests and Diseases* (pp. 189–195). Brighton.

Khashnobish, A., & Shearer, C. A., (1996). Phylogenetic relationships in some *Leptosphaeria* and *Phaeosphaeria* species. *Mycol. Res., 100*, 1355–1363.

Khatun, F., Hossain, M. A., Hossain, M. M., & Dey, D. K., (1996). Effect of culture media, temperature, pH and nitrogen sources on growth and sporulation of *Alternaria brassicciola*. Bangald. *J. Plant Pathol., 12*(1, 2), 29–32.

Kolte, S. J., (1985). *Diseases of Annual Edible Oilseed Crops: Rapeseed-Mustard and Sesame Diseases* (Vol. 2, p. 135). Boca Raton (FL): CRC Press Inc.

Kolte, S. J., Bardoloi, D. K., & Awasthi, R. P., (1991). The search of resistance to major diseases of rapeseed-mustard in India. *Proc. GCIRC 8ᵗʰ Int. Rapeseed Congr.*, (Vol. 1, pp. 60, 219–225). Saskatoon, Canada: (Abstract); Proc.

Konstantinova, P., Bonants, P. J. M., Van GentPelzer, M. P. E., Van der Zouwen, P., & Van den Bulk, R., (2002). Development of specific primers for detection and identification of Alternaria spp. in carrot material by PCR and comparison with blotter and plating assays. *Mycol. Res., 106*, 23–33.

Kubota, M., Abiko, K., & Yanagisawa, Y., (2006). Frequency of *Alternaria brassicicola* in commercial cabbage seeds in Japan. *J. Gen. Plant Pathol., 72*, 197–204.

Kumar, P. R., (1986). Rapeseed-mustard research in India. In: *Proceedings of the SAARC Member Countries Counterpart Scientists Meeting for Multi-Location Trial on Rapeseed-Mustard.* Kathmandu-Nepal.

Kumar, R., & Gupta, P. P., (1994). Survival of *Alternaria brassicae, A. brassicicola* and *A. alternata* in the seed of mustard (*B. juncea*) at different temperatures and relative humidities. *Ann. Biol., 10*, 55–58.

Kumar, S., Sangwan, M. S., Mehta, N., & Kumar, R., (2003). Pathogenic diversity in isolates of *Alternaria brassicae* infecting rapeseed and mustard. *J. Mycol. Pl. Path., 3*, 59–64.

Kumar, V., Haldar, S. P., Singh, K. P., Singh, K., & Singh, C., (2008). Cultural, morphological, pathogenic and molecular variability amongst tomato isolates of *Alternaria solani* in India. *World J. Microb. Biot., 24*, 1003–1009.

Kusaba, M., & Tsuge, T., (1995). Phylogeny of *Alternaria* fungi known to produce host-specific toxins on the basis of variation in internal transcribed spacers of ribosomal DNA. *Curr. Genet., 28*, 491–498.

Lamboy, W. F., (1994). Computing genetic similarity coefficients from RAPD data: The effects of PCR artifacts. *PCR Meth. Appl., 4*, 31–37.

Lecomte, P., Peros, J. P., Blancard, D., Bastien, N., & Delye, C., (2000). PCR-assays that identify the grapevine dieback fungus *Eutypa lata. Appl. Environ. Microbiol., 66*, 4475–4480.

Lederberg, J., (1992). Cellulases. In: *Encyclopaedia of Microbiology* (Vol. 1, A–C). Academic Press, Inc.

Leiminger, J., Bahnweg, G., & Hausladen, H., (2010). *Population Genetics-Consequences on Early Blight Disease* (Vol. 14, pp. 171–178). Lelystad, The Netherlands. PPO-Special Report.

Li, K. N., Rouse, D. I., Eyestone, E. J., & German, T. L., (1999). The generation of specific DNA primers using random amplified polymorphic DNA and its application to *Verticillium dahliae. Mycol. Res., 103*, 1361–1368.

Lynd, L. R., Weimer, P. J., Van, Z. W. H., & Pretorius, I. S., (2002). Microbial cellulose utilization: Fundamentals and biotechnology. *Microbiol. Mol. Biol. Rev., 66*, 506–577.

Mackinnon, S. L., Keifer, P., & Ayer, W. A., (1999). Components from the phytotoxic extract of *Alternaria brassicicola*, a black spot pathogen of canola. *Phytochemistry, 51*, 215–221.

Macris, B. J., (1984). Production and characterization of cellulose and β-glucosidase from a mutant of *Alternaria alternata. Appl. Environ. Microbiol., 47*(3), 560–565.

Marimuthu, T., Bhaskaran, R., Shanmugam, N., & Purushothaman, D., (1974). *In vitro* production of cell wall splitting enzymes by *Alternaria sesami*. Labd. *J. Sci. Technol., 12B*, 26–28.

Maude, R. B., & Humpherson-Jones, F. M., (1977). Dark leaf spot of *Brassicas* (*Alternaria brassicicola*). In: *27ᵗʰ Annual Report for 1976* (pp. 95, 96). National Vegetable Research Station, Wellesbourne, Warwick, U. K.

Maude, R. B., & Humpherson-Jones, F. M., (1980). Studies on the seed-borne phases of dark leaf spot (*Alternaria brassicicola*) and grey leaf spot (*Alternaria brassicae*) of brassicas. *Ann. Appl. Biol., 95*, 311–319.

Maude, R. B., Humpherson-Jones, F. M., & Shuring, C. G., (1984). Treatments to control *Phoma* and *Alternaria* infections of brassica seeds. *Plant Pathol., 33*, 525–535.

Mckay, G. J., Brown, E. A., Bjourson, A. J., & Mercer, P. C., (1999). Molecular characterization of *Alternaria lilonicola* and its detection in linseed. *Eur. J. Plant Pathol., 105*, 157–166.

Meena, P. D., Awasthi, R. P., Chattopadhyay, C., Kolte, S. J., & Kumar, A., (2010). Alternaria blight: A chronic disease in rapeseed-mustard. *J. Oilseeds Res., 1*, 1–11.

Meena, P. D., Chattopadhyay, C., Kumar, V. R., Meena, R. L., & Rana, U. S., (2005). Spore behavior in atmosphere and trends in variability of *Alternaria brassicae* population in India. *J. Mycol. Plant Pathol., 35*, 511.

Meena, P. D., Rani, A., Meena, R., Sharma, P., Gupta, R., & Chowdappa, P., (2012). Aggressiveness, diversity and distribution of *Alternaria brassicae* isolates infecting oilseed Brassica in India. *Afri. J. of Microbiol. Res., 6*, 5249–5258.

Mehta, N., Sangwan, M. S., Srivastava, M. P., & Kumar, R., (2003). Survival of *Alternaria brassicae* causing Alternaria blight in rapeseed mustard. *Indian J. Mycol. Pl. Path., 32*, 64–67.

Mercado, V. D., Renard, M. E., Duveiller, E., & Maraite, H., (2006). Identification of *Alternaria* spp. on wheat by pathogenicity assays and sequencing. *Plant Pathol., 55*, 485–493.

Michelmore, R. W., & Hulbert, S. H., (1987). Molecular markers for genetic analysis of phytopathogenic fungi. *Ann. Rev. of Phytopath., 25*, 383–404.

Milbraith, D. G., (1922). *Alternaria* from California. *Bot. Gaz., 74*, 320–324.

Milgroom, M. G., & Fry, W. E., (1997). Contribution of population genetics to plant disease epidemiology and management. *Adv. Bot. Res., 24*, 1–30.

Misaghi, I. J., Grogan, R. G., Duniway, J. M., & Kimble, K. A., (1978). Influence of environmental and culture media on spore morphology of *Alternaria alternata*. *Phytopathol., 68*, 29–34.

Morales, V. M., Jasalavich, C. A., Pelcher, L. E., Petrie, G. A., & Taylor, J. L., (1995). Phylogenetic relationship among several *Leptosphaeria* species based on their ribosomal DNA sequences. *Mycol. Res., 99*, 593–603.

Morris, P. F., Connolly, M. S., & St Clair, D. A., (2000). Genetic diversity of *Alternaria alternata* isolated from tomato in California assessed using RAPDs. *Myco. Res., 104*, 286–292.

Najafinia, M., & Sharma, P., (2011). Characterization of Indian isolates of Fusarium oxysporum f. sp. cucumerinum using vegetative compatibility groups and RAPD assay. Indian Phytopathol., 64, 12–18.

Nielson, O., (1933). Experiments in the control of the siliqua fungus. *Tidsskr. Planteavl., 39*, 437–452.

Nishimura, S., & Kohmato, K., (1983). Host-specific toxins and chemical structures from *Alternaria* species. *Annu. Rev. Phytopathol., 21*, 87–116.

Nozaki, K., Miyairi, K., Hizumi, S., Fukui, Y., & Okuno, T., (1997). Novel exo-polygalacturonases produced by *Alternaria mali*. *Biosci. Biotech. Bioch., 61*, 75–80.

Odds, F. C., Gow, N. A., & Brown, A. J. P., (2001). Fungal virulence studies come of age. *Genome Biol., 2*, 1009.1–1009.4.

Oliver, E. J., Thrall, P. H., Burdon, J. J., & Ash, J. E., (2001). Vertical transmission of disease in the *Cakile maritima-Alternaria brassicicola* interaction. *Aust. J. Bot., 49*, 561–569.

Panuthai, T., Sihanonth, P., Piapukiew, J., Sangvanich, P., & Karnchanatat, A., (2011). Screening and production of lipase from endophytic fungi. In: *Proceedings of the 12th Graduate Research Conference*. Khon Kaen University.

Park, M. S., Romanoski, C. E., & Pryor, B. M., (2008). A re-examination of the phylogenetic relationship between the causal agents of carrot black rot, *Alternaria radicina* and *A. carotiincultae. Mycologia, 100*, 511–527.

Patni, C. S., Kolte, S. J., & Awasthi, R. P., (2005). Cultural variability of *Alternaria brassicae*, causing Alternaria blight of mustard. *Annu. Rev. Plant Physiol., 19*, 231–242.

Pattanamahakul, P., & Strange, R. N., (1999). Identification and toxicity of *Alternaria brassicicola*, the causal agent of dark leaf spot disease of *Brassica species* grown in Thailand. *Plant Pathol., 48*, 749–755.

Philippidis, G. P., (1994). Cellulase production technology. In: Himmel, M. E., et al., (eds.), *Enzymatic Conversion of Biomass for Fuel Production* (p. 566). ACS symposium series.

Prasad, M. S., Sujatha, M., & Rao, S. C., (2009). Analysis of cultural and genetic diversity in *Alternaria helianthi* and determination of pathogenic variability using wild *A. helianthi* species. *J. Phytopathol., 157*, 609–617.

Prasad, R. D., Sharma, T. R., & Prameela, D. T., (2007). Molecular variability and detection of *Fusarium* species by PCR based RAPD, ISSR and ITS-RFLP analysis. *J. Mycol. Plant Path., 37*, 311–318.

Pryor, B. M., & Gilbertson, R. L., (2000). Molecular phylogenetic relationships amongst *Alternaria* species and related fungi based upon analysis of nuclear ITS and mt SSU rDNA sequences. *Mycol. Res., 104*, 1312–1321.

Pryor, B. M., & Michailides, T. J., (2002). Morphological, pathogenic and molecular characterization of *Alternaria* isolates associated with *Alternaria* late blight of pistachio. *Phytopathol., 92*, 406–416.

Pryor, B. M., Davis, R. M., & Gilbertson, R. L., (2002). Relationship and taxonomic status of *Alternaria radicina, A. carotiinclatae*, and *A. petroselini* based on morphological, biochemical and molecular characteristics. *Mycologia, 94*, 49–61.

Pusz, W., (2009). Morpho-physiological and molecular analyses of *Alternaria alternata* isolated from seeds of *Amaranthus. Phytopathol., 54*, 5–14.

Ram, R. S., & Chauhan, V. B., (1998). Assessment of yield loss due to *Alternaria* leaf spot in various cultivars of mustard and rapeseed. *J. Mycopathol. Res., 36*(2), 109–111.

Ramegowda, G., & Naik, M. K., (2008). Morphological, cultural and physiological diversity in isolates of *Alternaria* spp. Infecting *Bt*-cotton. *J. Mycol. Plant Path., 38*, 267–271.

Ramos, A. M., Gally, M., Garcia, M. C., & Levin, L., (2010). Pectinolytic enzyme production by *Colletotrichum truncatum*, causal agent of soybean anthracnose. *Revista Ibero Americana De Micología, 27*, 186–190.

Ramsey, G. B., & Smith, M. A., (1961). *Market Diseases of Cabbage, Cauliflower, Turnips, Cucumbers, Melons, and Related Crops* (pp. 184–249). U.S. Department of Agriculture, Agriculture Handbook.

Rangel, J. F., (1945). Two Alternaria diseases of cruciferous plants. *Phytopathol., 35*, 1002–1007.

Rathod, S. R., & Chavan, A. M., (2010). Extra-cellular hydrolytic enzymes action of *Alternaria* species under the influence of different nutritional sources. *J. Ecobiotechnol., 2*(6), 57–62.

Reis, A., & Boiteux, L. S., (2010). *Alternaria* species infecting *Brassicaceae* in the Brazilian neotropics: Geographical distribution, host range and specificity. *Plant Pathol. J., 92*, 661–668.

Rimmer, S. R., & Buchwaldt, H., (1995). Diseases. In: Kimber, D., & McGregor, D. I., (eds.), *Brassica Oilseeds-Production and Utilization* (pp. 111–140). CAB International, Alling ford, UK.

Roberts, R. G., Reymond, S. T., & Andersen, B., (2000). RAPD fragment pattern analysis and morphological segregation of small-spored *Alternaria* species and species groups. *Mycol. Res., 104*, 151–160.

Rogers, L. M., Kim, Y. K., Guo, W., Gonzalez-Candelas, L., & Li, D., (2000). Requirement for either a host or pectin-induced pectate lyase for infection of *Pisum sativum* by *Nectria hematococca*. *Proceedings of the National Academy of Sciences of the United States of America, 97*, 9813–9818.

Rotem, J., (1994). *The Genus Alternaria: Biology, Epidemiology and Pathogenicity*. American Phytopathological Society Press, St Paul, MN.

Saccardo, P. A., (1886). Hyphomyceteae. In: *Sylloge Fungorum Omninum Hucusque Cognitorum* (Vol. 4, p. 807). Pavia, Italy.

Saharan, G. S., & Kadian, A. K., (1983). Physiologic specialization in *Alternaria brassicae*. *Cruciferous News Letter, 8*, 32, 33.

Saharan, G. S., (1992a). Management of rapeseed and mustard diseases. In: Kumar, D., & Rai, M., (eds.), *Advances in Oilseeds Research* (Vol. 1, No. 7, pp. 152–188). Sci Pub., Jodhpur, India.

Saharan, G. S., (1992b). Disease resistance. In: Lagana, K. S., Banga, S. S., & Banga, S. K., (eds.), *Breeding Oilseeds Brassicas* (Vol. 12, pp. 181–205). Narosa Pub. House, New Delhi, India.

Saleem, A., El-Said, A. H. M., Maghraby, T. A., & Hussein, M. A., (2012). Pathogenicity and pectinase activity of some facultative mycoparasites isolated from *Vicia faba* diseased leaves in relation to photosynthetic pigments of plant. *J. Plant Pathol. Microb., 3*, 141.

Saleem, A., El-Said, A. H. M., Moharram, A. M., & Abdelnaser, E. G., (2013). Cellulolytic activity of fungi isolated from anise and cumin spices and potential of their oils as antifungal agents. *J. Med. Plants Res., 7*(17), 1169–1181.

Sangwan, M. S., & Mehta, N., (2007). Pathogenic variability in isolates of *Alternaria brassicae* (Berk.) Sacc. from different agro-climatic zones of India. *Plant Dis., 22*, 101–107.

Schilling, A. G., Moller, E. M., & Geiger, H. H., (1996). Polymerase chain reaction-based assays for species-specific detection of *Fusarium culmorum*, *F. graminearum*, and *F. avenaceum*. *Phytopathol., 86*, 515–522.

Schulz, B., Boyle, C., Draeger, S., Ro¨mmert, A. K., & Krohn, K., (2002). Endophytic fungi: A source of biologically active secondary metabolites. *Mycol. Res., 106*, 996–1004.

Sharma, A. K., Gupta, J. S., & Dixit, R. B., (1986). Incidence of pathogenic conidia on leaf surface of yellow sarson and taramira. *J. Indian Bot. Soc., 65*, 550–552.

Sharma, M., Deep, S., Bhati, D.S., Chowdappa, P., Selvamani, R., & Sharma, P., (2013a). Morphological, cultural, pathogenic and molecular studies of *Alternaria brassicae* infecting Cauliflower and Mustard in India. *Afr. J. Microbiol. Res., 7*(26), 3351–3363.

Sharma, P., Deep, S., Sharma, M., & Bhati, D. S., (2013b). Genetic variation of *Alternaria brassicae* (Berk) Sacc. causing dark leaf spot of cauliflower and mustard in India. *J. Gen. Plant Pathol., 79*(1), 41–45.

Sharma, P., Deep, S., Sharma, M., Singh, D., & Chowdappa, P., (2013c). PCR based assay for detecting pathogenic *Alternaria brassicicola* in Crucifers. *Indian Phytopathol., 66* (3), 263–268.

Sharma, P., Deep, S., Singh, D., Sharma, M., & Chowdappa, P., (2014). Penetration and infection processes of *Alternaria brassicicola* on cauliflower leaf and *A. brassicae* on mustard leaf: A histopathological study. *Plant Pathol. J., 13*(2), 100–111.

Sharma, T. R., & Tewari, J. P., (1995). Detection of genetic variation in *Alternaria brassicae* by RAPD fingerprints. *J. Plant Biochem Biot., 4*, 105–107.

Sharma, T. R., & Tiwari, J. P., (1998). RAPD analysis of three *Alternaria* species pathogenic to crucifers. *Mycol. Res., 102*, 807–814.

Sherf, A. F., & Macnab, A. A., (1986). Diseases of cruciferous. In: *Vegetable Diseases and Their Control* (2nd edn., pp. 251–306). The Ronald Press, New York.

Shurtleff, M. C., & Averre, C. W. III., (1997). *Glossary of Plant-Pathological Terms.* American Phytopathological Society, St. Paul, M. N.

Sigareva, M. A., & Earle, E. D., (1999). Camalexin induction in intertribal somatic hybrids between *Camelina sativa* and rapid cycling *Brassica oleracea. Theor. Appl. Genet., 98*, 164–170.

Silva, R. L. O., Luz, J. S., Silveira, E. B., & Cavalcante, U. M. T., (2006). Endophytic fungi of *Annona spp.*: Isolation, enzymatic characterization of isolates and plant growth promotion in *Annona squamosa* L. seedlings. *Acta Bot. Bras., 20*, 649–655.

Simmons, E. G., & Roberts, R. G., (1993). *Alternaria* themes and variations. *Mycotaxon, 48*, 109–140.

Simmons, E. G., (1992). *Alternaria* taxonomy: Current status, viewpoint, challenge. In: Chelkowski J., & Visconti, A., (eds.), *Alternaria Biology, Plant Diseases, and Metabolites* (pp. 1–35). Elsevier Science Publishers, Amsterdam.

Singh, A., & Bhowmik, T. P., (1985). Persistence and efficacy of some common fungicide against *Alternaria brassicae,* the causal agent of leaf blight of rapeseed and mustard. *Indian Phytopathol., 38*, 35–38.

Singh, D., Singh, R., Singh, H., Yadav, R. C., Yadav, N., Barbetti, M., Salisbury, P., et al., (2007). Cultural and morphological variability in *Alternaria brassicae* isolates of Indian mustard (*Brassica juncea* L. Czern & Coss.). In: *Proceeding of the 12th International Rapeseed Congress* (Vol. 4, pp. 158–160). Wuhan, China.

Singh, R., Mohan, C., & Sharma, V. K., (2012). Morphological and pathological variations in isolates of *Alternaria brassicae* causing leaf blight of rapeseed and mustard. *Plant Dis., 27*, 22–27.

Smith, I. M., Dunez, J., Philips, D. H., Lelliott, R. A., & Archer, S. A., (1988). *European Handbook of Plant Diseases.* Blackwell Sci. Pub., Oxford, London.

Smith, O. P., Peterson, G. L., Beck, R. J., Schaad, N. W., & Bonde, M. R., (1996). Development of a PCR-based method for identification of *Tilletia indica,* causal agent of Karnal bunt of wheat. *Phytopathology, 86*, 115–122.

Sofi, T. A., Beig, M. A., Dar, G. H., Ahmad, M., Hamid, A., Ahangar, F. A., Padder, B. A., & Shah, M. D., (2013). Cultural, morphological, pathogenic and molecular characterization of *Alternaria mali* associated with Alternaria leaf blotch of apple. *Afr. J. Biotechnol., 12*, 370–381.

Srivastava, A. K., & Gautam, N., (2013). Lipase activity of some fungi isolated from otitis media. *Adv. Biores., 4*(2), 63–66.

Suri, R. A., & Mandahar, C. L., (1982). Cytokinin-like substances and accumulation of labeled metabolites at infection sites of *Alternaria brassicae*. *Mycopathologia, 94*, 153–156.

Suri, R. A., & Mandahar, C. L., (1985). Involvement of cytokinin-like substances in the pathogenesis of *Alternaria brassicae* (Berk.) Sacc. *Plant Sci., 41*, 105–109.

Tahvonen, R., (1979). Seed-borne fungi on cruciferous cultivated plants in Finland and their importance in seedling raising. *Agr. Food Sci., 51*, 327–379.

Tan, R. X., & Zou, W. X., (2001). Endophyte a rich source of functional metabolites. *Nat. Prod. Rep., 18*, 448–459.

Taylor, J. L., (1993). A simple, sensitive, and rapid method for detecting seed contaminated with highly virulent *Leptosphaeria maculans*. *Appl. Environ. Microbiol., 59*, 3681–3685.

Tetarwal, M. L., Rai, P. K., & Shekhawat, K. S., (2008). Morphological and molecular variability of *Alternaria alternata*. *Indian J. Mycol. Pl. Pathol., 58*, 375.

Tewari, J. P., (1983). Cellular alterations in the black spot of rapeseed caused by *Alternaria brassicae*. *Phytopathol., 73*, 831.

Tewari, J. P., (1986). Subcuticular growth of *Alternaria brassicae* in rapeseed. *Can. J. Bot., 64*, 1227–1231.

Tigano, M. S., Aljanabi, S., & Marques, D. M. S. C., (2003). Genetic variability of Brazilian *Alternaria* spp. isolates as revealed by RAPD analysis. *Braz. J. Microbiol., 34*, 117–119.

Tivoli, B., Baranger, A., Avila, C. M., Banniza, S., & Barbetti, M., (2006). Screening techniques and sources of resistance to foliar diseases caused by major necrotrophic fungi in grain legumes. *Euphytica, 147*, 223–253.

Trail, F., & Koller, W., (1993). Diversity of cutinases from plant pathogenic fungi: Purification and characterization of two cutinases from *Alternaria brassicicola*. *Physiol Mol. Plant Pathol., 42*, 205–220.

Tripathi, N. N., & Kaushik, C. D., (1984). Studies on the survival of *Alternaria brassicae*, the causal organism of leaf spot of rapeseed and mustard. *Madras Agr. J., 71*, 237–241.

Tsuneda, A., & Skoropad, W. P., (1977). Formation of microsclerotia and chlamydospores from conidia of *Alternaria brassicae*. *Can. J. Bot., 55*, 1276–1281.

Tsuneda, A., & Skoropad, W. P., (1978a). Phylloplane fungal flora of rapeseed. *Trans. Br. Mycol. Soc., 70*, 329–334.

Tsuneda, A., & Skoropad, W. P., (1978b). Behavior of *Alternaria brassicae* and its mycoparasite *Nectria inventa* and intact and excised leaves of rapeseed. *Can. J. Bot., 56*, 1341–1345.

Varma, P. K., Singh, S., Gandhi, S. K., & Chaudhary, K., (2006). Variability among *Alternaria solani* isolates associated with early blight of tomato. *Comm. Agr. Appl. Biol. Sci., 71*, 37–46.

Vasudeva, R. S., (1958). Diseases of rape and mustard. In: Singh, D. P., (ed.), *Rape and Mustard* (pp. 77–86). Indian Central Oilseed Committee, Hyderabad, India.

Verma, P. R., & Saharan, G. S., (1994). *Monograph on Alternaria Disease of Crucifers* (p. 162). Research Branch, Agriculture and Agriculture Food Canada, Saskatoon Res. Centre, Canada.

Viswanath, K., & Kolte, S. J., (1997). Variability in *Alternaria brassicae:* Response to host genotypes, toxin production and fungicides. *Indian Phytopathol., 50*, 373–381.

Von, R. C., (1962). Histological studies of infection by *Alternaria longipes* on tobacco. *Phytopathol., 45*, 391–398.

Wakeham, A. J., Keane, G., & Kennedy, R., (2016). Field evaluation of a competitive lateral-flow assay for detection of Alternaria brassicae in vegetable Brassica crops. Plant Dis., 100 (9), 1831–1839.

Walsh, G., (2002). Industrial enzymes: Proteases and carbohydrases. In: *Proteins; Biochem., and Biotech.* John Wiley and Sons Ltd.

Walter, H. R., (1998). Microcrystalline cellulose technology. In: *Polysaccharide Association Structure in Food.* Marcel Dekker, Inc.

Walton, J. D., (1994). Deconstructing the cell wall. *Plant Physiol., 104,* 1113–1118.

Wang, H., & Zhang, T. Y., (2003). RAPD analysis on small spored *Alternaria* species. *Mycosystema, 22,* 35–41.

Warwick, S. I., Francis, A., & La Fleche, J., (2000). *Guide to Wild Germplasm of Brassica and Allied Crops (Tribe Brassicae, Brassicaceae)* (2nd edn). Agriculture and Agri-Food Research Branch Publication, ECORC Ottawa, Canada. Contribution No. 991475.

Weimer, J. L., (1924). Alternaria leaf spot and brown rot of cauliflower. *J. Agr. Res., 29,* 424–441.

Weir, T. L., Huff, D. R., Christ, B. J., & Romaine, C. P., (1998). RAPD-PCR analysis of genetic variations among isolates of *Alternaria solani,* and *Alternaria alternata* from potato, and tomato. *Mycologia, 90,* 813–821.

White, T. J., Bruns, T., Lee, S., & Taylor, J., (1990). Amplification and direct sequencing of fungal ribosomal genes for phylogenetics. In: Innis, M. A., Gelfand, D. H., Sninsky, J. J., & White, T. J., (eds.), *PCR Protocols: A Guide to Methods and Applications* (pp. 315–322). Academic Press, San Diego.

Wiltshire, S. P., (1947). *Species of Alternaria on Brassicae* (Vol. 20, pp. 8–15). Myco. Papers, CMI, England.

Xu, X. D., Dong, H. Y., Jilang, Y., & Bai, J. K., (2003). Analysis of genetic relationships among *Sporisonium relianum* isolates by RAPD. *Mycosystema, 22,* 56–61.

Yao, C., & Koller, W., (1995). Diversity of cutinases from plant pathogenic fungi: Different cutinases are expressed during saprophytic and pathogenic stages of *A. brassicicola. Mol. Plant Microbe. Interact, 8,* 122–130.

Zamani, M. R., Motallebi, M., & Roastamian, A., (2004). Characterization of Iranian isolates of *Fusarium oxysporum* on the basis of RAPD analysis, virulence and vegetative compatibility. *J. Phytopathol., 152,* 449–453.

Zur, G., Hallerman, E. M., Sharf, R., & Kashi, Y., (1999). Development of a Polymerase Chain Reaction-based assay for the detection of Alternaria fungal contamination in food products. *J. Food Prot., 62,* 1191–1197.

CHAPTER 14

Recent Advances in Diseases Management of Aonla (*Emblica officinalis*)

R. K. PRAJAPATI,[1] C. S. PANDEY,[2] P. K. GUPTA,[3] V. K. SINGH,[4] and
S. R. SINGH[5]

[1]*Scientist (Plant Protection), JNKVV, Krishi Vigyan Kendra, Tikamgarh,
Madhya Pradesh, India, E-mail: rkiipr@yahoo.com*

[2]*Assistant Professor (Horticulture), JNKVV, College of Agriculture,
Jabalpur, Madhya Pradesh, India*

[3]*Scientist (Plant Protection and Technical Officer), JNKVV,
Directorate of Extension Service, Jabalpur, Madhya Pradesh, India*

[4]*Associate Professor (Horticulture), JNKVV, College of Agriculture,
Tikamgarh, Madhya Pradesh, India*

[5]*Scientist (Plant Protection), CSUA&T, Krishi Vigyan Kendra,
Mahamayanagar, Uttar Pradesh, India*

ABSTRACT

Aonla (*Emblica officinalis*) considered as second high vitamin-C contain
fruit among the other fruits which contain vitamin-C. The disease of Aonla
affected its yield as well as orchard. Therefore, the studies of Aonla diseases
are compiled in this chapter. The Aonla affected by several diseases as
fungal, bacterial, viruses, and nematodes, but among them fungal diseases
are more important. The important fungal diseases are affected Aonla are
rust, fruit rot, moldes, anthracnose, leaf spots, wilt, and mildews. Lichens
important diseases causing algae in Aonla has been reported. Necrosis of
Aonla is also an important disorder caused by non-living factors. The study
has been carried out under different heads like economics importance of

diseases, symptoms, causal organism, diseases cycle epidemiology, and management. The different new causal organism of different fungal diseases has been review under this chapter.

14.1 INTRODUCTION

Aonla (*Emblica officinalis*) is also known as Indian gooseberry. Its origin is considered as India and around the Indian sub-continent area. Annual average cultivated area 93 (ha.000) with an annual average production of 1090 metric tons in India of Anola. Among the world, Anola growing countries like Cuba, Puerto Rico, the USA, Iran, China, Malaysia, Iraq, Bhutan, Thailand, Sri Lanka, Pakistan, the Philippines, Japan, Trinidad, Vietnam, and Panama, Indian rank first place in both area and production. Most cultivated states of Indian for Aonla are Uttar Pradesh, Maharashtra, Gujarat, Rajasthan, Andhra Pradesh, Karnataka, Tamil Nadu, and Himachal Pradesh (http://agricoop.gov.).

14.2 ECONOMIC IMPORTANCE

The Anola fruit is a good supply of diet vitamin C. The medicinal value of fruit known from ancient Indian history in Aruvedic literature its nature considered as laxative properties, cooling, acrid, diuretic, and beneficial for hemorrhages, diarrhea, dysentery, anemia, jaundice, dyspepsia, and cough. It has been used in Indin life as making Trifla and Chavanprash. Aonla has also been used since ancient times as Murabbas, pickles, candy, jelly, and jam. Fruits leaves, bark and even seeds were getting used for diverse purposes.

The importance of minor subtropical fruits is generally overlooked, although these fruits are very important and contribute to the fruit production and nutrition in our country. These fruits are generally of local importance and are within the reach of poor and common people. The yield of fruit loss of Aonla due to many fungal diseases, which are important and also causing severe incidence. Rhizopus rot is one of the most important and widespread diseases of Jack fruit. Loss up to 75% was estimated due to this rot (Pandey et al., 1979). Rusts, blue mold, leaf spot, powdery mildew, black leaf spot, and fruit rots were some of the very important diseases which cause severe damage.

Aonla was stricken by many diseases and disorders resulting from fungi, microorganisms, and viruses after harvesting (Morton, 1987). The post-harvest losses due to diseases were reported about 20–35% by means of (Sharma and Alam, 1998; Prakash, 2005; Rawal and Saxena, 2005). The Aonla fruit rots because of diverse fungi, i.e., *P. islandicum* Sopp. (Shetty, 1959), *Aspergillus* spp. (Srivastava et al., 1964), P. fruit rot (Tondon and Srivastava, 1964), *Cladosporium* (Jamaluddin, 1978), *P. phyllanthi* Punith (Lal et al., 1982), *C. gloeosporioides* (Penz) Sacc. (Mishra and Shivpuri, 1983), *A. alternata* (Fr.) Keissler (Pandey et al., 1984) and *P. funiculosum* Thom. (Yadav et al., 2009) have been the most essential as they have an effect on the fruit quality and amount in relation to the market place value (Bhardwaj and Sharma, 1999). Among the fungal pathogens which causing losses in the plant at field situation, the rust was discovered the most vital sickness which became caused extensive loss observed by using fruit rots because of a number of fungi (Om Prakash and Misra, 1993) Table 14.1.

TABLE 14.1   List of Various Disease of Aonla (*Emblica officinalis*)

Diseases	Causal Organism	References
Major Fungal Diseases		
Ring rust	*Ravenelia emblicae* Styd., *Phakopsora phyllanthi* Diet	Tyagi (1967); Zang et al. (1992)
Fruit Rot		
Phoma fruit rot	*Phoma putaminum* Speg. *Phoma emblicae* (Jamaluddin Tandon and Tandon)	Pandey et al. (1980); Jamaluddin (1978)
Pestalotia Fruit	*Pestalotia cruenta* Syd.	Tondon and Srivastava
Rot and leaf spot	*Pestalotiopsis versicolor, Pestalotiopsis heterocolor*	(1964); Zhang and Qi (2011)
Dry fruit rot	*Cladosporium tenuissimum* Cooke, *C. cladosporioidies* (Fries), *C. herbarum*	Pandey et al. (1980); Tadon and Verma (1964)
Nigarospora	*Nigrospora sphaerica* (Sacc.) Massan	Anonymous (2002)
Fruit rot		
Alternaria fruit rot	*Alternaria alternata* (Fr.) Keissler	Pandey et al. (1984)
Soft rot	*Phomopsis phyllanthi* Punith.	Lal et al. (1982)
Black soft rot, Cytospora Fruit Rot, Red fruit rot	*Syncephalastrum racemosum, Cytospora* sp, *C. herbarum*	Neelima Garg et al. (2004); Tadon and Verma (1964)
Molds		
Sooty mold	*Alternaria, Capnodium, Cladosporium, Fumago, Scorias, Capnodium* sp,	Chomnunti et al. (2014)

TABLE 14.1   *(Continued)*

Diseases	Causal Organism	References
Blue mold	*Penicillium citrinum P. oxalicum; P. islandicum, P. funiculosum* **Thom**	Shetty (1954); Om Prakash and Misra (1995); Yadav et al. (2012)
Black mold	*Aspergillus niger*	
Anthracnose	*Colletotrichum (Glomerella cingulata* Stoenm.) Spauld and Schrenk.	Mishra and Shivpuri (1983)
False anthracnose	*Kabatiella emblica*	Zhang and Qi (1996)
Leaf Spots		
Brown spot	*Phyllosticta emblica*	Zhang and Qi (1996)
Leaf spot	*Pestalotiopsis heterocolor*	Zhang and Qi (2011)
Wilt		
Fusarium wilt and	*Fusarium solani*	Soni and Verma (2010);
Powdery mildews	*Oidium sp.*	Zhang and Qi (1996)
Lichens	*Strigula elegans* (Fee.) Mull. Arg.	
Disorder		
Internal necrosis	**Disorder**	Ram et al. (1977)

14.3   MAJOR FUNGAL DISEASES

14.3.1   *RING RUST (RAVENELIA EMBLICAE STYD. PHAKOPSORA PHYLLANTHI DIET)*

14.3.1.1   *SYMPTOMS*

Rust was a chief and economically crucial disease in aonla. A most ailment occurrence of 21% became noted within the ultimate week of December and the improvement became higher when relative humidity multiplied and temperature decreased (Anonymous, 1989). It turned into also found in Rajasthan, Andhra Pradesh, Tamil Nadu, Haryana, etc. It was first found in Rajasthan via Tyagi (1967). In India, Rawal (1993) has observed that this causes significant loss in principal Aonla developing tracts of India. Black pustules develop on fruits, which later increase in a hoop pattern. In a complicated degree, many pustules coalesce together and increased in from a large place of the fruit surface. Pustules were covered with a papery masking which these were found black in color spores in the surface when the protecting ruptures. Initial signs seemed like pustules on fruit. The pustules which evolved collectively covered larger vicinity and be a part of collectively. The black colored spores rupturing exposed from a large paper-like overlaying. The appearance of fruits becomes dirty. Its spots group was formed on leaves also like pinkish to brown color spots both signs either on fruits or leaves were recorded in different in their appearance (Tyagi, 1667).

14.3.1.2   CAUSAL ORGANISM

Aonla rust (*Revenelia emblica* Syd. caused) an obligate parasite, which requires the stay host for contamination and establishment below beneficial conditions. The species changed originally defined by using Sydow and Butler (1901), who pronounced the occurrence of uredia and telia only. The pycnial, uredial, and telial stages arise in its existence cycle and on nuclear behavior in teliospore, basidium, and basidiospore were determined via Singh (1973).

14.3.1.3   DISEASE CYCLE

Ravenelia emblicae Styd was an autoceious rust fungi belong to Raveneli-aceae and complete its life cycle on a single host, i.e., Aonla.

14.3.1.4   EPIDEMIOLOGY

The first appears on the lower side of older leaves with brown, circular, and raised. These symptoms mostly started to be amperes from 2nd fortnight of October. The initial symptoms on fruits were observed from brown to black pustules, which later were developed in the form of ring form while not all times growth of fruit but first time in the season during 1st fortnight of October in both the years. The rust symptoms appeared at Sept first fortnight on fruits was observed (Pandey et al., 2006). The minimum symptoms were observed in August. The maximum and minimum temperature 31.7 and 250 C, Relative humidity morning and evening 92.40 and 69.70, Rainfall 24 mm, Sunshine hrs/day 1.4, Wind velocity 2.80 and evaporation was 4.1. The studies on correlation matrix resulted that, temperature (max. and min.) and evaporation were found highly significant negative correlation while evening RH% and rainfall too significant negative correlation with PDI. These results were also found in support with reports of Pandey et al. (2007); and Anonymous (2002–2003) (Devesh Anand Singh et al., 2015).

14.3.1.5 MANAGEMENT

The rust of Aonla may be managed by way of 3-four sprays of Wettable sulfur @ 5 g/liter of water at an interval of 1 month or Mancozeb @ 2.5 g/liter of water at duration of 15 days. Integrated approaches to be observed for effective management of Aonla rust, which includes all chemical control, cultural methods, resistant cultivator, and clean cultivation were critical and infected fruits and leaves need to be amassed and destroyed. Proper pruning to hold most excellent canopy density and avoiding humidity build-up consequences in much less incidence and intensity of infection of the pathogen. Three sprays of 0.5% wettable sulfur or Zineb (0.2%) at intervals of 1 month beginning from the month of July may be useful to manipulate rust (Tyagi, 1967). The commercially cultivated cultivars Chakaiya,' Krishna,' Kanchan,' NA-7, and NA-10 were relatively vulnerable to rust. Out of 9 cultivars screened underneath Faizabad (India) conditions, NA-6' was determined to be resistant. No cultivar was observed to be resistant to rust (Anonymous, 2002).

14.3.2 FRUIT ROTS

14.3.2.1 PHOMA FRUIT ROT (PHOMA PUTAMINUM SPEG, P. EMBLICAE JAMALUDDIN TANDON AND TANDON)

14.3.2.1.1 Symptoms

The above ground parts of the plants may be affecting by this diseases. Phoma fruit rot pathogen caused numerous, small dark brown to black spots on leaves which can also broaden concentric ring as they enlarge. Older leaves may be inflamed first, however, all leaves were susceptible, and defoliation can result when disease was intense. The leaf spot looks very similar to that as a result of early blight besides that the Phoma lesions contain several pycnidia like minute black fruiting bodies while on stem the formation of lesions dark brown with concentric earrings form on the stems, and both the inexperienced and ripe fruit may be inflamed. The lesions were on fruit lesions most broaden at the calyx in early stage but at the end as small sunken lesions that later change into sunken, black, leathery lesions with numerous pycnidia within the center. The signs of rotting on fruit were found at some point of Dec-Jan. The symptoms had been be gaining as a small pinkish brown necrotic spot which extends in the direction of both the ends of the fruit forming an eye-formed appearance. In intense cases, lesions coalesce, forming a larger pustule. The mature lesions were dark brown and critically infected fruits display wrinkling. The underlying tissues within the rotten fruits grow to be soft, the inflicting pathogen (*Phoma putaminum* Speg) said by way of Pandey et al. (1980). The Phoma emblicae become also suggested in Aonla causing dry fruit rot with the report of Jamaluddin et al. (1975). Around 20–25% Aonla fruit losses were reported from Allahabad in the market place (Srivastava et al., 1986).

14.3.2.1.2 Causal Organism

The Phoma putaminum Seg became first described through Seg (1980) and Phoma emblicae Jamaluddin Tandon and Tandon via Jamaluddin et al. (1975). The Phoma was identified with the aid of its mobile and greater or much less round fruiting structure (Pycnidia) containing hundreds of one-celled hyaline conidia. The conidia were born from inconspicuous peg like phialides lining the inner wall of pycnidium. The conidial fructification was a pycnidium which was a darkish flask-shaped opening typically through a single round ostiole and lined inside by means of hymenium of condiogenus cells which are enteroblastic, phialides, included or discrete, ampuliform to doliform and opening with the aid of minute aperture. The conidia of Phoma spp. were observed as in morphologically as thin-walled, various shaped, hyaline, aseptate or, regularly guttulate, and sometimes septate, i.e., Ellipsoid, cylindrical, fusiform, pyriform or globose. In addition to pycnidia, a few species produce dictyochlamydospores which might be darkish, irregularly ovate, with a number of transverse and longitudinal septa (Boerema and Vorhoeven, 1972; Sutton, 1980).

14.3.2.1.3 Disease Cycle

The disease cycle of fungus survived from one season to another season with the soil in inflamed plant particles of Aonla and closely associated weeds. Injury to the plant including pruning, insect feeding, the fungus caused mechanical harm or cracking to invasion from natural opening of plant. When the temperature (20°C) and humidity were optimum, loads of fungal conidia were exuded from the pycnidia. These were quite simply spread by means of rain, overhead irrigation, and on workers' clothing and equipment. Low soil nitrogen and phosphorus levels might also contribute to plant susceptibility.

14.3.2.1.4 Epidemiology

The rotting severity in papaya fruit through Phoma spp. Reported for the duration of rainy in India through Simmonds (1965). The fungal tiers inflicting damages in orchard reported with the aid of Chau and Alvarez (1979) that the sexual stage (perithecia) and the asexual stage (pycnidia) and can be chargeable for infections in the orchard. The optimum growth temperature requirement for the fungus was approximately 30°C. Phoma spp. differs of their rapidity to lessen the nutrition C of diverse end result. *Phoma psidii* and *Gloeosporium psidii* (Ghosh et al., 1965) reduced the total diet C content material after 10 days. Infection of mango culmination through *P. mangiferaa* and *P. emblica* via *P. exigua* resulted in a very rapid lower in vitamin C content compared with gradual lower all through garage of healthy culmination (Reddy and Laxminarayana, 1984).

14.3.2.1.5 Management

The orchard could be managed by using of Carbendazim @ 2 ml/liter of water has to be applied at the time of fruit setting whilst circumstance was wet in Aonla growing area. The other manipulate measures like post-harvest measures include a warm water bath, which might also have to be blended with Carbenazim spraying @ 2 ml/liter of water at 15 days interval and as per want based (Bolkan et al., 1976; Couey et al., 1984).

14.3.2.2 PESTALOTIA FRUIT ROT (PESTALOTIA CRUENTA SYD. PESTALOTIOPSIS VERSICOLOR)

14.3.2.2.1 Symptoms

The spots on the end result were usually irregular and brown. The disease commonly starts as a brownish discoloration on the fruit floor which grows slowly. Later, the spots end up medium-brown, and the skin around them develops mild brown coloration at fantastically later stage. The infected region becomes protected with white fluffy aerial increase of the fungus. The internal part of the diseased fruit suggests a dry dark brown area (Tandon and Srivastava, 1964). Pestalotiopsis leaf spot become a newly pronounced pathogen for China (Zhang and Qi, 1996).

14.3.2.2.2 Causal Organism

The first comprehensive treatment Pestalotia genera were by Saccardo (1884) who listed over 80 species. *Pestalotia cruenta* incidence in the rainy season, infection folicolous inciting brown, irregular spot with acervuli; acervuli scattered, punctiform, amphigenous, more on upper surface, subepidermal, erumpent at maturity with almost round opening, 100–170 micron in diameter; conidia 5-celled, fusiform, slightly constricted at septa, 17–24 × 5.5–7 micron, intermediate colored cells versicolored, upper two umber, lowest olivaceous, 11.9–15.5 p long, apical cell hyaline, short-cylindrical with a crest of 2 setulae, sometimes 1 or 3, rarely 4, basal cell hyaline, short, conic, pedicel hyaline, simple, short, 3–5 p; setulae simple, hyaline, divergent, 10–20.5 p long (Mahapatra, 1979). The fungus was constantly isolated from abnormal, brown spots which had evolved slowly on fruit of *Phyllanthus emblica*. This is a brand new host report for the fungus and a brand new record for India of five other end results inoculated; best guava became infected (Tandon and Srivastava, 1964). *Pestalotiopsis versicolor*, the fungus develops small, round, black, barely sunken spots on culmination to start with. Later on the spots increase hastily, became sunken, and ended in a smooth decay of the fruit flesh. The fungus changed into for the first time pronounced causing soft rot of Aonla (Verma and Verma, 2015).

14.3.2.2.3 Disease Cycle

Pestalotiopsis species rich genus going on as pathogens, endophytes, and saprophytes (Jeewan et al., 2004). However, maximum species of *Pestalotia* were plant pathogens (Zhang et al., 2003; Zhu et al., 1991). Pathogenic species of *Pestalotiopsis* initially make touch with the host in which the infection happens (inoculum), likely by using the conidia or fragmented spores (Espinoza et al., 2008). These inocula can also live on at some stage in harsh weather situations and may reason primary infections. Secondary inoculum produced on diseased tissue could additionally cause secondary infections and growth the severity of the sickness. The source of the inoculum can be wild plantations (Keith et al., 2006), flowers (Pandey, 1990), crop particles, sickness stock plants, used developing media, soil, and contaminated nursery equipment (McQuilken and Hopkins, 2004), splashed water droplets (Hopkins and McQuilken, 1997; Elliott et al., 2004) and additionally spores inside the air (Xu et al., 1999). Species of *Pestalotiopsis* have continuously been remoted as endophytes from plant tissues (Wei and Xu, 2004; Liu et al., 2006; Wei et al., 2005; Tejesvi et al., 2009; Watanabe et al., 2010). We suspect that many endophytic species stay as dormant symptomless population of flowers till the plant was stressed, and then dried.

14.3.2.2.4   Epidemiology

The spores germinated only at temperature from 15 to 30°C. While at 25°C the maximum spore germination and the highest length of germ tubes were recorded. Spores failed to germinate between 5 and 10°C as well as between 35 and 50°C. Optimal temperatures reported for other plant pathogens were 21°C–25°C with germination declining sharply at or just above 30°C (Bhagava and Khare, 1988; Maheswari et al., 2000). *Pestalotiopsis sp.* required dry weather (60–70% RH) Thus, 75% relative humidity seems optimum for conidial germination and gram tube development of Pestalotiopsis disseminate (Amy and Rowe, 1991; Surianchandraselvan et al., 1991).

14.3.2.2.5   Management

Leaf streak disease due to *Pestalotiopsis royenae* in black pepper and cardamom become controlled through spraying of schedules consistent with year of 3 rounds of 0.2% spray of Copper Oxychloride at fortnightly intervals, one in February-March and the opposite in September-October, can manage the unfold of the disorder (Nair, 2011). One spraying of Carbendazim (0.1%) must be carried out 15 days prior to fruit harvest. Harvesting must be achieved very cautiously to keep away from any injury to the culmination. Fruits need to be stored in smooth containers. Full sanitary measures need to be adopted throughout storage and transit. Sanitary situations in garage need to be maintained.

14.3.2.3 DRY FRUIT ROT (CLADOSPORIUM TENUISSIMUM COOKE, C. CLADOSPORI-OLDIES FRIES)

14.3.2.3.1 Symptoms

The dry fruit rot turned into recorded with the aid of Jamaluddin (1978) in Allahabad with an estimated lack of 2–5 in line with cent. *Cladosporium tenuissimum* Cooke: Infection turned into observed from November to February. The disorder began as colorless location, barely tender and eventually stepped forward in a circular manner. Light brown mycelial boom of the fungus became also obtrusive on infected regions. Jamaluddin (1978) reported that lesions might be increased 1–1.5 cm diameter within in 7 days. Again in 1978 Jamaluddin reported that for contamination injury was must. The rot shows darkish brown necrotic lesions. Slight boom of the organism seems in the necrotic cavity. The severity of infection turned into recorded simplest in mature and ripe culmination. Some freshly harvested fruits additionally showed infection. One-week-old lesion measured 0.7 to 1.2 cm in diameter (Jamaluddin, 1978). *C. herbarum* caused small spots of ochre red color on fruits Tandon and Verma (1964). Disease signs and symptoms started out as a gray fungus fashioned at the stigma, which led to the blossom blight and in the end to black rot and necrosis of the whole flower in blossom blight in Strawberry (Nam et al., 2015).

14.3.2.3.2 Causal Organism

Cladosporium was certainly one of the biggest and most heterogeneous genera of hyphomycetes (Ellis, 1971; Dugan et al., 2004). The genus *C. cladosporioides* was a very commonplace, cosmopolitan, saprobic species. The colony fashioned olivaceous-green to olivaceous-brown, velvet-like colonies with apically and laterally branched conidiophores and lemon-formed conidia, which had been generally clean, however, on occasion textured on PDA. The intercalary conidia were elliptical to lemon-form with sizes ranging $5.0{\sim}10.5 \times 2.5{\sim}3.0$ μm, and secondary ramoconidia had been cylindrical-oblong with sizes ranging $10.0{\sim}15.0 \times 2.5{\sim}3.7$ μm (Nam et al., 2011).

14.3.2.3.3   Disease Cycle

The fungus was big and lives on lifeless plant fabric within the soil. Conidia were disseminated within the air and were able to motive infection of fruit which has been broken through rain or through rough dealing with (English, 1945).

14.3.2.3.4   Epidemiology

The mycelia of isolates grew between 10°C and 30°C, while no growth happened after 7 days at 5°C on PDA, however, some isolates colonies have been considerably larger in radius than that of the other isolates after 7 days at each 20°C and 25°C. The optimal increase fee was attained after 7 days of incubation at 20°C. This end result became much like that pronounced via Tashiro et al. (2013), who stated optimal boom of *C. cladosporioides* at 20–22°C.

14.3.2.3.5   Management

Avoid fruit harm, at some stage in harvesting. Discard affected culmination from the orchard. Spray Dithane M-45 (zero.2%) or Carbendazim (0.1%) all through fruiting season.

14.3.2.4   NIGAROSPORA FRUIT ROT (NIGROSPORA SPHAERICA (SACC.) MASSAN)

14.3.2.4.1   Symptoms

The losses because of Nigarospora fruit rot become expected at 5–10% yearly it became because of *Nigrospora sphaerica* (Sacc.) Mason the prevalence of disease occurrence turned into recorded in moth of December from Kanpur for the first time. The symptoms of the disease described as rot initiation as a small dot of two mm diameter extending to a black ringspot as much as 7 mm. Several rings adjoining to each different coalesce resulting in whole rotting of the inflamed culmination (Kamthan et al., 1981).

14.3.2.4.2 Causal Organism

The fungus Nigarospora was belongs to the phylum Ascomycota and its mycelium filamentous. Fungal colonies were hairy to begin with white, turning into light to darkish gray with the onset of sporulation with black, round to sub-ground single-celled conidia (15 to 18 × 12 to fifteen μm), which have been born on a hyaline vesicle on the tip of the conidiophores. The morphological characters of the mycelium immersed and superficial, conidiophores micro-matous or semi-macronematous mononematous straight or flexuous, branched sub-hyline to faded brown or brown, smooth conidiophores cells discrete, determinate ampliform or sub-spherical hyaline, Conidia acrogenous fashioned singly, obligate spheroid, non-sepated, dark brown to blackish smooth. The pathogen saprophytic on many plant species, this fungal agent was also referred to as a leaf pathogen on several hosts worldwide (Mason, 1927; Ellis, 1971; Wright et al., 2008; Soylu et al., 2011; Zhang et al., 2011).

14.3.2.4.3 Disease Cycle

Nigrospora sphaerica was found an airborne, soil, plants, and leaf born (Webster, 1952). Nigrospora sphaerica was as an entophytic in nature where it produces antiviral and antifungal secondary metabolites (Zhang et al., 2009). The sporulation of N. sphaerica reasons its initial white colored colonies to unexpectedly turn black. N. sphaerica was often burdened with the closely associated species N. oryzae because of their morphological similarities. N. sphaerica reproduction was as spore in conidia through asexually. The conidia were ejected out forcefully at maximum horizontal distances of 6.7 cm, and a pair of cm vertically. Discharge of spores happens in all directions. The mechanism for projection is based at the conidiophores along with a flask-formed support cell that bears the conidium. Liquid from the support cell squirts via the assisting cellular projecting the spore outwards. This function of forcible spore discharge was hardly ever seen (Webster, 1952). N. sphaerica requires moisture to release spores into the air, consequently accu-mulation starts around 2:00 A.M. with height time of abundance happening round 10:00 A.M. Spore count unexpectedly decreases after 10:00 A.M. and remains low for the duration of the day (Tan et al., 1992).

14.3.2.4.4 Epidemiology

The rapid liberation of spores starting after 6 hr and reaching a peak between 8 and 10 hr. The concentration decreased rapidly during the afternoon and evening, with very few spores being trapped during the night. Liberation coincided with conditions of rising temperature and decreasing humidity, these usually occurring from 07.00 hr. onwards until about 14.00 hr. On damp mornings the following rainfall during the previous night, humidity often showed no marked decrease until about 11.00 hr. Rain and under-tree irrigation commonly resulted in large increases in the concentration of Nigrospora spores. This was probably a result of increased sporulation of the fungus after wetting of the spore-bearing substratum (Meredith, 1961).

14.3.2.4.5 Management

Carbendazim @ 2 ml or Propiconazole @ 1 ml/liter of water spraying found effective to control fungus. The garlic or cinnamon extract @ 100 ml/liter of water followed by *Bacillus subtilis* or *Pseudomonas fluorescens* market formulation product @ 4 gm/liter of water spraying at 15 days interval. Fruits protected with blue polythene bag directly reduced the inoculums, which resulted in much less disease beneath storage condition., bio-control sellers and bunch protected with blue polythene bag may be used as a potential source of sustainable eco-friendly IDM practices (Nath, 2014).

14.3.2.5   ALTERNARIA FRUIT ROT (ALTERNARIA ALTERNATA (FR.) KEISSLER)

14.3.2.5.1 Symptoms

The rot showed as a small, brownish, spherical necrotic spot, later changed in round fashion. In superior ranges, the spot became darkish brown to black and neighboring spots coalesced. The center parts of the inflamed tissues have become gentle and pulpy. In a survey of nearby fruit orchards (Allahabad) for the duration of January and February in 1981. Association of *Alternaria altanata* (Fr.) Keissler with dropped fruits turned into located in Allahabad. *A. altenata* became ubiquitous and has many hosts. By inoculating the Anola fruits in the lab by way of pinprick approach. Such culmination completely rots within 15–20 days of inoculation (Pandey et al., 1984).

14.3.2.5.2 Causal Organism

Colony gray to olivaceous, powdery or felty microscopy, conidiophores by and large unbranched with one or few conidial scars, conidia chain profusely branched. Conidia obclavate to ellipsoidal, with a quick cylindrical beak, medium brown rugulose with muriform septation, with an unmarried scare on the tip, arising in ordinarily unbranched chains of ten of greater one-of-a-kind (Hoog, 2000).

14.3.2.5.3 Disease Cycle

This fungus commonly survives on plant debris in soil as saprophyte from which conidia were picked up by way of wind and invade but has the potential to become pathogenic beneath most conducive environmental situations. The pathogen propagates itself through asexual spores known as conidia. These conidia were produced in lesions on mature or demise leaves (Timmer et al., 2003). The symptoms appeared in 50-days fully to manifestation after the first sign of ten days. During rain, the conidia of A. alternata dispersed through air currents, and their release from the lesions can be caused through rainfall, or even just a surprising drop in humidity (Dewdney, 2015). When the conidium lands on a leaf, it would wait until the nighttime dew, and then germinate. It can both enter through the stomata and penetrate immediately through the pinnacle of the leaf, the use of its appressorium, infecting the leaf inside 12 hours (Timmer et al., 2003).

14.3.2.5.4 Epidemiology

Alternaria spp. desires moist warm surroundings. It is often observed in areas with humid climates, or in which there has been enormous rainfall. The fungus lives in seeds and seedlings, and was likewise unfolding by spores. This sickness prospers in useless plant life which has been left in gardens over winter. Additionally, while dead infected particles is brought to compost pile it could unfold to other vegetation in the course of the garden.

14.3.2.5.5 Management

The chief means of management was prevention of harm and renovation of fruit vitality. Harvesting at correct adulthood was very vital. *In vitro* fungicides *viz.*, Mancozeb, Copper oxychloride, Carbendazim, and Bordeaux aggregate, etc. Inhibited the mycelial boom and spore germination of *A. altemata* (Mathur and Sarbhoy, 1983). Disease control was executed with the aid of pre-harvest fungicidal sprays coupled with a put up harvest dip treatment with Mancozeb (0.1%). The studies were carried out *in vitro* and *in vivo* conditions by Thakore et al. (1991) reported that sapota fruits could be saved from infection after harvesting by dipping in Mancozeb (0.1%).

14.3.2.6 SOFT ROTS (PHOMOPSIS PHYLLANTHI PUNITH)

14.3.2.6.1 Symptoms

The signs of fruit rot fruit in orchard and in marketplace of Allahabad and Fatehpur changed into first found in the course of December to February by means of (Lal et al., 1982). Isolations made from diseased end result yielded a nicely sporulating lifestyle of *Phomopsis phyllanthi* Punith. Originally, it changed into described as twigs on *P. reticulata* (Punithalingum, 1975). Isolations made from freshly inflamed Aonla end result yielded a properly sporulating pathogen of *Phomopsis phyllanthi* Punith (Lal et al., 1982). Earlier, it becomes described as *Phomopsis reticulata* infecting twigs of Aonla (Punithalingam, 1975). Smoke brown to black, rounded lesions develop inside 2–3 days of inoculation. The infected component later indicates olive-brown discoloration with water-soaked area and cover the complete fruit within eight days. The shape of the fruit additionally gets deformed.

14.3.2.6.2 Causal Organism

Phomopsis pathogens are nectro-trophic at the least for the latent segment of contamination and were consequently referred to as hemi-biotrophs (Rosskopf et al., 2000b). Species of *Phomopsis* were regularly occurring as endophytes of many hosts in each temperate and tropical area. The conidia known as gamma conidia were recorded (Rosskopf et al., 2000a, b; Cristescu, 2007). These conidia were hyaline, multi-guttulate, fusi-shape to subcylindrical with an acute or rounded apex, while the bottom was from time to time truncating (Mostert et al., 2001a; Punithalingam, 1974; Rodeva et al., 2009). Those species described, having a third type of spores were *Phomopsis hordei* Punith. *P. oryzae* Punith. *P. phyllanthi* Punith, *P. amaranthicola.* The

Diaporthe sexual nation is characterized by ascomata. Asci were unitunicate, clavate to clavate cylindrical, loosening from the ascogenous cells at an early stage and mendacity unfastened in ascoma. Punithalingam (1973) cited Phomopsis cotoneastri generating floccose, gray-white colony, with abundant aerial mycelium and numerous conidiomata disbursed in concentric circles on oat agar medium.

14.3.2.6.3   Disease Cycle

The genus *Phomopsis* contains more than 900 species names from an extensive range of hosts. In the month of December to February, the sickness was time-honored more at some stage. Although the fungus causes infection in younger and mature fruits but on inoculation, mature fruits were found to be more susceptible. Injury was important for the improvement of the disorder (Lal et al., 1982). The sickness occurs when spores were launched from a fungal fruiting body (pycnidia) and dispersed by using splashing rain, insects, and contaminated equipment. Spores germinate hastily when free moisture was present on leaves, stems, or leaves. The fungus survives among eggplant crops on and in crop debris, seeds, and soil.

14.3.2.6.4   Epidemiology

The sporulation of Phomopsis sp. was observed on dried petioles that continue to be attached to the tree. Conidia were discharged due to wet condition and deposited on fruit surfaces. The Phomopsis lacks cutinases and has the capability to penetrate through intact cuticle. It required wounds along with the peduncle broken at some stage in harvesting, fruit fly oviposition puncture wounds, and abrasions on the fruit to contaminate fruit. The disease was rarely visible in the discipline on inexperienced end result and was more commonly visible on fruits that have completely ripened. Phomopsis sp., like Rhizopus stolonifer, was capable of using anthracnose and Cercospora black spot lesions as courts of entry into ripe papaya fruits (http://www.Extento.Hawaii). Maximum sporulations of different Phomopsis spp. become observed at 22–25°C temperature. However, pH 5.6–6.8 turned into determined maximum appropriate for boom and sporulation of *Phomopsis* spp. (Luo et al., 2004).

14.3.2.6.5   Management

The pruning and destruction of debris, collectively with preventive fungicide sprays (Srivastava and Tandon, 1969). Dithane M-45 and @ (100 ppm) was reported effective in management of the diseases after the harvest of the fruit (Lal et al., 1982).

14.3.2.7   BLACK SOFT ROTS (SYNCEPHALASTRUM RACEMOSUM)

14.3.2.7.1   Symptoms

The first report of tender rot caused by Syncephalastrum racemosum changed into noticed on harvested and stored end result at Lucknow with the aid of Neelima Garg et al. (2004). These end result had numerous, minute brown necrotic lesions displaying white mycelial boom. A suggested halo of water-soaked, diminished tissue surrounded the lesion among the fringe of mycelium and healthy tissues. The black powder form spores layers were observed on the rotted floor of the fruit.

14.3.2.7.2   Causal Organism

The fungal colonies have been blackish gray, moderately, and dense. The fungus produced aseptate mycelium. The sporangial heads were 30 to 50 μm in diameter with sporangiospores determined linearly within cylindrical sacs (merosporangia) borne on spicules around the columella. Sporangiospores, round to cylindrical in form and borne in chains, measured 3.0 to 5.0 μm long. The fungus turned into morphologically and physiologically recognized as *Syncephalastrum racemosum* Schr. (Neelima Garg et al., 2004).

14.3.2.7.3   Disease Cycle

Syncephalastrum was the only genus inside the *Syncephalastraceae* circle of relatives in the order Mucorales. The soil, dung, plant debries may be naturally sources for the fungus *Syncephalastrum* and it cloud be isolated from these all sources. While it was also has been isolated from infected mycosis of the animals. *Syncephalastrum,* asexual reproduction happens through development of merosporangia leading to the manufacturing of sporangiospores. They also have a defined sexual cycle and broaden to form zygospores.

14.3.2.7.4   Epidemiology

Zygomycetes belong to mucorales. The mucorales were found as terrestrial fungi with characterized as ubiquitous; these molds were substantial found in soil, decaying vegetables, culmination, and seeds (Ribes, 2000). Some species of Mucorales were well known as meals contaminants, for this reason offering an essential supply of statistics concerning the hygiene condition of meals deposits, imperfection at the garage method, intense contamination of grain feedstock and presence of mycotoxins in substrata (Farias et al., 2000; Siqueira, 1995). Colonies of Syncephalastrum develop very hastily and fill the Petri dish or subculture tube. Maximum increase temperature was 40C. Non-dermatophyte molds were a fungus located in soil and decaying plant particles and was usually taken into consideration to be uncommon or secondary pathogens or diseased nails. Prevalence charges of onychomycoses because of non-dermatophyte molds range between 1.45% and 17.60%. The maximum not unusual nondermatophyte molds related to nail ailment were Scopulariopsis, Scytalidium, *Fusarium*, Aspergillus, and Onychocola canadensis. S. racemosum, a non-dermatophyte mold, belongs to the magnificence Zygomycetae.

14.3.2.7.5   Management

The fungicide like Carbendazim @2 ml/liter of water should be applied in the orchard at the time of fruit setting while condition is wet in Aonla growing area. Fruit rot can be managed by spraying of carbendazim 0.1% and need to be done 15 days prior to fruit harvest. The other control measures like post-harvest measures include a hot water bath, which may also should be combined with Carbenazim spraying @2 ml/liter of water at 15 days interval and as per need-based (Bolkan et al., 1976; Couey et al., 1984).

14.3.3 MOLDS

14.3.3.1   SOOTY MOLD (CAPNODIUM, LEPTOXYPHIUM FUMAGO, SCORIAS)

14.3.3.1.1 Symptoms

Sooty mold was a collective, self-descriptive time period for some of one-of-a-kind fungi; its miles a black, powdery coating adhering to flowers and their fruit or environmental objects. The mold blessings from both a sugary exudate produced by means of the plant or fruit, or when the plants were infected with honeydew-secreting insects or sapsuckers. Sooty mold itself could damage in little to the plant to the plant. Treatment was indicated when the mold was blended with insect infestation. Om Prakash (2012) reported that velvety covering of black fungal growth on floor of leaves, twigs, and flower were observed due to infected with sooty mold. The fundamental loss takes area at some stage in transit to the marketplace. The earliest symptom of contamination was visible as water-soaked lesion at the fruit floor, which enlarges in size accompanied by way of improvement of small pinhead size colonies of golden yellow color. The older colonies turn olive green. These were restricted only to the surface and do not penetrate into leaves Sooty molds were a remarkable, but poorly understood group of fungi. They coat fruits and leaves superficially with black mycelia, which reduce photosynthesis costs of host vegetation. Most sooty mold colonies comprise numerous species have affirm *Alternaria, Capnodium, Cladosporium, Fumago, Scorias, Capnodium* sp, with relationships fruits diseases (Chomnunti et al., 2014).

14.3.3.1.2 Causal Organism

The *Capnodium* spp. was common in sooty mold colonies, especially in the warmer parts of the arena and lots of this pycnidial paperwork had been assigned to the call Microxiohium. Mycelium moniliform, septate with subglobose, commonly darkish brown, bureaucracy subiculum on the leaves. Pycnidia flask like regularly branched, narrow at the upper component, and swollen on the middle element, dark brown. Pycnidiospores hyaline, spherical. Perithecia darkish brown, oval to spherical, but barely narrow on the tip and rise up normally from the base of the perithecia. Perithecial wall made up of polygonal cells. Asci elliptical, bitunnicate, and eight-spored, sessile, aparaphysate. Ascospores elliptical to elongated. Leptoxyphium was plant pathogenic fungi, the association occurs worldwide in the shape of sooty mildew and often in natural colonies (Hughes, 1976). *L. fumago* (Srivastava, 1982) on leaves of Kydia pinnata. The genus *Leptoxyphium* specifically characterized via the synnemata springing up from helically twisting hyphae or ropes of repent hyphae with the terminal conidiogenous area. Hyphae are composed of cylindrical cells and mucilaginous hyphal outer walls (Hughes, 1976). This genus has been elevated to encompass eighteen legitimate species consistent with mycobank. The genera *Scorias* were a genus of fungi within the Capnodiaceae family (Lumbsch, 2007). The genus becomes first defined Fries (1932). The fungus was referred to as sooty mold and was found developing on honeydew on leaves of many forms of bushes and flora (Tom, 2010).

14.3.3.1.3   Disease Cycle

When sap-sucking insects, aphids, whiteflies have deposited their honeydew, then the sooty molds was grow in thin black layers on leaves. It does not have longer grown parasitically, but it harms flora in a roundabout way and was likewise unsightly. The mold coats the leaves and this block out mild and makes photosynthesis much less powerful. Plant increase may be decreased, leaves blanketed in mold may also die upfront, and there can be a reduction in fruit yield. According to Hughes (1976), the term sooty mold inside the vernacular sense implemented for any saprophytic fungus generally with darkish colored hyphae which produces brown to black colonies superficially on living plant life. Such molds were regularly related to scale insects and different manufacturers of honeydew; however they can also arise without them. On smaller leaves sooty mildew may additionally form a skinny community of hyphae, a pellided a velutinous growth or a pseudoparenchymatous crust. Hyphae of many sooty molds have a markedly mucilaginous outer wall which absorbs water very conveniently acts as an adhesive and surely maintains a wet leaf floor for a longer length, on trunks and larger branches the growths are normally extra sturdy and could be in type of lumpy pseudo-parenchymatous stromata or spongy subicula composed of loosely interwoven hyphae. The sooty molds were seemed as deriving their nourishment from honeydew produced by means of diverse forms of bugs. The honeydew as 'the liquid droplet extraction from the alimentary track, as launched through the anus through aphids, coccids, and lots of different plant sucker bugs. The honey-dews were complex combos of a massive variety of chemical compounds together with sugars, free amino acids, proteins, minerals, etc. (Auclair, 1963).

14.3.3.1.4 *Epidemiology*

Sooty mildew increased the first grown on leaves, which lasts for the existence of the leaf. The second it persists on stems and twigs of woody plants and this kind was renewed from present parts of the fungus produced the preceding season. During the growing season, the fungi produce spores which could be blown and splashed to honeydew or plant exudate coated surfaces. When conditions were moist, the spores germinate and sooty mildew grows on the medium to be had. Warm temperatures and dry weather increase the prevalence of sooty molds. During dry periods, aphid populations and their honeydew manufacturing typically boom on foliage present process moisture stress. In addition, beneath dry situations, less rain was available to take away or dilute honeydew concentrations appropriate for sooty mold boom on leaves and other surfaces (Gillman, 2011). They were extra common in tropical, subtropical, and warm temperate regions and therefore their incidence in temperate regions was probably to boom with international warming. Sooty molds were rarely parasitized by using fungicolous taxa and those can also have biocontrol capacity. They seemingly grow in severe environments and can be xerophilic. This desires checking out as xerophilic taxa can be of interest for industrial applications. Sooty molds grow on sugars and seem to out-compete typical weed fungi and bacteria. They can also produce antibiotics for this reason and their biochemical capacity for acquiring novel bioactive compounds for medical application was underexplored.

14.3.3.1.5 *Management*

The presence of sooty mold could harm with a demonstration of insect pastime that might also purpose damage. Manage sooty mildew by reducing populations of sucking bugs, aphids, mealy bugs, leafhoppers, and smooth scales that excrete honeydew (Gillman, 2011). It was usually recommended that culmination showing such signs and symptoms were discarded for marketing. Bruising and damage at the time of harvesting need to be avoided. Treatment of culmination with borax and NaCl to control disease. Spray starch @2%, lambda-cyhalothrin @0.05% and wettable sulfur @ 0.2% be jumbled together starch if infection was more.

14.3.3.2 BLUE MOLD (PENICILLIUM CITRINUM)

14.3.3.2.1 Symptoms

The pathogen *Penicillium citrinum* Thom changed into first remoted by using Thom (1910) in the fruit of Aonla. Yield losses due to fruit rot, 5–50% has been mentioned to be sustained by way of growers within the fresh fruit industry or fruit processing enterprise within the Southeast Asia (Thom, 1930). Fruit rot was the maximum critical put up harvest decay of saved Aonla in all of the Aonla growing countries. The disease turned into first located in the United States (Janisiewicz, 1999) and it changed into further mentioned from South America, Southeast Asia, Central, and Southern India, Ceylon, Malaya, Singapore, Poland, Nigeria, Bangladesh, and China. In India, fruit rot fungi become first remoted from rotted Aonla fruit by using Chahal and Singh (1969) and citrinin become first time isolated from a cultural filtrate of *Penicillium citrinum* Thom. (Hetherington and Raistrick, 1931). It becomes first from U.P. pronounced from Varanasi (Shetty, 1954) described symptoms that because of *Penicillium citrinum* reasons brown patches and water-soaked areas at the culmination. As the disease progresses, 31 of a kind forms of colorations broaden, i.e., vivid yellow, purple-brown, and bluish green. There was exudation of drops of yellowish liquid at the fruit. The end result emits a bad odor. The whole fruit sooner or later offers a bluish-green pustulated or beaded look (Shetty, 1959).

The signs produced in Aonla culmination described by using Janisiewicz (1999) and Mishra and Hasib (2004) observed the blue or grayish-spore masses appear at the decayed area of fruit. The decayed place seems light tan to dark brown in color. Decayed tissue was tender, watery, and with ease separated from healthy tissue, leaving like a bowl. The presence of grayish-inexperienced spore hundreds at the decayed vicinity and related musty odor were the undoubtedly diagnostic indication of fruit rot (*P. citrinum*). *P. citrinum* was basically a wound pathogen. Wounds at the fruit skin, which includes punctures and bruises which might be created at harvest or in the course of submit-harvest coping with technique were the fundamental avenues of invasion via the fungus. *P. citrinum* can also reason decay via contamination of lenticels, but this sort of contamination commonly takes place on over mature fruit or when lenticels have been injured. Entire end result can also end up decayed after a few days of harvesting. An unmarried decayed fruit may additionally contain sufficient spores to contaminate water at the entire packing line (Rosenberger, 1990).

14.3.3.2.2   Causal Organism

Hooge et al. (2000) described in details of the fungus classification and anamorph or imperfect stage of pathogen causing fruit rot of Aonla was *P. citrinum* Thom. The cultural characters of *P. citrinum* were studied by using Thom (1914); Malmstrom et al. (2000); and Philip (2004). The colony of the fungus appeared as velvety, bluish-green in color, regularly generating yellow droplets at the surface and yellow in reverse. Conidiophores normally bi-verticillate, branched or seldom branched underneath the extent of metulae and every having flask-formed, asymmetrical sterigmata. Penicillin or conidia were not strongly divaricate, almost globose hyaline with barely brownish wall.

14.3.3.2.3   Disease Cycle

Penicillium spp. starts off evolved its life cycle as a spore which can sit down dormant for a few time. There were several species of *Penicillium* causing blue mold, every of which has a slightly special increase fee or infection speed. Few can have an effect on fruits with unbroken pores and skin, requiring a bruise or softening from over-ripeness to advantage a foothold. Once inside, the mold grows quickly. The mold has gear it makes use of to spread. First, it multiplies quickly, spreading thru the inflamed fruit very quickly. Secondly, it offers off a carcinogen known as mycotoxin patulin. This changes the flavor of the fruit ought to a person eat it, rotting, speeding their ripening, and resulting in gentle skins by which the *Penicillium* can spread.

14.3.3.2.4   Epidemiology

The higher germination and growth of *P. citrinum* changed into at 20°C however not at 40°C said with the aid of Halouat and Debevere (2003). Relative humidity 100% performs the vital role in the boom and sporulation of the fungus. RH control was required to provide the gold standard surround-ings for fresh end result for the duration of the garage. pH was a very vital element for the increase and sporulation of the *P. citrinum*. They determined *Aspergillus spp.* was more tolerant of alkaline pH however *Penicillium spp.* seemed to be extra tolerant of acidic pH Wheeler et al. (1991). Gopal (1985) observed the maximum vegetative boom and sporulation of *P. expansum* on pear fruit at neutral pH 7.

14.3.3.2.5   Management

The careful must be taken during handing of culmination. Any harm at the fruit surface throughout harvesting and storage make the Aonla fruits liable to blue mold. Avoid bruising or damage to fruits even as harvesting. Sanitary conditions in garage need to be maintained. Treatment of fruits with borax or sodium chloride (1%) exams the blue mold infection. Treatment with Carben-dazim 2 ml/liter of water after harvest. Sharma and Shukla (2003) stated the fruits packed in wood containers with iodine and di-phenylamine impreg-nated liners reduced the smooth rot infections by 80% and 70% according to percent. Rathod and Patel (2005) said the Carbendazim (500, 1000 ppm) and Mencozeb (2000, 4000 ppm) had been simplest in opposition to the *Penicil-lium* rot culmination in pre and submit-inoculation treatments. Jain (2006) observed that the culmination had been handled by Calcium Nitrate 1% and then packed in perforated polythene bags, reduced percentage physiological weight loss, pathological loss and lack of ascorbic acid (beneath 10%) in the course of garage Todkar et al. (2005) found the *T. harzianum* become superior as compared to *T. viride* in inhibiting the boom of *Aspergillus niger* and other mold fungi even as, *T. viride* become advanced.

14.3.3.3   BLACK MOLD (ASPERGILLUS NIGER)

14.3.3.3.1   Symptoms

Aspergillus awamori and *Aspergillus niger* van Tiegh causing surface rot of Aonla said by using Srivastava et al. (1964). Most of the 13 fungi isolated from saved fruits of Emblica officinalis in India by means of trendy blotter and agar plate techniques were Deuteromycetes. *Aspergillus spp.* was predominant (Misra and Misra, 1988). Symptoms had been described by way of Prakash and Srivastava (1987). Disease develops best in the injured fruits. The infection begins as stalk give up however the aspect of the fruit was not often involved. Grayish or pale brown spots may additionally appear on the fruit floor coalescing into dark brown or black lesions which can be tender and sunken, and later protected with a sooty mass of black spores. On mature lesions there may be formation of round resting bodies (sclerotia) about 1 mm in diameter, first of all white and later darkish brown (Snowdon, 1990).

14.3.3.3.2   Causal Organism

Aspergillus niger was a fungus and one of the most common species of the genus a set of fungi which might be generally taken into consideration asexual. It is ubiquitous in soil and was usually said from indoor environments, *A. niger* was commonly observed as a saprophyte growing on dead leaves, saved grains, compost piles and different decaying vegetation. Microscopically, its conidiophores were clean-walled, hyaline or turning darkish closer to borne on brown, regularly septate metulae. Conidia were globose to subglobose, dark brown to black and hard-walled (Kumari and Singh, 2016). *A. niger* was a filamentous fungus growing aerobically on natural matter. In nature, it's far determined in soil and litter, in compost and on rotten plant material Reiss (1986).

14.3.3.3.3   Disease Cycle

Aspergillus niger typically reproduces within the asexual kingdom, even though sexual replica has been located. In the typical asexual country, conidia (spores) were launched and disseminated through wind. Conidia germinated into hyphae. Following hyphal colonization of the substrate, aerial hyphae will emerge, producing conidiophores (stalks) and conidial heads that shape conidia. This fungus was spread via the air, soil, and water. It was normally a saprophyte, living off useless and decaying matter. Therefore, it's far commonly visible as a publish-harvest ailment. In the case of people and animals, a compromised immune system was commonly present when the disorder manifests. In vegetation, an irrigation practice which includes drip irrigation traces buried in soil and hot, humid growth situations were conducive to disease improvement (Sharma, 2012).

14.3.3.3.4   Epidemiology

A. niger was capable of growing in the extensive temperature variety of
6–47°C. The increase temperatures for *Aspergillus niger* was minimum,
6–8°C, most, 45–47°C and superior 35–37°C (Panasenko, 1967). *A. niger*
was a xerophile, *A. niger* was able to grow all the way down to pH 2.0 at
excessive. Some *A. niger* isolates can produce ochratoxin. *Aspergillus niger*
was used as a cellular manufacturing unit for the manufacturing of enzymes.
Aspergillus species chargeable for post-harvest decay of as an example
guava, litchis, mangoes, papaya, pineapples, pomegranates, apples, pears,
and grapes. Other food merchandise such as onions, rice, coffee, nuts, and
sunflower seeds also were substrates for *A. niger*. This fungus was present in
soil and inside the atmosphere, but it infects citrus culmination best at fairly
high temperature and best if they were weakened in a few way. The disease
was not unusual in the rainy season, while the relative humidity was high
and temperature ranges between 30°C and 35°C. Injury allows the pathogen
to purpose infection. Initially, small round water soaked spots seem (Khanna
and Chandra, 1975). Ghosh et al. (1966) discovered a fast decrease in
ascorbic acid content of the infected fruits. After 14 days of contamination by
using *Aspergillus sp.* The lack of ascorbic acid in inflamed aonla culmination
becomes 84.2% as compared with 41% in wholesome fruit (Jamaluddin et al.,
1974). Majumdar and Modi (1980) studied mango culmination infected with
A. flavus exhibited a higher degree of citric acid total acidity and sucrose. A
have a look at of metabolic pattern of the host plant revealed diverse challenges induced at some stage in pathogenesis. Aonla fruits contain different
sugar as known maltose, glucose, and fructose. During Incubation duration,
there was a sluggish loss in the quantity of maltose. The discount within the
total organic acid contents within the inflamed fruit may be attributed to the
host-pathogen interaction.

14.3.3.3.5   Management

The cautious managing was important in any respect levels to save you
mechanical harm to the fruit. Black mildew can be decreased by pre-harvest
sprays of fungicide (Subramanyam et al., 1972) and by means of publish
harvest remedies with warm water containing fungicide (Pandey et al.,
1990a). Storage at temperatures between 10°C and 15°C prevents improvement of black mold rot. Dipping of culmination inside the fungicide reduces

the possibilities of rot within the marketplace. Jadeja (1991) located that mango fruit rot due to *A. niger* there has been giant reduction in sickness. Intensity inside the hot water treatment of 55°C for 10 minutes and Carbendazim. Saini and Pathak (1991) attempted exclusive phyto-extracts as pre and post-inoculation treatments on fruits and minimum decay changed into discovered in Tulsi leaf extract.

14.3.4   ANTHRACNOSE (COLLETOTRICHUM (GLOMERELLA CINGULATA STOENM.) SPAULD AND SCHRENK.)

14.3.4.1   FALSE ANTHRACNOSE (KABATIELLA EMBLICA)

14.3.4.1.1   Symptoms

This pathogen was a huge trouble global wide, inflicting anthracnose and fruit rotting sicknesses on masses of economically essential hosts. The initial symptom of the disorder was leaflets with yellowish margin in the shape of minute, round, brown to gray spots. The crucial region of the spot stays grayish raised with dot like fruiting our bodies. The symptoms on culmination-depressed lesions broaden which later flip darkish in color bearing dot-like fruiting our bodies arranged in rings. Consequently the inflamed culmination become shrunk and rots. At high humidity on fruiting bodies, the lesion may additionally range in length and form with spore later the fruits grown to be shrunk and rot. The pathogen was favored through warm and humid weather conditions. Anthracnose (*C. gloeosporioides*) disease appears on leaflet and fruit at some stage in August-September (Mishra and Shivpuri, 1983). The false anthracnose caused by means of *Kabatiella emblica* was a newly mentioned pathogen from China (Zhang and Qi, 1996).

14.3.4.1.2   Causal Organisms

Glomerella cingulata was the sexual level (teleomorph) whilst the more normally noted asexual level (anamorph) was referred to as C, gloeosporioides. *Colletotrichum spp.* primarily based on their cultural characteristics, specifically whitish, grayish, and creamiest color and cottony/velvety mycelia on the pinnacle aspect of the culture and grayish cream with concentric zonation at the reverse aspect. *G. cingulata* spores were found non-septate and rounded stop.

14.3.4.1.3 Disease Cycle

The pathogen found in host in the form of teleomorph or anamorph stage. This distinction influences how the pathogen overwinters or survives durations without an inclined host. If the sexual stage (teleomorph) was a gift, the pathogen sexually reproduces to shape ascospores interior of asci, and ultimately packed into perithecia. This approach gives genetic variant and the convenience of perithecia that was found the reproduction as an asexual survival structure for next generation. Anamorph stage was recorded in plant tissues and alternate host for survival at 95% RH, 25–28°C, the ascospores were ejected and inflamed plant tissue sporulates. Ascospores infect immediately, even as the masses of conidia were produced from conidiophores and inflamed plant tissue produces acervuli. Sharma and Kulshrestha (2015) reported that masses of conidia were disseminated by using rain splash or wind onto new infection courts along with leaves, young fruit, or blossoms. When infection taken place the pathogen continues to supply conidia at some stage in the season resulting in a polycyclic sickness cycle. Once the host plant begins to senesce the teleomorph form of the disease (G. cingulata) will start to sexually produce ascospores in perithecia to restart the cycle.

14.3.4.1.4 Epidemiology

The fungus produces both the sexual stage (perithecia giving upward push to ascospores) and the asexual level (acervuli giving upward thrust to conidia) and the spore-bearing our bodies from in useless branches. The ascospores were not an idea to play an important function in infection (Fitzell and Peak, 1984). In moist seasons acervuli were produced in abundance on all components of tree and contamination of culmination was chiefly by way of conidia washed down by using rain: if moisture persists for numerous hours, the fungus was capable of direct penetration of the intact pores and skin of younger culmination (Wardlaw and Leonard, 1936). The pathogen survives in diseased debris or diseased twigs on the bushes (Sattar and Malik, 1939; Saaiman, 1997). The top of the line temperature for infection became observed to be 25°C. Cloudy and misty day favors the disease improvement. The pathogen enters fruits via pores of inexperienced end result and develops in flesh all through ripening.

14.3.4.1.5   Management

Removal and destruction of infected plant parts for the reduction of fruit rot (Sattar and Malik, 1939; Saniman, 1997). Spraying of Bordeaux mixture found effective for the management of the disease during occurrence at after harvest, inclusive of (Tandon and Singh, 1968). Sprays of Copper Oxychloride in mango orchard gave notably higher manipulate of Anthracnose (Lonsdale et al., 1994). Dithane M-45 gave the high quality management of anthracnose observed through Bordeaux combination (Ahmed et al., 1991). Various people have reported the efficacy of warm water remedy. Hot water remedies at 54.5–56.5°C for 10 minutes (Tandon and Singh, 1968). Disease can be controlled by dipping the culmination in diverse fungicides or by exposing them in Ammonia, Sulphur Dioxide, and Carbon Dioxide gases and additionally by hot air treatment (Tandon and Singh, 1968). Hot water treatments at 50–55°C for 5–15 minutes supply good manage of storage rot (Jacobs et al., 1973). Patel (1978) studies suggested that hyto-extracts of Kala Dhatura, ginger, and turmeric were discovered effective in inhibiting the boom of *C. gloeosporioides*

14.3.5   LEAF SPOTS

14.3.5.1   BROWN SPOT (PHYLLOSTICA EMBLICA, PESTALOTIOPSIS LEAF SPOT (P. HETEROCORNIS)

14.3.5.1.1   Symptoms

Phyllostica leaf spots with white centers, darkish brown margins and yellow halos. The spots had been circular or oval or elongated, typically isolated but occasionally confluent, causing intensive discoloration and the sickness changed into characterized via oval to lengthen or sometimes irregular spots white and papery at the center and had a dark brown margin with a yellowish halo. The spots were commonly remoted, but coalesced forming a bog lesion beneath humid situations, discoloration of leaves changed into evident from the distance in the subject, black, minute pycnidia inside the white papery area of roots were conspicuous in later ranges. The lesions of on leaves as numerous circular or abnormal with pycnidia later form brown and dry with quick-holes. The disease symptoms showed several circular, with white centers oval or elongated leaf spot, brown margins and yellowish halos which gave a standard yellowing appearance to the crop.

14.3.5.1.2 Causal Organism

These were brown spot (*P. emblicae*), false anthracnose (*K. emblicae*), Pestalotiopsis leaf spot (*P. heterocornis*), and powdery mildew (*Oidium sp.*). The first two species were described as new species and the third was a brand new report for China (Zhang et al., 1996).

14.3.5.1.3 Disease Cycle

The low temperature and humidity from June to September favored activity of the leaf spotting fungi such as Phyllosticta spp. but reduction in the rainy season (21–31°C) in July to October when the average temperature and RH were 26–30°C and 73–89%, respectively. Leaves were susceptible at 12–57 days after emergence under glasshouse conditions. They also discovered that pycnidia spore germination took place in 6–8 hours pycnidia were shaped extra rapidly at decrease air temperatures; leaf accidents shorten the incubation length through 2 days and leaves much less than 2 days vintage had been not inclined.

14.3.5.1.4 Epidemiology

A temperature range of 23–28°C with intermittent showers favored the disorder (Phyllosticta leaf spot) improvement and observed that the disease occurrence become most in the course of September, when temperature ranged from 22.6 to 29.20°C and relative humidity from 77–93 consistent with wide variety of wet days. The disorder spread through rain splashes. New foliage of ginger fantastically prone and the susceptibility decreased with growth in maturity of leaves. The number of pycnidia and pycnidiospores in line with the unit had been proportional to lesion vicinity. Disease reaction during peak ailment periods, i.e., October, and November in 1–9 scale. The minimal and maximum temperature (25.4°C to 34.6°C) and relative humidity (65–92%) prevailed for the duration of the disorder improvement with overall rainfall of 694.8 mm to 774.4 mm at some point of August to November.

14.3.5.1.5 Management

Bordeaux mixture (1%) spraying once or twice during the season checked *Phytlosticta*. Dithane M-45 proved effective in controlling the Phytlosticta leaf spot under field conditions.

14.3.6 WILT

14.3.6.1 FUSARIUM WILT (FUSARIUM SOLANI)

14.3.6.1.1 Symptoms

A new vascular wilt of Aonla was recorded in the course of the monsoon duration (July-August) in six months old grafted saplings at Balaghat and Seoni, Madhya Pradesh, India. The prevalence of disease various from 2.5 to 16% the nurseries, affecting saplings. Initially, the flowers confirmed yellowing of number one leaves followed by means of unexpected wilting and drying of seedlings inside some days. In older plants, signs of the disease were exhibited by the decrease most leaves which confirmed drooping, vein clearing and chlorosis. The disease affected eight kinds of Aonla *viz.,* Chakaiya, Francis, N-7, Kanchan, Anand-1, Anand-2, Hathizol, and Krishna. Repeated isolations from inflamed root areas yielded *Fusarium solani* (Soni and Verma, 2010).

14.3.6.1.2 Causal Organisms

F. solani produced colonies that were white, pink or violet blue-bluish brown center, faded, tea-with-milk-brown, or red-brown and cottony. Colonies were low floccose, loose, slimy, and sporadical on PDA, Aerial hyphae that given upward thrust to conidiophores. The conidia produced on conidiophores that branch into thin, elongated monophialides, macro-conidia, and micro-conidia was produced on phialides. The macro-conidia produced by means of *F. solani* were slightly curved, hyaline, and broad, frequently aggregating in fascicles. Typically the macro-conidia of this species have 3–5 septa. Thickened basal cells and tapered, rounded apical cells were characterized the micro-conidia. However, a few *F. solani* isolates have pointed, in place of rounded, macro-conidia. Micro-conidia are oval or cylindrical, hyaline, and easy. Some micro-conidia can be curved. Micro-conidia generally lack septa, but now and again have up to. *F. solani* also forms chlamydospores most generally underneath suboptimal growth situations. These can be produced in pairs or individually. They were plentiful, have difficult walls, and were 6–11 m. The chlamydospores were also spherical and brown.

14.3.6.1.3 *Disease Cycle*

F. solani was observed in soil global as per ecology and other complex, the fungus associated with the roots of vegetation and as deep inside the ground as eighty cm. It was regularly remoted in tropic, sub-tropic, and temperate locations, and less often isolated from alpine habitats. The pH of soil does not have a big impact on *F. solani*, however, soil fumigation causes an boom in occurrence. *F. solani* is usually sensitive to soil fungicides. *F. solani* has been discovered in ponds, rivers, sewage facilities, and water pipes.

14.3.6.1.4 *Epidemiology*

The temperature (15–35°C) and humidity (80–100%) for the spore germination of *F. solani*.

14.3.6.1.5 *Management*

In vitro tests had been made with six fungicides at 0.1% and 0.2%. Metalaxyl + Mancozeb (Ridomil) at and Carbendazim (Bavistin) at 0.2% have been located powerful against the disease. The disease became controlled in primary propagation nurseries of Balaghat and Seoni with applications of Ridomil (0.2%) at month-to-month intervals. Watering schedule was additionally monitored to avoid greater moisture in and round root zone of Aonla saplings and encourage clean air movement in nursery shade house (Soni and Verma, 2010).

14.4 LICHENS

14.4.1 *STRIGULA ELEGANS (FEE.) MULL. ARG.*

14.4.1.1 SYMPTOMS

Look for small (1–10 mm) gray-white spots at the higher leaf surfaces of plants that have leathery and continual leaves. Hand lens viewing might also screen black fungal fruiting structures within the leaf spots. Lichens were found on the surface of the trunk of the grown up trees. It was seen within the shape of whitish, pinkish, superficial patches of different shapes on the primary trunk and branches of the tree.

14.4.1.2   CAUSAL ORGANISM

Associations among algae and fungi called lichen. The genus *Strigula* contains some of species of plant-parasitic lichens which arise on leaves and very rarely on stems. The algal factor of *Strigula* was the well-known causal agent of algal leaf spot or inexperienced scurf, *C. virescens*. The parasitic capability of the lichen was attributable in particular to *C. virescens*. Fungal buddies of *Strigula* were acknowledged from numerous genera of fungi; the Ascomycetes Massaria and Microthyriella were most generally mentioned. Taxonomic classification in the lichens presently was primarily based primarily at the fungal aspect of the lichen (Hale, 1967), therefore upon taxonomic revision, the taxon *Strigula* may additionally within the destiny be separated into numerous new genera to accommodate every of the fungal components.

14.4.1.3   DISEASE CYCLE

The lifestyles cycle of the lichen on an algal thallus which colonization. *Strigula spp.* and typical, non-lichenized thalli of *Cephaleuros* can be discovered. The lichens are sub-cuticular gray-white crusts, 1–10 mm in diameter at the upper leaf surface, assisting black fungal fruiting structures. *Strigula* did not longer produce soredia. Reproduction of *Strigula* evidently relies upon risk encounters of the separate organisms which reproduce normally, *Cephaleuros* by zoospores from sporangia, and the fungus with the aid of conidia or ascospores from pycnidia or perithecia, respectively.

14.4.1.4   EPIDEMIOLOGY

Most species in temperate regions grown 0.5–8 mm at 12 months and that both seasonal and annual variation in Large availability of moisture and temperature have the greatest have an effect on boom little was regarded about life cycles besides that a brief juvenile following status quo of a propagules precedes a longer length of a rapid increase leading to a mature thallus. Senescence was reached because the thallus processes a most diameter or length that was species-specific. Growth rate sluggish and the crucial or older portions of the thallus sooner or later die away.

14.4.1.5   MANAGEMENT

Gunny rubbing, accompanied by using spraying of trunk and branches with commercial caustic soda (1%) earlier than onset of rains (Om Praksh, 2012).

14.5   DISORDER

14.5.1   INTERNAL NECROSIS (PHYSIOLOGICAL)

14.5.1.1   SYMPTOMS

The symptom starts with the browning of the innermost part of meso-carpic tissue on the time of endo-carp hardening and extends in the direction of the epi-carp resulting into brownish-black appearance of the flesh. In case of extreme incidence, those black spots turn out to be corky and gummy pockets increase (Om Prakash, 2012).

14.5.1.2   MANAGEMENT

Keep the orchards smooth and healthy to save you the infestation of this pest. Detect early infestation via periodically searching out for drying young shoots. In case of extreme infestation, dispose of webs and insert swab of cotton wool soaked in 0.025% dichlorvos or inject water emulsion of Chlorpyriphos (0.05%) and plug the holes. The larvae were parasitized by entomogenous fungus *B. bassiana* in nature. It may be used as an ability bio-manipulate agent (Om Prakash, 2012).

KEYWORDS

- aonla
- black mold
- blue mold
- *Emblica officinalis*
- leaf spot
- rhizopus rot

REFERENCES

Ahmed, H. U., Hossain, M. M., Alam, S. M. K., & Huq, H. M., (1991). Efficacy of different fungicides in controlling anthracnose and sooty mold of mango. *Bangladesh J. Agril. Res., 16*(1), 74–78.

Amy, C. J., & Rowe, R. C., (1991). Effects of temperature and duration of surface wetness on spore production and infection of cucumbers by *Didymella bryoniae*. *Phytopath., 81,* 206–209.

Anonymous, (1989). *Annual Report.* National Bureau of Plant Genetic Resources, New Delhi, India.

Anonymous, (2002). *Annual Report of Central Institute for Sub-Tropical Horticulture.* Lucknow, India.

Anonymous, (2002). *Final Report on Market Study of Aonla.* Submitted to UP Land Development Corporation (UPLDC) by M/s WIZMIN, Kanpur. www.wizmin.com (accessed on 12 February 2021).

Anonymous, (2002/2003). *Epidemiological Studies on Aonla Rust* (pp. 138, 139). Biennial report of AICRP on arid zone fruits (ICAR).

Auclair, J. L., (1963). Aphid feeding and nutrition. *Annual Rev. Entom., 8,* 439–490.

Bhardwaj, S. S., & Sharma, I. M., (1999). Diseases of minor fruits. In: Verma, L. R., & Sharma, R. C., (eds.), *Diseases of Horticultural Crops-Fruits* (pp. 540–562). Indus Publishing Co. New Delhi.

Bhargava, P. K., & Khare, M. N., (1988). Epidemiology of *Alternaria* blight of chickpea. *Indian Phytopath., 41,* 195–198.

Boerema, G. H., & Vorhoeven, A. A., (1972). Checklist for scientific names of common parasitic fungi. *Series a Fungus on Netherlendo Journal of Plant Aetho., 78,* 163.

Bolkan, H. A., Cupertino, F. P., Dianese, J. C., & Takatsu, A., (1976). Fungi associated with pre- and post-harvest fruit rots of papaya and their control in Central Brazil. *Pl. Dis. Reptr., 60,* 605–609.

Chahal, D. S., & Singh, N., (1969). Fruit rot of aonla. *FAO Plant Protection Bulletin., 14,* 67.

Chomnunti, P., Sinang, H., Aguirre-Hudson, B., Qing, T., Derek, P., Manpreet, K., Dhami, Aisyah, S., et al., (2014). The sooty molds. *Fungal Diversity, 66*(1), 1.

Couey, H. M., Alvarez, A. M., & Nelson, M. G., (1984). Comparison of hot water spray and immersion treatments for control of post-harvest decay of papaya. *Plant Dis., 68,* 436, 437.

Cristescu, C., (2007). The morphology and anatomy of structure somatic and reproductive of species of Phomopsis Sacc. *Bubak. Buletinul Grădinii Botanice Iaşi Tomul., 14*, 19–27.

Devesh, A. S., Sanjeev, K., Sukhvinder, S., & Govind, V., (2015). Epidemiology and management of rust disease (*Ravenelia emblicae* Syd.) of aonla. *Res. Environ. Life Sci., 8*(2), 213, 214.

Dewdney, M. M., (2015). *Alternaria Brown Spot.* EDIS New Publications RSS. Web.

Dugan, F. M., Schubert, K., & Braun, U., (2004). Check-list of *Cladosporium* names. *Schlechtendalia, 11*, 1–103.

Elliott, M. L., Broschat, T. K., Uchida, J. Y., & Simone, G. W., (2004). *Diseases and Disorders of Ornamental Palms.* American Phytopathological Society, St. Paul.

Ellis, M. B., (1971). *Dematiaceous Hyphomycetes.* Kew: CMI.

English, H., (1945). Fungi isolated from moldy sweet cherries in the Pacific Northwest. *Pl. Dis. Reptr., 29*, 559–566.

Espinoza, J. G., Briceno, E. X., Keith, L. M., & Latorre, B. A., (2008). Canker and twig dieback of blueberry caused by *Pestalotiopsis* spp. and a *Truncatella* sp. in Chile. *Plant Dis., 92*, 1407–1414.

Farias, A. X., Robbs, C. F., Bittencourt, A. M., Andersen, P. M., & Corrêa, T. B. S., (2000). Contaminação endógena por *Aspergillus* spp. em milho pós-colheita no estado do paraná Pesq. *Agropec. Bras., 35*, 617.

Fitzell, R. D., & Peak, C. M., (1984). The epidemiology of anthracnose disease of mango: Inoculum source. spore production and dispersal. *Ann. Appli. Biol., 104*, 53–59.

Fries, E. M., (1832). *Systema Mycologicum, 3*, 269, 290.

Ghosh, A. K., Bhargava, S. N., & Tandon, R. N., (1966). Studies on fungal diseases of some tropical fruits: IV. Post-infection change in ascorbic acid contents of mango and papaya. *Indian Phytopath., 19*(3), 262–268.

Ghosh, A. K., Tandon, R. N., Bhargava, S. N., & Srivastava, M. P., (1965). Vitamin C content of guava fruits after fungal infection. *Naturwissenschaften, 52*, 478.

Gillman, D. P., & Stipes, R. J., (2011). *Diseases of Woody Ornamentals and Trees.* APS Press and R. Scott Cameron, International Paper. https://ag.umass.edu/landscape/fact-sheets/sooty-mold (accessed on 12 February 2021).

Gopal, K. R., (1985). *Studies on Blue Mold rot of Pear Caused by Penicillium expansum Thom* (p. 71). MSc Thesis, Dr. Y.S. Parmar University of Horticulture and Forestry, Nauni, Solan (H.P.).

Hale, M. E., (1967). *The Biology of Lichens* (pp. 147–157). Edward Arnold Pub., Ltd., London.

Halouat, A., & Debevere, J. M., (2003). Effect of water activity and storage temperature on spore germination of molds isolated from prunes. *Int. J. Food Microbiol., 81*(2), 15–20.

Hetherington, J., & Raistrick, M., (1931). Isolation and identification of *Penicillium citrinum. Mycologia., 8*, 225–230.

Hoog, G. S. D., (2000). *Atlas of Clinical Fungi*, 1–1126.

Hooge, G. S., Guarro, J., & Figueras, M. J., (2000). *Atlas of Clinical Fungi* (Vol. 1, No. 2). Centralbureau voor Schimmelcultures, Utrecht, The Netherlands.

Hopkins, K. E., & McQuilken, M. P., (1997). *Pestalotiopsis on Nursery Stock, in HDC Project News No 39.* Horticultural Development Council, East Malling.

Hughes, S. J., (1976). Sooty molds. *Mycologia., 68*, 693–820.

Jacob, C. J., Brodrive, H. T., Swarts, H. D., & Mudler, N. J., (1973). Control of postharvest decay of mango fruit in South Africa. *Plant Disease Reporter, 57*(3), 173–176.

Jadeja, K. B., (1991). *Investigation on Postharvest Diseases of Mango Fruits and Their Management.* PhD thesis submitted TC Gujarat Agricultural University, S.K. Nagar (Junagadh Campus).

Jain, V., (2006). *Studies on Evaluation of Aonla (Emblica officinalis Gaertn.) Cultivars for Fruit Growth, Self-Life and Processing Technology of Beverages Under Chhattisgarh Condition.* PhD Thesis, Department of Horticulture, I.G.A.U., Raipur, (C.G.).

Jamaluddin, (1978). *Cladosporium* rot of fruits of *Phyllanthus emblica* caused by *Phomophsis phylanthi Punith* and its chemical control. *Nat. Acad. Sci. India., 5*(6), 183–185.

Jamaluddin, Tandon, M. P., & Tandon, R. N., (1974). Post-infection changes in ascorbic acid contents of aonla fruits caused by *Aspergillus niger* van. Tiegh. *Curro Sci., 43*(7), 218, 219.

Jamaluddin, Tandon, M. P., & Tandon, R. N., (1975). A fruit rot of aonla caused by *Phoma. Proc. Natl. Acad. Sci. India, 8*(45), 75–77.

Janisiewicz, W., (1999). Blue mold, *Penicillium* spp. In: *Compendium of Apple and Pear Diseases* (pp. 54, 55). Press, St. Paul, M.N.

Jeewon, R., Liew, E. C., & Hyde, K. D., (2004). Phylogenetic evaluation of species nomenclature of *Pestalotiopsis* in relation to host association. *Fungal Diversity, 17,* 39–55.

Joubert, J. J., & Rijkenberg, F. H. J., (1971). Parasitic green algae. *Ann. Rev. Phytopathol., 9,* 45–64.

Kamthan, K. P., Misra, R., & Shukla, A. K., (1981). *Nigrospora* fruit rot of *Emblica officinalis*: A new disease record. *Sci. Cult., 47,* 371, 372.

Keith, L. M., Velasquez, M. E., & Zee, F. T., (2006). Identification and characterization of *Pestalotiopsis* spp. causing scab disease of guava, *Psidium guajava* in Hawaii. *Plant Disease, 90,* 16–23.

Khanna, K. K., & Chandra, S., (1975). Studies on storage diseases of fruits and vegetables: III. Factors affecting storage rot of lime caused by *Aspergillus niger. Proc. Natl. Acad. Sci. India, 45,* 109, 110.

Kumari, M., & Singh, M., (2016). Morphological and pathogenic variabilities among *Aspergillus niger* isolates associated with groundnut (*Arachis htpogea.* L.). *Ann. Pl. Protec. Sci., 24*(2), 364–368.

Lal, B., Arya, A., & Rai, R. N., (1982). A new soft rot of aonla caused by *Phomopsis phyllanthi Punith* and its chemical control. *Nat. Acad. Sci. Letters., 5*(6), 183–185.

Liu, A. R., Wu, X. P., Xu, T., Guo, L. D., & Wei, J. G., (2006). Notes on endophytic *Pestalotiopsis* from Hainan, China. *Mycosystema, 25,* 389–397.

Lonsdale, J. H., Lonsdale, J. M., Greef, H., & Brooks, W., (1994). *The Efficacy of Prochloraz Chloramezal Sulphate and Guazatine on Post-Harvest Diseases of Mango* (Vol. 11, pp. 35–38). Yearbook South African Mango Growers Association.

Lumbsch, T. H., & Huhndorf, S. M., (2007). Outline of Ascomycota-2007. (Vol. 13, pp. 1–58). Myconet, Chicago, USA: The Field Museum, Department of Botany.

Luo, L., Xi, P., Jiang, Z., & Qi, P., (2004). Sporulation conditions of *Phomopsis* in pure culture. *Mycosystema, 23*(2), 219–225.

Maheshwari, S. K., Singh, D. V., & Saha, A. K., (2000). Role of environmental factors on *Alternaria* leaf spot of dolichous bean. *J. Mycopatho. Res., 38,* 81–83.

Majmudar, G., & Modi, V. V., (1980). Spoilage of mango by *Aspergillus flavus. Curr. Sci., 49*(21), 821, 822.

Malmstrom, J., Christophersen, C., & Frisvad, J. C., (2000). Secondary metabolites characteristics of *P. citrinum, P. steckii* and selected species. *Phytochemistry, 54,* 301–309.

Mason, E. W., (1927). On species of the genus *Nigro-spora* Zimmermann recorded on monocotyledons. *Transactions of the British Mycological Society, 12*(2/3), 152–165.

Mathur, S. B., & Sarbhoy, A. K., (1983). Chemical control of *Alternaria* leaf spot of sugar beet. *Indian Phytopath., 36*(4), 677–679.

McQuilken, M. P., & Hopkins, K. E., (2004). Biology and integrated control of *Pestalotiopsis* on container-grown ericaceous crops. *Pest Manag. Sci., 60,* 135–142.

Meredith, D. S., (1961). Atmospheric content of *Nigrospora* spores in Jamaican banana plantations. *J. Gen. Microbial., 26,* 343–349.

Mishra, A. K., & Hasib, M., (2004). Aonle main rog aur kit prabandh. *Phal. Phool.,* 17–20.

Mishra, A., & Shivpuri, A., (1983). Anthracnose: A new disease of aonla. *Indian Phytopath., 36,* 406–407.

Misra, N., & Misra, N., (1988). Studies on fungi deteriorating stored fruit of *Emblica officinalis* gaertn. *International J. Trop. Pl. Dis., 6*(1), 95–97.

Morton, J., (1987). *Emblica.* In: *Fruits of Warm Climates* (pp. 213–217). Miami, Florida.

Mostert, L., Crous, P. W., Kang, J. C., & Phillips, A. J. L., (2001a). Species of *Phomopsis* and a *Libertella* sp. occurring on grapevines with specific reference to South Africa: Morphological, cultural, molecular and pathological characterization. *Mycologia., 93,* 146–167.

Nair, K. P. P., (2011). Agronomy and economy of black pepper and cardamom, the king and queen of spices. *Mycobiology, 43*(3), 354–359.

Nam, M. H., Soo, M. P., Soo, H. K., Tae, K., & Kim, H. G., (2011). *Cladosporium cladosporioides* and *C. tenuissimum* v cause blossom blight in strawberry in Korea. *Agronomy and Economy of Black Pepper and Cardamom the King and Queen of Spices,* 109–366.

Nath, K., Solanky, K. U., & Kumawat, G. L., (2014). Effective approaches of potential bioagent, phytoextract, fungicide and cultural practice for management of banana fruit rot disease. *J. Plant Pathol. Microb., 5,* 6.

Neelima, G., Om, P., Pandey, B. K., Singh, B. P., & Pandey, G., (2004). First report of black Soft rot of Indian gooseberry caused by *Syncephalastrum racemosum, Plant Disease, 88*(5), 575.

Om, P. P. M., (2012). *Schedule for Aonla Pests Extension Bulletin No.05.* Horticulture Year, National Horticulture Mission Ministry of Agriculture Department of Agriculture and Cooperation Krishi Bhawan, New Delhi – 110001.

Om, P., & Mishra, A. K., (1993). Fungal disease of subtropical fruits. In: Chadha, K. L., & Pareek, O. P., (eds.), *Advances in Horticulture: Fruit Crops: Part 3* (Vol. 33, pp. 1275–1348). Malhotra Publishing House. New Delhi.

Om, P., & Mishra, A. K., (1995). Some new post-harvest diseases of aonla (*Phyllanthus emblica*) and management. *Annual Conf. and Nat. Sym. Recent Trends in Management of Biotic and Abiotic Stresses in Plants* (pp. 37, 38). HPKV, Palampur.

Panasenko, V. T., (1967). *Botanical Review, 33*(3), 189–215. https://www.jstor.org/stable/4353740 (accessed on 12 February 2021).

Pandey, A., Sukla, A. N., & Chandra, S., (2006). *Pestalotiopsis* stem canker of *Jatropha curcas. Indian Forester, 132,* 763–766.

Pandey, R. S., Bhargava, S. M., Shukla, D. M., & Dwivedi, D. K., (1984). Two new fruit disease caused by *Alternaria alternata. Int. J. Trop. Plant. Disease, 2*(5), 79, 80.

Pandey, R. S., Bhargava, S. N., Shukla, D. N., & Khati, D. V. S., (1979). Control of *Rhizopus* rot of jackfruits. *Indian Phytopath., 32,* 479, 480.

Pandey, R. S., Bhargave, S. N., Shukla, D. N., & Divedi, D. K., (1984). Two new fruit diseases of aonla caused by *Alternaria* sp. *Int. J. Trop. Pl. Dis., 2*, 79, 80.

Pandey, R. S., Dwivedi, O. K., Shukla, D. N., & Bhargava, S. N., (1990a). Two new fungicides for the control of *Aspergillus* rot of mango. *Natl. Acad. Sci. Lett., 3*, 263, 264.

Pandey, R. S., Shukla, D. N., Khati, D. K. S., & Bhargava, S. N., (1980). A new fruit rot of *Phyllanthus emblica. Indian Phytopath., 33*(3), 491.

Pandey, S., Singh, A. K., & Singh, A. K., (2007). Effect of weather parameters on aonla rust (*Ravenalia emblicae* Syd.) *Indian Phytopath., 61*, 412.

Pareek, S., & Kitinoja, L., (2011). *Aonla (Emblica officinalis Gaertn.) in Postharvest Biology and Technology of Tropical and Subtropical Fruits: Açai to Citrus.*

Patel, J. C., (1978). *Studies Regarding Anthracnose of Mango Caused by Colletotrichum Gloeosporio.* des Penz. MSc (AgrL) thesis. Gujarat Agricultural University S.K. Nagar.

Philip, B. M., (2004). *Training in Mold Isolation, Identification, Handling and Evaluation of Conditions Leading to Mycotoxin Production* (p. 82). UNDP, FAO, U.S. Food and Drug Administration.

Philip, S., (1979). Aspergillus rot of pomegranate fruit. *Indian Phytopath., 32*(2), 332.

Prakash, O., & Srivastava, D. N., (1987). *Mango Diseases and Their Management: A World Review (book).* Today and Tomorrow's Print and Pub., New Delhi.

Prakash, O., (2005). Progress towards integrated management of disease of mango. *Abstract in Second Global Conference, Plant Health-Global Wealth* (pp. 133, 134). Udaipur, India.

Punithalingam, E., (1973). *Transactions of the British Mycological Society, 60*, 157–160.

Punithalingam, E., (1974). Studies on *Sphaeropsidales* in culture: II. *Mycol. Pap., 136*, 1–63.

Punithalingam, E., (1975). Some new species and combinations in phomopsis. *Trans. Brit. Mycol. Soc., 64*, 427–435.

Qi-Hui, Z., Tian, L., Lian-Di, Z., Zhang, Y., Zhi-Feng, L., Hui-Ming, H., & Yue-Hu, P., (2009). Two new compounds from the marine *Nigrospora spherica. Journal of Asian Natural Products Research., 11*(11), 962–966.

Ram, S., Dwivedi, T. S., & Bist, L. D., (1976). Internal fruit necrosis in aonla (*Emblica officinalis* Gaertn.). *Progressive. York, 8*, 5–12.

Rathod, R. S., & Patel, J. G., (2005). Effect of pre- and post-inoculation treatment on post-harvest rots of aonla fruits. *J. Myco. Pl. Pathol., 35*(3), 525.

Rawal, R. D., & Saxena, A. K., (2005). Management of post-harvest diseases of tropical fruits with reference to India. *Abstract in Second Global Conference, Plant Health-Global Wealth* (p. 133). Udaipur, India.

Rawal, R. D., (1993). Fungal disease of tropical fruits spp. In: Chadha, K. L., & Pareek, O. P., (eds.), *Advances in Horticulture, Fruit Crops* (Vol. 3, pp. 1255–1273). Malhotra Publ. House, New Delhi.

Reddy, S. M., & Laxminaray, P., (1984). Post-infection changes in ascorbic acid contents of mango and aonla caused by two fruit rot fungi. *Curro. Sci., 53*(17), 927–928.

Reiss, J., (1986). *Schimmelpilze: Lebensweise, Nutzen, Schaden, Bekämpfung* (pp. 33–41). Springer, Berlin Heidelberg New York.

Rodeva, R., Stoyanova, Z., & Pandeva, R., (2009). A new fruit disease of pepper in Bulgaria caused by *Phomopsis capsici. Acta Hortic., 830*, 551–556.

Rosenberger, D. A., (1990). Blue mold. In: Jones, A. L., & Aldwinckle, H. S., (eds.), *Compendium of Apple and Pear Diseases* (pp. 54, 55). APS Press, St. Paul, MN.

Rosskopf, E. N., Charudattan, R., DeValerio, J. T., & Stall, W. M., (2000b). Field evaluation of *Phomopsis amaranthicola*, a biological control agent of *Amaranthus* spp. *Plant Dis., 84,* 1225–1230.

Rosskopf, E. N., Charudattan, R., Shabana, Y. M., & Benny, G. L., (2000a). *Phomopsis amaranthicola*: A new species from *Amaranthus* sp. *Mycologia., 92,* 114–122.

Saaiman, W. E., (1997). *Orchard Sanitation as Q Means of Controlling Postharvest Diseases* (Vol. 17, pp. 73, 74). Yearbook South African Mango Growers Association.

Saccardo, P. A., (1906). *Sylloge Fungorum, 18,* 572.

Saini, P. K., & Pathak, V. N., (1991). *Management of Aspergillus Rot of Mango Fruits (Abs.).* National Symposium on Plant rust-The shifty enemies at HAU, Hisar.

Sattar, A., & Malik, S. A., (1939). Studies on anthracnose of mango caused by *Colletotrichum glosporioides* in Punjab. *Indian J. Agric. Sci., 9,* 511, 512.

Sharma, M., & Kulshrestha, S., (2015). *Colletotrichum gloeosporioides*: An anthracnose causing pathogen of fruits and vegetables. *Biosci. Biotechnol. Res. Asia., 12*(2).

Sharma, N., & Mashkoor, A. M., (1998). In: *Post-Harvest Diseases of Horticultural Perishables* (pp. 226–253). International Books Distributing Co., Lucknow.

Sharma, R. L., & Shukla, A., (2003). Effect of different packaging on soft rot (*Fusarium* spp.) of banana in storage. *J. Mycol. Pl. Pathol., 33*(1), 299, 300.

Sharma, R., (2012). Pathogenicity of *Aspergillus niger* in plants. *Cibtech Journal of Microbiology, 1.*

Shetty, H. A. H., (1954). Blue mold of aonla (*Phyllanthus emblica* L.). *Curr. Sci., 28,* 208.

Shetty, K. G. H., (1959). Blue mold of amla (*Phyllamhus emblica* L.). *Curr. Sci., 28,* 208.

Singh, U. P., (1973). Morphology and cytology of *Ravenelia emblicae* Sydow. *Journal of Basic Microbiology, 13*(3), 251–258.

Siqueira, R. S., (1995). *Manual De Tecnologia De Alimentos.* Rio de Janeiro: EMBRAPA-CTAA.

Snowdon, A. L., (1990). *A Color Atlas of Post-Harvest Diseases and Disorders of Fruits and Vegetables* (Vol. 1, p. 133). General Introduction and Fruits. University of Cambridge, Wolfe Scientific.

Soni, K. K., & Verma, R. K., (2010). A new vascular wilt of aonla (*Emblica officinalis*) and its management. *Journal of Mycology and Plant Pathology, 40*(2), 187–191.

Soylu, S., Dervis, & Soylu, E. M., (2011). First report of *Nigrospora sphaerica* causing leaf spots on Chinese wisteria: A new host of the pathogen. *Plant Disease, 95*(02), 219.

Srivastava, M. P., & Tandon, R. N., (1969). *Aspergillous niger* chemical management. *Plant Dis. Reptr., 53,* 206–208.

Srivastava, M. P., Chandra, S., & Tandon, R. N., (1964). Post-harvest diseases of some fruits and vegetables. *Proc. Natl. Acad. Sci. India, 34*(4), 339–342.

Srivastava, R. C., (1982). Notes on two interesting fungi from India. *Archiv Für Protistenkunde 125*(1–4), 331–333.

Subramaniyam, H., Narayana, M. N. V., Lakshminarayana, S., & Dalal, V. B., (1972). Control of fungal spoilage *Alphanso* mangoes by preharvest application of fungicides. *Acta Hort., 24,* 224–226.

Surianchandraselvan, M., Bhaskaran, R., & Ramadoss, M., (1991). Epidemiology of grey leaf spot disease on coconut caused by *Pestalotiopsis palmarum* (Cooke). *Indian Coconut J. (Cochin), 21,* 19, 20.

Sutton, B. C., (1980). *The Coelomycetes* (pp. 1–696). Fungi imperfecti with pycnidia, acervuli and Saccardo's Syll. fung., Published by CMI, Kew England.

Sydow, H., & Butler, E. J., (1901). Fungi *Indiae orientalis*: Part I. *Annual. Mycology, 4,* 424–445.

Tan, T. K., Teo, T. S., Han, H., Lee, B. W., & Chong, A., (1992). Variations in tropical airspora in Singapore. *Mycological Research, 96*(3), 221–224.

Tandon, I. N., & Singh, B. B., (1968). Control of mango anthracnose by fungicides. *Indian Phytopath., 21,* 212–216.

Tandon, R. N., & Shrivastava, M. P. (1964). Fruit rot of aonla caused by *Pestalotia cruenta* Syd. in India. *Curr. Sci., 33*(3), 86–87.

Tandon, R. N., & Srivastava, M. P., (1964). Fruit rot of *Emblica officinalis* Gaertn. caused by *Pestalotia cruenta* Syd. in India. *Journal article: Current Science, 33*(3), 86, 87.

Tandon, R. N., & Verma, A., (1964). Some new storage diseases of fruits and vegetables. *Curro. Sci. 33,* 625–627.

Tashiro, N., Noguchi, M., Ide, Y., & Kuchiki, F., (2013). Sooty spot caused by *Cladosporium cladosporioides* in postharvest *Satsuma mandarin* grown in heated greenhouses. *J. Gen. Plant Pathol., 79,* 158–161.

Tejesvi, M. V., Tamhankar, S. A., Kini, K. R., Rao, V. S., & Prakash, H. S., (2009). Phylogenetic analysis of endophytic *Pestalotiopsis* species from ethno-pharmaceutically important medicinal trees. *Fungal Divers., 38,* 167–183.

Thom, C., (1910). *Cultural Studies of Species of Penicillium citrinum* (Vol. 118, pp. 1–109). U.S. Dept. Agri. Bur. Anim. Ind. Bul.

Thom, C., (1914). Conidium production in *Penicillium. Mycologia, 6,* 211–215.

Thom, C., (1930). Blue mold of aonla (*Phyllanthus emblica* L.). *Mycologia, 6,* 210, 211.

Timmer, Lavern, M., Tobin, L., Peever, Zvi, S., & Kazuya, A., (2003). Alternaria diseases of citrus-novel pathosystems. *Phytopathology Mediterranea, 42,* 99–112. Citrus Research and Education Center.

Todkar, L., Sapkal, R. T., & Sawant, D. M., (2005). Antagonistic effect of bioagents against the organisms causing post-harvest losses in grapes. In: *Proceeding of Second Global Conference Plant.*

Tyagi, R. N. S., (1967). *Morphological and Taxonomical Studies on the Genus Ravenelia Berk.* Occurring in Rajasthan. PhD thesis, University of Rajasthan, Jaipur.

Verma, P., & Verma, R. K., (2015). A new pre-harvest fruit rot of aonla (*Emblica officinalis*) caused by *Pestalotiopsis* versicolor from central India. *World Journal of Pharmaceutical Research, 4*(10), 2461–2465.

Wardlaw, C. W., & Leonard, E. R., (1936). *The Storage of West Indian Mangoes* (Vol. 2, p. 47). Memoir of the Low Temperature Research Station, Trinidad.

Watanabe, K., Motohashi, K., & Ono, Y., (2010). Description of *Pestalotiopsis pallidotheae*: A new species from Japan. *Mycoscience, 51,* 82–188.

Webster, J., (1952). Spore projection in the hyphomycete *Nigrospora sphaerica*. *The New Phytologist, 51*(2), 229–235.

Wei, J. G., & Xu, T., (2004). *Pestalotiopsis kunmingensis* sp. Nov., an endophyte from *Podocarpus macrophyllus. Fungal Diversity, 15,* 247–254.

Wei, J. G., Xu, T., Guo, L. D., & Pan, X. H., (2005). Endophytic *Pestalotiopsis* species from southern China. *Mycosystema, 24,* 481–493.

Wei, J. G., Xu, T., Guo, L., D, Liu, A. R., Zhang, Y., & Pan, X. H., (2007). Endophytic *Pestalotiopsis* species associated with plants of *Podocarpaceae, Theaceae* and Taxaceae in southern China. *Fungal Divers, 24,* 55–74.

Wheeler, K. A., Hurdman, B. F., & Pitt, J. J., (1991). Influence of pH on the growth of some toxigenic species of *Aspergillus, Penicillium,* and *Fusarium* (Vol. 12, No. 2, 3, pp. 141–149). CSIRO Division of Food Processing. Australia.

Wright, E. R., (2008). *Plant Dis., 92,* 171.

Xu, L., Kusakari, S., Hosomi, A., Toyoda, H., & Ouchi, A., (1999). Postharvest disease of grape caused by *Pestalotiopsis* species. *Annals of the Phytopathological Society of Japan, 65,* 305–311.

Yadav, S. M., Patil, R. K., Saurabh, S., Balai, L. P., & Rai, A. K., (2012). Ecochemical management of a new fungal rot (*P. funiculosum* Thom.) of aonla. *The Bioscan, 7*(4), 649–651.

Yadav, S. M., Waghunde, R. R., Patil, R. K., & Pandey, R. N., (2009). A new fruit rot of aonla incited by *Penicillium funiculosum* and its management. *J. Plant Disease Science, 4*(1), 132, 133.

Zhang, C. F., & Qi Pei, K., (1996). Identification of fungal diseases in amla (*Phyllanthus emblica* L.). *Journal of South China Agricultural University, 17*(4), 6–10. Chinese language.

Zhang, J. X., Xu, T., & Ge, Q. X., (2003). Notes on *Pestalotiopsis* from Southern China. *Mycotaxon, 85,* 91, 92.

Zhu, P. L., Ge, Q. X., & Xu, T., (1991). The perfect stage of *Pestalotiopsis* from China. *Mycotaxon, 40,* 129–140.

CHAPTER 15

Exploitation of Biofumigation and Biocontrol Agents for the Management of Soil-Borne Diseases

G. BINDUMADHAVI[1] and R. GOPI[2]

[1]Regional Agricultural Research Station, Lam, Acharya N.G. Ranga Agricultural University, Guntur, Andhra Pradesh, India, E-mail: bindugopireddy@gmail.com

[2]ICAR-Sugarcane Breeding Institute Research Center, Kannur – 670002, Kerala, India

ABSTRACT

The yield potential of different agricultural crops is influenced by several biotic and abiotic factors of which the biotic factors cause severe yield losses, among these diseases caused by soil-borne pathogens are major concern. The management of soil-borne diseases is very challenging due to their highly distinct nature of incidence and production of dormant structures like chlamydospores, conidia, and sclerotia during offseason, and wide range of alternate and collateral hosts including weeds, use of chemical control measures are generally not effective, not ecologically viable, too costly and application of chemicals cause ill-effects to non-target soil microorganisms. Several alternate strategies like crop rotation, soil amendments, improved cultural practices and biological control agents were available for the management of diseases. Nowadays management of diseases by biocontrol strategies has become an interesting tool (Cook, 1993). An effective biocontrol strategy is feasible only when the biocontrol agents survive, establish, proliferate, and combat the pathogen. However, the success rate of biological control is less because of several reasons, especially environment, soil pH, organic matter and moisture content. In the present scenario use of mustard biofumigation in combination with antagonists may be most agreeable, cost effective, ecofriendly viable option. Biofumigation is an agronomic practice, where different crops are cultivated and incorporated into the soil at the time of flowering like green manuring. Addition of more plant biomass, especially roots, can offer the gains of green manure crops, and volatile compounds released during crushing of biofumigant plants can suppress soil-borne pathogens, pests, and weeds. The full potential of bioagents can be exploited by using in combination of biofumigation is convenient as it delivers the dual benefits of biocontrol and creates favorable conditions for the growth of biocontrol agents by addition of organic matter to the soil.

15.1   SOIL-BORNE DISEASES IMPORTANCE

Diseases caused by soil-borne organisms greatly reduce yield and quality in crop produce. Management of these pathogens is difficult because they usually survive in soil as sclerotia, microsclerotia, chlamydospores, oospores, etc., for many years, even in the absence of host plant. Simultaneous infections from different soil-borne pathogens usually result in complex diseases that may further causes huge losses to the crop. Several diseases caused by soil-borne microorganisms are challenging to predict, detect, and diagnose. Further, the extremely complex soil environment is making strenuous to understand the ecology of soil-borne microorganisms. The term *soil-borne pathogens,* therefore, can be outlined as *pathogens* that cause plant diseases by the agent that reaches the plant through the soil.

The most important diseases caused by soil-borne pathogens are usually rots that affect belowground parts (including seed decay, damping-off of seedlings, root rots, and crown rots) and vascular wilts started with root infections. Rarely soil-borne pathogens incite foliar diseases which can show signs and symptoms on above-ground parts of the plants. Soil-borne pathogens can generally be categorized into *soil inhabitants* (able to harbor in the soil for a long time) and *soil invaders* or *soil transients* (able to reside in soil for a short time). Under certain condition several soil-borne plant pathogens also can live as non-pathogenic soil microorganisms such as *saprophytes* (organisms that live on dead and decaying organic matter).

15.2   MAJOR PATHOGEN GROUPS

15.2.1   *FUNGI*

Multicellular microorganisms cause most of the soil-borne diseases hence considered the very important pathogen group. Plant-pathogenic fungi categorized into five major taxonomic classes depending on morphological and biological attributes, Plasmodiophoromycetes, Zygomycetes, Oomycetes, Ascomycetes, and Basidiomycetes. Few species of Ascomycetes and Basidiomycetes produce asexual spores and these asexual stages are put in a new separate class, the Fungi Imperfecti. Notable oomycetes pathogens include *Aphanomyces, Bremia, Phytophthora,* and *Pythium.* Significant Ascomycetes are *Monosporascus* and *Sclerotinia.* Examples of soil-borne Fungi Imperfecti pathogens are *Fusarium (wilt causing), Rhizoctonia (induce* root rots), *Verticillium (*incites wilts)*. Plasmodiophora brassicae* (manifests clubroot disease of brassicas) and *Spongospora subterranea* (instigates powdery scab of potato) are the conspicuous soil-borne pathogens. Several soil-borne fungi live in soil for longer periods as dormant survival structures like melanized mycelium, chlamydospores, oospores, and sclerotia, whereas the thin-walled mycelium of numerous fungi survives for only a brief time frame in the soil.

15.2.2   *BACTERIA*

Single-celled organisms that have rigid cell walls but lack a membrane bound nucleus. Soil-borne plant pathogenic bacteria causes not many diseases when compared to soil-borne fungi. Examples are *Erwinia, Rhizomonas, Streptomyces, Pseudomonas,* and *Xanthomonas* and generally bacteria reside in the soil for just a brief period.

15.2.3   *VIRUS*

Soil-borne viruses causing diseases to crops are very less. Viruses are subcellular entities containing nucleic acid surrounded by a protein coat and replicate only in the living cells of the organism. Generally, viral disease symptoms manifest as stunted growth of the plant, tissue malformations, yellowing, mosaic patterns, and necrosis of foliar parts. Soil-borne viruses are obligatory parasites that survive only in the living tissues of the host plant or in the soil microorganisms such as nematode or fungi that transmit virus to the host crops. Lettuce big vein disease, caused by the Mirafiori lettuce virus survive in soil fungus *(Olpidium brassicae)* that moves in soil water, anchors to lettuce roots, and transmits the virus.

15.2.4   NEMATODES

Soil-borne plant-parasitic nematodes are tiny, non-segmented roundworms. Spend most of their life in soil, either as ectoparasitic nematodes which feed on the external surface of plant roots or as endooparasitic which live inside roots. Nematode infected plants show reduced growth and vigor. In soil, during offseason, plant-parasitic nematodes live as free-living organisms or as eggs or durable cysts. Ex: Root-knot nematodes (*Meloidogyne* species) infection manifest symptoms like reduction in vigor in several crops and can induce severe malformations and bulging of roots.

For the management of nematodes, different strategies have been developed and implemented; however, none of the strategies are significantly effective in the complete elimination of nematodes from the soil.

Chemical control is one of the effective strategies for the management of diseases. However, their misuse and abuse may also lead to environmental pollution and development of fungicide resistance (Christopher et al., 2010). Because of several environmental and health hazards caused by pesticides, researchers have focused on developing alternatives to chemical pesticides for management of soil-borne pathogenic microorganisms and nematodes (Larkin et al., 1998). In this context, alternative approaches including crop rotation, use of soil amendments, solarization, biofumigation, biological soil disinfestations, grafting, and application of antagonists or organic amendments like composts, are of paramount concern among researchers and farmers (Kirkegaard et al., 2000; Ryckeboer, 2001; Bailey and Lazarovits, 2003; Louws et al., 2010). Soil fumigant methyl bromide was banned after Montreal protocol, and the search for the non-chemical approaches for the management of soil-borne pathogens was intensified by researchers and found that biofumigation is one of the viable and eco-friendly option. Hence the aim of this paper was to outline current knowledge on biofumigation concept and uses of biofumigation and integrated approach of combining biofumigation with biocontrol agents.

15.3 BIOFUMIGATION

The earliest concept of biofumigation was documented by Theophastrus in 300BC when he found that the odors of cabbage were causing harmful effects on vines (Willis, 1985). After the ban of notoriously toxic chemicals for soil fumigation, including methyl bromide, substitutes to pesticides have been intensively explored (Duniway, 2002; Porter and Mattner, 2002).

15.3.1 CONCEPT OF BIOFUMIGATION

Biofumigation is the agronomic practice where soil-borne pests are suppressed by the incorporation of glucosinolate-containing plant parts as green manures into the soil (Angus et al., 1994). It is the important concept of using volatile chemicals (allelochemicals) released from macerating plant tissues to suppress pests (Brown and Morra, 1997; Rosa et al., 1997). The range of pests suppressed includes germinating weed seeds, nematodes, bacteria, fungi, viruses, and insects. The term biofumigation was coined by Australian scientists to describe the suppression of soil-borne pests by compounds released by brassica species (Kirkegaard et al., 1993). Glucosinolate-containing *Brassica* spp. generally release volatalic isothiocyanates (ITCs), which are toxic to different pathogens (Kirkegaard et al., 1993; Matthiessen and Kirkegaard, 2006). The chemistry involved in the biofumigation can be accredited to the reaction of myrosinase enzyme on the glucosinolates (GLS) thereby releasing ITCs, thiocyanates, nitriles, oxalidine, dimethyl sulphide, methanethiol among several compounds (Matthiessen and Kirkegaard, 2006; Gimsing and Kirkegaard, 2008).

Approximately 20 diverse GLS have been recorded based on the side organic chain. Their concentrations vary with the age of the plant and conditions in which they are grown (Sarwar and Kirkegaard, 1998). GLS are generally found in members of Tovariaceae, Resedaceae, Capparaceae, Moringaceae, and Brassicaceae (Brown and Morra, 1997), however ITCs remains the prime choice of interest for study because they are the major hydrolytic products of GLS compared to thiocyanates or nitriles (Gimsing and Kirkegaard, 2008). Concentrations of ITCs have been shown to decrease by 90% within 24 hours of incorporation of *Brassica* residues (Brown et al., 1991). Their persistence up to 45 days has also been demonstrated (Gimsing and Kirkegaard, 2008; Poulsen et al., 2008). Plants of the family Brassicaceae contain GSLs that are hydrolyzed into various products upon tissue degradation. Crop plants of the *Brassica* species include cabbage (*Brassica oleracea*), broccoli, radish (*Raphanus sativus*), canola (*Brassica napus*), cauliflower, mustard (*Brassica juncea*) and white mustard (*Sinapsis alba*) (Larkin and Griffin, 2007). These GSLs are hydrolyzed by the enzyme myrosinase (thioglucoside glucohydrolase) (Brown and Morra, 1997; Kirkegaard et al., 2000; Matthiessen and Kirkegaard, 2006) into ITCs, thiocyanates, nitriles, epithionitriles, oxazolidinethiones (Fahey et al., 2001). These products, especially ITCs, are highly toxic to various microorganisms and efficiently used as biofumigants (Fahey et al., 2001; Yulianti et al., 2007). Effective mycelium inhibition by *B. juncea* plant species tested can be associated to high levels of GLS in their plant tissue. As the amount of *B. juncea* plant tissue added into soil increases, AITC emission enhances and therefore antifungal activity becomes stronger. AITC emission from *B. juncea* added soil was completed within 72 h after amendment, even up to 1.0% (wt/wt) application rate to soil. Disease reduction is depending on the efficiency at which GLS are converted into isothiocyanate (Mazzola and Zhao, 2010).

Omirou et al. (2013) reported that the rapid dissipation of GLS in brassicaceous plant materials added in the natural soil is accredited mainly to myrosinase activity. ITCs are harmful to several types of microorganisms (Walker et al., 1937); they react with sulfur-containing proteins by nonspecific and irreversible reactions (Brown and Morra, 1997). The biocidal compounds released during biofumigation suppress pathogens, weeds, and alter rhizosphere microbial populations (Matthiessen and Kirkegaard, 2006; Hoagland et al., 2008). Biofumigation have been studied against soil-borne fungal pathogens like *Rhizoctonia, Verticillium, Colletotrichum, Fusarium, Pythium, Phytophthora* spp. (Steffek et al., 2006; Mattner et al., 2008; Friberg et al., 2009).

15.3.2   EFFECT ON SOIL MICROBIAL COMMUNITY

Brassica sp. as plant material or its seed meal has been tested by several researchers for green manuring and was found to influence microbial community structures (Vera et al., 1987; Williams-Woodward et al., 1997; Mazzola et al., 2001; Cohen and Mazzola, 2006; Hoagland et al., 2008; Friberg et al., 2009; Omirou et al., 2010). The addition of brassica plant material for biofumigation has been found to enhance or reduce the count of the rhizosphere microorganisms such as *Trichoderma* spp., *Pythium* spp., fluorescent pseudomonads, *Streptomyces* spp., actinomycetes, and other antagonists of soil-borne pathogens depending on the plant species and soil type (Mazzola et al., 2001; Cohen and Mazzola, 2006; Perez et al., 2007; Mazzola and Zhao, 2010). The added plant materials can aid the native microbial community with competition, parasitism, antagonism, and predatory trait against the soil-borne microorganisms (Raaijmakers et al., 2009). ITCs react with sulfhydryl groups, amine groups, and the disulfide bonds of proteins, causing the deterioration of enzymes and the suppression of microbial growth (Brown and Morra, 1997). Bacteria and actinomycetes are more resistant to toxicants than fungi and oomycetes (Kreutzer, 1963). Smith and Kirkegaard (2002) studied the growth sensitivity of fungi, bacteria, and oomycetes to 2-phenylethyl ITC as it is the main ingredient of the taproot GLS of canola, and found that the susceptibility of fungi and oomycetes are higher for aromatic ITCs than for aliphatic ITCs. The long-term use of ITCs may directly suppress soil diseases or indirectly cause changes in the indigenous soil microbial community. Smith and Kirkegaard (2002) demonstrated the suppressive effects of 2-phenylethyl ITC on a variety of pathogenic and non-pathogenic microorganisms *in vitro*. Bacteria were found to be usually more tolerant of 2-phenylethyl ITC when compared to eukaryotic organisms tested, and *Trichoderma* spp. were found to be more tolerant than the other eukaryotes tested, including *Aphanomyces, Gaeumannomyces, Phytophthora* spp., *Thielaviopsis, Rhizoctonia solani,* and *Pythium* spp. Despite variations in lethal doses, 2-phenylethyl ITC was found to be suppressive to all microorganisms tested, indicating its potential in biofumigation. Smith and Kirkegaard (2002) also observed that *Trichoderma* spp. were tolerant to a particular isothiocyanate produced from Brassica tissues than several genera of plant-pathogenic fungi. Larkin and Griffin (2007) reported that, volatiles released from chopped leaf material of Brassica crops inhibited growth of a variety of soil-borne pathogens of potato, including *Rhizoctonia solani, Phytophthora erythroseptica, Pythium ultimum, Sclerotinia sclerotiorum,* and *Fusarium sambucinam* very low when compared with Indian mustard which reported nearly complete inhibition (80–100%) *in vitro. Trichoderma virens* and *Pseudomonas chrysogenum*, were tolerant to inhibition effects from Indian mustard than any of the potato pathogens. Soil amendment with *Brassica* plants increases organic matter, improves soil structure, and enhances useful microbial populations as well as their biocidal effects on soil-borne microorganisms. Galletti et al. (2008) investigated the tolerance of the forty isolates of *Trichoderma* spp. to the biofumigant substances in view to integration with *Brassica carinata* seed meal (BCSM) *in vitro* and found that *Trichoderma* spp. are generally less inhibited than the tested pathogens (*Pythium ultimum, Rhizoctonia solani, Fusarium oxysporum*), however fungistatic effect was found at the highest dose (10 μmole of sinigrin). BCSM addition enhanced pathogen population, but decreased disease incidence, apparently due to indirect mechanisms. Yulianti et al. (2007) reported that *Rhizoctonia solani* pathogen was reduced through the incorporation of *Brassica* green manure into the soil. Incorporation takes place at a depth of 10–20 cm (Bellostas et al., 2007). The GSL products are mostly released after maceration of the plant material (Motisi et al., 2010).

15.3.3   FACTORS EFFECTING BIOFUMIGATION

Various factors influence the successful hydrolysis of GSLs to ITCs including the soil physical and chemical properties (Morra and Kirkegaard, 2002), water availability, temperature fluctuations, the GSL content (Matthiessen et al., 2004) and the amount of tissue disruption by freezing, drying or maceration to increase the contact between the enzyme and the GSLs (Morra and Kirkegaard, 2002). Increasing the soil moisture content helps the GSL hydrolysis process and reduces the loss of volatile ITCs. This implies a higher ITC concentration released into the soil. It was also found that most of the ITCs will be released within four days after tissue incorporation, depending on the soil temperature, water content, and the physical and chemical properties (Zasada and Ferris, 2004). The ITC concentration in the soil decreases rapidly within a few days due to microbial degradation. It has been reported that after 1–2 weeks no ITC could be detected (Gimsing and Kirkegaard, 2006). Soil type can also influence the degradation rate of ITCs (Gimsing et al., 2006). Degradation is fast in clay soil when compared to the sandy soil. Clayey soil contains more nutrients, have more microbial populations, and stabilize cellular enzymes through the adsorption to clay particles.

Matthiessen and Shackleton (2005) demonstrated that the ITCs derived from aliphatic GSLs (allyl-ITCs) are more active in the soil than those from aromatic GSLs (2-phenylethyl-ITCs). It was observed that *Brassicas* with high number of short-chain aliphatic ITCs are highly efficient in pest suppression and those with long-chain aromatic ITCs have a low biofumigation capacity. It was concluded that aliphatic GSLs degrade much easier than aromatic GSLs. Aliphatic ITCs are generally present in shoots and the aromatic ITCs are commonly found in roots (Kirkegaard and Sarwar, 1998). Earlier *in vitro* bioassays have reported contrary results on the effect of *Brassica* plant tissue on growth inhibition of many plant pathogenic fungi (Smith and Kierkegaard, 2002; Sarwar et al., 1998; Charron and Sams, 1999; Yulianti et al., 2006) especially when advanced to field evaluations. Such inconsistencies between *in vitro* bioassays and field efficacy may be due to the fact that the methods used for screening plant materials as biofumigants exclude the role of soil properties and soil microbes. Hydrolysis of GLS in soil depends not only on the concentration of glucosionlates but soil factors, extent of plant tissue disruption, soil temperature and water content (Morra and Kirkegaard, 2002; Hanschen et al., 2015).

15.4 USES OF BRASSICA BIOFUMIGANT CROPS

15.4.1 ROTATION CROPS

Brassica (and other) rotation crops also decreases disease incidence, by altering the soil microbial populations and improvement of suppressive conditions, that are different from the biofumigation response. Larkin and Honeycutt (2006) recorded that canola and rapeseed rotations (*Brassica napus*) enhanced diverse group of microbial populations when compared to non-Brassica rotations, and these rotations culminate in suppressed incidence and severity of Rhizoctonia disease in potato, even when the crops were not added to the soil as green manures. Further, Larkin and Griffin (2007) found that disease suppression was not constantly related with high glucosinolate-containing crops, and recorded that the rotations with barley and ryegrass suppressed several soil-borne diseases of potato that was proportionate to that of Brassica green manures.

15.4.2 COVER CROPS

Cover crops can also associated with suppression of disease and pathogen population through non-host and trap-crop effects, enhancement of beneficial microorganism communities, and the release of non-ITC decomposing substances (Matthiessen and Kirkegaard, 2006). Cover cropping contributes several benefits to the crop, such as enhancing the nutrient, moisture availability, and altered soil microbial populations, which may have also attributed to the improved yields (Clark, 2007).

15.4.3   GREEN MANURE

In a study performed by Friberg et al. (2009) the incorporation of mustard (*Brassica juncea*) as a green manure decreased the inoculum level of *Rhizoctonia solani* in the soil. High concentrations of GSL hydrolysis are needed to suppress pathogen growth. *R. solani* produce pseudosclerotia or thick-walled hyphae as survival bodies. These survival structures are less susceptible to GSL hydrolysis products than young hyphae, restricting the biocidal effect of green manure to suppress fungal pathogens (Yulianti et al., 2006). Canola (*Brassica napus*) reduces soil-borne pathogen population when it is incorporated into the soil as a green manure crop. Roots contributed as the major source of biocidal ITCs when *B. napus* green-manure crops are incorporated into the soil (Gardiner et al., 1999). Inhibition of potential potato pathogens, including *Pythium ultimum, R. solani, F. sambucinam* and *V. dahliae* (Charron and Sams, 1999; Mayton et al., 1996) was recorded with compounds released from certain Brassica cultivars. Mayton et al. (1996) demonstrated that allyl isothiocyanate released from macerated *Brassica juncea* cv. Cutlass tissue completely suppressed *in vitro* growth of five common plant pathogens; *Pythium ultimum, Rhizoctonia solani, Verticillium dahliae, Verticillium alboatrum,* and *Colletotrichum coccodes.* In addition, *Brassica napus* (winter rapeseed) had the highest above ground biomass (7.7 $\pm$ 3.4 kg m^{-2}) of ten Brassica species tested.

15.4.4   SEED MEAL

Smolinska et al. (1997) evaluated the effect of seed meal of *Brassica napus* which produced volatile fungitoxic compounds against *Aphanomyces* causing root rot of pea. Bioassays of pea (*Pisum sativum*) seed inoculated with zoospore suspensions and incubated for 24 hours in the presence of volatiles from rapeseed meal recorded less than 50% of root rot disease severity when compared with the absence of meal. The results indicated that *B. napus* allelochemicals released by enzymatic hydrolysis of GLS are attributed for harmful effects toward *Aphanomyces euteiches* f. sp. *pisie.* The impact of *Brassica napus* seed meal on the microbial complex causing apple replant disease was evaluated by Mazzola et al. (2001) in greenhouse trials. Regardless of glucosinolate content, seed meal amendment at a rate of 0.1% (vol/vol) significantly improved the growth of apple and decreased infections by *Rhizoctonia* spp. and *Pratylenchus penetrans* to apple root. Application of a *B. napus* seed meal with less GLS at the rate of 1.0% (vol/vol) significantly enhanced the recovery of *Pythium* sp. from apple roots and reduced the apple seedling root biomass. Application of seed meal attributed to increased soil populations of total bacteria and actinomycetes were phytotoxic to apple when applied at a rate of 2% (vol/vol), and phytotoxicity was not reduced. Smolinska et al. (2003) conducted bioassays with four *Fusarium oxysporum* isolates using sealed containers in which 0.3 μl of 2-propenyl, ethyl, butyl phenylethyl, and benzyl or phenyl isothiocyanates inhibited mycelial growth completely and suppressed conidial and chlamydospore germination of all isolates. The concentration of propenyl and ethyl isothiocyanates that inhibited mycelial growth completely, also suppressed conidial and chlamydospore germination of all isolates. Cohen et al. (2005) demonstrated that suppression of *R. solani* by *Brassica napus* seed meal was associated with specific changes in soil microbial communities and was unrelated to levels of glucosinolate. Larkin and Griffin (2006) recorded significant *R. solani* inoculum decrease in greenhouse assays, and also disease reductions in field tests followed by addition of brassica plant tissue, with *B. juncea* leading to the highest disease reductions. Brassicas in particular are known to be efficient nutrient scavengers. Fayzalla et al. (2009) observed the differences in the sensitivity of the pathogens to the mustard seed meal at all concentrations and found that *Rhizoctonia solani* as the most susceptible fungus under *in vitro* conditions. The least values of linear growth (3.8, 1.2, and 0.7 cm) were documented with *R. solani* at 5, 10, and 25 mg of seed meal, respectively. Results of field experiment indicated that mustard seed meal reduced the disease incidence over the control by 69.7% at four months after planting.

15.5 LIST OF FUNGAL/BACTERIAL/NEMATODE DISEASES

Inhibition of mycelial growth as a result of exposure to ITCs was documented for several plant pathogenic fungi and nematodes. There are number of reports on effect of isothiocyanates of brassica plants under *in vitro* studies for different pathogens like *Gaeumannomyces graminis* (Angus et al., 1994), *Leptosphaeria maculans* (Mithen et al., 1986), *Rhizoctonia solani, Fusarium graminearum, Bipolaris sorokiniana,* and *Pythium irregulare* (Sarwar et al., 1998), *Helminthosporium solani* and *Verticillium dahliae* (Olivier et al.1999), *F. sambucinum* (Mayton et al., 1996), etc., Plant pathogenic nematodes and fungi show varying degree of sensitivity to specific ITCs depending on the stages and structures such as spores, mycelium, and sclerotia (Brown and

Morra, 1997; Smith and Kirkegaard, 2002; Kirkegaard et al., 1996; Manici et al., 1997). The pathogens like *Aphanomyces, Fusarium, Gaumanno-myces, Phytophthora, Pythium, Rhizoctonia, Sclerotinia,* and *Verticillium* and endoparasitic and semi-endoparasitic nematodes such as *Globodera, Meloidogyne, Pratylenchus,* and *Tylenchus* were most studied in biofumigation. Below is the list of management of fungal and nematode pathogens using biofumigation (Table 15.1).

TABLE 15.1   List of Fungal and Nematode Diseases Managed by Using Biofumigaiton

SL. No.	Name of Pathogen	Diseases	Crop	References
1.	*Xiphinema index*	Grapevine fanleaf virus vector	Grapevine	Bello et al. (2004)
2.	*Rhizoctonia solani* Kühn;	Canker and black scurf	Potato	Larkin et al. (2010)
3.	*Meloidogyne javanica* and *Tylenchulus semipenetrans*	Plant-parasitic nematodes	–	Zasada and Ferris (2004)
4.	*Meloidogyne* spp.	Root-knot nematode	Pepper	Bello et al. (2001); Roubtsova et al. (2007)
5.	*Meloidogyne incognita*	Root-knot nematode	Tomato	Charles et al. (2015)
6.	*Streptomyces scabiei*	potato scab	Potato	Gouws and Mienie (2000); Larkin et al. (2010)
7.	*Pythium aphanidermatum*	Damping-off	Cucumbers	Deadman et al. (2006)
8.	*Globodera pallida*	Cyst nematode	Potato	Lord et al. (2011)
9.	*R. solani*	Canker and black scurf	Potato	Larkin and Griffin (2006)
10.	*Meloidogyne incognita, Sclerotium rolfsii, and Pythium ultimum*	–	–	Stapleton and Duncan (1998)
11.	Gaeumannomyces *(take-all),* Rhizoctonia, *and Fusarium*	Soil-borne cereal pathogens	Wheat	Smith et al. (http://www.regional.org.au/au/gcirc/2/334.htm)
12.	*S. cepivorum*	white rot	onion	Smolinska (2003)
13.	*R. solani*	Stem and bulb infection	Tulip	Van Os et al. (2004)

TABLE 15.1 *(Continued)*

SL. No.	Name of Pathogen	Diseases	Crop	References
14.	*R. solani*	black scurf of potato		Nazf (2006)
15.	*Gaeumannomyces graminis*	Take all disease	wheat	Angus et al. (1994); Kirkegaard et al. (1998)
16.	*F. oxysporum* f. sp. *lycopersici*	wilt	tomato	Smolinska (2000)
17.	*Meloidogyne chitwoodi*	Root-knot nematode	potato	Mojtahedi et al. (1993); Henderson et al. (2009); Lazzeri et al. (2009)
18.	*Aphanomyces euteiches*	Root rot	peas	Muehlchen et al. (1990), Chan and Close (1987)
19.	*Fusarium oxysporum* f. sp. *conglutinans*	–	–	Ramirez-Villapudua (1988)
20.	*Thielaviopsis basicola*	Root rot	Bean, sesame	Adams (1971); Papavizas (1968)
21.	*Fusarium oxysporum* f. sp. *cepae*	*Fusarium* basal rot (FBR)	Onion	Sintayehu (2014)
22.	Fungal complex (Cylindrocarpon, Phytophthora, Pythium, and Rhizoctonia)	Replant disease	Apple	Mazzola et al. (2001)
23.	*R. solanacearum*	bacterial wilt	tobacco	Akiew and Trevorrow (1999)

15.6 BIOFUMIGATION BY FUNGUS

15.6.1 *BIOFUMIGATION WITH FUNGUS MUSCODOR ALBUS*

An endophytic fungus *Muscodor albus* Worapong, Strobel, and Hess was first isolated from the cinnamon tree in a rainforest in Honduras (Strobel et al., 2001; Worapong et al., 2001). The fungal hyphae is white in color with marked features of lacking spores and spore-producing structures. An important characteristic feature of this fungus is its ability to produce volatile organic compounds of antifungal and antibacterial nature and able to kill fungi and bacteria pathogenic to both plants and human (Strobel et al., 2001; Worapong et al., 2001). Grimme et al. (2007) reported that a *M. albus* produces different compounds *viz.*, 1-butanol, 2-methyl, 1, butanol, 2-methyl acetat; propanoic acid; and 2-methyl-, 2-methyl propyl ester, whereas Strobel et al. (2001) revealed that the *M. albus* produce 28 organic volatile compounds which were capable of inhibiting fungi, bacteria, and oomycetes. Mercier and Jimenez (2007) narrated the delineation of production of the important antifungal volatiles, isobutyric acid, 2-methyl-1-butanol and isobutanol, produced by *M. albus* (Table 15.2).

TABLE 15.2   List of Plant Pathogens Controlled by Biofumigation with *M. albus*

SL. No.	Plant-Pathogen/Disease	References
1.	Smut on grain	Strobel et al. (2001)
2.	Soil-borne pathogens	Stinson et al. (2003); Mercier and Manker (2005)
3.	Post-harvest pathogens of fruit	Mercier and Jimenez (2004); Mercier and Smilanick (2005); Schnabel and Mercier (2006)
4.	Potato tuber moth	Lacey and Neven (2006); Lacey et al. (2008)
5.	Root-knot nematode	Grimme et al. (2007)
6.	Brown rot of peaches Gray mold Blue mold of apple	Mercier and Jiménez (2004)
7.	*Meloidogyne chitwoodi, Meloidogyne hapla, Paratrichodorus allius,* and *Pratylenchus penetrans*	Riga et al. (2008)
8.	Green mold and sour rot of lemons	Mercier and Smilanick (2004)

15.7   BIOFUMIGATION BY ESSENTIAL OIL

In a greenhouse, tomatoes transplanted to soils biofumigated with essential oils of thyme, palmarosa, lemongrass, Greek oregano and tea tree exhibited no symptoms of the disease (Pradhanang et al., 2003), the same as tomato and geranium plants in soils treated with clove oil (Huang and Lakshman, 2010).

Biofumigation is the most common method for applying essential oils to control diseases caused by soil-inhabiting pathogens, such as the agents of bacterial wilt of tomato (Huang and Lakshman, 2010; Ji et al., 2005;

Pradhanang et al., 2003), geranium (Huang and Lakshman, 2010) and edible ginger (Paret et al., 2010). In greenhouse-grown tomatoes use of Palmarosa and thyme as biofumigants resulted in 100% control of bacterial wilt, whereas use of only palmarosa oil decreased the wilt incidence up to 20% in another experiment (Pradhanang et al., 2003). However, Ji et al. (2005) reported that tomato wilt incidence was lower in thyme biofumigated field (33.1%) than with palmarosa (48.1%) oils.

In conclusion, soil biofumigation with palmarosa essential oil reduced the severity of bacterial wilt of sweet peppers in both the greenhouse and field; it was not detrimental to plant growth and increased the number of fruits per plant. Furthermore, it did not affect the pH nor the characteristics of biofumigated soil, and resulted in decreased growth of *Ralstonia solanasiarum* thus proved as the viable management option for bacterial wilt of sweet peppers. Dhingra et al. (2004) also found that essential oil extracted from mustard seeds (*Brassica rapa*) significantly reduced the growth of *R. solani* both *in vitro* and in field conditions, but higher concentrations of mustard oil is required for field situation. It was reported that different cultivars of *Brassica* species have varying levels of efficacy due to diverse types of GLS in their plant tissue in various concentrations (Kirkegaard et al., 1996; Kirkegaard and Sarwar, 1998). The degree of fungal suppression by *Brassica* plant materials was determined by the category and quantity of antimicrobial volatiles released (Kirkegaard et al., 1996). Blok et al. (2000) reported that there was a trend of higher yield in the biofumigant treatments, although it was not apparently related to disease reductions induced by the introduction of ITCs into the soil. This result is probably related to the enhanced anaerobic soil environment created following the incorporation of organic matter into the soil compared to the fallow (+) VIF treatment.

15.8 INTEGRATED USE OF BIOFUMIGANTS AND BIOCONTROL AGENTS FOR THE MANAGEMENT OF SOIL-BORNE DISEASES

The strategies adopted for pest management over the last 50 years has been influenced by several factors, including the possibility of pest management strategies, concerns of governments to provide sufficient food supplies, the growth of agribusiness, and the value placed on scientific achievements. These factors and their impact have been alleviated by increasing public concern for food safety and the sustainable use of resources. IPM is essentially a holistic approach for pest management by optimum use of a combination of methods to manage different pests within a particular cropping system.

A major constraint for the successful field application of biocontrol agents is their irreconcilable performance under field conditions (Siddiqui et al., 2014). However, integrated use of biofumigation and antagonists may pave the way for this complication (Talibi et al., 2014).

Combined use of biological control agents with biofumigation resulted in enhancement of carbon source that may modify or even stimulate the growth of native or introduced microorganisms, including both useful and disease causing organisms. Till date, the information regarding the effect of biocidal compounds, especially ITCs, on the useful inhabitant soil microorganisms or artificially inoculated antagonists is limited. In particular, the effect of rapeseed meal on Trichoderma is reported to be ambiguous, hence incorporation of the meal prior to Trichoderma application is suggested, as the direct combination seemed uncompatible (Dandurand et al., 2000).

Deadman et al. (2006) reported that under commercial conditions, solarization and biofumigation (solarization following cabbage residue incorporation) both reduced *Pythium aphanidermatum* inoculum levels in soil relative to untreated controls under greenhouse conditions. Both treatments also reduced the level of damping-off disease in greenhouse seedlings.

Galletti et al. (2008) tested Trichoderma spp. for sensitivity to antifungal volatiles released from BCSM and in direct contact with the meal. They reported that all Trichoderma spp. tested were less sensitive to BCSM than the assayed pathogens (*Pythium ultimum, Rhizoctonia solani,* and *Fusarium oxysporum*). Moreover, most of the isolates have shown tolerance by recording moderate growth on seed meal when compared to the pathogens tested. BCSM incorporation increased pathogen population, but reduced sugar beet damping-off disease incidence, probably due to indirect mechanisms.

Ojaghian et al. (2012) who found that Brassica crops used as green manure cover crops were able to significantly reduce potato stem rot caused by *Sclerotinia sclerotiorum* in field experiments.

Neubauer et al. (2014) found that amendments of *Brassica juncea* shoot tissue reduced the number of viable microsclerotia of *Verticillium dahliae* significantly with efficiencies from 69.3 to 81.3%. A similar type of result was also found by Dohroo et al. (2009), who reported that the total microbial population in rhizosphere soil was influenced by cabbage biofumigation followed by mustard and PGPR consortia.

An experiment was conducted to study the effect of soil solarization and Ethiopian mustard as biofumigant to manage bacterial spot disease and yield improvement in tomato and found that solarization decreased the population of bacterial pathogen from 10.68 to 8.79 CFU g^{-1}, total bacterial count from 11.27 to 9.86 CFU g^{-1}, and total actinomycete from 11.69 to 9.44 CFU g^{-1}. It can be concluded that the solarization and bio-fumigation cannot be used as biorational possibility for effective management of bacterial leaf spot however, these two ways can be accustomed to enhance tomato yield in the presence of pathogen (Misrak et al., 2014).

Handiseni et al. (2015) studied the effectiveness of a *Brassica juncea* cover crop integrated with use of a tolerant rice cultivar and fungicide application 2011, 2012, and 2013 and found that *B. juncea* cover crop along with use of a tolerant rice cultivar and half the label rate of azoxystrobin (0.08 kg a.i.ha^{-1}) can be an efficient approach for the control of sheath blight in rice.

In ginger it was found that the use of cabbage residues as biofumigant in the soil resulted in less incidence of soft rot (9.97%) and bacterial wilt diseases (5.92%) with maximum yield of 15.16 t/ha. The highest bacterial wilt disease incidence of 22.11% and lowest yield of 7.70 t/ha was recorded by Control.

It was concluded that soil sterilization was more beneficial than seed treatment alone, and both seed as well as soil treatment should be practiced to manage them efficiently (Sekhar and Surajit, 2016).

At the early growth stages of tomato, the integrated use of *Streptomyces rubrogriseus* HDZ-9-47 and biofumigation had important effects on the soil microbial and nematode populations, which leads to management of *Meloidogyne incognita* through direct and indirect mechanisms. Denaturing gradient gel electrophoresis (DGGE) study revealed that the combination of *S. rubrogriseus* HDZ-9-47 and biofumigation increased beneficial microorganisms and reduced soil-borne fungal pathogens resulted in effective management against *M. incognita* (Jin et al., 2019).

Li et al. (2014) recorded that integrated use of soil biofumigation with biocontrol agent *Pseudomonas reinekei* SN21 was highly successful against *M. incognita* in field conditions. Biofumigation of soil with cabbage residues, irrigation to the soil up to saturation, and covering of soil with plastic foil, in combination with the antagonistic strain *Streptomyces rubrogriseus* HDZ-9-47 was more effective against *M. incognita* than the application of nematicide fosthiazate in a field trial (Jin et al., 2017).

Biocontrol agents use the nutrients carbon and nitrogen-enriched by biofumigation, thereby enhancing their activity and combativeness in the soil (Handerson et al., 2009). Integrated use of bioagent *Bacillus amyloliquefaciens* strain BS211 with biofumigation influenced microbial population, enhanced diversity of soil bacteria, and reduced Phytophthora blight in pepper (Wang et al., 2014). Valdes et al. (2012) found that soil biofumigation reduced plant-parasitic nematode densities while improving the population of those of bacterivorous nematodes.

Biofumigation with Brassica revealed that the soil actinomycetes population was negatively correlated with damping-off disease (Ascencion et al., 2015).

Wang et al. (2009) reported that Brassica biofumigants decreased the fungivores in the soil. Gruver et al. (2010) recorded that biofumigation did not impact the count of omnivores/predators.

Madhavi et al. (2015) studied the biofumigation effect of different *Brassica* spp. and onion against *Rhizoctonia solani* f. sp. *sasakii in vitro* and found that among different brassica species tested, macerated mustard plant parts are effective in inhibition of radial growth and sclerotial production of *R. solani* f. sp. *sasakii*. In all tested concentrations, there was a positive association between increase of volumes and the inhibition of radial growth. The growth of *R. solani* f. sp. *sasakii*, *T. harzianum* and *P. fluorescens* were monitored for 72 hours of continuous exposure to volatiles produced from different quantities of hydrated mustard powder *in vitro* and found that with the increase in dose of mustard seed powder there was a continuous reduction in the growth of *R. solani* whereas the inhibition of growth of *T. harzianum* was less compared to *R. solani*. Fungistatic effect was not observed for *P. fluorescens* count and this shows the tolerance of *P. fluorescens* to mustard seed meal (Madhavi et al., 2016). Banded leaf and sheath blight disease of maize can be effectively controlled by biofumigation of soil with mustard plant material in combination with potential *P. fluorescens* and *Trichoderma harzianum*. Maximum reduction in the viability of sclerotia was recorded in combination of biofumigation with mustard and seed treatment + soil drenching + foliar spray with *P. fluorescens*. Total bacteria, *Trichoderma* spp. and *Pseudomonas* spp. colony-forming units were high in bio-fumigant with mustard treatments in both greenhouse and field studies.

Prasad and Kumar (2017) recorded that the soil amendment with brassicas might be a practical non-chemical approach to control *Fusarium* wilt of chickpea. Among different Brassica species tested, *Brassica alba* L. was found to be the most effective as it recorded maximum inhibition of mycelia growth of the pathogen followed by *B. nigra L.* and *B. juncea L.* tissues. Lowest wilt incidence was observed in *B. juncea*. Biofumigation whereas maximum grain yield was produced in *B. alba* L. amended plots under field conditions.

15.9   FUTURE PROSPECTS

After the Montreal protocol, withdrawal of soil fumigants, especially methyl bromide, from markets and increasing concerns toward environmental safety and health hazards to humans and animals forced researchers to develop alternative strategies for the management of soil-borne diseases. Biofumigation with brassica and non-brassica crops is one of the alternative approaches which can enrich the soil and controls the soil-borne pathogens. In several countries over the past few years, several trials have been carried out to evaluate the effectiveness of biofumigant crops as rotation crops, green manure crops, and seed meal for containing soil-borne pathogens. The use of biofumigation strategy for the management of soil-borne diseases is rapidly increasing in the USA, Australia, Italy, The Netherlands, and South Africa, thus stimulated the interest of researchers for developing cultivars with high glucosinolate content. New potential has also been explored for the dehydrated plant tissues and/or for defatted seed meal pellets production and use. Another important point need to be fully explored is the effect of biocidal compounds released during incorporation of plant tissue, especially ITCs on the native and introduced beneficial microorganisms.

Few areas which also need attention of scholars are as follows:

- Breeding of varieties/hybrids with high glucosinolate content;
- Developing standard protocols for incorporation of biofumigant crop as plant extract, green manure crop, intercropping, etc., in the soil;
- research projects to evaluate the effect of biofumigation on soil rhizosphere microorganism especially beneficial microflora;
- Ecological studies on the plant and pathogen interaction with biofumigants including the non-target microbial population in the soil.

15.10   CONCLUSION

By reviewing the current literature on the biofumigation concept, methods of application, and advantages of adoption and management of soil-borne pathogens, it could be concluded as an alternative method to combat the soil-borne plant pathogens with several direct and indirect benefits.

KEYWORDS

- bacteria fungi
- biofumigation
- denaturing gradient gel electrophoresis
- glucosinolates
- isothiocyanates
- soil-borne pathogens

REFERENCES

Adams, P. B., (1971). Effect of soil temperature and soil amendments on *Thielaviopsis* root rot of sesame. *Phytopath., 61*, 93–97.

Akiew, S., & Trevorrow, P., (1999). Biofumigation of bacterial wilt of tobacco. In: Magarey, R. C., (ed.), *Proceedings 1st Australasian Soilborne Disease Symposium* (pp. 207, 208). Bureau of Sugar Experiment Stations, Brisbane.

Angus, J. F., Gardner, P. A., Kirkegaard, J. A., & Desmarchelier, J. M., (1994). Biofumigation: isothiocyanates released from *Brassica* roots inhibit growth of the take-all fungus. *Plant and Soil, 162*, 107–111.

Ascencion, L. C., Liang, W. J., & Yen, T. B., (2015). Control of *Rhizoctonia solani* damping-off disease after soil amendment with dry tissues of *Brassica* results from an increase in actinomycetes population. *Biological Control, 82*, 21–30.

Bailey, K. L., & Lazarovits, (2003). Suppressing soil-borne diseases with residue management and organic amendments. *Soil and Tillage Research, 72*, 169–180.

Bello, A. M., Arias, J. A., López-Pérez, A., & García-Álvarez, J., (2004). Biofumigation, fallow, and nematode management in vineyard replant. *Nematropica, 34*, 53–64.

Bello, A., Lopez, J., Perez, J. A., & Arias, A., (2001). *Biofumigation and Grafting in Pepper as Alternatives to Methyl Bromide.* Centrolciencias Edioambirntales.

Bellostas, N., Sorensen, J. C., & Sorensen, H., (2007). Profiling glucosinolates in vegetative and reproductive tissues of four *Brassica* species of the U-triangle for their biofumigation potential. *J. Sci Food and Agr., 87*, 1586–1594.

Blok, W. J., Lamers, J. G., Termorshuizen, A. J., & Bollen, G. J., (2000). Control of soil-borne plant pathogens by incorporating fresh organic amendments followed by tarping. *Phytopathology, 90*, 253–259.

Brown, P. D., & Morra, M. J., (1997). Hydrolysis products of glucosinolates in *Brassica napus* tissues as inhibitors of seed germination. *Plant and Soil, 181*, 307–316.

Brown, P. D., Tokuhisa, J. G., Reichelt, M., & Gershenzon, J., (1991). Variation of glucosinolate accumulation among different organs and development stages of *Arabidopsis thaliana. Phytochemistry, 62*, 471–481.

Chan, M. K. Y., & Close, R. C., (1987). Aphanomyces root rot of peas: Control by the use of cruciferous amendments. *New Zeal. J. Agr. Res., 30*, 225–233.

Charles, K., Agathar, K., Ronald, M., Cosmas, P., Ignitius, M., & Blessing, M., (2015). Nematicidal effects of brassica formulations against root-knot nematodes (*Meloidogyne javanica*) in tomatoes (*Solanum Lycopersicum* L.). *Pak. J. Phytopathol., 27*(02), 109–114.

Charron, C. S., & Sams, C. E., (1999). Inhibition of *Pythium ultimum* and *Rhizoctonia solani* by shredded leaves of brassica species. *J. Am. Soc. Hortic. Sci., 124*(5), 462–467.

Christopher, D. J., Raj, T. S., Rani, S. U., & Udhayakumar, R., (2015). Role of defense enzymes activity in tomato as induced by *Trichoderma virens* against *Fusarium* wilt caused by *Fusarium oxysporum* f sp. *lycopersici*. *Journal of Biopesticides., 1,* 158–162.

Clark, A., (2007). *Managing Cover Crops Profitably* (3rd edn.). Sustainable Agriculture Network, Beltsville, MD. 248.

Cohen, M. F., & Mazzola, M., (2006). Resident bacteria, nitric oxide emission and particle size modulate the effect of *Brassica napus* seed meal on disease incited by *Rhizoctonia solani* and *Pythium* spp. *Plant and Soil, 286*(1/2), 75–86.

Cohen, M. F., Yamasaki, H., & Mazzola, M., (2005). *Brassica napus* seed meal soil amendment modifies microbial community structure, nitric oxide production and incidence of *Rhizoctonia* root rot. *Soil Biol. Biochem., 37,* 1215–1227.

Cook, R. J., (1993). Making greater use of introduced microorganisms for biological control of plant pathogens. *Phytopathology, 31,* 5340.

Deadman, M. A., Hasani, H. A., & Sa'di, A., (2006). Solarization and biofumigation reduce *Pythium aphanidermatum* induced damping-off and enhance vegetative growth of greenhouse cucumber in Oman. *J. Plant Pathol., 88*(3), 335–337.

Dhingra, O. D., Costa, M. L. N., & S. J. G. J., (2004). Potential of allyl isothiocyanate to control *Rhizoctonia solani* seedling damping-off and seedling blight in transplant production. *J. Phytopathol., 152,* 352–357.

Dohroo, N. P., Gupta, M., Shanmugam, V., & Kumar, A., (2009). Status and non-chemical management of ginger rhizome diseases in Himachal Pradesh. *Pl. Dis. Res., 24*(1), 71.

Duniway, J., (2002). Chemical alternatives to methyl bromide for soil treatment particularly in strawberry production. *Proceedings of International Conference on Alternatives to Methyl Bromide* (Vol. 432). Seville, Spain.

Fahey, J. W., Zalcmann, A. T., & Talalay, P., (2001). The chemical diversity and distribution of glucosinolates and isothiocyanates among plants. *Phytochemistry, 56,* 5–51.

Fayzalla, E. A., El-Barougy, E., & El-Rayes, M. M., (2009). Control of soil-borne pathogenic fungi of soybean by biofumigation with mustard seed meal. *J. Appl. Sci. Res., 9,* 272–279.

Friberg, H., Edel-Herman, V., Faivre, C., Gautheron, N., Fayolle, L., Faloya, V., Montfort, F., & Steinberg, C., (2009). Cause and duration of mustard incorporation effects on soil-borne plant pathogenic fungi. *Soil Biol. and Biochem., 41,* 2075–2084.

Galletti, S., Sala, E., Leoni, B. O., & Cerato, P. L. C., (2008). *Trichoderma spp.* tolerance to *Brassica carinata* seed meal for a combined use in biofumigation. *Biol. Control, 45,* 319–327.

Gardiner, J. B., Morra, M. J., Eberlein, C. V., Brown, P. D., & Borek, V., (1999). Allelochemicals released in soil following incorporation of rapeseed (*Brassica napus*) green manures. *J. Agric. Food Chem., 47,* 3837–3842.

Gimsing, A. L., & Kirkegaard, J. A., (2006). Glucosinolate and isothiocyanate concentration in soil following incorporation of *Brassica* biofumigants. *Soil Biol. Biochem., 38,* 2255–2264.

Gimsing, A. L., & Kirkegaard, J. A., (2008). Glucosinolates and biofumigation: Fate of glucosinolates and their hydrolysis products in soil. *Phytochem. Rev., 8*(1), 299–310.

Gimsing, A. L., Sorensen, J. C., Tovgaard, L., Jorgensen, A. M. F., & Hansen, H. C. B., (2006). Degradation kinetics of glucosinolates in soil. *Environ. Toxicol. Chem., 25,* 2038–2044.

Grimme, E., Zidack, N. K., Sikora, R. A., Strobel, G. A., & Jacobsen, B. J., (2007). Comparison of *Muscodor albus* volatiles with a biorational mixture for control of seedlings diseases of sugar beet and root-knot nematode on tomato. *Plant Disease, 91,* 220–224.

Gruver, L. S., Weil, R. R., Zasada, I. A., Sardanelli, S., & Momen, B., (2010). Brassicaceous and rye cover crops altered free-living soil nematode community composition. *Applied Soil Ecology, 45,* 1–12.

Handiseni, M., Young-Ki, J., Kyung-Min, L., & Xin-Gen, Z., (2015). Screening brassicaceous plants as Biofumigants for management of *Rhizoctonia solani* AG1-IA. *Plant Disease, 100.* 10.1094/PDIS-06-15-0667-RE.

Hanschen, F. S., Yim, B., Winkelmann, T., Smalla, K., & Schreiner, M., (2015). Degradation of biofumigant isothiocyanates and allyl glucosinolate in soil and their effects on the microbial community composition. *Plos One, 7,* 1–18.

Henderson, D. R., Riga, E., Ramirez, R. A., Wilson, J., & Snyder, W. E., (2009). Mustard biofumigation disrupts biological control by *Steinernema* spp. nematodes in the soil. *Biological Control, 48,* 316–322.

Hoagland, L., Carpenter-Boggs, L., Reganold, J. P., & Mazzola, M., (2008). Role of native soil biology in brassicaceous seed meal-induced weed suppression. *Soil Biol. Biochem., 40,* 1689–1697.

Huang, Q., & Lakshman, D. K., (2010). Effect of clove oil on plant pathogenic bacteria and bacterial wilt of tomato and geranium. *Journal of Plant Pathology, 92,* 701–707.

Jin, N., Lu, X., & Wang, X., (2019). The effect of combined application of *Streptomyces rubrogriseus* HDZ-9–47 with soil biofumigation on soil microbial and nematode communities. *Sci Rep., 9,* 16886. https://doi.org/10.1038/s41598-019-52941-9.

Jin, N., Xue, H., Wen-Jing, L., Xue-Yan, W., Liu, Q., Shu-Sen, L., Liu, P., Jian-Long, Z., & Jian, H., (2017). Field evaluation of *Streptomyces rubrogriseus* HDZ-9-47 for biocontrol of *Meloidogyne incognita* on tomato. *Journal of Integrative Agriculture, 16*(06), 1347–1357.

Kirkegaard, J. A., & Sarwar, M., (1998). Biofumigation potential of Brassicas I. variation in glucosinolate profiles of diverse field-grown Brassicas. *Plant and Soil, 201,* 71–89.

Kirkegaard, J. A., Gardener, A. P., Desmarchelier, M. J., & Angus, F. J., (1993). Biofumigation-using *Brassica* species to control pests and diseases in horticulture and agriculture. In: *9th Australian Research Assembly on Brassicas* (pp. 77–82).

Kirkegaard, J. A., Sarwar, M., & Matthiessen, J. N., (1998). Assessing the biofumigation potential of crucifers. *Acta Horticulture, 459,* 105–111.

Kirkegaard, J. A., Sarwar, M., Wong, P. T. W., Mead, A., Howe, G., & Newell, M., (2000). Field studies on the biofumigation of take-all by *Brassica* break crops. *Aust. J. Agric. Res., 51,* 445–456.

Kirkegard, J. A., Wong, P. T. W., & Desmarchelier, J. M., (1996). *In vitro* suppression of fungal root pathogens of cereals by Brassica tissues. *Plant Pathol., 45,* 593–603.

Kreutzer, W. A., (1963). Selective toxicity of chemicals to soil microorganisms. *Annu. Rev. Phytopathol., 1,* 101–126.

Lacey, L. A., & Neven, L. G., (2006). The potential of the fungus, *Muscodor albus* as a microbial control agent of potato tuber moth (Lepidoptera: Gelechiidae) in stored potatoes. *Journal of Invertebrate Pathology, 91,* 195–198.

Lacey, L. A., Horton, D. R., & Jones, D. C., (2008). The effect of temperature and duration of exposure of potato tuber moth (Lepidoptera: Gelechiidae) in infested tubers to the biofumigant fungus *Muscodor albus*. *Journal of Invertebrate Pathology, 97,* 159–164.

Larkin, R. P., & Griffin, T. S., (2007). Control of soil-borne potato diseases using Brassica green manures. *Crop Prot., 26,* 1067–1077.

Larkin, R. P., & Honeycutt, C. W., (2006). Effects of different 3-year cropping systems on soil microbial communities and *Rhizoctonia* diseases of potato. *Phytopathology, 96,* 68–79.

Larkin, R. P., & Honeycutt, C.W., (1999 & 2000). Crop rotation effects on *Rhizoctonia* canker and black scurf of potato in central Maine. *Biological and Cultural Tests (Online).* Report 17, PT06. doi: 10.1094/ BC17. The American Phytopathological Society, St. Paul, MN.

Larkin, R. P., Griffin, T. S., & Honeycutt, C. W., (2010). Rotation and cover crop effects on soil-borne potato diseases, tuber yield, and soil microbial communities. *Plant Dis., 94,* 1491–1502.

Lazzeri, L., Curto, G., Leoni, O., Dallavalle, E. D., Avino, L., Malaguti, L., Santi, R., & Patalano, G., (2009). Nematicidal efficacy of biofumigation by defatted *Brassicaceae* meal for control of *Meloidogyne incognita* (Kofoid *et* White) Chitw. on a full field zucchini crop. *J. of Sus. Agril., 33,* 349–358.

Li, G. J., Dong, Q. E., Ma, L., Huang, Y., Zhu, M. L., Ji, Y. P., Wang, Q. H., Mo, M. H., & Zhang, K. Q., (2014). Management of *Meloidogyne incognita* on tomato with endophytic bacteria and fresh residue of *Wasabia japonica*. *Journal of Applied Microbiology, 117,* 1159–1167.

Lord, J. S., Lazzeri, L., Atkinson, H. J., & Urwin, P. E., (2011). Biofumigation for control of pale potato cyst nematodes: Activity of brassica leaf extracts and green manures on *Globodera pallida in vitro* and in soil. *J. Agric. Food Chem., 59,* 7882–7890.

Louise-Marie, D., Mosher, R., & Knudsen, G., (2000). Combined effects of *Brassica napus* seed meal and *Trichoderma harzianum* on two soil-borne plant pathogens. *Canadian Journal of Microbiology, 46,* 1051–1057. 10.1139/cjm-46-11-1051.

Louws, F., Rivard, C., & Kubota, C., (2010). Grafting fruiting vegetables to manage soil-borne pathogens, foliar pathogens, arthropods, and weeds. *Scientia Horticulturae, 127,* 127–146. 10.1016/j.scienta.2010.09.023.

Madhavi, G. B., Umadevi, G., Kumar, K. V. K., Babu, T. R., & Naidu, T. C. M., (2015). Evaluation of different *Brassica* species and onion for their biofumigation effect against *Rhizoctonia solani* f. sp *Sasakii in vitro. J. Res. ANGRAU, 43*(3/4), 22–28.

Madhavi, G. B., Umadevi, G., Kumar, K. V. K., Babu, T. R., & Naidu, T. C. M., (2016). Effect of volatiles produced by hydrolysis of mustard seed powder against *Rhizoctonia solani* f. sp. *sasakii, Trichoderma harzianum* and *Pseudomonas fluorescens. Progre. Res., 11*(VII), 4821–4823.

Madhavi, G. B., Umadevi, G., Kumar, K. V. K., Babu, T. R., & Naidu, T. C. M., (2018). Effect of combined application of biofumigant, *Trichoderma harzianum* and *Pseudomonas fluorescens* on *Rhizoctonia solani* f. sp. *sasakii. Indian Phytopath.* https://doi.org/10.1007/ s42360-018-0039-6.

Manici, L. M., Lazzeri, L., & Palmieri, S., (1997). *In vitro* fungitoxic activity of some glucosinolates and their enzyme-derived products toward plant pathogenic fungi. *J. Agric. Food Chem., 45,* 2768–2773.

Matthiessen, J. N., & Kirkegaard, J. A., (2006). Biofumigation and enhanced biodegradation: Opportunity and challenge in soil-borne pest and disease management. *Critical Reviews in Plant Sciences. 25,* 235–265.

Matthiessen, J. N., & Shackleton, M. A., (2005). Biofumigation: Environmental impacts on the biological activity of diverse pure and plant-derived isothiocyanates. *Pest Manag. Sci., 61*, 1043–1051.

Matthiessen, J. N., Warton, B., & Shackleton, M. A., (2004). The importance of plant maceration and water addition in achieving high *Brassica*-derived isothiocyanate levels in soil. *Agroindustria., 3*(3), 277–280.

Mattner, S. W., Porter, I. J., Gounder, R. K., Shanks, A. L., Wren, D. J., & Allen, D., (2008). Factors that impact on the ability of biofumigants to suppress fungal pathogens and weeds of strawberry. *Crop Prote., 27*, 1165–1173.

Mayton, S. H., Olivier, C., Vaughn, F. S., & Loria, R., (1996). Correlation of fungicidal activity of brassica species with allyl isothiocyanate production in macerated leaf tissue. *Phytopathol., 86*, 267–271.

Mazzola, M., & Zhao, X., (2010). *Brassica juncea* seed meal particle size influences the chemistry but not soil biology-based suppression of individual agents inciting apple replant disease. *Plant Soil., 337*, 313–324.

Mazzola, M., Granatstein, D. M., Elfving, D. C., & Mullinix, K., (2001). Suppression of specific apple root pathogens by *Brassica napus* seed meal amendment regardless of glucosinolate content. *Phytopathol., 91*, 673–679.

Mercier, J., & Jime´nez, J. I., (2004). Control of fungal decay of apples and peaches by the biofumigant fungus *Muscodor albus*. *Postharvest Biology and Technology, 31*, 1–8.

Mercier, J., & Jime´nez, J. I., (2007). Potential of the volatile-producing fungus *Muscodor albus* for control of building molds. *Canadian Journal of Microbiology, 53*, 404–410.

Mercier, J., & Manker, D. C., (2005). Biocontrol of soil-borne diseases and plant growth enhancement in glasshouse soilless mix by the volatile producing fungus *Muscodor albus*. *Crop Protection, 24*, 355–362.

Mercier, J., & Smilanick, J. L., (2005). Control of green mold and sour rot of stored lemon by biofumigation with *Muscodor albus*. *Bio. Control., 32*, 401–407.

Misrak, K., Amare, A., & Dechassa, N., (2014). Evaluation of soil solarization and biofumigation for the management of bacterial spot of tomato. *African Journal of Food, Agriculture, Nutrition and Development, 14*(4).

Morra, M. J., & Kirkegaard, J. A., (2002). Isothiocyanate release from soil-incorporated *Brassica* tissues. *Soil Biol. Biochem., 34*, 1683–1690.

Motisi, N., Dore, T., Lucas, P., & Montfort, F., (2010). Dealing with the variability in biofumigation efficacy through an epidemiological framework. *Soil Biol. and Biochem., 42*, 2044–2057.

Muehlchen, A. M. R., Rand, R. E., & Parke, J. L., (1990). Evaluation of crucifer green manures for controlling *Aphanomyces* root rot of peas. *Plant Dis., 74*, 651–654.

Nazf, (2006). *Integrated Nutrient Management of Black Scurf of Potato* (Vol. 161). Thesis submitted to Dept. of Plant pathology. Univ of Arid Agriculture, Rawalpindi.

Neubauer, C., Heitmann, B., & Muller, C., (2014). Biofumigation potential of Brassicaceae cultivars to *Verticillium dahliae*. *Eur. J. Plant Pathol., 140*, 341–352.

Ojaghian, M. R., Cui, Z. Q., Xie, G. L., Li, B., & Zhang, J., (2012). Brassica green manure rotation crops reduce potato stem rot caused by *Sclerotinia sclerotium*. *Australas. Plant Pathol., 41*(4), 347–349.

Olivier, C., Vaughn, S. F., & Mizubuti, E. S. G., (1999). Variation in allyl isothiocyanate production within *Brassica* species and correlation with fungicidal activity. *J. Chem. Ecol., 25*, 2687–2701.

Omirou, M., Karpouzas, D. G., Papadopoulou, K. K., & Ehaliotis, C., (2013). Dissipation of pure and broccoli released glucosinolate in soil under high and low moisture content. *Eur. J. Soil Biol., 56*, 49–55.

Omirou, M., Rousidou, C., Bekris, F., Papadopoulou, K. K., Menkissoglou-Spiroudi, U., Ehaliotis, C., & Karpouzas, D. G., (2010). The impact of biofumigation and chemical fumigation methods on the structure and function of the soil microbial community. *Microb. Ecol., 61*, 201–213.

Papavizas, G. C., (1968). Survival of root-infecting fungi in soil: VI. Effect of amendments on bean root rot caused by *Thielaviopsis basicola* and on inoculum density of the causal organism. *Phytopathol., 58*, 421–428.

Paret, M. L., Kratky, B. A., & Álvarez, A. M., (2010). Effect of plant essential oils on *Ralstonia solanacearum* race 4 and bacterial wilt of edible ginger. *Plant Disease, 94*, 521–527.

Perez, C., Dill-Macky, R., & Kinkel, L. L., (2007). Management of soil microbial communities to enhance populations of *Fusarium graminearum*-antagonists in soil. *Plant and Soil, 302*, 53–69.

Porter, I. J., & Mattner, S., (2002). Non chemical alternatives to methyl bromide for soil treatment in strawberry production 39–48. *Proceedings of International Conference on Alternatives to Methyl Bromide, 432*.

Poulsen, J. L., Gimsing, A. L., Halkier, B. A., Bjarnholt, N., & Hansen, H. C. B., (2008). Mineralization of benzyl glucosinolate and its hydrolysis product the biofumigant benzyl isothiocyanate in soil. *Soil Biol. Biochem., 40*, 135–141.

Pradhanang, P. M., Momol, M. T., Olson, S. M., & Jones, J. B., (2003). Effects of plant essential oils on *Ralstonia solanacearum* population density and bacterial wilt incidence in tomato. *Plant Disease., 87*, 423–427.

Prasad, P., & Kumar, J., (2017). Management of *Fusarium* wilt of chickpea using brassicas as biofumigants. *Legume Research: An International Journal, 40*, 178–182. 10.18805/lr.v0i0.7022.

Raaijmakers, J. M., Paulitz, T. C., Steinberg, C., Alabouvette, C., & Moënne-Loccoz, Y., (2009). The rhizosphere: A playground and battlefield for soil-borne pathogens and beneficial microorganisms. *Plant Soil., 321*, 341–361.

Ramirez-Villapudua, J., & Munnecke, D. E., (1988). Effect of solar heating and soil amendments of cruciferous residues on *Fusarium oxysporum* f. sp. *conglutinant* and other organisms. *Phytopath., 78*, 289–295.

Riga, E., Lawrence, A., Lacey, B., & Neussa, G., (2008). *Muscodor albus*, a potential biocontrol agent against plant-parasitic nematodes of economically important vegetable crops in Washington State, USA. *Biological Control, 45*, 380–385.

Rosa, A. S. E., (1997). Daily variation in glucosinolate concentrations in the leaves and roots of cabbage seedlings in two constant temperature regimes. *J. Sci. Food Agric., 73*, 364–368.

Roubtsova, T., Jose-Antonio, L. P., Scott, E., & Antoon, P., (2007). Effect of Broccoli (*Brassica oleracea*) Tissue, incorporated at different depths in a soil column, on *Meloidogyne incognita*. *J. Nematol., 39*, 111–117.

Ryckeboer, J., (2001). *Biowaste and Yard Waste Composts: Microbiological and Hygienic Aspects-Suppressiveness to Plant Diseases, 53*.

Sarwar, M., & Kirkegaard, J. A., (1998). Biofumigation potential of brassicas: II. Effect of environment and ontogeny on glucosinolate production and implications for screening. *Plant and Soil, 201*, 91–101.

Sarwar, M., Kirkegaard, J. A., Wong, P. T. W., & Desmarchelier, J. M., (1998). Biofumigation potential of brassicas: III. *In vitro* toxicity of isothiocyanates to soil-borne fungal pathogens. *Plant and Soil, 201,* 103–112.

Schnabel, G., & Mercier, J., (2006). Use of a *Muscodor albus* pad delivery system for the management of brown rot of peach in shipping cartons. *Postharvest Biology and Technology, 42,* 121–123.

Sekhar, B., & Surajit, K., (2016). Biofumigation: An eco-friendly approach for managing bacterial wilt and soft rot disease of ginger. *Indian Phytopath., 69*(1), 53–56.

Siddiqui, I. A., & Shaukat, S. S., (2004). *Trichoderma harzianum* enhances the production of nematicidal compounds *in vitro* and improves biocontrol of *Meloidogyne javanica* by *Pseudomonas fluorescens* in tomato. *Letters in Applied Microbiology, 38,* 169–175.

Sintayehu, A., Ahmed, S., Fininsa, C., & Sakhuja, P. K., (2014). Evaluation of green manure amendments for the management of *Fusarium* basal rot (*Fusarium oxysporum* f. sp. *cepae*) on shallot. *International J. of Agron.* http://dx.doi.org/10.1155/2014/150235.

Smith, B. J., & Kirkegaard, J. A., (2002). *In vitro* inhibition of soil microorganisms by 2-phenylethyl isothiocyanate. *Plant Pathol., 51,* 585–593.

Smith, B. J., Sarwar, M., Wong P. T. W., & Kirkegaard, J. A. (1999). http://www.regional.org.au/au/gcirc/2/334.htm (accessed on 12 February 2021).

Smolinska, U., (2000). Survival of *Sclerotium cepivorum* sclerotia and *Fusarium oxysporum* chlamydospores in soil amended with cruciferous residues. *J. Phytopathol., 148,* 343–349.

Smolinska, U., Knudsen, G. R., Morra, M. J., & Borek, V., (1997). Inhibition of *Aphanomyces euteiches* f. sp. *Pisi* by volatiles produced by hydrolysis of *Brassica napus* seed meal. *Plant Dis., 81,* 288–292.

Smolinska, U., Morra, M. J., Knudsen, G. R., & James, R. L., (2003). Isothiocyanates produced by *Brassicaceae* species as inhibitors of *Fusarium oxysporum. Plant Dis., 87,* 407–412.

Stapleton, J. J., & Duncan, R. A., (1998). Soil disinfestation with cruciferous amendments and sublethal heating: Effects on *Meloidogyne incognita, Sclerotium rolfsii,* and *Pythium ultimum. Plant Pathol., 47,* 737–742.

Steffek, R., Spornberger, A., & Altenburger, J., (2006). Detection of microsclerotia of *Verticillium dahliae* in soil samples and prospects to reduce the inoculum potential of the fungus in the soil. *Agriculturae Conspectus Scientificus, 71,* 145–148.

Stinson, A. M., Zidack, N. K., Strobel, G. A., & Jacobson, B. J., (2003). Mycofumigation with *Muscodor albus* and *Muscodor roseus* for control of seedling diseases of sugar beet and *Verticillium wilt* of eggplant. *Plant Disease, 87,* 1349–1354.

Strobel, G. A., Dirkse, E., Sears, J., & Markworth, C., (2001). Volatile antimicrobials from *Muscodor albus,* a novel endophytic fungus. *Microbiology Reading, 147,* 2943–2950.

Talibi, I., Boubaker, H., Boudyach, E. H., & Ait, B. A. A., (2014). Alternative methods for the control of postharvest citrus diseases. *Journal of Applied Microbiology, 117,* 1–17.

Valdes, Y., Viaene, N., & Moens, M., (2012). Effects of yellow mustard amendments on the soil nematode community in a potato field with focus on *Globodera rostochiensis. Applied Soil Ecology, 59,* 39–47.

Van, O. G. J., Bijman, V., Boer, M. D., Breeuwsma, S. J., Van, D. B., & Lazzeri, L., (2004). Biofumigation against soil borne fungal diseases in flower bulbs. In: *Proceedings of the First International Symposium Biofumigation: A Possible Alternative to Methyl bromide* (Vol. 21). A way for optimizing potential for an integrated management of soil borne pests and diseases.

Vera, C., McGergor, D., & Downey, R., (1987). Detrimental effects of volunteer *Brassica* on production of certain cereal and oilseed crops. *Can J. Plant Sci., 67,* 983–995.

Walker, C. J., Morell, S., & Foster, H. H., (1937). Toxicity of mustard oils and related Sulphur compounds to certain fungi. *Am. J. Bot., 24,* 241–536.

Wang, D., Rosen, C., Kinkel, L., Cao, A., Tharayil, N., & Gerik, J., (2009). Production of methyl sulfide and dimethyl disulfide from soil-incorporated plant materials and implications for controlling soil-borne pathogens. *Plant Soil, 324*(1), 185–197.

Wang, Q., Ma, Y., Yang, H., & Chang, Z., (2014). Effect of biofumigation and chemical fumigation on soil microbial community structure and control of pepper Phytophthora blight. *World J. Microbiol. Biotechnol., 30*(2), 507–518.

Williams, W. J. L., Pfleger, F. L., Fritz, V. A., & Allmaras, R. R., (1997). Green manures of oat, rape and sweet corn for reducing common root rot in pea (*Pisum sativum*) caused by *Aphanomyces euteiches. Plant and Soil, 188,* 43–48.

Willis, R. J., (1985). The historical bases of the concept of allelopathy. *J. Hist. Biol., 18,* 71–102.

Worapong, J., Strobel, G., Ford, E. J., Li, J. Y., Baird, G., & Hess, W. M., (2001). *Muscodor albus* anam. gen. et sp. nov., an endophyte from *Cinnamomum zeylanicum. Mycotaxon., 79,* 67–79.

Yulianti, T., Sivasithamparam, K., & Turner, D. W., (2007). Saprophytic and pathogenic behavior of *R. solani* AG2-1 (ZG-5) in a soil amended with *Diplotaxis tenuifolia* or *Brassica nigra* manures and incubated at different temperatures and soil water content. *Plant and Soil, 294,* 277–289.

Yulianti, Y., Sivasithamparam, K., & Turner, W. D., (2006). Response of different forms of propagules of *Rhizoctonia solani* AG2-1 (ZG5) exposed to the volatiles produced in soil amended with green manures. *Ann. Appl. Biol., 148,* 105–111.

Zasada, I. A., & Ferris, H., (2004). Nematode suppression with brassicaceous amendments: Application based upon glucosinolate profiles. *Soil. Biol. Biochem., 36,* 1017–1024.

Index